Lecture Notes in Computer Science 85

Commenced Publication in 1973
Founding and Former Series Editors:
Gerhard Goos, Juris Hartmanis, and Jan v

Editorial Board

Jack Snoeyink   Pinyan Lu   Kaile Su
Lusheng Wang (Eds.)

Frontiers in Algorithmics and Algorithmic Aspects in Information and Management

Joint International Conference, FAW-AAIM 2012
Beijing, China, May 14-16, 2012
Proceedings

 Springer

Volume Editors

Jack Snoeyink
University of North Carolina, Chapel Hill, NC, USA
E-mail: snoeyink@cs.unc.edu

Pinyan Lu
Microsoft Research Asia, Shanghai, China
E-mail: pinyanl@microsoft.com

Kaile Su
Peking University, Beijing, China
E-mail: isskls@zsu.edu.cn

Lusheng Wang
City University of Hong Kong, Kowloon, Hong Kong, SAR
E-mail: lwang@cs.cityu.edu.hk

ISSN 0302-9743   e-ISSN 1611-3349
ISBN 978-3-642-29699-4   e-ISBN 978-3-642-29700-7
DOI 10.1007/978-3-642-29700-7
Springer Heidelberg Dordrecht London New York

Library of Congress Control Number: 2012935871

CR Subject Classification (1998): F.2, G.2, I.3.5, E.1, F.1, J.1, I.2

LNCS Sublibrary: SL 1 – Theoretical Computer Science and General Issues

Typesetting: Camera-ready by author, data conversion by Scientific Publishing Services, Chennai, India

Printed on acid-free paper

Springer is part of Springer Science+Business Media (www.springer.com)

Preface

This volume contains the papers presented at FAW-AAIM 2012: the 6th International Frontiers of Algorithmics Workshop (FAW 2012) and the 8th International Conference on Algorithmic Aspects of Information and Management (AAIM 2012), jointly held during May 14–16, 2012, at Peking University, Beijing, China.

The joint conference provides a focused forum on current trends of research on algorithms, discrete structures, operation research, combinatorial optimization and their applications, and brings together international experts at the research frontiers in these areas to exchange ideas and to present significant new results.

There were 81 submissions to this edition of the conference, of which 33 papers were accepted. All papers were rigorously reviewed by the Program Committee members and/or external referees; almost all papers received at least three detailed reviews. The papers were evaluated on the basis of their significance, novelty, soundness and relevance to the conference.

We were pleased to deliver the best paper award to Kazuhide Nishikawa, Takao Nishizeki and Xiao Zhou for their paper "Algorithms for Bandwidth Consecutive Multicolorings of Graphs" and the best student paper award to Bryan He for his paper "Optimal Binary Representation of Mosaic Floorplans and Baxter Permutations."

Besides the regular talks, the program also included two invited talks by Tao Jiang (University of California - Riverside, USA) and Joseph S.B. Mitchell (State University of New York at Stony Brook, USA).

We are very grateful to all the people who made this meeting possible: the authors for submitting their papers, the Program Committee members and external reviewers for their excellent work, and the two invited speakers. In particular, we would like to thank Peking University for hosting the conference and providing organizational support.

We also acknowledge EasyChair, a powerful and flexible system for managing all stages of the paper handling process, from the submission stage to the preparation of the final version of the proceedings.

May 2012

Jack Snoeyink
Pinyan Lu
Kaile Su
Lusheng Wang

Organization

General Chairs

John Hopcroft   Cornell University, USA
Hong Mei   Peking University, China

Program Committee Co-chairs

Jack Snoeyink   University of North Carolina at Chapel Hill,
     USA
Pinyan Lu   Microsoft Research Asia, China
Kaile Su   Peking University, China
Lusheng Wang   City University of Hong Kong, Hong Kong

Program Committee: FAW Track

Andrej Bogdanov   Chinese University of Hong Kong, China
Leizhen Cai   Chinese University of Hong Kong, China
Xin Chen   Nanyang Technological University, Singapore
Yijia Chen   Shanghai Jiao Tong University, China
Zhi-zhong Chen   Tokyo Denki University, Japan
Miklós Csürös   Université de Montréal, Canada
Bin Fu   University of Texas-Pan American, USA
Ming-Yang Kao   Northwestern University, USA
Guohui Lin   University of Alberta, Canada
Tian Liu   Peking University, China
Xiaoming Sun   Institute of Computing Technology, CAS,
     China
Haitao Wang   University of Notre Dame, USA
Jianxin Wang   Central South University, China
David Woodruff   IBM Almaden Research Center, USA
Yi Wu   IBM Almaden Research Center, USA
Mingji Xia   Software Institute, CAS, China
Jinhui Xu   University at Buffalo, the State University of
     New York, USA
Yitong Yin   Nanjing University, China
Shengyu Zhang   Chinese University of Hong Kong, China

Program Committee: AAIM Track

Ning Chen	Nanyang Technological University, Singapore
Xi Chen	Columbia University, USA
Yongxi Cheng	Xi'an Jiao Tong University, China
Giorgos Christodoulou	University of Liverpool, UK
Hao Yuan	City University of Hong Kong, China
Martin Hoefer	RWTH Aachen University, Germany
Hon Wai Leong	National University of Singapore, Singapore
Guojun Li	Shandong University, China
Julian Mestre	University of Sydney, Australia
Karthik Natarajan	City University of Hong Kong, China
Anthony Man-Cho So	Chinese University of Hong Kong, China
Periklis Papakonstantinou	Tsinghua University, China
Zhiyi Tan	Zhejiang University, China
Chung-Piaw Teo	National University of Singapore, Singapore
Yajun Wang	Microsoft Research Asia, China
Ke Xu	Beihang University, China
Ke Yi	Hong Kong University of Science and Technology, China
Guochuan Zhang	Zhejiang University, China
Jian Zhang	Software Institute, CAS, China
Louxin Zhang	National University of Singapore, Singapore
Lu Zhang	Peking University, China

Local Organizing Committee

Hanpin Wang	Peking University, China
Tian Liu	Peking University, China

Additional Reviewers

Bei, Xiaohui	Li, Shuguang
Burcea, Mihai	Lin, Bingkai
Cai, Yufei	Liu, Yang
Chen, Shiteng	Liu, Yangwei
Deng, Yuxin	Lopez-Ortiz, Alejandro
Frati, Fabrizio	Ma, Tengyu
Guo, Chengwei	Mak, Yan Kei
Halim, Steven	Megow, Nicole
Hu, Haiqing	Narodytska, Nina
Huang, Ziyun	Ng, Yen Kaow
Jiang, Minghui	Qiao, Youming
Li, Guojun	Shi, Zhiqiang
Li, Jian	Srihari, Sriganesh

Van Zuylen, Anke
Wahlstrom, Magnus
Wang, Jiun-Jie
Wang, Xiangyu
Yang, Guang
Ye, Nan
Yin, Minghao

Yin, Yitong
Zhang, Chihao
Zhang, Jinshan
Zheng, Changwen
Zhou, Yuan
Zhu, Shanfeng

Computational Geometry Approaches to Some Algorithmic Problems in Air Traffic Management

Joseph S.B. Mitchell

Department of Applied Mathematics and Statistics, Stony Brook University, USA
joseph.mitchell@stonybrook.edu

Abstract. The next generation of air transportation system will have to use technology to be able to cope with the ever increasing demand for flights. Several challenging optimization problems arise in trying to maximize efficiency while maintaining safe operation in air traffic management (ATM). Constraints and issues unique to air transportation arise in the ATM domain, including weather hazards, turbulence, no-fly zones, and three-dimensional routing. The challenge is substantially compounded when the constraints vary in time and are not known with certainty, as is the case with weather hazards. Human oversight is provided by air traffic controllers, who are responsible for safe operation within a portion of airspace known as a sector.

In this talk we discuss algorithmic methods that can be used in modeling and solving air traffic management problems, including routing of traffic flows, airspace configuration into load-balanced sectors, and capacity estimation in the face of dynamic and uncertain constraints and demands. We highlight several open problems.

Keywords: computational geometry, geometric flow, air traffic management, load balancing, sectorization.

Acknowledgments. This research has been supported by grants from the National Science Foundation (CCF-0729019, CCF-1018388), NASA Ames, and Metron Aviation. The talk is based on collaborative work with many, including Anthony D. Andre, Dominick Andrisani, Estie Arkin, Amitabh Basu, Jit-Tat Chen, Nathan Downs, Moein Ganji, Robert Hoffman, Joondong Kim, Victor Klimenko, Irina Kostitsyna, Shubh Krishna, Jimmy Krozel, Changkil Lee, Tenny Lindholm, Anne Pääkkö, Steve Penny, Valentin Polishchuk, Joseph Prete, Girishkumar Sabhnani, Robert Sharman, Philip J. Smith, Amy L. Spencer, Shang Yang, Arash Yousefi, Jingyu Zou.

Combinatorial Methods for Inferring Isoforms from Short Sequence Reads

Tao Jiang[1,2]

[1]Department of Computer Science and Engineering,
University of California, Riverside, CA
[2] School of Information Science and Technology, Tsinghua University, Beijing, China
jiang@cs.ucr.edu

Abstract. Due to alternative splicing, a gene may be transcribed into several different mRNA transcripts (called *isoforms*) in eukaryotic species. How to detect isoforms on a genomic scale and measure their abundance levels in a cell is a central problem in transcriptomics and has broad applications in biology and medicine. Traditional experimental methods for this purpose are time consuming and cost ineffective. Although deep sequencing technologies such as RNA-Seq provide a possible effective method to address this problem, the inference of isoforms from tens of millions of short sequence reads produced by RNA-Seq has remained computationally challenging. In this talk, I will first briefly survey the state-of-the-art methods for inferring isoforms from RNA-Seq short reads including Cufflinks, Scripture and IsoInfer, and then describe the algorithmic framework behind IsoInfer in more detail. The design of IsoInfer exhibits an interesting combination of combinatorial optimization techniques (*e.g.*, convex quadratic programming) and statistical concepts (*e.g.*, maximum likelihood estimation and p-values). Finally, I will introduce our recent improvement of IsoInfer, called IsoLasso. The new method incorporates the well-known LASSO regression method into the quadratic program of IsoInfer and is likely to deliver isoform solutions with both good accuracy and sparsity. Our extensive experiments on both simulated and real RNA-Seq data demonstrate that this addition could help IsoLasso to filter out lowly expressed isoforms (which are often noisy) and achieve higher sensitivity and precision simultaneously than the existing transcriptome assembly tools.

This is a joint work with Wei Li (UC Riverside) and Jianxing Feng (Tongji University).

Table of Contents

Optimal Binary Representation of Mosaic Floorplans and Baxter Permutations

Bryan He

Department of Computer Science, California Institute of Technology,
1200 E. California Blvd., 91126 Pasadena, California, United States of America
bryanhe@caltech.edu

Abstract. A *floorplan* is a rectangle subdivided into smaller rectangular *blocks* by horizontal and vertical line segments. Two floorplans are considered equivalent if and only if there is a bijection between the blocks in the two floorplans such that the corresponding blocks have the same horizontal and vertical boundaries. *Mosaic floorplans* use the same objects as floorplans but use an alternative definition of equivalence. Two mosaic floorplans are considered equivalent if and only if they can be converted into equivalent floorplans by sliding the line segments that divide the blocks. The *Quarter-State Sequence* method of representing mosaic floorplans uses $4n$ bits, where n is the number of blocks. This paper introduces a method of representing an n-block mosaic floorplan with a $(3n - 3)$-bit binary string. It has been proven that the shortest possible binary string representation of a mosaic floorplan has a length of $(3n - o(n))$ bits. Therefore, the representation presented in this paper is asymptotically optimal. *Baxter permutations* are a set of permutations defined by prohibited subsequences. There exists a bijection between mosaic floorplans and Baxter permutations. As a result, the methods introduced in this paper also create an optimal binary string representation of Baxter permutations.

Keywords: Binary Representation, Mosaic Floorplan, Baxter Permutation.

1 Introduction

In this section, the definitions of mosaic floorplans and Baxter permutations are introduced, previous work in the area and their applications are described, and the main result is stated.

1.1 Floorplans and Mosaic Floorplans

Definition 1. *A* floorplan *is a rectangle subdivided into smaller rectangular subsections by horizontal and vertical line segments such that no four subsections meet at the same point.*

J. Snoeyink, P. Lu, K. Su, and L. Wang (Eds.): FAW-AAIM 2012, LNCS 7285, pp. 1–12, 2012.
© Springer-Verlag Berlin Heidelberg 2012

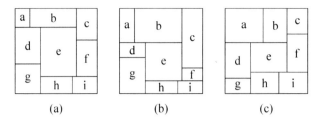

Fig. 1. Three example floorplans

The smaller rectangular subsections are called *blocks*. Figure 1 shows three floorplans, each containing 9 blocks. Note that the horizontal and vertical line segments do not cross each other. They can only form *T-junctions* (⊢, ⊥, ⊣, and ⊤).

The definition of equivalent floorplans does not consider the size of the blocks in the floorplan. Instead, two floorplans are considered equivalent if and only if their corresponding blocks have the same relative position relationships. The formal definition of equivalent floorplans follows.

Definition 2. *Let F_1 be a floorplan with R_1 as its set of blocks. Let F_2 be another floorplan with R_2 as its set of blocks. F_1 and F_2 are considered* equivalent *floorplans if and only if there is a bijection $g : R_1 \to R_2$ such that the following conditions hold:*

1. *For any two blocks $r, r' \in R_1$, r and r' share a horizontal line segment as their common boundary with r above r' if and only if $g(r)$ and $g(r')$ share a horizontal line segment as their common boundary with $g(r)$ above $g(r')$.*
2. *For any two blocks $r, r' \in R_1$, r and r' share a vertical line segment as their common boundary with r to the left of r' if and only if $g(r)$ and $g(r')$ share a vertical line segment as their common boundary with $g(r)$ to the left of $g(r')$.*

In Figure 1, (a) and (b) have the same number of blocks and the position relationships between their blocks are identical. Therefore, (a) and (b) are equivalent floorplans. However, (c) is not equivalent to either.

The objects of *mosaic floorplans* are the same as the objects of floorplans. However, mosaic floorplans use a different definition of equivalence. Informally, two mosaic floorplans are considered equivalent if and only if they can be converted to each other by sliding the horizontal and vertical line segments one at a time. The equivalence of the mosaic floorplans is formally defined by using the *horizontal constraint graph* and the *vertical constraint graph* [9]. The horizontal constraint graph describes the horizontal relationship between the vertical line segments of a floorplan. The vertical constraint graph describes the vertical relationship between the horizontal line segments of a floorplan. The formal definitions of horizontal constraint graphs, vertical constraint graphs, and equivalence of mosaic floorplans follow.

Definition 3. *Let F be a floorplan.*

1. *The* horizontal constraint graph $G_H(F)$ *of F is a directed graph. The vertex set of $G_H(F)$ has a bijection with the set of the vertical line segments of F. For two vertices u_1 and u_2 in $G_H(F)$, there is a directed edge $u_1 \rightarrow u_2$ if and only if there is a block b in F such that the vertical line segment v_1 corresponding to u_1 is on the left boundary of b and the vertical line segment v_2 corresponding to u_2 is on the right boundary of b.*
2. *The* vertical constraint graph $G_V(F)$ *of F is a directed graph. The vertex set of $G_V(F)$ has a bijection with the set of the horizontal line segments of F. For two vertices u_1 and u_2 in $G_V(F)$, there is a directed edge $u_1 \rightarrow u_2$ if and only if there is a block b in F such that the horizontal line segment h_1 corresponding to u_1 is on bottom boundary of b and the horizontal line segment h_2 corresponding to u_2 is on the top boundary of b.*

The graphs in Figure 2 are the constraint graphs of all three floorplans shown in Figure 1. Note that the top, bottom, right, and left boundaries of the floorplan are represented by the north, south, east, and west vertices labeled by N, S, E, and W, respectively, in the constraint graphs. Also note that each edge in $G_H(F)$ and $G_V(F)$ corresponds to a block in the floorplan.

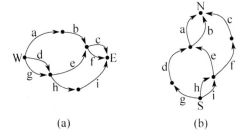

(a)   (b)

Fig. 2. The constraint graphs representing all three mosaic floorplans in Figure 1. (a) is the horizontal constraint graph. (b) is the vertical constraint graph.

Definition 4. *Two mosaic floorplans are* equivalent mosaic floorplans *if and only if they have identical horizontal constraint graphs and vertical constraint graphs.*

Thus, in Figure 1, (a), (b), and (c) are all equivalent mosaic floorplans. Note that (c) is obtained from (b) by sliding the horizontal line segment between blocks d and g downward, the horizontal line segment between blocks c and f upward, and the vertical line segment between blocks a and b to the right.

1.2   Applications of Floorplans and Mosaic Floorplans

Floorplans and mosaic floorplans are used in the first major stage in the physical design cycle of VLSI (Very Large Scale Integration) circuits [10]. The blocks in a floorplan correspond to the components of a VLSI chip. The floorplanning stage is used to plan the relative position of the circuit components. At this stage, the blocks do not have specific sizes assigned to them yet, so only the position relationship between the blocks are considered.

For a floorplan, the wires between two blocks run cross their common boundary. In this setting, two equivalent floorplans provide the same connectivity between blocks. For a mosaic floorplan, the line segments are the wires. Any block with a line segment on its boundary can be connected to the wires represented by the line segment. In this setting, two equivalent mosaic floorplans provide the same connectivity between blocks.

Binary representations of floorplans and mosaic floorplans are used by various algorithms to generate floorplans in order to solve various VLSI layout optimization problems.

Floorplans are also used to represent rectangular cartograms [15,17]. Rectangular cartograms provide a visual method of displaying statistical data about a set of regions.

1.3 Baxter Permutations

Baxter permutations are a set of permutations defined by prohibited subsequences. They were first introduced in [3]. It was shown in [8] that the set of Baxter permutations has bijections to all objects in the *Baxter combinatorial family*. For example, [4] showed that *plane bipolar orientations* with n edges have a bijection with Baxter permutations of length n. [5] established a relationship between Baxter permutations and pairs of alternating sign matrices.

In particular, it was shown in [1,6,20] that mosaic floorplans are one of the objects in the Baxter combinatorial family. A simple and efficient bijection between mosaic floorplans and Baxter permutations was established in [1,6]. As a result, any binary representation of mosaic floorplans can also be converted to a binary representation of Baxter permutations.

1.4 Previous Work on Representations of Floorplans and Mosaic Floorplans

Because of their applications in VLSI physical design, the representations of floorplans and mosaic floorplans have been studied extensively by mathematicians, computer scientists and electrical engineers. Although their definitions are similar, the combinatorial properties of floorplans and mosaic floorplans are quite different. The following is a list of research on floorplans and mosaic floorplans.

Floorplans

There is no known formula for calculating $F(n)$, the number of n-block floorplans. The first few values of $F(n)$ are $\{1, 2, 6, 24, 116, 642, 3938, \ldots\}$. Researchers have been trying to bound the range of $F(n)$. In [2], it was shown that there exists a constant $c = \lim_{n \to \infty} (F(n))^{1/n}$ and $11.56 < c < 28.3$. This means that

$11.56^n \leq F(n) \leq 28.3^n$ for large n. The upper bound of $F(n)$ was reduced to $F(n) \leq 13.5^n$ in [7].

Algorithms for generating floorplans are presented in [12]. In [18], a $(5n-5)$-bit representation of n-block floorplans is shown. A different $5n$-bit representation of n-block floorplans is presented in [19]. The shortest known binary representation of n-block floorplans uses $(4n-4)$ bits [16].

Since $F(n) \geq 11.56^n$ for large n [2], any binary string representation of n-block floorplans must use at least $\log_2 11.56^n = 3.531n$ bits. Closing the gap between the known $(4n-4)$-bit binary representation and the $3.531n$ lower bound remains an open research problem [16].

Mosaic Floorplans

It was shown in [6] that the set of n-block mosaic floorplans has a bijection to the set of Baxter permutations, and the number of n-block mosaic floorplans equals to the nth *Baxter number* $B(n)$, which is defined as the following:

$$B(n) = \frac{\sum_{r=0}^{n-1} \binom{n+1}{r} \binom{n+1}{r+1} \binom{n+1}{r+2}}{\binom{n+1}{1} \binom{n+1}{2}}$$

In [14], it was shown that $B(n) = \Theta(8^n/n^4)$. The first few Baxter numbers are $\{1, 2, 6, 22, 92, 422, 2074, \ldots\}$.

There is a long list of papers on representation problem of mosaic floorplans. [11] proposed a *Sequence Pair* (SP) representation. Two sets of permutations are used to represent the position relations between blocks. The length of the representation is $2n \log_2 n$ bits.

[9] proposed a *Corner Block List* (CB) representation for mosaic floorplans. The representation consists of a list S of blocks, a binary string L of $(n-1)$ bits, and a binary string T of $2n-3$ bits. The total length of the representation is $(3n + n \log_2 n)$ bits.

[21] proposed a *Twin Binary Sequences* (TBS) representation for mosaic floorplans. The representation consists of 4 binary strings $(\pi, \alpha, \beta, \beta')$, where π is a permutation of integers $\{1, 2, \ldots, n\}$, and the other three strings are n or $(n-1)$ bits long. The total length of the representation is $3n + n \log_2 n$.

A common feature of above representations is that each block in the mosaic floorplan is given an explicit name (such as an integer between 1 and n). They all use at least one list (or permutation) of these names in the representation. Because at least $\log_2 n$ bits are needed to represent every integer in the range $[1, n]$, the length of these representations is inevitably at least $n \log_2 n$ bits.

A different approach using a pair of *Twin Binary Trees* was introduced in [20]. The blocks of the mosaic floorplan are not given explicit names. Instead, the shape of the two trees are used to encode the position relations of blocks. In this representation, each tree consists of $2n$ nodes. Each tree can be encoded by using $4n$ bits, so the total length of the representation is $8n$ bits.

In [13], a representation called *Quarter-State-Sequence* (QSS) was presented. It uses a Q sequence that represents the configuration of one of the corners of the mosaic floorplan. The length of the Q sequence representation is $4n$ bits. This is the best known representation for mosaic floorplans.

The number of n-block mosaic floorplans equals the nth Baxter number, so at least $\log_2 B(n) = \log_2 \Theta(8^n/n^4) = 3n - o(n)$ bits are needed to represent mosaic floorplans.

1.5   Main Result

Theorem 1. *The set of n-block mosaic floorplans can be represented by $(3n-3)$ bits, which is optimal up to an additive lower order term.*

Most binary representations of mosaic floorplans discussed in section 1.4 are complex. In contrast, the representation introduced in this paper is very simple.

By using the bijection between mosaic floorplans and Baxter permutations described in [1], the methods in this paper also work on Baxter permutations. Hence, the optimal representation of mosaic floorplans results in an optimal representation of all objects in the Baxter combinatorial family.

2   Optimal Representation of Mosaic Floorplans

In this section, an optimal representation of mosaic floorplans is described.

2.1   Standard Form of Mosaic Floorplans

Let M be a mosaic floorplan. Let h be a horizontal line segment in M. The *upper segment set* of h and the *lower segment set* of h are defined as the following:

ABOVE(h) = the set of vertical line segments above h that intersect h.
BELOW(h) = the set of vertical line segments below h that intersect h.

Similarly, for a vertical line segment v in M, the *left segment set* of v and the *right segment set* of h are defined as the following:

LEFT(v) = the set of horizontal segments on the left of v that intersect v.
RIGHT(v) = the set of horizontal segments on the right of v that intersect v.

Definition 5. *A mosaic floorplan M is in* standard form *if the following hold:*

1. *For every horizontal segment h in M, all vertical segments in ABOVE(h) appear to the right of all vertical segments in BELOW(h). (Figure 3(a))*
2. *For every vertical segment v in M, all horizontal segments in RIGHT(v) appear above all horizontal segments in LEFT(v). (Figure 3(b))*

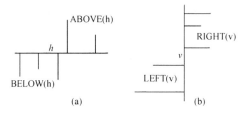

Fig. 3. Standard form of mosaic floorplans

The mosaic floorplan shown in Figure 1 (c) is the standard form of mosaic floorplans shown in Figure 1 (a) and Figure 1 (b).

The standard form $M_{standard}$ of a mosaic floorplan M can be obtained by sliding its vertical and horizontal line segments. Because of the equivalence definition of mosaic floorplans, $M_{standard}$ and M are considered the same mosaic floorplans. For a given M, $M_{standard}$ can be obtained in linear time by using the horizontal constraint graphs and vertical constraint graphs described in [9]. From now on, all mosaic floorplans are assumed to be in standard form.

2.2 Staircases

Definition 6. *A staircase is an object that satisfies the following conditions:*

1. *The border is formed by a line segment on the positive x-axis starting from the origin and a line segment on the positive y-axis starting from the origin connected by non-increasing vertical and horizontal line segments.*
2. *The interior is divided into smaller rectangular subsections by horizontal and vertical line segments.*
3. *No four subsections meet at the same point.*

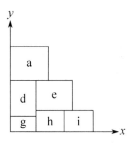

Fig. 4. A staircase with $n = 6$ blocks and $m = 3$ steps that is obtained from the mosaic floorplan in Figure 1 (c) by deleting blocks b, c and f

A *step* of a staircase S is a horizontal line segment on the border of S, excluding the x-axis. Figure 4 shows a staircase with $n = 6$ blocks and $m = 3$ steps. Note that a mosaic floorplan is a staircase with $m = 1$ step.

2.3   Deletable Rectangles

Definition 7. *A* deletable rectangle *of a staircase S is a block that satisfies the following conditions:*

1. *Its top edge is completely contained in the border of S.*
2. *Its right edge is completely contained in the border of S.*

In the staircase shown in Figure 4, the block a is the only deletable rectangle. The concept of deletable rectangles is a key idea for the methods introduced in this paper. This concept was originally defined in [16] for their $(4n-4)$-bit representation of floorplans. However, a modified definition of deletable rectangles is used in this paper to create a $(3n-3)$-bit representation of mosaic floorplans.

Lemma 1. *The removal of a deletable rectangle from a staircase results in another staircase unless the original staircase contains only one block.*

Proof. Let S be a staircase with more than one block and let r be a deletable rectangle in S. Define S' to be the object that results when r is removed from S. Because the removal of r still leaves S' with at least one block, the border of S' still contains a line segment on the x-axis and a line segment on the y-axis, so condition (1) of a staircase holds for S'. Removing r will not cause the remainder of the border to have an increasing line segment because the right edge of r must be completely contained in the border, so condition (2) of a staircase also holds for S'. The removal of r does not form new line segments, so the interior of S' will still be divided into smaller rectangular subsections by vertical and horizontal line segments, and no four subsections in S' will meet at the same point. Thus, conditions (3) and (4) of a staircase hold for S'. Therefore, S' is a staircase.

The basic ideas of the representation can now be outlined. Given a mosaic floorplan M, the deletable rectangles of M are removed one by one. By Lemma 1, this results in a sequence of staircases, until only one block remains. The necessary location information of these deletable rectangles are recorded so that the original mosaic floorplan M can be reconstructed. However, if there are multiple deletable rectangles for these staircases, many more bits will be needed. Fortunately, the following key lemma shows that this does not happen.

Lemma 2. *Let M be a n-block mosaic floorplan in standard form. Let $S_n = M$, and let S_{i-1} $(2 \leq i \leq n)$ be the staircase obtained by removing a deletable rectangle r_i from S_i.*

1. *There is a single, unique deletable rectangle in S_i for $1 \leq i \leq n$.*
2. *r_{i-1} is adjacent to r_i for $2 \leq i \leq n$.*

Proof. The proof is by reverse induction.

$S_n = M$ has only one deletable rectangle located in the top right corner.

Assume that S_{i+1} $(i \leq n-1)$ has exactly one deletable rectangle r_{i+1}. Let h be the horizontal line segment in S_{i+1} that contains the bottom edge of r_{i+1}, and let v be the vertical line segment in S_{i+1} that contains the left edge of r_{i+1} (Figure 5). Let a be the uppermost block in S_{i+1} whose right edge aligns with v, and let b be the rightmost block in S_{i+1} whose top edge aligns with h. Note that either a or b may not exist, but at least one will exist because $2 \le i$. After r_{i+1} is removed from S_{i+1}, a and b are the only candidates for deletable rectangles of the resulting staircase S_i. There are two cases:

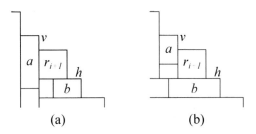

(a)　　　　　　　　(b)

Fig. 5. Proof of Lemma 2

1. The line segments h and v form a $\vdash$-junction (Figure 5 (a)) Then, the bottom edge of a must be below h because M is a standard mosaic floorplan, and a is not a deletable rectangle in S_i. Thus, the block b is the only deletable rectangle in S_i.
2. The line segments h and v form a $\perp$-junction (see Figure 5 (b)) Then, the left edge of b must be to the left of v because M is a standard mosaic floorplan, and b is not a deletable rectangle in S_i. Thus, the block a is the only deletable rectangle in S_i.

In both cases, only one deletable rectangle r_i (which is either a or b) is revealed when the deletable rectangle r_{i+1} is removed. There is only one deletable rectangle in $S_n = M$, so all subsequent staircases contain exactly one deletable rectangle, and (1) is true. In both cases, r_{i+1} is adjacent to r_i, so (2) is true.

Let S be a staircase and r be a deletable rectangle of S whose top side is on the k-th step of S. There are four types of deletable rectangles.

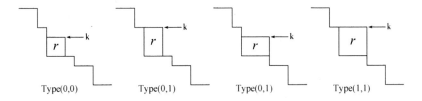

Type(0,0)　　　　Type(0,1)　　　　Type(0,1)　　　　Type(1,1)

Fig. 6. The four types of deletable rectangles

1. Type $(0,0)$:
   (a) The upper left corner of r is a $\vdash$-junction.
   (b) The lower right corner of r is a $\perp$-junction.
   (c) The deletion of r decreases the number of steps by one.
2. Type $(0,1)$:
   (a) The upper left corner of r is a $\top$-junction.
   (b) The lower right corner of r is a $\perp$-junction.
   (c) The deletion of r does not change the number of steps.
3. Type $(1,0)$:
   (a) The upper left corner of r is a $\vdash$-junction.
   (b) The lower right corner of r is a $\dashv$-junction.
   (c) The deletion of r does not change the number of steps.
4. Type $(1,1)$:
   (a) The upper left corner of r is a $\top$-junction.
   (b) The lower right corner of r is a $\dashv$-junction.
   (c) The deletion of r increases the number of steps by one.

2.4   Optimal Binary Representation

This binary representation of mosaic floorplans depends on the fact that a mosaic floorplan M is a special case of a staircase and the fact that the removal of a deletable rectangle from a staircase results in another staircase. The binary string used to represent M records the unique sequence of deletable rectangles that are removed in this process. The information stored by this binary string enables the original mosaic floorplan M to be reconstructed.

A 3-bit binary string is used to record the information for each deletable rectangle r_i. The string has two parts: the type and the location of r_i. To record the type of r_i, the bits corresponding to its type is stored directly. To store the location, note that, by Lemma 2, two consecutive deletable rectangles r_i and r_{i-1} are adjacent. Thus, they must share either a horizontal edge or a vertical edge. A single bit can be used to record the location of r_i with respect to r_{i-1}: a 1 if they share a horizontal edge, and a 0 if they share a vertical edge.

Encoding Procedure

Let M be the n-block mosaic floorplan to be encoded. Starting from $S_n = M$, remove the unique deletable rectangles r_i, where $2 \le i \le n$, one by one. For each deletable rectangle r_i, two bits are used to record the type of r_i, and one bit is used to record the type of the common boundary shared by r_i and r_{i-1}.

Decoding Procedure

The process starts with the staircase S_1, which is a single rectangle. Each staircase S_{i+1} can be reconstructed from S_i by using the 3-bit binary string for the deletable rectangle r_{i+1}. The 3-bit string records the type of r_{i+1} and the type of edge shared by r_i and r_{i+1}, so r_{i+1} can be uniquely added to S_i. Thus, the decoding procedure can reconstruct the original mosaic floorplan $S_n = M$.

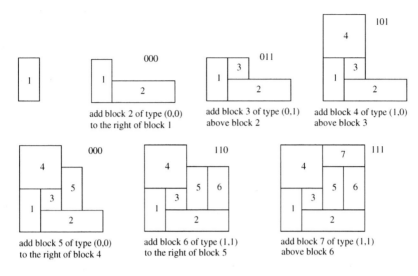

Fig. 7. The decoding of the binary representation (000 011 101 000 110 111)

The lower left block of the mosaic floorplan M (which is the only block of S_1) does not need any information to be recorded. Each of the other blocks of M needs three bits. Thus, the length of the representation of M is $(3n - 3)$ bits.

3 Conclusion

In this paper, a binary representation of n-block mosaic floorplans using $(3n - 3)$ bits was introduced. Since any representation of n-block mosaic floorplans requires at least $(3n - o(n))$ bits [14], this representation is optimal up to an additive lower term. This representation is very simple and easy to implement.

Mosaic floorplans have a bijection with Baxter permutations, so the optimal representation of mosaic floorplans leads to an optimal $(3n - 3)$ bit representation of Baxter permutations and all objects in the Baxter combinatorial family.

References

1. Ackerman, E., Barequet, G., Pinter, R.Y.: A bijection between permutations and floorplans, and its applications. Discrete Applied Mathematics 154, 1674–1684 (2006)
2. Amano, K., Nakano, S., Yamanaka, K.: On the number of rectangular drawings: Exact counting and lower and upper bounds. IPSJ SIG Notes 2007-AL-115-5C, 33–40 (2007)
3. Baxter, G.: On fixed points of the composite of commuting functions. Proceedings American Mathematics Society 15, 851–855 (1964)
4. Bonichon, N., Bousquet-Mélou, M., Fusy, É.: Baxter permutations and plane bipolar orientations. Séminaire Lotharingien de Combinatoire 61A (2010)

5. Canary, H.: Aztec diamonds and baxter permutations. The Electronic Journal of Combinatorics 17 (2010)
6. Dulucq, S., Guibert, O.: Baxter permutations. Discrete Mathematics 180, 143–156 (1998)
7. Fujimaki, R., Inoue, Y., Takahashi, T.: An asymptotic estimate of the numbers of rectangular drawings or floorplans. In: Proceedings 2009 IEEE International Symposium on Circuits and Systems, pp. 856–859 (2009)
8. Giraudo, S.: Algebraic and combinatorial structures on baxter permutations. Discrete Mathematics and Theoretical Computer Science, DMTCS (2011)
9. Hong, X., Huang, G., Cai, Y., Gu, J., Dong, S., Cheng, C.-K., Gu, J.: Corner-block list: An effective and efficient topological representation of non-slicing floorplan. In: Proceedings of the International Conference on Computer Aided Design (ICCAD 2000), pp. 8–12 (2000)
10. Lengauer, T.: Combinatorial Algorithms for Integrated Circuit Layout. John Wiley & Sons (1990)
11. Murata, H., Fujiyoshi, K.: Rectangle-packing-based module placement. In: Proceedings of the International Conference on Computer Aided Design (ICCAD 1995), pp. 472–479 (1995)
12. Nakano, S.: Enumerating Floorplans with n Rooms. In: Eades, P., Takaoka, T. (eds.) ISAAC 2001. LNCS, vol. 2223, pp. 107–115. Springer, Heidelberg (2001)
13. Sakanushi, K., Kajitani, Y., Mehta, D.P.: The quarter-state-sequence floorplan representation. IEEE Transactions on Circuits and Systems - I: Fundamental Theory and Applications 50(3), 376–386 (2003)
14. Shen, Z.C., Chu, C.C.N.: Bounds on the number of slicing, mosaic, and general floorplans. IEEE Transactions on Computer-Aided Design of Integrated Circuits and Systems 22(10), 1354–1361 (2003)
15. Speckmann, B., van Kreveld, M., Florisson, S.: A linear programming approach to rectangular cartograms. In: Proceedings 12th International Symposium on Spatial Data Handling (SDH), pp. 527–546 (2006)
16. Takahashi, T., Fujimaki, R., Inoue, Y.: A $(4n-4)$-Bit Representation of a Rectangular Drawing or Floorplan. In: Ngo, H.Q. (ed.) COCOON 2009. LNCS, vol. 5609, pp. 47–55. Springer, Heidelberg (2009)
17. van Kreveld, M., Speckmann, B.: On rectangular cartograms. Computational Geometry: Theory and Applications 37(3), 175–187 (2007)
18. Yamanaka, K., Nakano, S.: Coding floorplans with fewer bits. IEICE Transactions Fundamentals E89(5), 1181–1185 (2006)
19. Yamanaka, K., Nakano, S.: A Compact Encoding of Rectangular Drawings with Efficient Query Supports. In: Kao, M.-Y., Li, X.-Y. (eds.) AAIM 2007. LNCS, vol. 4508, pp. 68–81. Springer, Heidelberg (2007)
20. Yao, B., Chen, H., Cheng, C.-K., Graham, R.: Floorplan representation: Complexity and connections. ACM Transactions on Design Automation of Electronic Systems 8(1), 55–80 (2003)
21. Young, E.F.Y., Chu, C.C.N., Shen, Z.C.: Twin binary sequences: A nonredundant representation for general nonslicing floorplan. IEEE Transactions on Computer-Aided Design of Integrated Circuits and Systems 22(4), 457–469 (2003)

Succinct Strictly Convex Greedy Drawing of 3-Connected Plane Graphs

Jiun-Jie Wang and Xin He*

Department of Computer Science and Engineering,
University at Buffalo, Buffalo, NY, 14260, USA
{jiunjiew,xinhe}@buffalo.edu

Abstract. Geometric routing by using virtual locations is an elegant way for solving network routing problems. Greedy routing, where a message is simply forwarded to a neighbor that is closer to the destination, is a simple form of geometric routing. Papadimitriou and Ratajczak conjectured that every 3-connected plane graph has a greedy drawing in the $\mathcal{R}^2$ plane [10]. Leighton and Moitra settled this conjecture positively in [9]. However, their drawings have two major drawbacks: (1) their drawings are not necessarily planar; and (2) $\Omega(n \log n)$ bits are needed to represent the coordinates of their drawings, which is too large for routing algorithms for wireless networks. Recently, He and Zhang [8] showed that every triangulated plane graph has a succinct (using $O(\log n)$ bit coordinates) greedy drawing in $\mathcal{R}^2$ plane with respect to a metric function derived from Schnyder realizer. However, their method fails for 3-connected plane graphs. In this paper, we show that every 3-connected plane graph has drawing in the $\mathcal{R}^2$ plane, that is succinct, planar, strictly convex, and is greedy with respect to a metric function based on parameters derived from Schnyder wood.

1 Introduction

As communication technology progresses, traditional wired communication networks are rapidly replaced by wireless networks (such as sensor networks). The nodes of such networks are equipped with very limited memory and computing power. Thus the traditional network communication protocols are not suitable for them. *Geometric routing* is an interesting class of routing algorithms for wireless communication which uses the geographic location to determine routing paths. The simplest geometric routing is *greedy routing*: to send a message from a source node s to a destination node t, s simply forwards the message to a neighbor that is closer to t.

However, greedy routing has drawbacks: the nodes need to be equipped GPS devises in order to determine their geographic location, which are too expensive and power consuming. Even worse, a node s might be located in a *void position*, (namely s has no neighbor that is closer to the destination). In this

---

* Research supported in part by NSF Grant CCR-0635104.

J. Snoeyink, P. Lu, K. Su, and L. Wang (Eds.): FAW-AAIM 2012, LNCS 7285, pp. 13–25, 2012.

case, the greedy routing completely fails. As a solution, Papadimitriou et. al. [10] introduced the concept of *greedy drawing*: Instead of using real geographic coordinates, one could use graph drawing to compute the drawing coordinates for the nodes of a network G. Then geometric routing algorithms rely on drawing coordinates to determine the routing paths. Simply speaking, a *greedy drawing* is a drawing of G for which the greedy routing works. More precisely:

Definition 1. [10] Let S be a set and $H(*, *)$ a metric function over S. Let $G = (V, E)$ be a graph.

1. A *drawing* of G into S is a 1-1 mapping $d : V \to S$.
2. The drawing d is a *greedy drawing* with respect to H if for any two vertices u, w of G ($u \neq w$), u has a neighbor v such that $H(d(u), d(w)) > H(d(v), d(w))$.
3. The drawing d is a *weakly greedy drawing* with respect to H if for any two vertices u, w of G ($u \neq w$), u has a neighbor v such that $H(d(u), d(w)) \geq H(d(v), d(w))$.

The following conjecture was posed in [10]:

Greedy Embedding Conjecture: Every 3-connected plane graph has a greedy drawing in the Euclidean plane $\mathcal{R}^2$.

Leighton and Moitra [9] recently settled this conjecture positively. A similar result was obtained by Angelini et al. [3]. However, their drawing algorithms have drawbacks. First, as pointed out in [7], their drawings are not necessarily planar nor convex. Second, $\Omega(n \log n)$ bits are needed to represent the drawing coordinates produced by their algorithms. This is the same space usage as traditional routing table approaches, which is not practical for sensor networks. To make this routing scheme work in practice, we need a *succinct* greedy drawing: Namely, the drawing coordinates can be represented by using $O(\log n)$ bits.

Two recent papers made some progresses toward this goal. Goodrich et al. [7] used a set of virtual coordinates from the structure of *Christmas cactus* [9] to represent the vertex positions. However, their succinct representation of each vertex is totally different from the real underlying geometric embedding. Hence it is not a true greedy drawing. He and Zhang [8] showed that the classical Schnyder drawing of triangulated plane graphs is a succinct greedy drawing in $\mathcal{R}^2$ with respect to a simple and natural metric function H. They showed that the Schnyder drawing for 3-connected plane graphs is succinct, planar and convex, and *weakly greedy* with respect to the same function H. With a greedy drawing, the greedy routing algorithm is very simple: the source node s just sends the message to a neighbor that is strictly closer to the destination t. With a weakly greedy drawing, the routing algorithm is more complicated. Since s might only have a neighbor u whose distance to t is the same, s might have to send the message to u. Hence the message might be sent back and forth among nodes with equal distances to t, but never reaches t.

Papadimitriou and Ratajczak also posed another related conjecture [10]:

Convex Greedy Embedding Conjecture: Every 3-connected plane graph has a convex greedy embedding in the Euclidean plane.

Cao et al. [5] recently showed that there exists a 3-connected plane graph G such that any convex greedy drawing of G in the $\mathcal{R}^2$ Euclidean plane must use $\Omega(n)$-bit coordinates. Thus it is impossible to find a succinct, convex, greedy drawing on $\mathcal{R}^2$ Euclidean plane for 3-connected plane graphs. In other words, in order to find a drawing in $\mathcal{R}^2$ plane that is succinct, greedy, and convex, one must give up Euclidean distance. In this paper, we describe a drawing for 3-connected plane graphs in $\mathcal{R}^2$ plane that is succinct, strictly convex, and greedy with respect to a metric function based on *Schnyder woods*.

The classical Schnyder drawing of 3-connected plane graphs [11] is based on three *Schnyder coordinates*, which are obtained by counting the number of faces in the three regions divided by a Schnyder wood. It was shown in [8] that this drawing is only weakly greedy with respect to the metric function H. Our new drawing algorithm is also based on Schnyder woods. However, in addition to the three Schnyder coordinates, our algorithm also uses other information obtained from Schnyder woods, which we call *Schnyder parameters*. (There are totally 9 parameters, including three Schnyder coordinates). Schnyder parameters are used to calculate the drawing coordinates and to define the metric function.

The present paper is organized as follows. In section 2, we give the definition and basic properties of Schnyder woods. In section 3, we describe the metric function $H(u, v)$ used in our greedy drawing. We also show how to transform the Schnyder parameters into a set of equations which can be used to determine the relative locations of the vertices. In section 4, we show our drawing has the greedy property with respect to the the metric function $H(u, v)$. In section 5, we describe how to obtain drawing coordinates from Schnyder parameters and show that the drawing is planar and strictly convex.

2   Preliminaries

Definition 2. [1,6] Let G be a 3-connected plane graph with three external vertices v_1, v_2, v_3 in counterclockwise (ccw) order. A *Schnyder wood* of G is a triple of rooted spanning trees $\{T_1, T_2, T_3\}$ of G with the following properties:

- For $i \in \{1, 2, 3\}$, the root of T_i is v_i, the edges of G are directed from children to parent in T_i.
- Each edge e of G is contained in at least one and at most two trees. If e is contained in two spanning trees, then it has different directions in the two trees.
- For each vertex $v \notin \{v_1, v_2, v_3\}$ of G, v has exactly one edge leaving v in each of T_1, T_2, T_3. The ccw order of the edges incident to v is: leaving in T_1, entering in T_3, leaving in T_2, entering in T_1, leaving in T_3, and entering in T_2. Each entering block may be empty. An edge with two opposite directions is considered twice. The first and the last incoming edges are possibly coincident with the outgoing edges.

Fig 1 (1) and (2) show two examples of edge pattern around a vertex v. (In the second example, the edge leaving v in T_3 and an edge entering v in T_2 are the same edge). Fig 1 (3) shows an example of Schnyder wood. The edges in T_1, T_2, T_3 are drawn as red solid, blue dashed, and green dotted lines respectively. Each edge of G belongs to 1 or 2 trees, and is said to be *unidirectional* or *bidirectional*, respectively.

Let $\{T_1, T_2, T_3\}$ be a Schnyder wood of G. We assume a cyclic structure on the set $\{1, 2, 3\}$ so that $i - 1$ and $i + 1$ are always defined. Namely if $i = 3$ then $i + 1 = 1$ and if $i = 1$ then $i - 1 = 3$. For each vertex u and $i \in \{1, 2, 3\}$, let $P_i(u)$ denote the path in T_i from u to the root v_i of T_i. If u is the parent of w in T_i, then w is an *i-child* of u. If u is an ancestor of w in T_i, then u is an *i-ancestor* of w, and w is an *i-descendant* of u. $R_i(u)$ denotes the region of G bounded by the paths $P_{i-1}(u), P_{i+1}(u)$ and the exterior path between the vertex v_{i-1} and v_{i+1}, excluding the vertices on the path $P_{i-1}(u)$. $x_i(u)$ denotes the number of faces of G in $R_i(u)$. $n_i(u)$ denotes the ccw pre-order number of the node u in T_i, and $s_i(u)$ denotes the number of descendants of u in T_i.

We call $x_i(u), n_i(u), s_i(u)$ $(i = 1, 2, 3)$ the *Schnyder parameters* of u. They will be used to define the drawing coordinates and the distance between vertices.

We further partition the vertices in the region $R_i(u)$ into four subsets:

1. The *i-Descendant*: $D_i(u) = \{v \mid v \text{ is an } i\text{-descendent of } u\}$.
2. The *i-Boundary*: $B_i(u) = \{v \mid v \text{ is a vertex} \in P_{i+1}(u)\}$.
3. The *i-Left-Cousin*: $LC_i(u) = \{v \mid v \text{ is an } i\text{-descendent of } w \in P_{i+1}(u) \text{ where } w \neq u\}$.
4. The *i-Right-Cousin*: $RC_i(u) = \{v \mid v \text{ is an } i\text{-descendent of } w \in P_{i-1}(u) \text{ where } w \neq u\}$.

In Fig 1 (3), we have: $P_2(b) = \{b, a, v_2\}$, $P_3(b) = \{b, e, v_3\}$, $R_1(b) = \{v_2, a, d, g, h\}$. $B_1(b) = \{v_2, a\}$, $D_1(b) = \{d, g\}$, $LC_1(b) = \{h\}$ and $RC_1(b) = \emptyset$.

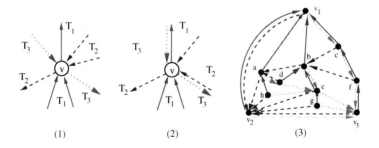

(1)   (2)   (3)

Fig. 1. (1) and (2) Two examples of edge pattern around an internal vertex v; (3) A 3-connected graph G with its Schnyder wood

3   Metric Function $H(u, v)$

The metric function H used for greedy drawing is a hierarchical function with two components $H_1(u, v)$ and $H_2(u, v)$. Let $\mathcal{Q} = 2n - 4$ (which is strictly greater than the number of internal faces in G). For two vertices u and v, define:

$$H_1(u, v) = |x_1(u) - x_1(v)| + |x_2(u) - x_2(v)| + |x_3(u) - x_3(v)| \tag{1}$$

$$H_2(u, v) = \begin{cases} 0 & \text{if } u = v \\ \mathcal{Q} + \min\{|x_i(u) - x_i(v)|, |x_{i-1}(u) - x_{i-1}(v)|\} & \text{if } v \in B_i(u) \\ \mathcal{Q} + \min\{|x_i(u) - x_i(v)|, |x_{i+1}(u) - x_{i+1}(v)|\} & \text{if } v \in LC_i(u) \\ \mathcal{Q} + \min\{|x_{i-1}(u) - x_{i-1}(v)|, |x_{i+1}(u) - x_{i+1}(v)|\} & \text{if } v \in D_i(u) \\ \mathcal{Q} + \min\{|x_i(u) - x_i(v)|, |x_{i-1}(u) - x_{i-1}(v)|\} & \text{if } v \in RC_i(u) \end{cases} \tag{2}$$

We pack the two components of $H(u, v)$ into a single integer as follows:

$$H(u, v) = H_1(u, v) \times 2\mathcal{Q} + H_2(u, v) \tag{3}$$

Note that $\min\{|x_j(u) - x_j(v)|, |x_k(u) - x_k(v)|\} < \mathcal{Q}$ $(1 \leq j, k \leq 3)$. Hence $H_2(u, v) < 2\mathcal{Q}$. So, for any four vertices u, v, w, z, $H(u, v) < H(w, z)$ if and only if $H_1(u, v) < H_1(w, z)$; or $H_1(u, v) = H_1(w, z)$ and $H_2(u, v) < H_2(w, z)$.

Lemma 1. *Let u and v be any two vertices of G. Then: (1) $v \in D_i(u)$ if and only if $u \in B_{i-1}(v)$. (2) $v \in LC_i(u)$ if and only if $u \in RC_{i+1}(v)$.*

Proof. We prove the lemma for $i = 1$. The other cases are symmetric.
  (1) From Fig 2 (1), it is easy to see that $v \in D_1(u)$ if and only if $u \in B_3(v)$.
  (2) From Fig 2 (2), it is easy to see that $v \in LC_1(u)$ if and only if $u \in RC_2(v)$. □

Theorem 1. *$H(u, v)$ is a metric function.*

Proof. Since $H_1(u, v)$ is the Manhattan distance in R^3, it is a metric function. In the following we show $H_2(u, v)$ is also a metric function. Since $H(u, v)$ is a

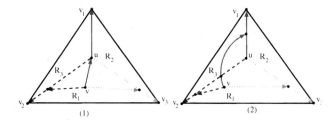

Fig. 2. The proof of Lemma 1

linear combination of $H_1(u,v)$ and $H_2(u,v)$, this implies that $H(u,v)$ is a metric function.

First, by the definition of $H_2(u,v)$, we have $H_2(u,v) \geq 0$, and $H_2(u,v) = 0$ if and only if $u = v$. Next, we show $H_2(u,v) = H_2(v,u)$ for all u,v. Assume $v \in R_1(u)$. Other cases are symmetric.

Case 1: $v \in D_1(u)$. Then $H_2(u,v) = Q + \min\{|x_3(u) - x_3(v)|, |x_2(u) - x_2(v)|\}$. By Lemma 1, $u \in B_3(v)$. Thus $H_2(v,u) = Q + \min\{|x_3(v) - x_3(u)|, |x_2(v) - x_2(u)|\}$. Hence $H_2(u,v) = H_2(v,u)$.

Case 2: $v \in B_1(u)$. Then $H_2(u,v) = Q + \min\{|x_1(u) - x_1(v)|, |x_3(u) - x_3(v)|\}$. By Lemma 1, $u \in D_2(v)$. Thus $H_2(v,u) = Q + \min\{|x_1(v) - x_1(u)|, |x_3(v) - x_3(u)|\}$. Hence $H_2(u,v) = H_2(v,u)$.

Case 3: $v \in LC_1(u)$. Then $H_2(u,v) = Q + \min\{|x_1(u) - x_1(v)|, |x_2(u) - x_2(v)|\}$. By Lemma 1, $u \in RC_2(v)$. Thus: $H_2(v,u) = Q + \min\{|x_2(v) - x_2(u)|, |x_1(v) - x_1(u)|\}$. Hence $H_2(u,v) = H_2(v,u)$.

Case 4: $v \in RC_1(u)$. Then $H_2(u,v) = Q + \min\{|x_1(u) - x_1(v)|, |x_3(u) - x_3(v)|\}$. By Lemma 1, $u \in LC_3(v)$. Thus: $H_2(v,u) = Q + \min\{|x_3(v) - x_3(u)|, |x_1(v) - x_1(u)|\}$. Hence $H_2(u,v) = H_2(v,u)$.

Next we show that the triangle inequality holds for H_2. For any u,v,w:

$$
\begin{aligned}
& H_2(u,v) + H_2(v,w) \\
=\ & Q + \min\{|x_i(u) - x_i(v)|, |x_{i+1}(u) - x_{i+1}(v)|\} \\
& + Q + \min\{|x_i(v) - x_i(w)|, |x_{i+1}(v) - x_{i+1}(w)|\} \\
\geq\ & 2Q > Q + \min\{|x_i(u) - x_i(w)|, |x_{i+1}(u) - x_{i+1}(w)|\} = H_2(u,w)
\end{aligned}
$$

This completes the proof of the Theorem. □

For two vertices u and v, in order to calculate the distance $H(u,v)$, we need to know the relative locations between u and v. More specifically: First, we need to know if v is in $R_1(u)$, $R_2(u)$ or $R_3(u)$. Second, we need to know if v is an i-Descendant, i-Boundary, i-Left-Cousin or i-Right-Cousin of u. In the following, we describe how to determine these information by using Schnyder parameters of u and v (namely, $x_i(u), x_i(v), n_i(u), n_i(v), s_i(u), s_i(v)$).

We use an index function to represent the relative locations of u and v. It has two components. The first one is used to decide which of regions $R_1(u)$, $R_2(u)$, $R_3(u)$ contains v. The second one is used to determine if v is in $D_i(u)$, $LC_i(u)$, $RC_i(u)$, or $B_i(u)$.

Define a function $I(x)$ as follows: $I(x) = 1$ if $x \geq 0$ and $I(x) = 0$ if $x < 0$.

Observation 1. *Let u,v be any two vertices of G.*

1. *$v \in R_1(u)$ if and only if $n_3(v) > n_3(u)$ and $n_2(u) > n_2(v)$.*
2. *$v \in R_2(u)$ if and only if $n_1(v) > n_1(u)$ and $n_3(u) > n_3(v)$.*
3. *$v \in R_3(u)$ if and only if $n_2(v) > n_2(u)$ and $n_1(u) > n_1(v)$.*

For simplicity, we define the following items:

- $\mathcal{R}^1(u,v) = I(n_3(v) - n_3(u)) \times I(n_2(u) - n_2(v))$.
- $\mathcal{R}^2(u,v) = I(n_1(v) - n_1(u)) \times I(n_3(u) - n_3(v))$.
- $\mathcal{R}^3(u,v) = I(n_2(v) - n_2(u)) \times I(n_1(u) - n_1(v))$.

Lemma 2. *For any two vertices u, v of G, $v \in R_i(u)$ if and only if $\mathcal{R}^i(u,v) = 1$.*

Proof. Immediate from Observation 1. □

Observation 2. *Let u, v be any two vertices of G. The necessary and sufficient conditions that $v \in D_i(u)$, $v \in B_i(u)$, $v \in LC_i(u)$, or $v \in RC_i(u)$ are given in Table 1. (The first two rows in the table are the conditions for $v \in R_i(u)$).*

Table 1. Necessary and sufficient conditions for deciding relative locations of u and v

$v \in B_i(u)$	$v \in LC_i(u)$	$v \in D_i(u)$	$v \in RC_i(u)$
$n_{i+1}(u) > n_{i+1}(v)$	$n_{i+1}(u) > n_{i+1}(v)$	$n_{i+1}(u) > n_{i+1}(v)$	$n_{i+1}(u) > n_{i+1}(v)$
$n_{i-1}(v) > n_{i-1}(u)$	$n_{i-1}(v) > n_{i-1}(u)$	$n_{i-1}(v) > n_{i-1}(u)$	$n_{i-1}(v) > n_{i-1}(u)$
$n_i(u) > n_i(v)$	$n_i(u) > n_i(v)$	$n_i(u) < n_i(v)$	$n_i(u) + s_i(u) < n_i(v)$
$n_{i+1}(v) + s_{i+1}(v)$ $\geq n_{i+1}(u)$	$n_{i+1}(u) > n_{i+1}(v)$ $+ s_{i+1}(v)$	$n_i(v) \leq n_i(u) + s_i(u)$	

Note: In Lemma 1 (1), it is stated that $v \in D_i(u)$ if and only if $u \in B_{i-1}(v)$. For $i = 1$, it states that $v \in D_1(u)$ if and only if $u \in B_3(v)$. From Table 1, $v \in D_1(u)$ if and only if the following conditions hold: (i) $n_2(u) > n_2(v)$; (ii) $n_3(v) > n_3(u)$; (iii) $n_1(u) < n_1(v)$ and (iv) $n_1(u) + s_1(u) \geq n_1(v)$.

On the other hand, from Table 1, $u \in B_3(v)$ if and only if: (i') $n_1(v) > n_1(u)$; (ii') $n_2(u) > n_2(v)$, (iii') $n_3(u) < n_3(v)$ and (iv') $n_1(u) + s_1(u) \geq n_1(v)$. So the conditions for $v \in D_1(u)$ and the conditions for $u \in B_3(v)$ are identical. Similarly, from Table 1, the conditions for $v \in LC_i(u)$ and the conditions for $u \in RC_{i+1}(v)$ are also identical.

For simplicity, we define the following terms ($i = 1, 2, 3$):

- $\mathcal{B}^i(u,v) = I((n_i(u) - n_i(v)) \times (n_{i+1}(v) + s_{i+1}(v) - n_{i+1}(u)))$.
- $\mathcal{LC}^i(u,v) = I((n_i(u) - n_i(v)) \times (n_{i+1}(u) - (n_{i+1}(v) + s_{i+1}(v))))$.
- $\mathcal{D}^i(u,v) = I((n_i(v) - n_i(u)) \times (n_i(u) + s_i(u) - n_i(v)))$.
- $\mathcal{RC}^i(u,v) = I(n_i(v) - (n_i(u) + s_i(u)))$.

Lemma 3. *Let u and v be two vertices of G where $v \in R_i(u)$ ($i = 1, 2, 3$):*

1. *$v \in B_i(u)$ if and only if $\mathcal{B}^i(u,v) = 1$.*
2. *$v \in LC_i(u)$ if and only if $\mathcal{LC}^i(u,v) = 1$.*
3. *$v \in D_i(u)$ if and only if $\mathcal{D}^i(u,v) = 1$.*
4. *$v \in RC_i(u)$ if and only if $\mathcal{RC}^i(u,v) = 1$.*

Proof. Immediate from Observation 2. □

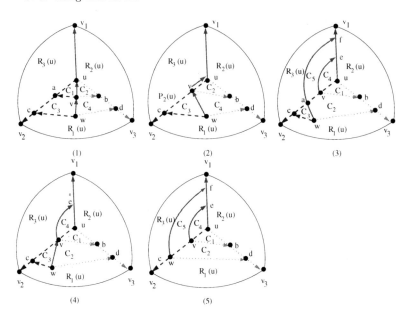

Fig. 3. The proof of Theorem 2

Let $H_2^{i\mathcal{B}}(u, v)$ denote the function $H_2(u, v)$ for $v \in B_i(u)$, $H_2^{i\mathcal{LC}}(u, v)$ denote the function $H_2(u, v)$ for $v \in LC_i(u)$, $H_2^{i\mathcal{D}}(u, v)$ denote the function $H_2(u, v)$ for $v \in D_i(u)$ and $H_2^{i\mathcal{RC}}(u, v)$ denote the function $H_2(u, v)$ for $v \in RC_i(u)$. From above lemmas, the explicit definition of the distance function $H(u, v)$ defined in equation (3) can be written as:

$$H(u, v) = 2\mathcal{Q} \times H_1(u, v) + \frac{1}{2} \sum_{i=1}^{3} \mathcal{R}^i(u, v) \times \sum_{\alpha \in \{\mathcal{B}, \mathcal{LC}, \mathcal{D}, \mathcal{RC}\}} \alpha^i(u, v) \times H_2^{i\alpha}(u, v)$$

4 Greedy Drawing Property

In this section, we show that the drawing defined by Schnyder parameters has greedy drawing property with respect to the function $H(u, v)$.

Theorem 2. *For any two vertices u, w, u has a neighbor v such that $H(u, w) > H(v, w)$.*

Proof. Without loss of generality, we assume $w \in R_1(u)$. Other cases are symmetric. Due to space limitation, we only prove the Case 2a (the most complicated case). The complete proof will be given in the full paper.

Case 2: $w \in LC_1(u)$. Let v be the neighbor of u on $P_2(u)$.

Case 2a: $w \in LC_1(v)$ (see Fig 3 (3)).

Let $a \in P_2(u)$ be the first common vertex of $P_2(u)$ and $P_1(w)$, $b \in P_3(u)$ be the first common vertex of $P_3(u)$ and $P_3(v)$, $c \in P_2(u)$ be the first common vertex of $P_2(u)$ and $P_2(w)$, $d \in P_3(u)$ be the first common vertex of $P_3(u)$ and $P_3(w)$, $e \in P_1(u)$ be the first common vertex of $P_1(u)$ and $P_1(v)$ and $f \in P_1(u)$ be the first common vertex of $P_1(u)$ and $P_1(w)$. Let C_1, C_2, C_3, C_4, C_5 be the regions bounded by the paths $P_i(u), P_i(v)$ and $P_i(w)$ as shown in Fig 3 (3). (Some of these regions might be empty.) Then:

$$
\begin{aligned}
H_1(u, w) &= (\#(C_1) + \#(C_2) + \#(C_3)) + (\#(C_1) + \#(C_2) + \#(C_4) + \#(C_5)) \\
&\quad + |\#(C_3) - (\#(C_4) + \#(C_5))| \\
H_1(v, w) &= (\#(C_2) + \#(C_3)) + (\#(C_2) + \#(C_5)) + |\#(C_3) - \#(C_5)|
\end{aligned}
$$

Case 2a1: $\#(C_5) > \#(C_3)$. In this case, we have:

$$
\begin{aligned}
H_1(u, w) &= (\#(C_1) + \#(C_2) + \#(C_3)) + (\#(C_1) + \#(C_2) + \#(C_4) + \#(C_5)) \\
&\quad + (\#(C_4) + \#(C_5)) - \#(C_3)) \\
H_1(v, w) &= (\#(C_2) + \#(C_3)) + (\#(C_2) + \#(C_5)) + (\#(C_5) - \#(C_3))
\end{aligned}
$$

Clearly $H_1(u, w) > H_1(v, w)$ is equivalent to $2(\#(C_1) + \#(C_4)) > 0$. Although either C_1 or C_4 can be empty, they cannot be both empty. (Otherwise the edge (u, v) will have to be in all three trees.) So $2(\#(C_1) + \#(C_4)) > 0$ and $H_1(u, w) > H_1(v, w)$. This implies $H(u, w) > H(v, w)$.

Case 2a2: $\#(C_5) \leq \#(C_3)$ and $\#(C_4) + \#(C_5) > \#(C_3)$. In this case, we have:

$$
\begin{aligned}
H_1(u, w) &= (\#(C_1) + \#(C_2) + \#(C_3)) + (\#(C_1) + \#(C_2) + \#(C_4) + \#(C_5)) \\
&\quad + (\#(C_4) + \#(C_5)) - \#(C_3)) \\
H_1(v, w) &= (\#(C_2) + \#(C_3)) + (\#(C_2) + \#(C_5)) + (\#(C_3) - \#(C_5))|
\end{aligned}
$$

Clearly $H_1(u, w) > H_1(v, w)$ is equivalent to $2(\#(C_1) + \#(C_4) + \#(C_5)) > 2\#(C_3)$. This is true because we assumed $\#(C_4) + \#(C_5) > \#(C_3)$ in this case. So $H_1(u, w) > H_1(v, w)$. This implies $H(u, w) > H(v, w)$.

Case 2a3: $\#(C_4) + \#(C_5) \leq \#(C_3)$. In this case we have:
$$
H_1(u, w) = 2 \times (\#(C_1) + \#(C_2) + \#(C_3)) \quad H_1(v, w) = 2 \times (\#(C_2) + \#(C_3))
$$
If $\#(C_1) > 0$, then $H_1(u, w) > H_1(v, w)$ and hence $H(u, w) > H(v, w)$.

If $\#(C_1) = 0$, then $H_1(u, w) = H_1(v, w)$. We must consider $H_2(u, w)$ and $H_2(v, w)$:

$$
\begin{aligned}
H_2(u, w) &= Q + \min\{|x_1(u) - x_1(w)|, |x_2(u) - x_2(w)|\} \\
&= Q + \min\{[\#(C_1) + \#(C_2) + \#(C_3)], \\
&\qquad\qquad [\#(C_1) + \#(C_2) + \#(C_4) + \#(C_5)]\} \\
&= Q + [\#(C_2) + \#(C_4) + \#(C_5)]
\end{aligned}
$$

The last equality is because we assumed $\#(C_4) + \#(C_5) \leq \#(C_3)$ and $\#(C_1) = 0$.

$$H_2(v,w) = Q + \min\{|x_1(v) - x_1(w)|, |x_2(v) - x_2(w)|\}$$
$$= Q + \min\{[\#(C_2) + \#(C_3)], [\#(C_2) + \#(C_5)]\}$$
$$= Q + [\#(C_2) + \#(C_5)]$$

The last equality is because $\#(C_5) \leq \#(C_4) + \#(C_5) \leq \#(C_3)$. Since $\#(C_1) = 0$, C_4 cannot be empty and $\#(C_4) > 0$. Thus $H_2(u,w) > H_2(v,w)$. This implies $H(u,w) > H(v,w)$.

5 Strictly Convex Embedding

In this section, we describe a drawing of a 3-connected plane graph G by using Schnyder parameters. Our algorithm is based on an elegant algorithm in [4], which constructs a strictly convex drawing of G by using Schnyder coordinates. This algorithm has three steps:

Step 1: Draw each vertex u of G at the coordinates $(x_1(u), x_2(u))$. Recall that $x_i(u)$ is the number of internal faces in the region $R_i(u)$.

Step 2: Enlarge the drawing by a factor of n^h (where h is a constant integer). Namely, draw each vertex u at the new coordinates $(x_1(u) \times n^h, x_2(u) \times n^h)$.

Step 3: Perform a *fine perturbation* for each vertex $u \in G$: Each vertex u is moved $f_x(u)$ units along x-direction and $f_y(u)$ units along y-direction where $f_x(u)$ and $f_y(u)$ are two integers such that $|f_x(u)| < n^h$ and $|f_y(u)| < n^h$. So the final coordinates of u is:

$$(x_1(u) \times n^h + f_x(u), x_2(u) \times n^h + f_y(u)) \tag{4}$$

It is well known that the drawing obtained in Step 1 is a convex drawing of G, but might not be strictly convex. In Step 2, this convex drawing is enlarged by a large factor n^h. During Step 3, each vertex u is moved to a new location by a small distance, this results in a strictly convex drawing [4]. Let $F_x = \max\{f_x(u)|\forall u \in G\}$ and $F_y = \max\{f_y(u)|\forall u \in G\}$. Let $\mathcal{F} = \max\{2n - 5, F_x, F_y\} + 1$. If we replace the term n^h in Equation (4) by $\mathcal{F}^h$, the arguments in [4] still work. Thus we have:

Claim. The mapping $\overline{d} : u \to (\overline{X}(u), \overline{Y}(u))$ where:

$$\overline{X}(u) = x_1(u) \times \mathcal{F}^h + f_x(u) \quad \overline{Y}(u) = x_2(u) \times \mathcal{F}^h + f_y(u) \tag{5}$$

defines a strictly convex drawing of G on $\mathcal{R}^2$ plane.

Next we prove some properties of the drawing produced by the above algorithm.

Lemma 4. *Consider the drawing of G defined by $\overline{d} : u \to (\overline{X}(u), (\overline{Y}(u))$. Let F be a face of G. Let v, u, w be three consecutive vertices on F. Let L be the line segment connecting v and w. Then the distance between u and L is at least $\frac{1}{\sqrt{2}\mathcal{F}^{2(h+1)}}$.*

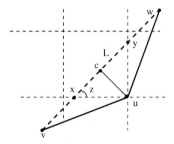

Fig. 4. The proof of Lemma 4

Proof. Let s_L be the slope of the line L. Then $s_L = \frac{a}{b}$ for some integers a, b such that $0 \le |a|, |b| \le \mathcal{F}^{h+1}$ (see Fig 4). Let x (respectively, y) denote the intersection point of L with the horizontal (respectively, vertical) line that passes u.

The distance between u and y is: $D_y = \overline{Y}(v) + [\overline{X}(u) - \overline{X}(v)] \times s_L - \overline{Y}(u) = \overline{Y}(v) + \frac{[\overline{X}(u) - \overline{X}(v)] \times a}{b} - \overline{Y}(u)$. Since u is not on L, D_y is a non-zero integer. Because $0 \le |a|, |b| \le \mathcal{F}^{h+1}$, we have $D_y \ge \frac{1}{\mathcal{F}^{h+1}}$. Similarly, the distance between u and x is $D_x \ge \frac{1}{\mathcal{F}^{h+1}}$.

Let c be the intersection point of L with the line perpendicular to L that passes u. Clearly c is between x and y. Denote the angle $\angle yxu$ by z. Without loss of generality, we assume $z \le \pi/4$. (Otherwise switch the roles of x and y.) Since the slope of L is $\ge \frac{1}{\mathcal{F}^{h+1}}$, $\tan z \ge \frac{1}{\mathcal{F}^{h+1}}$. Since we assumed $z \le \pi/4$, $\cos z \ge 1/\sqrt{2}$. This implies $\sin z \ge \frac{\tan z}{\sqrt{2}} \ge \frac{1}{\sqrt{2}\mathcal{F}^{h+1}}$. Thus, the distance between u and c is: $D_x \times \sin z \ge \frac{1}{\sqrt{2}\mathcal{F}^{2(h+1)}})$. $\qquad\square$

In the following, we describe how to encode other Schnyder parameters $(n_i(u)$ and $s_i(u))$ of a vertex u into drawing coordinates $(X(u), Y(u))$, while still maintaining the strictly convex property of the drawing. First, we enlarge the drawing defined by the coordinates $(\overline{X}(u), \overline{Y}(u))$ by a factor $4 \times \mathcal{F}^{2(h+1)+3}$. Then we perform *fine perturbation* to the enlarged drawing by using other Schnyder parameters. By extending the ideas in [4], we are able to show the resulting drawing is still strictly convex. More precisely, we define:

$$X(u) = \overline{X}(u) \times 4\mathcal{F}^{2(h+1)+3} + n_1(u) \times \mathcal{F}^2 + s_1(u) \times \mathcal{F} + n_3(u) = 4 \times x_1(u) \times$$
$$\mathcal{F}^{2(h+1)+h+3} + 4 \times f_x(u) \times \mathcal{F}^{2(h+1)+3} + n_1(u) \times \mathcal{F}^2 + s_1(u) \times \mathcal{F} + n_3(u).$$

$$Y(u) = \overline{Y}(u) \times 4\mathcal{F}^{2(h+1)+3} + n_2(u) \times \mathcal{F}^2 + s_2(u) \times \mathcal{F} + s_3(u) = 4 \times x_2(u) \times$$
$$\mathcal{F}^{2(h+1)+h+3} + 4 \times f_y(u) \times \mathcal{F}^{2(h+1)+3} + n_2(u) \times \mathcal{F}^2 + s_2(u) \times \mathcal{F} + s_3(u).$$

It's straightforward to recover the Schnyder parameters from $(X(u), Y(u))$.

Theorem 3. *1. The drawing defined by $d : u \to (X(u), Y(u))$ is strictly convex.*

2. For each vertex u, the number of bits needed to represent $(X(u), Y(u))$ is $O(\log n)$.

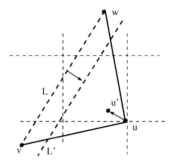

Fig. 5. The proof of Theorem 3

Proof. (1) Let F be any face of G. Let v, u, w be three consecutive vertices on F. Consider the angle $\angle vuw$. It is enough to show $\angle vuw < \pi$ in the drawing.

Let L be the line connecting v and w. By Lemma 4, in the drawing defined by $\overline{d} : u \to (\overline{X}(u), \overline{Y}(u))$, the distance between u and L is at least $\delta \geq \frac{1}{\sqrt{2}\mathcal{F}^{2(h+1)}}$. After we enlarge the drawing $\overline{d}$ by a factor of $4 \times \mathcal{F}^{2(h+1)+3}$, the distance between u and L is at least $\Delta \geq 2\sqrt{2} \times \mathcal{F}^3$.

In the drawing defined by $d : u \to (X(u), Y(u))$, the vertices are moved by a distance strictly less than $\sqrt{2}\mathcal{F}^3$. The line L is moved to a new location L'. The vertex u is moved to a new location u' (see Fig 5). In the worst case, L and u are moved toward each other in the direction perpendicular to L. Since the vertices are moved by a distance less than $\sqrt{2}\mathcal{F}^3$, u' is still located at the same side of L' as before the movement. Thus $\angle vuw < \pi$ as to be shown.

(2) By the definition, $\overline{X}(u) \leq \mathcal{F}^{h+1}$. Thus $X(u) \leq 4 \times \mathcal{F}^{3(h+1)+4}$. Similarly $Y(u) \leq 4 \times \mathcal{F}^{3(h+1)+4}$. Since $\mathcal{F}$ is an integer that is polynomial in n, $X(u) = O(n^p)$ and $Y(u) = O(n^p)$ (for some constant p). So we need only $O(\log n)$ bits to represent $(X(u), Y(u))$. $\square$

References

1. Bonichon, N., Felsner, S., Mosbah, M.: Convex drawings of 3-connected planar graph. Algorithmica 47, 399–420 (2007)
2. Angelini, P., Di Battista, G., Frati, F.: Succinct Greedy Drawings Do Not Always Exist. In: Eppstein, D., Gansner, E.R. (eds.) GD 2009. LNCS, vol. 5849, pp. 171–182. Springer, Heidelberg (2010)
3. Angelini, P., Frati, F., Grilli, L.: An algorithm to construct greedy drawing of triangulations. Journal of Graph Algorithms and Applications 14(1), 19–51 (2010)
4. Rote, G.: Strictly Convex Drawings of Planar Graphs. In: Proc. 16th Annual ACM-SIAM Symp. on Discrete Algorithms, SODA 2005, pp. 728–734 (2005)
5. Cao, L., Strelzoff, A., Sun, J.Z.: On succinctness of geometric greedy routing in Euclidean plane. In: Proc. ISPAN, pp. 326–331 (2009)
6. Felsner, S.: Convex Drawings of Planar Graphs and the Order Dimension of 3-Polytopes. Order 18, 19–37 (2001)

7. Goodrich, M.T., Strash, D.: Succinct Greedy Geometric Routing in the Euclidean Plane. In: Dong, Y., Du, D.-Z., Ibarra, O. (eds.) ISAAC 2009. LNCS, vol. 5878, pp. 781–791. Springer, Heidelberg (2009)
8. He, X., Zhang, H.: Succinct Convex Greedy Drawing of 3-Connected Plane Graphs. In: SODA 2011 (2011)
9. Leighton, T., Moitra, A.: Some results on greedy embeddings in metric spaces. Discrete Comput. Geom. 44(3), 686–705 (2010)
10. Papadimitriou, C.H., Ratajczak, D.: On a conjecture related to geometric routing. Theoretical Computer Science 334(1), 3–14 (2005)
11. Schnyder, W.: Planar graphs and poset dimension. Order 5, 323–343 (1989)
12. Schnyder, W.: Embedding planar graphs on the grid. In: in Proc. 1st ACM-SIAM Symp. Discrete Algorithms, pp. 138–148 (1990)

Weighted Inverse Minimum Cut Problem under the Sum-Type Hamming Distance[*]

Longcheng Liu[1,**], Yong Chen[2], Biao Wu[3], and Enyu Yao[3]

[1] School of Mathematical Sciences, Xiamen University, Xiamen 361005, P.R. China
longchengliu@xmu.edu.cn
[2] Institute of Operational Research & Cybernetics, Hangzhou Dianzi University,
Hangzhou 310018, P.R. China
[3] Department of Mathematics, Zhejiang University, Hangzhou 310027, P.R. China

Abstract. An inverse optimization problem is defined as follows: Let S denote the set of feasible solutions of an optimization problem P, let c be a specified cost (capacity) vector , and $x^0 \in S$. We want to perturb the cost (capacity) vector c to d such that x^0 becomes an optimal solution of P with respect to the cost (capacity) vector d, and to minimize some objective functions. In this paper, we consider the weighted inverse minimum cut problem under the sum-type Hamming distance. First, we show the general case is NP-hard. Second we present a combinatorial algorithm that run in strongly polynomial time to solve a special case.

Keywords: Minimum cut, Inverse problem, Hamming distance, Strongly polynomial algorithm.

1 Introduction

Let $N(V, A, c)$ be a connected and directed network, where $V = \{1, 2, \ldots, n\}$ is the node set, A is the arc set ($|A| = m$) and c is the capacity vector for arcs. Each component c_{ij} of c is called the *capacity* of arc (i, j). There are two special nodes in V: the source node s and the sink node t. Let X and $\overline{X} = V \setminus X$ be a partition of all vertices such that $s \in X$ and $t \in \overline{X}$. An $s - t$ cut, denoted by $\{X, \overline{X}\}$, is the set of arcs with one endpoint in X and another endpoint in $\overline{X}$. We further use $(X, \overline{X})$ to express the set of all forward arcs from a vertex in X to a vertex in $\overline{X}$ and use $(\overline{X}, X)$ to express the set of all backward arcs from a vertex in $\overline{X}$ to a vertex in X in an $s - t$ cut. As we know the capacity of an $s - t$ cut $\{X, \overline{X}\}$, denoted by $c(\{X, \overline{X}\})$, is the sum of the capacities of all forward arcs:

$$c(\{X, \overline{X}\}) = \sum_{(i,j) \in (X, \overline{X})} c_{ij}.$$

---

[*] This research is supported by the National Natural Science Foundation of China (Grant No. 11001232), Fundamental Research Funds for the Central Universities (Grant No. 2010121004) and Department of Education of Zhejiang Province of China (Grant No. Y200909535).
[**] Corresponding author.

J. Snoeyink, P. Lu, K. Su, and L. Wang (Eds.): FAW-AAIM 2012, LNCS 7285, pp. 26–35, 2012.

The *minimum-cut problem* is to determine an $s - t$ cut of minimum capacity. It is a classical network optimization problem that has many applications. It is well-known that minimum cut problem can be solved in strongly polynomial time.

Conversely, an inverse minimum cut problem is to modify the arc capacity vector as little as possible such that a given $s - t$ cut can form a minimum cut. Yang et al. [8] showed that the inverse minimum cut problem under l_1 norm is strongly polynomial time solvable, where the modification cost is measured by l_1 norm. Liu and Yao [5] showed that the weighted inverse minimum cut problem under the bottleneck type Hamming distance is also strongly polynomial time solvable, where the modification cost is measured by the bottleneck type Hamming distance. In this paper, we consider the weighted inverse minimum cut problem under the sum-type Hamming distance, in which we measure the modification cost by the weighted sum-type Hamming distance.

Let each arc (i, j) have an associated capacity modification cost $w_{ij} \geq 0$, and let w denote the arc modification cost vector. Let $\{X^0, \overline{X}^0\}$ be a given cut in the network $N(V, A, c)$. Then for the general weighted inverse minimum cut problem under the sum-type Hamming distance, we look for an arc capacity vector d such that

(a) $\{X^0, \overline{X}^0\}$ is a minimum cut of the network $N(V, A, d)$;
(b) for each $(i, j) \in A$, $-l_{ij} \leq d_{ij} - c_{ij} \leq u_{ij}$, where $l_{ij}, u_{ij} \geq 0$ are respectively given bounds for decreasing and increasing capacity c_{ij};
(c) the total modification cost for changing capacities of all arcs is minimized, i.e., $\sum_{(i,j)\in A} w_{ij} H(c_{ij}, d_{ij})$ is minimized, where $H(c_{ij}, d_{ij})$ is the Hamming distance between c_{ij} and d_{ij}, i.e., $H(c_{ij}, d_{ij}) = 0$ if $c_{ij} = d_{ij}$ and 1 otherwise.

In general, in an inverse combinatorial optimization problem, a feasible solution is given which is not optimal under the current parameter values, and it is required to modify some parameters with minimum modification cost such that the given feasible solution becomes an optimal solution. A lot of such problems have been well studied when the modification cost is measured by (weighted) l_1, l_2, and l_∞ norms. Readers may refer to the survey paper Heuberger [2] and papers cited therein. Recently, inverse problems under the weighted Hamming distance also received attention. In fact the weighted Hamming distance corresponds to the situation in which we might care about only whether the parameter of an arc is changed, but without considering the magnitude of its change as long as the adjustment is restricted to a certain interval. Noting that not like the l_1, l_2 and l_∞ norms which are all convex and continuous about the modification, the Hamming distance $H(\cdot, \cdot)$ is discontinuous and nonconvex, which makes the known methods for l_1, l_2 and l_∞ norms unable to be applied directly to the problems under such distance measure.

Zhang et al. [10] considered the inverse center location improvement problem under the weighted Hamming distance. For the bounded case, they showed that even under the unweighted sum-type Hamming distance, achieving an algorithm with a worst-case ratio $O(\log n)$ is strongly NP-hard, but under the weighted bottleneck-type Hamming distance, a strongly polynomial algorithm with a time complexity $O(n^2 \log n)$ is available. He et al. [1] discussed the inverse minimum spanning tree problem under the weighted sum-type Hamming distance. For both unbounded and bounded cases, they presented strongly polynomial algorithms with a time complexity $O(n^3 m)$. Zhang et al. [11] further discussed the inverse minimum spanning tree problem under the weighted bottleneck-type Hamming distance. For the unbounded case, they presented algorithms with a time complexity $O(nm)$, and for the constrained case, they presented an algorithm with a time complexity $O(n^3 m \log m)$. Liu and He [3] discussed the inverse minimum spanning tree problem and reverse shortest-path problem with discrete values. They presented strongly polynomial algorithms for the inverse minimum spanning tree problems with discrete values and they showed the reverse shortest-path problem with discrete values is strongly NP-complete. Liu and Zhang [7] discussed the inverse maximum flow problems under the weighted Hamming distance, for both sum-type and bottleneck-type, they presented strongly polynomial algorithms. Yang and Zhang [9] discussed inverse sorting problems under the weighted sum-type Hamming distance. For both unbounded and bounded cases, they presented strongly polynomial algorithms. Liu and Yao [6], Liu and Wang [4] discussed the inverse min-max spanning tree problem under the weighted Hamming distance. They presented strongly polynomial algorithms for all the problems they discussed. The above n and m are the numbers of nodes and edges (arcs), respectively, in a given undirected (directed) network.

The paper is organized as follows. Section 2 contains some preliminary results. Section 3 shows that the general case of weighted inverse minimum cut problem under the sum-type Hamming distance is NP-hard. Section 4 consider a polynomial solvable case and a strongly polynomial algorithm is presented. Some final remarks are made in Section 5.

2   Preliminary Results

The following result is well known.

Lemma 1. *An $s-t$ cut $\{X, \overline{X}\}$ of the network $N(V, A, c)$ is a minimum cut if and only if there exists a feasible flow f from node s to node t that "saturates" the cut $\{X, \overline{X}\}$, i.e., there exists a feasible flow f such that*

$$f_{ij} = c_{ij}, \quad if \ (i, j) \in (X, \overline{X}),$$
$$f_{ji} = 0, \quad if \ (j, i) \in (\overline{X}, X).$$

In such case the flow f must be a maximum flow of network $N(V, A, c)$.

The general weighted inverse minimum cut problem under the sum-type Hamming distance can be formulated as follows.

$$\min \sum_{(i,j) \in A} w_{ij} H(c_{ij}, d_{ij})$$
s.t. Cut $\{X^0, \overline{X}^0\}$ is a minimum cut of $N(V, A, d)$;   (1)
$$-l_{ij} \leq d_{ij} - c_{ij} \leq u_{ij}, \text{ for each } (i, j) \in A.$$

Lemma 2. *If problem (1) has a feasible solution, then there exists an optimal solution d^* such that*

(I) $d_{ij}^* \leq c_{ij}$ *for* $(i,j) \in (X^0, \overline{X}^0)$.

(II) $d_{ij}^* = c_{ij}$ *for* $(i,j) \in (\overline{X}^0, X^0)$.

(III) $d_{ij}^* \geq c_{ij}$ *for other* (i,j).

Proof. (a) We first prove the validity of (I). Let d^* be an optimal solution of problem (1). If (I) is not true, then we can find an arc $(x, y) \in (X^0, \overline{X}^0)$ with $d_{xy}^* > c_{xy}$. Define $\overline{d}$ as

$$\overline{d}_{ij} = \begin{cases} c_{ij}, & \text{if } (i,j) = (x, y), \\ d_{ij}^*, & \text{otherwise.} \end{cases}$$

We say the cut $\{X^0, \overline{X}^0\}$ is a minimum cut of the network $N(V, A, \overline{d})$. In fact, if there exists another $s - t$ cut $\{X, \overline{X}\}$ such that

$$\overline{d}(\{X, \overline{X}\}) < \overline{d}(\{X^0, \overline{X}^0\}),$$

then from the definition of $\overline{d}$, we have the following two cases:

Case 1: If $(x, y) \in (X, \overline{X})$, then we have

$$d^*(\{X, \overline{X}\}) - \overline{d}(\{X, \overline{X}\}) = d_{xy}^* - c_{xy} = d^*(\{X^0, \overline{X}^0\}) - \overline{d}(\{X^0, \overline{X}^0\}),$$

i.e.,

$$d^*(\{X^0, \overline{X}^0\}) - d^*(\{X, \overline{X}\}) = \overline{d}(\{X^0, \overline{X}^0\}) - \overline{d}(\{X, \overline{X}\}) > 0.$$

Case 2: If $(x, y) \notin (X, \overline{X})$, then we have

$$d^*(\{X, \overline{X}\}) = \overline{d}(\{X, \overline{X}\}) < \overline{d}(\{X^0, \overline{X}^0\}) < d^*(\{X^0, \overline{X}^0\}).$$

Both of the two cases above conflict the fact that the cut $\{X^0, \overline{X}^0\}$ is a minimum cut of the network $N(V, A, d^*)$.

And from the definition of $\overline{d}$, we have

$$-l_{ij} \leq \overline{d}_{ij} - c_{ij} \leq u_{ij}, \text{ for each } (i,j) \in A.$$

So $\overline{d}$ is a feasible solution of problem (1). However, we have

$$\sum_{(i,j) \in A} w_{ij} H(c_{ij}, d_{ij}^*) - \sum_{(i,j) \in A} w_{ij} H(c_{ij}, \overline{d}_{ij}) = w_{xy} \geq 0.$$

If $w_{xy} > 0$, then d^* cannot be an optimal solution of problem (1), a contradiction. Otherwise, $\overline{d}$ is another optimal solution of problem (1), but it satisfies $\overline{d}_{xy} = c_{xy}$. Hence, by repeating the above argument, we can conclude that there exists an optimal solution d^* of problem (1) such that $d_{ij}^* \leq c_{ij}$ for $(i,j) \in (X^0, \overline{X}^0)$.

(b) By a similar argument as in (a), we can get $d_{ij}^* \geq c_{ij}$ for $(i,j) \in A \setminus (X^0, \overline{X}^0)$. Let f^* be a maximum flow in network $N(V, A, d^*)$. From Lemma 1 we know that

$$f_{ij}^* = d_{ij}^*, \quad if \ (i,j) \in (X^0, \overline{X}^0),$$

$$f^*_{ji} = 0, \quad if \ (j,i) \in (\overline{X}^0, X^0).$$

If there exists an arc $(x,y) \in (\overline{X}^0, X^0)$ such that $d^*_{xy} > c_{xy}$, then we define $\overline{d}$ as follows:

$$\overline{d}_{ij} = \begin{cases} c_{ij}, & if \ (i,j) = (x,y), \\ d^*_{ij}, & \text{otherwise.} \end{cases}$$

It is easy to know that f^* and $\{X^0, \overline{X}^0\}$ are also a pair of maximum flow and minimum cut in the network $N(V, A, \overline{d})$. Combining with the fact that

$$-l_{ij} \le \overline{d}_{ij} - c_{ij} \le u_{ij}, \qquad \text{for each}(i,j) \in A$$

we know $\overline{d}$ is a feasible solution of problem (1). And by a similar argument as in the last part of (a) we can get $d^*_{ij} = c_{ij}$ for $(i,j) \in (\overline{X}^0, X^0)$.
    Combining (a) and (b), the lemma holds.

Let the network $N'(V, A', c')$ be obtained from $N(V, A, c)$ by the following way: $A' = A \setminus (\overline{X}^0, X^0)$, $c'_{ij} = c_{ij}$. Then we have the following lemmas:

Lemma 3. *The cut $\{X^0, \overline{X}^0\}$ is a minimum cut of $N(V, A, d)$ if and only if $\{X^0, \overline{X}^0\}$ is a minimum cut of $N'(V, A', d')$.*

Proof. Suppose $\{X^0, \overline{X}^0\}$ is a minimum cut of $N(V, A, d)$, then from Lemma 1 we know there exists a maximum flow f of the network $N(V, A, d)$ such that

$$f_{ij} = d_{ij}, \quad if \ (i,j) \in (X^0, \overline{X}^0),$$

$$f_{ji} = 0, \quad if \ (j,i) \in (\overline{X}^0, X^0).$$

And from the definition of network $N'(V, A', d')$, the cut $\{X^0, \overline{X}^0\}$ only has the forward arcs in the network $N'(V, A', d')$. Define a flow f' of the network $N'(V, A', d')$ by setting $f'_{ij} = f_{ij}$ for each arc $(i,j) \in A'$. It is clear that f' is a feasible flow of the network $N'(V, A', d')$ and it saturates the cut $\{X^0, \overline{X}^0\}$, i.e., $\{X^0, \overline{X}^0\}$ is a minimum cut of the network $N'(V, A', d')$.
    Conversely, suppose $\{X^0, \overline{X}^0\}$ is a minimum cut of the network $N'(V, A', d')$, then from Lemma 1 we know there exists a maximum flow f' of the network $N'(V, A', d')$ such that

$$f'_{ij} = d'_{ij}, \quad if \ (i,j) \in (X^0, \overline{X}^0).$$

Define a flow f of the network $N(V, A, d)$ by setting $f_{ij} = f'_{ij}$ for each arc $(i,j) \in A \setminus (\overline{X}^0, X^0)$ and $f_{ij} = 0$ for each $(i,j) \in (\overline{X}^0, X^0)$. It is clear that f is a feasible flow of the network $N(V, A, d)$ and it saturates the cut $\{X^0, \overline{X}^0\}$, i.e., $\{X^0, \overline{X}^0\}$ is a minimum cut of the network $N(V, A, d)$.

Lemma 4. *If the capacity vector $d^{'}$ is an optimal solution for the inverse minimum cut problem under the weighted sum-type Hamming distance for the network $N^{'}(V, A^{'}, c^{'})$, then the capacity vector d^* is an optimal solution for the inverse minimum cut problem under the weighted sum-type Hamming distance for the network $N(V, A, c)$, where $d^*_{ij} = d^{'}_{ij}$ for each arc $(i, j) \in A \setminus (\overline{X}^0, X^0)$ and $d^*_{ij} = c_{ij}$ for each arc $(i, j) \in (\overline{X}^0, X^0)$.*

Proof. Suppose that the arc capacity vector $d^{'}$ is an optimal solution for the inverse minimum cut problem under the weighted sum-type Hamming distance for the network $N^{'}(V, A^{'}, c^{'})$. Define d^* as follows:

$$d^*_{ij} = \begin{cases} d^{'}_{ij}, & \text{if } (i, j) \in A \setminus (\overline{X}^0, X^0), \\ c_{ij}, & \text{otherwise.} \end{cases}$$

Since $\{X^0, \overline{X}^0\}$ is a minimum cut of the network $N^{'}(V, A^{'}, d^{'})$, from Lemma 3 we know $\{X^0, \overline{X}^0\}$ is a minimum cut of the network $N(V, A, d^*)$. And from the definition of d^*, we have

$$-l_{ij} \le d^*_{ij} - c_{ij} \le u_{ij}, \text{ for each } (i, j) \in A,$$

and

$$\sum_{(i,j)\in A} w_{ij} H(c_{ij}, d^*_{ij}) = \sum_{(i,j)\in A'} w_{ij} H(c_{ij}, d^{'}_{ij}). \tag{2}$$

So d^* is a feasible solution for the inverse minimum cut problem for the network $N(V, A, c)$.

Now suppose $\overline{d}$ is an optimal solution for the inverse minimum cut problem for the network $N(V, A, c)$. It is obvious $\overline{d}^{'}$ obtained by setting $\overline{d}^{'}_{ij} = \overline{d}_{ij}$ for each $(i, j) \in A^{'}$ is a feasible solution for the inverse minimum cut problem for the network $N^{'}(V, A^{'}, c^{'})$. Hence we have

$$\sum_{(i,j)\in A'} w_{ij} H(c_{ij}, d^{'}_{ij}) \le \sum_{(i,j)\in A'} w_{ij} H(c_{ij}, \overline{d}^{'}_{ij}) \le \sum_{(i,j)\in A} w_{ij} H(c_{ij}, \overline{d}_{ij}). \tag{3}$$

From the Equations (2) and (3), we have

$$\sum_{(i,j)\in A} w_{ij} H(c_{ij}, d^*_{ij}) \le \sum_{(i,j)\in A} w_{ij} H(c_{ij}, \overline{d}_{ij}).$$

Combining with the fact that d^* is a feasible solution and $\overline{d}$ is an optimal solution for the inverse minimum cut problem under the weighted sum-type Hamming distance for the network $N(V, A, c)$, we know the capacity vector d^* is an optimal solution for the inverse minimum cut problem for the network $N(V, A, c)$.

3   Complexity of the General Case

In this section, we will show problem (1) is NP-hard. To show this result, we transfer a well-known NP-hard problem-Knapsack Problem-into a special case of the decision version of problem (1).

Knapsack Problem(KP)
Given a knapsack of capacity $C > 0$ and I items. Each item has value $p_i > 0$ and weight $w_i > 0$. Find the selection of items ($\theta_i = 1$ if item i be selected and 0 otherwise) that fit $\sum_{i=1}^{I} \theta_i \cdot w_i \leq C$, and the total value, $\sum_{i=1}^{I} \theta_i \cdot p_i$ is maximized.

The Decision Version of Knapsack Problem(DVKP)
Given a knapsack of capacity $C > 0$ and I items. Each item has value $p_i > 0$ and weight $w_i > 0$. For a given threshold K, whether there is a selection of items ($\theta_i = 1$ if item i be selected and 0 otherwise) satisfying the following two consitions:

(a) $\sum_{i=1}^{I} \theta_i \cdot w_i \leq C$;

(b) $\sum_{i=1}^{I} \theta_i \cdot p_i \geq K$.

Theorem 1. *Even if the modified arcs are restricted in $(X^0, \overline{X}^0)$, i.e., $l_{ij} = u_{ij} = 0$ if $(i,j) \notin (X^0, \overline{X}^0)$, problem (1) is NP-hard.*

Proof. The decision version of the problem (1) is as follows:
   For a given threshold C, whether there is a solution d satisfying the following conditions:
   (a) $\{X^0, \overline{X}^0\}$ is a minimum cut of the network $N(V, A, d)$;
   (b) for each $(i,j) \in A$, $-l_{ij} \leq d_{ij} - c_{ij} \leq u_{ij}$;
   (c) the total modification cost for changing capacities of all arcs is not greater than C, i.e., $\sum_{(i,j) \in A} w_{ij} H(c_{ij}, d_{ij}) \leq C$.
   For a given instance of DVKP $\{n, w_i, p_i, C, K\}$, we construct a network N and an instance of the decision version of problem (1) as follows:
   The network N has $2n+3$ nodes and $3n+1$ arcs $\{a_1, a_2, \ldots, a_n, b_1, b_2, \ldots, b_n, c_1, c_2, \ldots, c_n, d\}$. An illustration of network N is shown in Figure 1.
   Set the $\{w_{ij}, c_{ij}, l_{ij}, u_{ij}\}$ as follows:

(1) If $(i,j) \in \{a_1, a_2, \ldots, a_n, c_1, c_2, \ldots, c_n\}$, then $l_{ij} = u_{ij} = 0$, $c_{ij} = w_{ij} = M$, where M is a sufficiently large number.

(2) If $(i,j) = d$, then $l_{ij} = u_{ij} = 0$, $c_{ij} = \sum_{i=1}^{n} p_i - K$, $w_{ij} = +\infty$.

(3) For the arc b_i in $\{b_1, b_2, \ldots, b_n\}$, we set the capacity equals the item's value p_i, set the upper bound u_{ij} equals 0, set the lower bound l_{ij} equals the item's value p_i, set the modification cost equals the item's weight w_i.

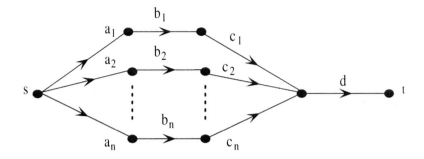

Fig. 1. An illustration of network N

For a sufficiently large number M, the minimum cut of the constructed network N is $\{X, \overline{X}\} = \{d\}$.

At last, we set the given cut as $\{X^0, \overline{X}^0\} = \{b_1, b_2, \ldots, b_n\}$.

It is clear that the construction of the network N can be done in polynomial time.

Combining the property of the Hamming distance with the definition of $\{w_{ij}, c_{ij}, l_{ij}, u_{ij}\}$ above, the constructed instance of the decision version of problem (1) is whether there is an arc set $\Omega \subseteq \{b_1, b_2, \ldots, b_n\}$ satisfying the following conditions:

(a) $\sum\limits_{i \in \Omega} p_i \geq K$;

(b) $\sum\limits_{i \in \Omega} w_i \leq C$ (where C is the capacity of the knapsack).

It is clear that the answer of the above decision problem is *Yes* if and only if the answer of the given instance of DVKP is *Yes*.

4   A Polynomial Solvable Case

From the results stated in Section 2, the weighted inverse minimum cut problem under the sum-type Hamming distance can be formulated as follows:

$$
\begin{aligned}
\min \quad & \sum_{(i,j) \in A'} w_{ij} H(c'_{ij}, d'_{ij}) \\
\text{s.t.} \quad & \text{Cut } \{X^0, \overline{X}^0\} \text{ is a minimum cut of } N'(V, A', d'); \\
& -l_{ij} \leq d'_{ij} - c'_{ij} \leq u_{ij}, \qquad \text{for each } (i, j) \in A'.
\end{aligned}
\tag{4}
$$

By a similar argument as in Lemma 2, we have the following lemma:

Lemma 5. *Problem (4) is feasible if and only if d' is a feasible solution of problem (4), where d' is defined as:*

$$
d'_{ij} = \begin{cases} c'_{ij} - l_{ij}, & \text{if } (i, j) \in (X^0, \overline{X}^0), \\ c'_{ij} + u_{ij}, & \text{otherwise.} \end{cases}
\tag{5}
$$

In section 3, we have shown even restricted the modified arcs in $(X^0, \overline{X}^0)$, the weighted inverse minimum cut problem under the sum-type Hamming distance is NP-hard. So in this section, we consider a polynomial solvable case. We not only restrict the modified arcs in $(X^0, \overline{X}^0)$, but also set $w_{ij} = w$ for all $(i,j) \in A$, which can be formulated as follows (for simplicity, we set $w_{ij} = 1$):

$$\min \sum_{(i,j) \in A'} H(c_{ij}', d_{ij}')$$

s.t. Cut $\{X^0, \overline{X}^0\}$ is a minimum cut of $N'(V, A', d')$;

$$-l_{ij} \le d_{ij}' - c_{ij}' \le u_{ij}, \text{ for each } (i,j) \in (X^0, \overline{X}^0);$$

$$l_{ij} = u_{ij} = 0, \text{ for each } (i,j) \in A' \setminus (X^0, \overline{X}^0). \qquad (6)$$

We are going to give an algorithm for solving problem (6) in strongly polynomial time. But before stating the algorithm, it is helpful to explain the main motivation of the algorithm.

First, by Lemma 5, we find out whether the problem (6) is feasible or not. If problem (6) is infeasible, then we stop and do nothing any more.

Second, if problem (6) is feasible, we want to find a new vector d' to make the cut $\{X^0, \overline{X}^0\}$ become a minimum cut of the network $N'(V, A', d')$. From Lemma 2 and the property of the Hamming distance we know the new vector d' satisfies the following property: if $d_{ij}' \ne c_{ij}'$ then $d_{ij}' = c_{ij}' - l_{ij}$.

Third, to meet the objective request, we change the arcs with the largest modification first. By Lemma 1, when the value of the maximum flow of the current network is equal to the capacity of the cut $\{X^0, \overline{X}^0\}$, we stop.

Now we are ready to state our algorithm.

Algorithm 1

Step 0: Check whether problem (6) is feasible or not. (By Lemma 1 and Lemma 5, we can run the maximum flow algorithm to see whether $\{X^0, \overline{X}^0\}$ is a minimum cut of the network $N'(V, A', d')$, where d' is defined as in (5).) If problem (6) is infeasible, stop. Otherwise, go to Step 1.

Step 1: Let $\{Y, \overline{Y}\}$ be the minimum cut of the network $N'(V, A', c')$, $C = c'(\{X^0, \overline{X}^0\}) - c'(\{Y, \overline{Y}\})$, $\Omega = \emptyset$, $q = 0$, $\Gamma = (X^0, \overline{X}^0) \setminus (Y, \overline{Y})$. For $(i,j) \in \Gamma$, denote $\Delta_{ij} = l_{ij}$. We rearrange all the Δ_{ij} in an nonincreasing order, express them as: $\Delta_1 \ge \Delta_2 \ge \ldots \ge \Delta_k$ and denote the associated arcs as: $\{(i_1, j_1), (i_2, j_2), \ldots, (i_k, j_k)\}$.

Step 2: $q = q + 1$, $\Omega = \Omega \cup \{(i_q, j_q)\}$.

Step 3: If $\sum_{i=1}^{q} \Delta_i \le C$, go back to Step 2. Otherwise, output an optimal solution d' of the problem (6) as:

$$d_{ij}' = \begin{cases} c_{ij}' - l_{ij}, & \text{if } (i,j) \in \Omega, \\ c_{ij}', & \text{otherwise.} \end{cases} \qquad (7)$$

The following theorem is straightforward.

Theorem 2. *Algorithm 1 is a strongly polynomial time algorithm for problem (6).*

5 Concluding Remarks

In this paper we studied the weighted inverse minimum cut problem under the sum-type Hamming distance. With some preliminary results we first proved the general case is NP-hard. We also consider a polynomial solvable case.

As a future research topic, it will be meaningful to consider other inverse combinational optimization problems under Hamming distance. Studying computational complexity results and proposing optimal/approximation algorithms are promising.

References

1. He, Y., Zhang, B.W., Yao, E.Y.: Wighted inverse minimum spanning tree problems under Hamming distance. Journal of Combinatorial Optimization 9, 91–100 (2005)
2. Heuberger, C.: Inverse Optimization: A survey on problems, methods, and results. Journal of Combinatorial Optimization 8, 329–361 (2004)
3. Liu, L.C., He, Y.: Inverse minimum spanning tree problem and reverse shortest-paht problem with discrete values. Progress in Natural Science 16(6), 649–655 (2006)
4. Liu, L.C., Wang, Q.: Constrained inverse min-max spanning tree problems under the weighted Hamming distance. Journal of Global Optimization 43, 83–95 (2009)
5. Liu, L.C., Yao, E.Y.: A weighted inverse minimum cut problem under the bottleneck type Hamming distance. Asia-Pacific Journal of Operational Research 24, 725–736 (2007)
6. Liu, L.C., Yao, E.Y.: Inverse min-max spanning tree problem under the weighted sum-type Hamming distance. Theoretical Computer Science 396, 28–34 (2008)
7. Liu, L.C., Zhang, J.Z.: Inverse maximum flow problems under the weighted Hamming distance. Journal of Combinatorial Optimization 12, 395–408 (2006)
8. Yang, C., Zhang, J.Z., Ma, Z.F.: Inverse maximum flow and minimum cut problems. Optimization 40, 147–170 (1997)
9. Yang, X.G., Zhang, J.Z.: Inverse sorting problem by minimizing the total weighted number of changes and partial inverse sorting problems. Computational Optimization and Applications 36, 55–66 (2007)
10. Zhang, B.W., Zhang, J.Z., He, Y.: The center location improvement problem under the Hamming distance. Journal of Combinatorial Optimization 9, 187–198 (2005)
11. Zhang, B.W., Zhang, J.Z., He, Y.: Constrained inverse minimum spanning tree problems under the bottleneck-type Hamming distance. Journal of Global Optimization 34, 467–474 (2006)

Voronoi Diagram with Visual Restriction*

Chenglin Fan[1], Jun Luo[1,3], Wencheng Wang[2], and Binhai Zhu[3,4]

[1] Shenzhen Institutes of Advanced Technology, Chinese Academy of Sciences, China
{cl.fan,jun.luo}@siat.ac.cn
[2] Institute of Software, Chinese Academy of Sciences, China
whn@ios.ac.cn
[3] Top Key Discipline of Computer Software and Theory, Zhejiang Provincial Colleges,
Zhejiang Normal University, China
[4] Department of Computer Science, Montana State University, Bozeman, MT 59717, USA
bhz@cs.montana.edu

Abstract. In a normal Voronoi diagram, each site is able to see all the points in the plane. In this paper, we study the case such that each site is only able to see a visually restricted region in the plane and construct the so-called Visual Restriction Voronoi Diagram (VRVD). We show that the visual restriction Voronoi cell of each site is not necessary convex and it could consist of many disjoint regions. We prove that the combinatorial complexity of the VRVD on n sites is $\Theta(n^2)$. Then we give an $O(n^2 \log n)$ time and $O(n^2)$ space algorithm to construct VRVD.

1 Introduction

Voronoi diagram is a fundamental structure in computational geometry and plays important roles in other fields such as GIS and physics [2]. One of the major applications of Voronoi diagram is to answer the *nearest-neighbor* queries efficiently. Much has been done on the variants of Voronoi diagrams and the algorithms for computing Voronoi diagrams in various fields [1,4,5,7–10], Aurenhammer [3] considered a special kinds of Voronoi Diagram: visibility is constrained to a segment on a line avoiding the convex hull of the sites, he also presented a quadratic complexity and construction time.

In some situations, we want to find not only the closest site but also the site which is visible to the query point. For example, in a football match, each player has his/her own vision area at any given time. If we want to find which player is the closest to the ball and can also see the ball efficiently, we can construct the so-called Visual Restriction Voronoi Diagram. In this paper, we consider the case when each player (or site) has some visible angle (which may not be $180°$) in the plane.

In the following paragraphs we define some notations, given a set of n sites/points $P = \{p_1, p_2, ..., p_n\}$ in the plane. For each site/point p_i, associate it with two rays $r_1(p_i)$ and $r_2(p_i)$ which have the same endpoint p_i, and the point p_i is only visible to the region from $r_1(p_i)$ to $r_2(p_i)$ in clockwise order. We use $VR(p_i)$ to denote the region which p_i is visible to (see Figure 1).

---

* This research has been partially funded by the International Science & Technology Cooperation Program of China (2010DFA92720), Shenzhen Fundamental Research Project (grant no. JC201005270342A) and by NSF of China under project 60928006.

J. Snoeyink, P. Lu, K. Su, and L. Wang (Eds.): FAW-AAIM 2012, LNCS 7285, pp. 36–46, 2012.

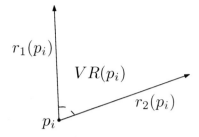

Fig. 1. Illustration of $VR(p_i), r_1(p_i)$, and $r_2(p_i)$

A point q in the plane is said to be *controlled* by p_i if it meets two conditions below:

1. $q \in VR(p_i)$.
2. Among all the sites in P which are visible to q, p_i is the closest one.

A region in the plane is said to be *controlled* by p_i if every point in the region is controlled by p_i and this region is called the visual restriction Voronoi cell of p_i. Note that different from the normal Voronoi diagram, in our setting a member in P may control more than one connected region, and each connected region is not necessary convex (see Figure 2). For the special case when the angle of each site could see is $180°$, every site is visible to a half-plane. We call that specific Voronoi diagram as Half-Plane Voronoi diagram (HPVD for short). If we use Theorem 2.2(i) by Edelsbrunner *et al.* [6] and compute the upper envelope of a set of n triangles in three dimensions in $O(n^2\alpha(n))$ time and storage, then project it to the two dimensional plane to get the VRVD. We can easily get an $O(n^2\alpha(n))$ combinatorial complexity and build VRVD in $O(n^2\alpha(n))$ time and space. In this paper, we obtain an optimal combinatorial complexity $O(n^2)$ for VRVD. We also design an algorithm using $O(n^2)$ space to compute VRVD, as the algorithm in [6] does not work in practice.

In this paper, we assume that there are no two sites at the same position, and no two sites whose boundary lines overlap with each other. Section 2 gives the analysis of combinatorial complexity of VRVD, while the algorithm to construct VRVD is presented in section 3. We give the conclusion in section 4.

2   Combinatorial Complexity of VRVD

In this section, we investigate the combinatorial complexity (i.e., the size) of VRVD. Let us first consider the edges of VRVD. It is not difficult to find that the edges of VRVD have only two types: either it belongs to a perpendicular bisector of two sites or it belongs to a boundary line $l_1(p_i)$ or $l_2(p_i)$ ($1 \le i \le n$, $l_1(p_i), l_2(p_i)$ are the extended line of rays $r_1(p_i), r_2(p_i)$ respectively). Therefore, there are two types of edges of VRVD:

1. Those edges lie on a boundary line,
2. Those edges do not not lie on a boundary line;

and there are three types of vertices of VRVD:

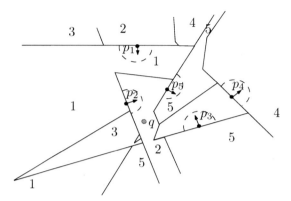

Fig. 2. An example of Visual Restriction Voronoi diagram of five sites p_1, p_2, p_3, p_4, p_5. In this special example, we assume the angle between two rays of point is $180°$, and each point is visible to a half-plane, and the boundary of that half-plane is perpendicular to the arrow of that point. The number in each cell is the site ID which controls that cell.

1. V_1: the intersections between two edges of the second type,
2. V_2: the intersections between two edges of the first type,
3. V_3: the intersections between two edges: one of first type and another one of the second type.

Now we want to bound the number of vertices of VRVD of n sites P, denoted as $VRVD(P)$. Since there are only n boundary lines, there are only $O(n^2)$ intersections between them. Therefore $|V_2| = O(n^2)$. However, what is the number of V_1 and V_3?

Lemma 1. *In $VRVD(P)$, there are at most $O(n)$ edges of the second type intersecting with each boundary line.*

Proof. Consider the boundary line $l(p_k)$ (without loss of generality, we just assume $l(p_k)$ is horizontal), which represents any one of line $l_1(p_k), l_2(p_k)$ and the intersections $I_1, I_2, \ldots, I_m$ between $r_1(p_i), r_2(p_i)$ $(1 \leq i \leq n, i \neq k)$ and $l(p_k)$ (see Figure 3). These intersections separate the line $l(p_k)$ into $2m + 1$ intervals (if we consider each intersection point as an interval): $(-\infty, I_1), [I_1, I_1], (I_1, I_2), \ldots, (I_m - 1, I_m), [I_m, I_m], [I_m, +\infty)$ such that each site p_i is either visible to all points in an interval or invisible to any point of a given interval. We use n_i to denote the number of sites in P that is visible to the i-th interval. Note that one site may be visible to two different intervals but the sets of visible sites of two different intervals cannot be identical. For a site p_i, if p_i is visible to the i-th interval but is not visible to the $(i + 1)$-th interval, then there are two cases: (1) p_i is not visible to all intervals to the right of the i-th interval. (2) p_i is not visible to those intervals to the right of the i-th interval until the point I_r or I_l, where I_r and I_l are the right and left intersection points between $l(p_k)$ and rays $r_1(p_i)$ and $r_2(p_i)$ respectively. We use D_i to denote the set of those sites and $|D_i|$ to denote the number of those sites.

Similarly, if p_i is not visible to the i-th interval but visible to the $(i + 1)$-th interval, then there are two cases: (1) p_i is visible to all intervals to the right of i-th interval; (2) p_i is visible to those intervals to the right of the i-th interval until the point I_r or I_l.

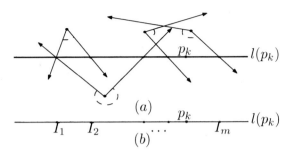

Fig. 3. The boundary line $l(p_k)$ is visible by many sites in P

We use A_i to denote the set of those sites and $|A_i|$ to denote the number of those sites. Hence we have the following:

$$\sum_{i=1}^{2m} |D_i| \le n$$

$$n_1 + A_1 + A_2 + A_3 + \cdots + A_{2m} \le 2n$$

We first consider the leftmost interval $(-\infty, I_1)$. There are n_1 sites which are visible to this interval. Hence there are at most $n_1 - 1$ second kind of edges of VRVD (perpendicular bisector edge) which intersect with the interval $(-\infty, I_1)$. Suppose the number of intersections in this interval is $k_1 \le n_1 - 1$. There are at most $n_1 - k_1$ sites that do not control the leftmost interval (or in other words, do not contribute any perpendicular bisector edges intersecting with the leftmost interval) and they may contribute perpendicular bisector edges intersecting with the intervals to the right later. For two sites p_i, p_j, they create a perpendicular bisector b_{ij}. If p_i is to the left of b_{ij}, then p_i can not control some intervals to the right of b_{ij} unless p_j is not visible to some intervals right of b_{ij}. That means if $p_j \in D_i$, then p_i could control some intervals to the right of the i-th interval again.

We then consider the second interval. From the above discussion, we know that at most $(n_1 - k_1) + |D_1|$ sites may control the second interval. Moreover, the sites in A_1 which are not visible to the first interval and visible to the second interval may also control the second interval. Therefore, totally at most $(n_1 - k_1) + |D_1| + |A_1|$ sites may control the second interval and there are at most $(n_1 - k_1) + |D_1| + |A_1| - 1 \ge k_2$ third type intersections in the second interval. We continue considering the third interval and so on using a similar analysis. For the $(i+1)$-th interval, there are at most $|D_i| + |A_i| + |D_{i-1}| + |A_{i-1}| + \cdots + |D_1| + |A_1| + |n_1| - k_1 - k_2 - \cdots - k_i) - 1 \ge k_{i+1}$ third type intersections. Then we have:

$$k_1 + k_2 + \cdots + k_{2m+1}$$

$$\le k_1 + k_2 + \cdots + k_{2m} + (|D_{2m}| + |A_{2m}| + |D_{2m-1}| + |A_{2m-1}| + \cdots$$

$$+ |D_1| + |A_1| + n_1 - k_1 - k_2 - \cdots - k_{2m}) - 1 \le 3n - 1$$

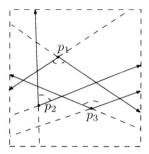

Fig. 4. The boundary lines $l_1(p_i), l_2(p_i)(i = 1, 2, .., n)$ separate the big rectangle plane into many convex polygons

Therefore we can conclude that there are at most $O(n)$ second kind of edges intersecting with each boundary line. What is more, there are at most $2(n - 2)$ intersections between other boundary lines intersecting with $l(p_k)$. Hence $l(p_k)$ is divided into at most $O(n)$ disjoint intervals from left to right such that each interval is controlled by at most one site.   □

Actually, from Lemma 1, we know that $|V_3| = O(n^2)$. Then the question is: how many second kind of edges of VRVD are there?

We draw a big rectangle containing all the sites of P and all intersections of boundary lines. Then the rectangle is decomposed by the (arrangement of the) boundary lines $l_1(p_i), l_2(p_i)(i = 1, 2, .., n)$ into many convex polygons. Each convex polygon CP_j is visible to some sites of P. In fact, each site is either visible to the whole area of CP_j, or not visible to any point in CP_j (see Figure 4). Moreover, for any convex polygon CP_j, all the sites are either outside of CP_j, or on the boundary of CP_j. Let the set of sites which are visible to CP_j be VP_j. Then VRVD(P) inside CP_j is the same as the Voronoi diagram of VP_j inside CP_j (see Figure 5). Let the part of VRVD(P) inside CP_j be $VRVD_j$. We now study the property of $VRVD_j$.

Lemma 2. *Let the number of edges of $VRVD_j$ which intersect with polygon CP_j be x (the third type of intersections). The number of edges of $VRVD_j$ is at most $2x - 1$ and the first type of intersections in $VRVD_j$ is at most $x - 1$.*

Proof. First we prove that $VRVD_j$. is a tree. If $VRVD_j$. is not a tree, then it contains cycle. This cycle must contain a site which contradicts the fact there are no sites inside CP_j. Since $VRVD_j$. is actually the normal Voronoi diagram of VP_j inside CP_j, the degree of inner node of $VRVD_j$. is at least three. The intersection edges of $VRVD_j$ with polygon CP_j are actually the leaves of $VRVD_j$.. For a tree with x leaves and internal nodes with degree at least three, the number of edges of this tree is at most $2x - 1$, and the inner nodes of the tree is at most $x - 1$, which is equal to the number of the first type of intersections in $VRVD_j$.   □

According to Lemma 2 and the fact that there are $O(n^2)$ third type of intersections, we know $|V_1| = O(n^2)$ and the total number of bisector edges of VRVD is also $O(n^2)$. Therefore we have the following theorem:

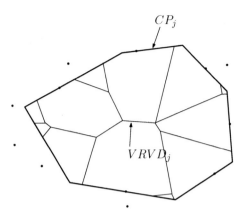

Fig. 5. Illustration of $VRVD_j$ inside convex polygon CP_i (thick line) where all sites are outside of CP_j or on the boundary of CP_j

Theorem 1. *The combinatorial complexity of VRVD of n sites is $\Theta(n^2)$.*

Proof. Since $|V_1| = O(n^2), |V_2| = O(n^2)$ and $|V_3| = O(n^2)$, the total number of vertices is $O(n^2)$. From Lemma 1 and Lemma 2, we know that the total number of bisector edges of VRVD is also $O(n^2)$ and the boundary edges are cut by other boundary lines and bisector edges which add at most $O(n^2)$ edges. Hence the total number of edges is $O(n^2)$. According to the Euler formula for planar graphs, the number of faces of VRVD is also $O(n^2)$.

It is easy to construct an example such that VRVD does contain $\Omega(n^2)$ vertices, edges and cells. For example, in Figure 6, all sites' boundaries are horizontal. Therefore all their visual angles are $180°$ and normal vectors of those boundaries are either up or down vertically. There are $n/2$ points on line l_1 and the distance between two adjacent points is d_0, and $n/2$ points on l_2 such that l_2 is crossing the center point on l_1 and l_2 is perpendicular to l_1. All points on l_2 are on one side of l_1. The distance between the leftmost point on l_1 and l_2 is d. The first point on l_2 whose distance to line l_1 is $d_1 = d + \epsilon$. The distance between the second point and line l_1 is $d_2 = 2 * d_1 + \epsilon, ...,$ the *ith* point whose distance to l_1 is $d_i = 2 * d_{i-1} + \epsilon, ...$

Therefore, the combinatorial complexity of VRVD of n sites is $\Theta(n^2)$.   □

3   Algorithm for Computing Visual Restriction Voronoi Diagram

The basic idea for computing VRVD is to first compute all $VRVD_i$. for all CP_i (see Figure 5) and then merge them together. If we know the sites that control the cells inside CP_i, then we can compute the normal Voronoi diagram of those sites to obtain $VRVD_i$.. Next, we show how to compute the sites that control the cells inside CP_i.

Since $VRVD_i$. is a tree, all cells inside CP_i have edges on the boundary of CP_i. That is to say, the sites that control a part of the boundary of CP_i are the same as the sites that control cells inside CP_i. Thus, we only need to focus on which site controls which part of boundary line. For a boundary line $l_1(p_k)$ (the same as $l_2(p_k)$), we need to compute the intervals that are controlled by different sites. Without loss of generality, assume

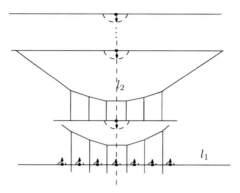

Fig. 6. Example of $\Omega(n^2)$ vertices, edges and cells in the Visual-Restriction Voronoi Diagram of set P with parallel boundaries

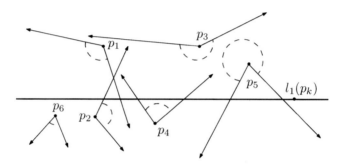

Fig. 7. Illustration of three groups of sites whose directions are to the left, to the right, or vertical

that $l(p_k)$ is a horizontal line. The n sites are separated into four groups according to the number of intersection between their rays and $l(p_k)$ (see Figure 7):

A. Those sites p_i whose rays $r_1(p_i), r_2(p_i)$ intersect with $l(p_k)$ with just one intersections (see p_1, p_2 in Figure 7). We assume the total number of points of this group is m_l.
B. Those sites p_i whose rays $r_1(p_i), r_2(p_i)$ intersect with $l(p_k)$ with just two intersections (see p_4, p_5 in Figure 7). We assume the total number of points of this group is m_r.
C. Those sites p_i which are visible to all points on $l(p_k)$ (see p_3 in Figure 7). We assume the total number of points of this group is m_v. Note that those sites invisible to any points on $l(p_k)$ can be ignored (see p_6 in Figure 7).
D. The site p_k is only visible to a half-line of $l(p_k)$, what is more, only one side of $r_1(p_k), r_2(p_k)$.

We compute the intervals controlled by each group of sites independently and then merge them together later. First we show how to compute the intervals controlled by the first group of sites A. There are two kinds of sites for the first group of sites: (A.1) those sites which is invisible to left part of $l(p_k)$ and visible to the right part of $l(p_k)$;

(A.2) those sites which is visible to left part of $l(p_k)$ and invisible to the right part of $l(p_k)$. They are symmetrical to each other. We compute the two cases (A.1 and A.2) respectively and merger them together. The first case A.1 is covered as follows (the second case A.2 can be covered similarly):

The boundary lines of m_r sites intersect with $l(p_k)$ at points $I_1, I_2, ..., I_{m'_r}$ from left to right ($m'_r \leq m_r$ since some boundary lines intersect with $l(p_k)$ at the same point). These intersections separate $l(p_k)$ into m'_r visible intervals: $[I_1, I_1), [I_2, I_3)$, $..., [I_{m'_r}, +\infty)$. Note that the sites of this group are not visible to any point on the interval $(-\infty, I_1)$. We deal with each interval on $l(p_k)$ from left to right. For each interval, it could be separated into several subintervals such that each subinterval is controlled by one site and we need to compute those subintervals.

Suppose that we have already processed all intervals to the left of the j-th interval $[I_j, I_{j+1})$ and now we are dealing with the j-th interval. Those sites that are visible to the intervals to the left of I_j arc visible to interval $[I_j, +\infty)$. Suppose that those sites that control x subintervals on $[I_j, +\infty)$ are $q_1, q_2, ..., q_x$ from left to right which have already been computed in the previous steps. Note that the order of x subintervals on $[I_j, +\infty)$ are in the same order of $q_1, q_2, ..., q_x$ from left to right. There could be several sites whose boundary line intersect with $l(p_k)$ at I_j and they are only visible to the interval $[I_j, +\infty)$. For each site p of them, we search $q_1, q_2, ...q_x$ to find the position of p among them according to their x-coordinates. There are three cases:

1. p is to the right of q_x, that means p controls the rightmost subinterval. We can sweep from site q_x to the left (until reaching I_j since p is not visible to the left of I_j) to find the left endpoint of the subinterval that p controls. First, we compute the intersection $I_{p,x}$ between the bisector $b_{q_x,p}$ (between site q_x and p) and $l(p_k)$. Assume that the endpoint separating two consecutive subintervals which are controlled by sites q_{x-1} and q_x is a. If $I_{p,x}$ is to the right of a, then stop sweeping and $I_{p,x}$ is the left endpoint of the subinterval that p controls. Else if $I_{p,x}$ is to the left of a or $I_{p,x} = a$, site q_x does not control any point in $[I_j, +\infty)$, and q_x will never be considered later on. Then we treat q_{x-1} as q_x and continue sweeping as above until we find the left endpoint of the subinterval that p controls or reach I_j.

2. p is to the left of q_1. We sweep from q_1 to the right. First we compute the intersection $I_{q_1,p}$ between the bisector $b_{q_1,p}$ (between site q_1 and p) and $l(p_k)$. If $I_{q_1,p}$ is to the left of I_j, p does not control any point on line $l(p_k)$. Else the left endpoint of the subinterval that p controls is I_j and we need to find the right endpoint of the subinterval that p controls. Assume that the endpoint separating two consecutive subintervals that are controlled by sites q_1 and q_2 is b. If $I_{q_1,p}$ is to the left of b, then $I_{q_1,p}$ is the right endpoint of the subinterval that p controls; else q_1 does not control any point in $[I_j, +\infty)$, and we continue sweeping as above from q_2 to the right until we find the right endpoint. Note that the right endpoint could be $+\infty$.

3. The neighbor sites of p from left to right are q_i and q_{i+1}. First, we compute the intersection $I_{p,i}$ between the bisector $b_{p,i}$ (between q_i and p) and $l(p_k)$, and the intersection $I_{p,i+1}$ between the bisector $b_{p,i+1}$ (the bisector between q_{i+1} and p) and $l(p_k)$. If $I_{p,i}$ is not to the left of $I_{p,i+1}$, then p does not control any point in $l(p_k)$; else sweep from q_i toward left to find the left endpoint of the subinterval that

p controls as in step 1, and sweep from q_{i+1} toward right to find the right endpoint of the subinterval that p controls as in step 2.

Note that if the bisector $b_{q_j,p}$ $(1 \le j \le x)$ is parallel to $l(p_k)$, that means there is no intersection between $b_{q_j,p}$ and $l(p_k)$. In this case, we can set $I_{p,j}$ as $+\infty$ or $-\infty$ according to the positions of p, q_j, $l(p_k)$ and $b_{q_j,p}$. If $b_{q_j,p}$ is above $l(p_k)$ and q_j is above $b_{q_j,p}$, then we set $I_{p,j}$ as $-\infty$ since p controls the interval which is originally controlled by q_j. If $b_{q_j,p}$ is above $l(p_k)$ and q_j is below $b_{q_j,p}$, then we set $I_{p,j}$ as $+\infty$. Actually, that means we can just ignore p since it does not control any interval in $[I_j, +\infty)$. Similarly, we can set $I_{p,j}$ as $+\infty$ if $b_{q_j,p}$ is below $l(p_k)$ and q_j is above $b_{q_j,p}$ and set $I_{p,j}$ as $-\infty$ if $b_{q_j,p}$ is below $l(p_k)$ and q_j is below $b_{q_j,p}$. There is still one special case when the bisector $b_{q_j,p}$ overlaps with $l(p_k)$. In this case, we can just set the interval controlled by q_j to be controlled by both q_j and p.

When the above steps are finished, we compute the sites control subintervals in interval $[I_j, I_{j+1})$, which is easy as all sites $q_1, q_2, ..., q_x$ are already sorted and controlled one subinterval from left to right by now for $[I_j, +\infty)$, we just sweep from the q_1 to right, to find the site q_y which control at least one point to the right of I_j, then all the sites on the left of q_y are deleted and never need to be considered later. After the j-th interval $[I_j, I_{j+1})$ is processed, we can process all other intervals to the right of the j-th interval as above from left to right.

The computation of the intervals controlled by the second kinds of sites A.2 is similar to the computation of the intervals controlled by the first kinds of sites A.1.

We now show how to merge intervals generated by sites of A.1 and A.2. We sort the endpoints of two types of intervals from left to right and create some new intervals. For a new interval that originally belongs to one type, we do not need to take care of it. Otherwise if we have a new interval $[a, b]$ that originally belongs to two types, we need to decide whether $[a, b]$ should be split into two subintervals that are controlled by the two type of sites separately. Suppose that $[a, b]$ is controlled by sites p_l and p_r before merging, where p_l points left and p_r points right. We compute the intersection I_{lr} between $l(p_k)$ and the bisector b_{lr} (bisector of p_l and p_r), the intersection I_{lk} between $l(p_l)$ and $l(p_k)$, and the intersection I_{rk} between $l(p_r)$ and $l(p_k)$. According to the positions of I_{lr}, I_{lk} and I_{rk}, we can split $[a, b]$ easily. Details are omitted here due to space limitation. After the merge of the first and the second type of intervals, we obtain the intervals in $l(p_k)$ determined by the first group of sites.

For the second groups of sites B, we use the divide and conquer algorithm to solve it, we just separate the m_r points into two sets of almost the same size, and compute the their intervals controlled on $l(p_k)$ from left to right respectively, then just merge it using the above method. Then we have $T(n) = 2T(n/2) + O(n)$, in which $T(n)$ denotes the time to compute n sites' intervals. The number of intervals controlled by n sites on $l(p_k)$ is $O(n)$ according to Lemma 1. The merging procedure needs only $O(n)$ time for two sets (each with $n/2$ sites), hence $T(n) = O(n \log n)$.

For the third group of sites C, since their boundary lines have no intersections with $l(p_k)$, we can not use the above algorithm directly. However, we can treat the intersections between rays and $l(p_k)$ to be $+\infty$ or $-\infty$. Then we can use the above algorithm again.

We can merge the intervals decided by the sites in A and B in a similar way. Finally, if two newly created consecutive intervals are controlled by the same site, we just merge them together.

At last, we merge the intervals controlled by the four groups of sites respectively. The merging procedure is similar to the procedure of merging intervals determined by A.1 and A.2.

Lemma 3. *All the intervals on line $l(p_k)$ that are controlled by the corresponding sites can be computed in $O(n \log n)$ time.*

Proof. Computing the intersections between $n-1$ boundary lines and $l(p_k)$, and sorting all of them from left to right can be achieved in $O(n \log n)$ time. Before processing the interval $[I_j, +\infty)$, its subintervals and the sites $q_1, q_2, ..., q_x$ that control them are in order from left to right. We use a balanced binary search tree to store them. The sites $q_1, q_2, ..., q_x$ are stored in the leaves of tree. Searching the left neighbor and right neighbor of site p needs $O(\log n)$ time. Then sweeping one subinterval toward left or right takes $O(\log n)$ time. Note that a site that is deleted from the balanced search tree will never be considered again in $[I_j, +\infty)$. Since there are at most $O(n)$ sites, the total time complexity of sweeping is $O(n \log n)$.

Before the merge of different types of $O(n)$ intervals, those intervals are already sorted from left to right. Merging them together takes $O(n)$ time and it produces $O(n)$ new intervals. Splitting each new interval takes only constant time. Hence these steps of merge and splitting take $O(n)$ time. The total running time for computing intervals on line $l(p_k)$ that correspond to sites on one side of $l(p_k)$ is $O(n \log n)$.   □

From the above lemma, we know that we can compute all intervals on all n boundary lines in $O(n^2 \log n)$ time. For a convex polygon CP_i, we can traverse the edges of CP_i to collect the set S_i of sites that control the intervals on edges of CP_i. Then we compute the normal Voronoi diagram VD_i of S_i and use CP_i to cut off VD_i to obtain VP_i. Note that if p_k controls some interval on $l(p_k)$ which belongs to one edge of CP_i, then p_k may or may not be visible to CP_i according to the direction of p_k. If p_k is visible to CP_i, we can include p_k in S_i. Otherwise, p_k is not counted.

Lemma 4. *VP_i can be computed in $O(h_i \log h_i + c_i)$ time, where h_i is the number of sites which control cells inside CP_i and c_i is the number of edges of CP_i.*

Proof. After the arrangement of lines $l(p_k), i = 1, 2, \cdots, n$, is computed and stored in a doubly-connected edge list, the set of sites that controls the edges of CP_i can be obtained in $O(h_i + c_i)$ time through a traversal in the arrangement. Computing the normal Voronoi diagram VD_i takes $O(h_i \log h_i)$ time. The trim of VD_i to obtain HP_i takes $O(h_i + c_i)$ time. So the total running time is $O(h_i \log h_i + c_i)$.   □

Note that some intervals on edges of CP_i are not on the final VRVD since the two sites that control two sides of those intervals are the same. If the two sides of an interval are controlled by the same site (or neither sides is controlled by any sites), then that interval is not edge of final VRVD, else it is an edge of final VRVD. We can traverse the edges of all CP_i and check whether each interval is an edge of final VRVD or not.

Theorem 2. *The VRVD of n sites can be computed in $O(n^2 \log n)$ time.*

Proof. Based on Lemma 4, each VP_i can be computed in $O(h_i \log h_i + c_i)$ time. Then all VP_i can be calculated in $O(\sum_{i=1}^{X}(h_i \log h_i + c_i))$ time where $X = O(n^2)$ is the number of CP. For each boundary line, it is divided into $O(n)$ intervals and each interval is controlled by at most one site according to the proof of lemma 1. Therefore, $\sum_{i=1}^{X} h_i = O(n^2)$. The total number of edges for all convex polygon CP_j $(j = 1, 2, ..., n)$ is $O(n^2)$, as n lines intersecting with each other produces at most $O(n^2)$ line segments, that means $\sum_{i=1}^{X} c_i = O(n^2)$. Hence $O(\sum_{i=1}^{X}(h_i \log h_i + c_i)) = O(n^2 \log n)$. The procedure of merging the edges between adjacent part of Voronoi diagram VP_i can be done by sweeping every boundary line from left to right, which needs $O(n^2)$ time in total, as all intervals are already sorted. Hence the theorem is proved. $\square$

4   Concluding Remarks

In this paper, we propose a new variant of Voronoi diagram such that each site has its own visual area. We study the corresponding Visual Restriction Voronoi Diagram (VRVD) with the visual area of each site being a sector. First, we prove that the combinatorial complexity of VRVD of n sites is $\Theta(n^2)$. Then an algorithm with $O(n^2 \log n)$ running time is given to construct the VRVD with n sites. Whether we can remove the additional log factor and design an $O(n^2)$ time algorithm to construct the VRVD is still an open problem. In the future, we could consider solving the VRVD problem in an output-sensitive fashion because we expect the complexity of VRVD is usually small, possibly only $O(n)$ in real applications. What is more, the VRVD in higher dimension is also an interesting problem.

References

1. Aggarwal, A., Guibas, L., Saxe, J., Shor, P.: A linear time algorithm for computing the voronoi diagram of a convex polygon. In: Proceedings of the Nineteenth Annual ACM Symposium on Theory of Computing, STOC 1987, pp. 39–45. ACM, New York (1987)
2. Aurenhammer, F.: Voronoi diagrams–a survey of a fundamental geometric data structure. ACM Comput. Surv. 23, 345–405 (1991)
3. Aurenhammer, F., Stöckl, G.: On the Peeper's Voronoi diagram. SIGACT News 22, 50–59 (1991)
4. Aurenhammer, F., Edelsbrunner, H.: An optimal algorithm for constructing the weighted voronoi diagram in the plane. In: Pattern Recognition, pp. 251–257 (1984)
5. de Berg, M., van Kreveld, M., Overmars, M., Schwarzkopf, O.: Computational Geometry Algorithms and Applications. Springer (1997)
6. Edelsbrunner, H., Guibas, L.J., Sharir, M.: The upper envelope of piecewise linear functions: Algorithms and applications. Discrete & Computational Geometry 4, 311–336 (1989)
7. Fan, C., He, J., Luo, J., Zhu, B.: Moving network voronoi diagram. In: ISVD, pp. 142–150 (2010)
8. Fortune, S.: A sweepline algorithm for voronoi diagrams. Algorithmica 2, 153–174 (1987)
9. Gowda, I.G., Kirkpatrick, D.G., Lee, D.T., Naamad, A.: Dynamic voronoi diagrams. IEEE Transactions on Information Theory 29(5), 724–730 (1983)
10. Guibas, L.J., Knuth, D.E., Sharir, M.: Randomized incremental construction of delaunay and voronoi diagrams. Algorithmica 7(4), 381–413 (1992)

Minimization of the Maximum Distance between the Two Guards Patrolling a Polygonal Region[⋆]

Xuehou Tan[1,2] and Bo Jiang[2]

[1] School of Information Science and Technology,
Dalian Maritime University, Linghai Road 1, Dalian, China
[2] School of Information Science and Technology,
Tokai University, 4-1-1 Kitakaname, Hiratsuka 259-1292, Japan
`tan@wing.ncc.u-tokai.ac.jp`

Abstract. The *two-guard* problem asks whether two guards can walk to detect an unpredictable, moving target in a polygonal region P, no matter how fast the target moves, and if so, construct a walk schedule of the guards. For safety, two guards are required to always be mutually visible, and thus, they move on the polygon boundary. Specially, a *straight walk* requires both guards to monotonically move on the boundary of P from beginning to end, one clockwise and the other counterclockwise.

The objective of this paper is to find an optimum straight walk such that the *maximum* distance between the two guards is minimized. We present an $O(n^2 \log n)$ time algorithm for optimizing this metric, where n is the number of vertices of the polygon P. Our result is obtained by investigating a number of new properties of the *min-max* walks and converting the problem of finding an optimum walk in the min-max metric into that of finding a shortest path between two nodes in a graph. This answers an open question posed by Icking and Klein.

1 Introduction

Motivated by the relations to the well-known *Art Gallery* and *Watchman Route* problems, much attention has recently been devoted to the problem of detecting an unpredictable, moving target in an n-sided polygon P by a group of mobile guards. Both the target and the guards are modeled by points that can continuously move in P. The goal of the guards is to eventually "see" the target, or to verify that no target is present in the polygon, no matter how fast the target moves. Many types of polygon shapes and the vision sensors of the guards have been studied in the literature [7, 10–17, 19].

In this paper, we focus on the two-guard model, in which two guards move on the polygon boundary and are always kept to be mutually visible [10, 11].

[⋆] Work by Tan was partially supported by Grant-in-Aid (23500024) for Scientific Research from Japan Society for the Promotion of Science, and work by Jiang was partially supported by National Natural Science Foundation of China under grant 61173034.

J. Snoeyink, P. Lu, K. Su, and L. Wang (Eds.): FAW-AAIM 2012, LNCS 7285, pp. 47–57, 2012.

The goal is to patrol P by two guards so that at any instant, the line segment connecting the guards partitions P into a "clear" region (not containing the target) and an "uncleared" region (it may contain the target). In the end, we would like to know whether the whole polygon P is clear or the target is detected, if it is ever possible. This target-finding model may have applications in adversarial settings, as it has obvious advantages for safety and communication between the guards.

Icking and Klein were the first to study the *two-guard problem* [11]. Suppose that two vertices u and v on P are given. The problem of walking two guards in P then asks the guards to start at the entrance u and force the target out of the region through the exit v. Icking and Klein have shown that the solution of the two-guard problem consists of a number of instances of straight walks and counter walks [11]. A walk is said to be *straight* if two guards monotonically move on P from u to v, or *counter* if both guards move on P clockwise, one from u to v and one from v to u. They gave an $O(n \log n)$ time *decision* algorithm for determining whether the polygon P is straight, counter, or general walkable, where n denotes the number of vertices of P. Later, a linear-time decision algorithm was presented by Heffernan [10]. Tseng *et al.* gave an $O(n \log n)$ time algorithm to determine all pairs of boundary points which admit straight, general walks [20]. Recently, Bhattacharya *et al.* improved this time bound to $O(n)$ [3].

The two-guard problem also involves giving a walk, if it exists. For the general walk problem, Icking and Klein have given a *schedule reporting* algorithm with $O(n \log n + k)$ running time, where k is the size of the walk schedule reported. For the straight and counter walk versions, the schedule reporting algorithms of Icking and Klein take $O(n \log n)$ time [11]. Later, Heffernan showed that a straight or counter walk schedule can be reported in $\Theta(n)$ time, if it exists [10].

The objective of this paper is to find an optimum straight walk such that the *maximum* distance between the two guards is minimized. The study of the two-guard problem in this *min-max* metric is motivated by the fact that for a police patrol in a dangerous midtown street, the closer the two guards are kept the safer they are. Actually, finding a min-max walk was left as an open problem in [11]. Our work is also stimulated by its relationship to the well-known Fréchet distance problem [1, 2, 5]. The Fréchet distance for two curves is usually illustrated by a person walking a dog on a leash. Both the person and the dog walk forward on their respective curves. The Fréchet distance of these two curves (they may intersect each other or even self-intersect) is defined as the length of the *shortest* leash that makes it possible for the person and the dog to walk from beginning to end. An $O(N^2 \log N)$ time algorithm has been developed to compute the Fréchet distance between polygonal curves A and B in arbitrary dimensions for obstacle-free environments, where N is the larger of the complexities of A and B [1]. In our straight walk problem, we place one more restriction that the person and the dog should always be mutually visible inside the given region. The $O(n^2 \log n)$ time solution obtained in this paper answers an open question posed by Icking and Klein [11], and matches with the best known result for computing the Fréchet distance between two polygonal curves [1]. Moreover, our algorithm can automatically find the starting and ending vertices such that the maximum distance over all pairs of the starting and ending points is minimized, provided that the given polygon is straight walkable.[1]

The rest of this paper is organized as follows. In Section 2 of this paper, we give basic definitions used in the paper. In Section 3, we show that an optimum straight walk can be decomposed into a sequence of basic motions, called the *atomic walks*. The atomic walks are so defined that their solutions in the min-max metric can easily be found. In Section 4, we introduce a data structure that records all possible atomic walks in the given polygon and a transition relation among the atomic walks. By applying Dijkstra's algorithm to the obtained diagram, we can then give our $O(n^2 \log n)$ time algorithm for finding an optimum straight walk such that the maximum distance between the two guards is minimized. Finally, we pose some open problems for further research in Section 5.

2   Preliminary

Let P denote a simple polygon of n vertices and ∂P the boundary of P. Just for convenience, we assume that P is in a general position in the plane. That is, no three vertices of P are collinear, and no three edge extensions have a common point. Two points p, $q \in P$ are *visible* from each other if the line segment connecting them, denoted by $\overline{pq}$, does not intersect the exterior of P. We denote by $|xy|$ the length of $\overline{xy}$.

Let g_1, g_2 be two point guards on ∂P. Let $g_1(t)$ and $g_2(t)$ denote the positions of g_1 and g_2 on ∂P at time t; we require that $g_1(t)$ and $g_2(t)$: $[0, \infty) \to \partial P$ be two continuous functions. A point $p \in P$ is said to be *detected* or *illuminated* at t if p is contained in the line segment $\overline{g_1(t)g_2(t)}$. A *configuration* of g_1 and g_2 at time t is a pair of the points $g_1(t)$ and $g_2(t)$ such that the line segment $\overline{g_1(t)g_2(t)}$ lies in the interior of P. Assume that the initial positions of two guards are located at a same vertex of P. The configuration of g_1 and q_2 at a time thus divides P into a *clear* region that does not contain the target, and an uncleared (or a contaminated) region that may contain the target. Any line segment $\overline{g_1(t)g_2(t)}$ at a time t is called a *walk segment*. The point $g_1(t)$ is the *walk partner* of $g_2(t)$, and vice versa.

A *walk instruction* can be given by a pair of functions $g_1(t)$, $g_2(t)$ such that either of $g_1(t)$ and $g_2(t)$ specifies an algebraic path, i.e., an edge of P along which the guard g_1 or g_2 moves. More specifically, the following two types of walk instructions are considered: Two guards g_1 and g_2 move along segments of single edges such that (i) no intersections occur among all line segments $\overline{g_1(t)g_2(t)}$ during the movement, or (ii) any two segments $\overline{g_1(t)g_2(t)}$ intersect each other. (Probably, one guard stands still, while the other moves.) See [11].

Let the area of P be one, and let $P(t)$ denote the fraction of the clear area at time t. Initially, $P(0) = 0$. We say a *walk schedule* exists for P, or equally, P

---

[1] Without loss of generality, we assume that a walk schedule always starts (ends) at a polygon vertex.

allows a *walk* if $P(t) = 1$ for some $t > 0$. The *complexity* of a walk schedule is the total number of walk instructions it consists of.

Theorem 1. *(See [3]) It takes $O(n)$ time to determine whether the given polygon P is straight walkable, where n is the number of vertices of P.*

For the remainder of the paper, we assume that the polygon P is straight walkable.

3   Properties of the Min-max Walks

In this section, we study the straight walk probem in the min-max metric. We first describe a method to assign some *critical* walk partners with every vertex of P. These critical walk segments are then used to define a basic motion of the two guards, whose solution in the min-max metric can easily be given.

First, we briefly review an important concept of ray shots. A vertex of P is *reflex* if its interior angle is strictly greater than 180°. For a reflex vertex r, its forward (backward) *ray shot*, denote by $F(r)$ $(B(r))$ is defined as the first point of P hit by a "bullet" shot at r along the edge adjacent to r, in clockwise (counterclockwise) direction. See Fig. 1. Clearly, the motion of the two guards are restricted by these ray shots inside P [10, 11, 18]. Assume that the shaded region below the line segment $\overline{ab}$ in Fig. 1 is currently clear. Assume also that there is a reflex vertex r in the contaminated region such that four points a, $F(r)$, r and b (a, r, $B(r)$ and b) are in clockwise order. See Fig. 1. Then, the shot $F(r)$ requires that the guard at a should move to $F(r)$ clockwise by the time the guard at b moves to pass through r; otherwise, the contaminated region can never be cleared by any straight walk, which starts at $\overline{ab}$ and ends at some vertex of the contaminated region. Analogously, $B(r)$ requires that the guard at b should move to $B(r)$ counterclockwise by the time the guard at a moves to pass through r.

In order to assign the walk partners for every vertex of P, we compute the ray shots for all reflex vertices of P. A portion of an edge of P is called a "pseudo-edge" if its endpoints are either a polygon vertex or a ray shot, and its interior does not contain any other ray shots. Next, we compute for each vertex x its

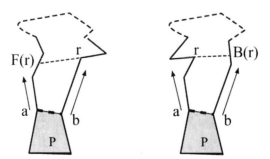

Fig. 1. Ray shots $F(r)$ or $B(r)$ place the restrictions on the motion of the guards

visible region inside P. Denote by $vis(x)$ the visible region of x, and denote by $S(x)$ the set of the pseudo-edges contained in $vis(x)$. For each pseudo-edge e of $S(x)$, we let the point of e closest to x be a *critical* walk partner for x. The vertex x is called the *defining vertex* of the pseudo-edge e. Clearly, several critical partners may be assigned with x. We call the segments connecting x and its critical partners, the *critical* walk segments. So, a critical walk segment always has a polygon vertex (i.e., the defining vertex) as one of its endpoints.

Two critical walk segments, together with at most two pseudo-edges, or some portions of the pseudo-edges between them, form a *walk-quadrilateral* (it may degenerate into a triangle) if the walk segments do not intersect (but they may share an endpoint). A walk-quadrilateral is *minimal*, if it does not contain any other walk-quadrilaterals. We call the walk of the two guards in a minimal walk-quadrilateral, an *atomic walk*. Clearly, an atomic walk does not contain any other atomic walks.

We present below several properties of the atomic walks, which are the key to our solution.

Lemma 1. *An atomic walk can be made such that the maximum distance between the guards is the larger of the lengths of two critical walk segments.*

Proof. Observe that any walk-quadrilateral, from its definition, is straight walkable from one critical walk segment to the other. So, if the underlying walk-quadrilateral degenerates into a triangle, then only one guard needs to move from one critical walk segment to the other, while the other stands still. The lemma simply follows.

Suppose that the walk-quadrilateral is not a triangle. Assume that $\overline{x_1 y_1}$ and $\overline{x_2 y_2}$ are two critical walk segments, which define the walk-quadrilateral for the atomic walk, and their defining vertices are x_1 and x_2. Our method is then to introduce at most two (walk) segments, each of the length smaller than the larger of $|x_1 y_1'|$ and $|x_2 y_2|$, to divide the walk-quadrilateral into a smaller quadrilateral with two parallel segments and at most two triangles.

Without loss of generality, assume that two guards are currently located at x_1 and y_1, i.e., the segment connecting the two guards is required to move from $\overline{x_1 y_1}$ to $\overline{x_2 y_2}$. If either critical walk segment, say, $\overline{x_1 y_1}$, happens to have two vertices x_1, y_1 as its endpoints, then the segment $\overline{x_1 y_1}$ is perpendicular to both the edge containing x_1 and the edge containing y_1. This is because otherwise the angle $\angle x_1 y_1 y_2$ is strictly larger than $\pi/2$, and thus the atomic walk defined by x_1 and x_2 contains the one defined by x_1 and y_1, a contradiction (Fig. 2(a)). In this case, the segment connecting the guards can first move parallel to $\overline{x_1 x_2}$ till it touches x_2 or y_2, and then one guard further moves to reach the other endpoint of $\overline{x_2 y_2}$. Clearly, the distance function between the two guards is not changed in the former movement, and monotonically increasing the latter movement. The lemma is true.

Assume now that neither of y_1 and y_2 is a vertex. If neither y_1 nor y_2 is a ray shot, then both $\overline{x_1 y_1}$ and $\overline{x_2 y_2}$ are perpendicular to the line containing y_1 and y_2. See Fig. 2(b). The segment connecting the two guards can be moved, parallel to $\overline{x_1 y_1}$ (or $\overline{x_2 y_2}$), from one critical walk segment to the other. Again,

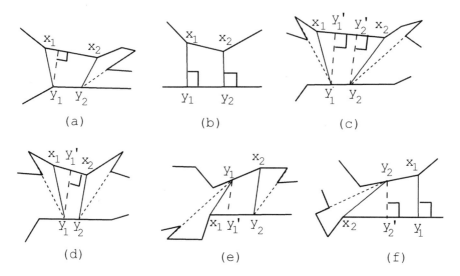

Fig. 2. Illustrating the proof of Lemma 1

the lemma is true. If y_1 (y_2) is a ray shot, then $\angle x_1 y_1 y_2$ ($\angle x_2 y_2 y_1$) is at least $\pi/2$. In this case, we find the point y_1' (y_2') of the segment $\overline{x_1 x_2}$, which is closest to y_1 (y_2). See Fig. 2(c)-(d). Note that one of y_1' and y_2' may not exist (Fig. 2(d)). The atomic walk can then be made by at most three following motions of the guards: from $\overline{x_1 y_1}$ to $\overline{y_1 y_1'}$, from $\overline{y_1 y_1'}$ to $\overline{y_2 y_2'}$, and from $\overline{y_2 y_2'}$ to $\overline{x_2 y_2}$. See Fig. 2(c) or 2(d). Also, the lemma follows.

Finally, consider the situation in which x_1 and x_2 are on different pseudo-edges. If the edges containing $\overline{x_1 y_2}$ and $\overline{x_2 y_1}$ are parallel, the segment connecting the guards can simply be moved from $\overline{x_1 y_1}$ to $\overline{x_2 y_2}$ such that the distance function between the two guards is monotone. If $\overline{x_1 y_2}$ and $\overline{x_2 y_1}$ are not parallel, then as discussed above, two angles $\angle x_1 y_1 x_2$, $\angle x_2 y_2 x_1$ are at least $\pi/2$. If the angle at x_1 is the smallest in the quadrilateral $x_1 y_1 x_2 y_2$, we let y_1' the point of $\overline{x_1 y_2}$ such that $\overline{y_1 y_1'}$ and $\overline{x_2 y_2}$ are parallel (Fig. 2(e)). Otherwise, the angle at x_2 is the smallest and we let y_2' the point of $\overline{x_2 y_1}$ such that $\overline{y_2 y_2'}$ and $\overline{x_1 y_1}$ are parallel (Fig. 2(f)). We have $|y_1 y_1'| < |x_1 y_1|$, $|y_1 y_1'| < |x_2 y_2|$ in the former case, and $|y_2 y_2'| < |x_1 y_1|$, $|y_2 y_2'| < |x_2 y_2|$ in the latter case. In either case, the motion of the guards can simply be arranged such that the maximum distance between the two guards is the larger of $|x_1 y_1'|$ and $|x_2 y_2|$. The proof is complete.   □

Next, we show that there is an optimum walk in the min-max metric such that it consists of only atomic walks. Suppose that W is an optimum walk in the min-max metric, and u, v are its starting, ending vertices, respectively. All polygon vertices are touched, in the walk W, by the segment connecting the guards one by one. We call the first walk segment reached a vertex x, the W-*walk segment* of x. Also, we consider u (v) as the very first (last) W-walk segment. Since no two W-walk segments properly intersect, all W-walk segments can be ordered from u to v. Consider the motion of the guards between two consecutive

W-walk segments. By an argument similar to the proof of Lemma 1, it can also be arranged so that the maximum distance between the guards is the larger of the lengths of two W-walk segments. Thus, we focus below our attention on the lengths of W-walk segments.

Observe that a walk W is globally optimum if it is locally optimum, i.e., the portion of the walk between any two W-walk segments is optimum, too. An immediate result is the following.

Lemma 2. *Suppose that W is an optimum straight walk in the min-max metric. Then, one can always assume that the portion of W between any two W-walk segments is optimum. Moreover, all the locally longest W-walk segments belong to the set of critical walk segments.*

Proof. If the portion of W between some two W-walk segments is not optimum, then we can rearrange the walk between the W-walk segments such that that portion of W is optimum. The first part of the lemma simply follows.

Turn to the second part of the lemma. Let S_i, ..., S_j, ..., S_k ($i < j < k$) be a sequence of consecutive W-walk segments such that the length of S_j, i.e., $|S_j|$ is the maximal among $|S_i|$, ..., $|S_k|$. Assume now that S_j is not a critical walk segment. Let x be the defining vertex of S_j, and let y be the other endpoint of S_j. Then, we can move the point y to a new position y', on the edge containing y, by an arbitrary small distance such that $|S_j| > |xy'|$ and $|S_h| \leq |xy'|$, $h \neq j$. This implies that the portion of W between S_i and S_j is not optimum, a contradiction. Therefore, S_j is a critical walk segment. The proof is complete. □

The above result is still a little away from what we need. Let us now consider a *discrete* version of this min-max problem, in which not only the maximum distance but also the distance between the guards on every W-walk segment is minimized. (So, any solution to the discrete, min-max problem also minimizes the sum of the lengths of all W-walk segments.) Then, we can simply obtain following results.

Observation 1. *Any optimum straight walk in the discrete, min-max metric gives the same answer (i.e., the maximum distance between the two guards) as that in the min-max metric.*

Lemma 3. *Suppose that W is an optimum straight walk in the discrete, min-max metric. Then, one can always assume that the portion of W between any two W-walk segments is optimum. Moreover, all the W-walk segments belong to the set of critical walk segments.*

Corollary 1. *Suppose that W is an optimal straight walk W in the discrete, min-max metric. Then, all the motions of the guards between two consecutive W-walk segments are the atomic walks.*

4   Algorithm

In this section, we introduce a data structure, called the *atomic walk diagram*, which records all possible atomic walks in the given polygon P and a transition relation among these walks (see also [19]). Next, we show that any optimum solution in the discrete, min-max metric can be represented by a path between two special nodes of the diagram. This makes it possible to apply Dijkstra's algorithm to the atomic walk diagram, so as to find an optimum solution.

Suppose that all vertices of P and their assigned walk partners are ordered on the boundary of P clockwise. Let us number all vertices and walk partners in the sorted order using integers $0, 1, \ldots, m - 1$. For ease of presentation, we slightly modify the description of the polygon P. If x is a polygon vertex, we let x' be a copy of x, and consider x and x' as two *different* vertices of P. All the original and copied vertices appear on the boundary of P alternately, and thus, an edge has an original vertex and a copied vertex, and (One may also consider that x' and x have two consecutive integers, but only the original vertices have their walk partners.)

The atomic walk diagram G is constructed as follows. First, we put into the set $V(G)$ all the nodes (i, j) $(0 \le i, j \le m - 1)$, where i and j are two endpoints of a critical walk segment. For all vertices x, we also put into $V(G)$ the nodes (x, x') and (x', x). Note that any walk schedule is assumed to start (end) at a polygon vertex. The nodes (x, x') will be considered as the possible starting points of walk schedules, and thus, called the *starting nodes*. The nodes (x', x) are considered as the possible ending points of walk schedules, and called the *ending nodes*. Finally, we add two special nodes s (called the *source* node), t (called the *target* node) to $V(G)$. Since no more than n critical walk segments can be defined by a vertex, the total number of nodes of $V(G)$ is clearly $O(n^2)$.

Let us describe how to construct the arc set of the diagram G, which is denoted by $E(G)$. Note that the starting point (vertex) of a straight walk is always contained in the clear region. Although the exact position of this starting point is not known, we can assume that the vertices of the clear region are in

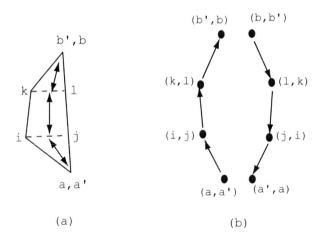

Fig. 3. Illustrating the construction of the diagram G

clockwise order, from the starting point. Two nodes (i, j) and (k, l) are then connected by a *single* arc from (i, j) to (k, l) if and only if (i) there is an atomic walk between them and (ii) the points i, k, l and j are in clockwise order. See Fig. 3. Clearly, if there is an arc from (i, j) to (k, l) in G, then the (symmetric) arc from (l, k) to (j, i) exists. Specially, if there is an arc from (x, x') (the starting node) to (k, l) $(k \neq l)$, then the arc from (l, k) to (x', x) (the ending node) exists. Finally, for every starting node (x, x'), we add an arc from s to (x, x'). And, for every ending node (x', x), we add an arc from (x', x) to t. See Fig. 3 for an example, which shows the arcs representing three atomic walks. (Only a portion of G is shown in Fig. 3.) Since a polygon vertex x can define at most $O(n)$ atomic walks, the total number of arcs of the obtained set $E(G)$ is also $O(n^2)$. Observe that the transition relation among the atomic walks is implicitly represented by all arcs of $E(G)$.

Lemma 4. *The atomic walk diagram G can be constructed in $O(n^2)$ time and space.*

Proof. First, all ray shots can be computed in $O(n \log n)$ time using the ray-shooting query algorithm [4]. For each vertex x, its visible region $vis(x)$ as well as the pseudo-edge set $S(x)$ can be computed in $O(n)$ time [8]. All critical walk segments can then be computed in $O(n^2)$ time. As shown in the proof of Lemma 1, an atomic walk occurs in the following situations: (i) two adjacent (defining) vertices have their walk partners on a pseudo-edge, including the special case that the edge degenerates into a vertex, and (ii) a defining vertex and the partner of the other defining vertex are on a pseudo-edge such that no other ray shots are contained between them; otherwise, it contradicts the definition of atomic walks. Thus, all the atomic walks defined by a vertex x can be found in $O(n)$ time. The time required to compute all atomic walks is $O(n^2)$, too. After all atomic walks are found, the diagram Q can simply be constructed in $O(n^2)$ time and space. $\quad\square$

Lemma 5. *Suppose that P is straight walkable. Any optimum walk W in the discrete, min-max metric can then be represented as an st-path in the diagram G. Also, an st-path in G always correspoonds to a valid walk of the two guards.*

Proof. First, we claim that not only the vertices but also the ray shots have to be considered in a straight walk (i.e., both polygon vertices and ray shots are needed to give the walk instructions of the straight walk). Let $W_{u,v}$ denote a straight walk from a vertex u to the other v. With respect to $W_{u,v}$, the boundary of P can be divided into two polygonal chains, with common endpoints u and v. Then, for a vertex x on a chain, all of its possible walk partners in $W_{u,v}$ belong to a non-empty interval $[lo(x), hi(x)]$ on opposite chain, where each of $lo(x)$ and $hi(x)$ is either a polygon vertex or a ray shot [11]. Since u and v may be any two vertiecs of P, the claim simply follows.

Besides the vertices and ray shots, the critical walker partners (defined for all polygon vertices) are also used to define the atomic walks. It follows from Corollary 1 that the optimum walk W in the discrete, min-max metric can be decomposed into a sequence of atomic walks. From our construction of the diagram G, all possible atomic walks in P are represented by arcs of G. (Remember that an atomic walk is represented by two arcs in G.) Thus, all the atomic walks in W can be mapped to a sequence of arcs in G. Next, add two additional arcs to the obtained sequence; one connects s to the first node of the arc sequence representing W, and the other connects the last node of the arc sequence. Since W gives a continuous motion of the two guards, the resulting sequence is a directed st-path in G.

Assume now that G contains a directed st-path. Except for the very first and last arcs, any other arc of the st-path corresponds an atomic walk of the two guards. Since the very first (last) arc actually indicates the starting (ending) vertex, the directed st-path in G can thus be transformed into a valid walk of the two guards. This completes the proof.   □

To solve the discrete, min-max problem, we assign a weight with each arc of $E(G)$. First, the weight of an arc connected to the node s or t is defined as zero. Since all other arcs of $E(G)$ represent the atomic walks, the weight of an arc is defined as the larger of the lengths of two critical walk segments. We denote by G_m this weighted diagram of G.

It follows from Lemma 5 that an optimum walk W corresponds to such a directed st-path in G_m that the maximum of the weights of the arcs in the st-path is minimized. Such a min-weight st-path in G_m can be computed in $O(n^2 \log n)$ time using Dijkstra's algorithm [6]. (If two computed paths happen to have the same maximum distance, we can further compare the sum of the lengths of the critical walk segments encountered by them, and select the path having the smaller sum.) Note also that Dijkstra's algorithm can find two symmetric optimum walks (see Fig. 3).

By now, we obatin the main result of this paper.

Theorem 2. *Suppose that P is straight walkable. One can compute in $O(n^2 \log n)$ time an optimum walk such that the maximum distance between the two guards is minimized.*

5   Concluding Remarks

In this paper, we study the problem of finding an optimum straight walk such that the maximum distance between the two guards is minimized. We presented an $O(n^2 \log n)$ time algorithm for optimizing this metric. (Actually, the discrete version of the min-max walk problem is also solved.) The key point is to decompose an optimum straight walk into a sequence of atomic walks, whose solutions in the min-max metric can easily be found. Because of its relationship to the Fréchet distance problem, our method may find new applications in computing various Fréchet distances.

We pose several open questions for further research. Suppose that P is known to be straight walkable from the entrance u to the exit v. In this special case, our algorithm still needs $O(n^2 \log n)$ time to compute an optimmum walk from u to v. It is thus an interesting work to develop a more efficient solution for this special case. Second, is it possible to extend our method to counter (or general) walks? Note that all walk segments intersect each other in a counter walk. We find it difficult to define the concept of the *atomic counter walks*, analogous to that of the atomic walks. In contrast, an $O(n^2)$ time algorithm for computing an optimum *general* walk such that the sum of the distances travelled by the two guards is minimized has been developed by the authors [19].

References

1. Alt, H., Godau, M.: Computing the Fréchet distance between two polygonal curves. Int. J. Comput. Geom. & Appl. 5, 75–91 (1995)
2. Bespamyatnikn, B.: An optimal morphing between polylines. Int. J. Comput. Geom. & Appl. 12, 217–228 (2002)
3. Bhattacharya, B.K., Mukhopadhyay, A., Narasimhan, G.: Optimal Algorithms for Two-Guard Walkability of Simple Polygons. In: Dehne, F., Sack, J.-R., Tamassia, R. (eds.) WADS 2001. LNCS, vol. 2125, pp. 438–449. Springer, Heidelberg (2001)
4. Chazelle, B., Guibas, L.: Visibility and intersection problem in plane geometry. Discrete Comput. Geom. 4, 551–581 (1989)
5. Cook, A.F., Wenk, C.: Geodesic Fréchet distance inside a simple polygon. ACM Trans. Algo. 7(1) (2010)
6. Corman, T.H., Leiserson, C.E., Rivest, R.L., Stein, C.: Introdution to algorithms, 2nd edn. The MIT Press (2001)
7. Efrat, A., Guibas, L.J., Har-Peled, S., Lin, D.C., Mitchell, J.S.B., Murali, T.M.: Sweeping simple polygons with a chain of guards. In: Proc., ACM-SIAM Sympos. Discrete Algorithms, pp. 927–936 (2000)
8. Ghosh, S.K.: Visibility algorithms in the plane. Cambridge University Press (2007)
9. Guibas, L.J., Latombe, J.C., Lavalle, S.M., Lin, D., Motwani, R.: Visibility-based pursuit-evasion in a polygonal environment. IJCGA 9, 471–493 (1999)
10. Heffernan, P.J.: An optimal algorithm for the two-guard problem. Int. J. Comput. Geom. & Appl. 6, 15–44 (1996)
11. Icking, C., Klein, R.: The two guards problem. Int. J. Comput. Geom. & Appl. 2, 257–285 (1992)
12. LaValle, S.M., Simov, B., Slutzki, G.: An algorithm for searching a polygonal region with a flashlight. Int. J. Comput. Geom. & Appl. 12, 87–113 (2002)
13. Lee, J.H., Park, S.M., Chwa, K.Y.: Searching a polygonal room with one door by a 1-searcher. Int. J. Comput. Geom. & Appl. 10, 201–220 (2000)
14. Suzuki, I., Yamashita, M.: Searching for mobile intruders in a polygonal region. SIAM J. Comp. 21, 863–888 (1992)
15. Tan, X.: A unified and efficient solution to the room search problem. Comput. Geom. Theory Appl. 40(1), 45–60 (2008)
16. Tan, X.: An efficient algorithm for the three-guard problem. Discrete Appl. Math. 158, 3312–3324 (2008)
17. Tan, X.: Sweeping simple polygons with the minimum number of chain guards. Inform. Process. Lett. 102, 66–71 (2007)
18. Tan, X.: The Two-Guard Problem Revisited and Its Generalization. In: Fleischer, R., Trippen, G. (eds.) ISAAC 2004. LNCS, vol. 3341, pp. 847–858. Springer, Heidelberg (2004)
19. Tan, X., Jiang, B.: Optimum Sweeps of Simple Polygons with Two Guards. In: Lee, D.-T., Chen, D.Z., Ying, S. (eds.) FAW 2010. LNCS, vol. 6213, pp. 304–315. Springer, Heidelberg (2010)
20. Tseng, L.H., Heffernan, P.J., Lee, D.T.: Two-guard walkability of simple polygons. Int. J. Comput. Geom. & Appl. 8(1), 85–116 (1998)

On Covering Points with Minimum Turns

Minghui Jiang

Department of Computer Science, Utah State University, Logan, UT 84322-4205, USA
mjiang@cc.usu.edu

Abstract. We point out mistakes in several previous FPT algorithms for k-LINK COVERING TOUR and its variants in $\mathbb{R}^2$, and show that the previous NP-hardness proofs for MINIMUM-LINK RECTILINEAR COVERING TOUR and MINIMUM-LINK RECTILINEAR SPANNING PATH in $\mathbb{R}^3$ are incorrect. We then present new NP-hardness proofs for the two problems in $\mathbb{R}^{10}$.

1 Introduction

The problem of covering a set of points by a minimum number of lines is one of the oldest problems in computational geometry. Megiddo and Tamir [16] proved that the line cover problem is NP-hard even in $\mathbb{R}^2$. For the rectilinear version of the problem in which the lines must be axis-parallel, Hassin and Megiddo [12] observed the problem in $\mathbb{R}^2$ reduces to vertex cover in bipartite graphs and hence is solvable in polynomial time, and then proved that the problem in $\mathbb{R}^3$ in NP-hard. On the other hand, Gaur and Bhattacharya [10] presented a $(d-1)$-approximation algorithm for the problem in $\mathbb{R}^d$ for all $d \geq 3$. Also, Langerman and Morin [14] presented FPT algorithms for the general problem of covering n points in $\mathbb{R}^d$ by k hyperplanes with both d and k as parameters; see also [11,19].

Instead of using lines, we can cover the points using a polygonal chain of line segments, with the goal of minimizing the number of links or turns in the chain. Given a set of n points in $\mathbb{R}^d$, a chain of line segments that covers all n points is called a *covering tour* if the chain is closed, and is called a *spanning path* if the chain is open. A covering tour (or a spanning path) is *rectilinear* if all segments in the tour (or the path) are axis-parallel. Thus we have four optimization problems MINIMUM-LINK COVERING TOUR, MINIMUM-LINK RECTILINEAR COVERING TOUR, MINIMUM-LINK SPANNING PATH, and MINIMUM-LINK RECTILINEAR SPANNING PATH (and correspondingly, four decision problems k-LINK COVERING TOUR, k-LINK RECTILINEAR COVERING TOUR, k-LINK SPANNING PATH, and k-LINK RECTILINEAR SPANNING PATH). These problems have been extensively studied in terms of both computational complexity and combinatorial bounds [15,6,17,2,5,18,4]; see also [9,1,3] for related results.

We now review previous results on these problems. On the negative side, Kranakis et al. [15] noted that MINIMUM-LINK SPANNING PATH in $\mathbb{R}^2$ is NP-hard (they credited Clote for this result), and Arkin et al. [2] proved that MINIMUM-LINK COVERING TOUR in $\mathbb{R}^2$ is NP-hard. On the positive side, Stein and Wagner [17] presented an $O(\log n)$-approximation for MINIMUM-LINK COVERING TOUR in $\mathbb{R}^2$ (using the approximation for set cover). Moreover, for the rectilinear versions of the problems,

J. Snoeyink, P. Lu, K. Su, and L. Wang (Eds.): FAW-AAIM 2012, LNCS 7285, pp. 58–69, 2012.

Stein and Wagner [17] presented a 2-approximation for MINIMUM-LINK RECTILIN-
EAR COVERING TOUR in $\mathbb{R}^2$, and Bereg et al. [4] presented a 2-approximation for
MINIMUM-LINK RECTILINEAR SPANNING PATH in $\mathbb{R}^2$ and a d^2-approximation for
MINIMUM-LINK RECTILINEAR SPANNING PATH in $\mathbb{R}^d$ for all $d \geq 3$.

In contrast to the NP-hardness of (non-rectilinear) MINIMUM-LINK COVERING
TOUR and MINIMUM-LINK SPANNING PATH in $\mathbb{R}^2$, the complexities of MINIMUM-
LINK RECTILINEAR COVERING TOUR and MINIMUM-LINK RECTILINEAR SPAN-
NING PATH in $\mathbb{R}^d$ for any fixed $d \geq 2$ remain unknown, even for the simplest case
that all points are in the plane and no two points share the same x or y coordinate. Interestingly, for this simplest case, Stein and Wagner [17] presented an approximation
algorithm for MINIMUM-LINK RECTILINEAR COVERING TOUR that uses at most 2
more turns than the optimal, and Bereg et al.'s constructive proof for a related combinatorial bound [4, Theorem 5] implies an approximation algorithm for MINIMUM-LINK
RECTILINEAR SPANNING PATH that uses at most 1 more turn than the optimal.

Recently, Estivill-Castro et al. [7,8] reported NP-hardness proofs for MINIMUM-
LINK RECTILINEAR COVERING TOUR (they called it the RECTILINEAR MINIMUM
LINK TRAVELING SALESMAN PROBLEM) and for MINIMUM-LINK RECTILINEAR
SPANNING PATH (they called it the RECTILINEAR MINIMUM LINK SPANNING PATH
PROBLEM) in $\mathbb{R}^3$ [7], and presented several FPT algorithms for k-LINK RECTILINEAR
COVERING TOUR (they call it the RECTILINEAR k-BENDS TRAVELING SALESMAN
PROBLEM) and its variants in $\mathbb{R}^2$ [8]. The NP-hardness proofs in [7] are reminiscent
of the NP-hardness proof of Hassin and Megiddo [12] for the related problem of line
cover in $\mathbb{R}^3$. The FPT algorithms in [8] are based on standard kernelization and bounded
search tree techniques as in the previous work of Langerman and Morin [14] on line
cover. In this paper, we point out mistakes in several FPT algorithms in [8], and show
that the NP-hardness proofs in [7] are incorrect. We then present new NP-hardness
proofs for MINIMUM-LINK RECTILINEAR COVERING TOUR and MINIMUM-LINK
RECTILINEAR SPANNING PATH in $\mathbb{R}^{10}$.

2   Mistakes in Previous Algorithms

In this section, we point out mistakes in several previous FPT algorithms for k-LINK
COVERING TOUR and its variants in $\mathbb{R}^2$ [8].

We first point out a mistake in the FPT algorithm for the problem k-LINK COVERING
TOUR [8, Section 2.1]. This algorithm uses a kernelization procedure to find a set L_{k+1}
of at most k lines, where each line covers at least $k + 1$ points, then obtains a reduced
instance of at most $2k+k^2$ points, including the two extreme points of each line in L_{k+1},
and the at most k^2 points not covered by any line in L_{k+1}. The algorithm then finds the
set R of all lines through at least two points in the reduced instance, and enumerates
tours based on the lines in R. The choice of the set R depends on a lemma [8, Lemma 6]
which states that "If a tour T has the minimum number of turns, then every line segment
in T covers at least two points." The mistake in this algorithm is that it neglected the
possibility that a segment in T may cover at least two points but either none or only
one of these points is included in the reduced instance: in particular, a point may not
be included in the reduced instance if it is on a line in L_{k+1} but is not one of the two

extreme points. Consequently, the line supporting this segment may not be included in R and hence may not be used in enumerating tours. We remark that the algorithm for the related problem k-LINK RECTILINEAR COVERING TOUR [8, Section 3.1] is more careful in composing the set R; see [8, Lemma 10 and Lemma 11].

We next point out a mistake in the FPT algorithm for the problem k-LINK COVERING TOUR with the constraint that "one line-segment covers all points on the same line" [8, Section 2.2]. This algorithm depends on the same lemma [8, Lemma 6], which was proved by a translation/rotation argument. While the argument is valid for the problem without constraint, it does not hold for the problem with constraint. We refer to Figure 1 for a counterexample. It is clear that no tour can cover these 11 points with less than 4 turns. On the other hand, these 11 points can be covered by a tour with 4 turns at $(3,0)$, $(-5,0)$, $(0,5)$, and $(0,-1)$, where each segment in the tour covers all points on the line supporting the segment. The following proposition disproves the lemma [8, Lemma 6] for the problem with constraint:

Proposition 1. *In any tour with 4 turns that covers the 11 points in Figure 1 under the constraint, there is at least one segment that covers either a single point or no point at all.*

Proof. Observe that each of the three lines $y = 0$, $x = 0$, and $y - x = 5$ covers four points. We claim that the tour must include one segment from each of these three lines. Suppose the contrary that one of these three lines does not host a segment, then each of the four points on this line must be covered by a distinct segment in the tour. It is easy to verify (by a case analysis) that the four segments through the four points cannot cover the other seven points.

Now, the segment on the horizontal line $y = 0$ must contain the four points $(-5,0)$, $(1,0)$, $(2,0)$, and $(3,0)$ because of the constraint, so it cannot be consecutive with the segment on the vertical line $x = 0$. Thus we can assume, up to symmetry, that the segment on $y = 0$ is followed by the segment on $y - x = 5$, which is then followed

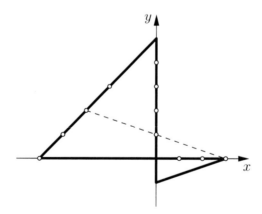

Fig. 1. A counterexample for the general problem with constraint. The point set consists of 11 points: $(1,0)$, $(2,0)$, $(3,0)$, $(0,1)$, $(0,2)$, $(0,3)$, $(0,4)$, $(-5,0)$, $(-4,1)$, $(-3,2)$, $(-2,3)$. The number of turns is 4.

by the segment on $x = 0$. Now consider the fourth segment that closes the tour. This segment cannot be the segment between the two points $(0, 1)$ and $(3, 0)$ because then it would miss the point $(-3, 2)$ which is on the same line. Similarly, it cannot be on the horizontal line $y = 0$. Indeed this fourth segment can contain at most one point, either $(0, 1)$ or $(3, 0)$.   □

After reducing the point set to a kernel of size at most k^2, the algorithm for the problem with constraint simply enumerates tours based on the lines through at least two points in the kernel as justified by this lemma; see [8, page 199]. Note that the point set in our counterexample is already a kernel by their criteria because no line can covered more than $k = 4$ points. Since the lemma is no longer valid for the problem with constraint, the algorithm fails to find a solution to our counterexample illustrated in Figure 1. (Another mistake in this algorithm is that the enumeration of tours in the kernel is done with no regard to the points outside the kernel. Consequently, there may be segments through at least two points in the kernel that miss points on the same line but outside the kernel. This may lead to invalid tours that violate the constraint.)

There is a similar mistake in the FPT algorithm for the problem k-LINK RECTILIN-EAR COVERING TOUR with the constraint that "one line-segment covers all points on the same line" [8, Section 3.2.1]. This algorithm only considers "tours where every line-segment covers at least one point." To justify this, [8, page 205, footnote e] argues that "If a line-segment in the tour covers no points, it can be translated in parallel until it is placed over one point." Again, this argument is valid for the problem without constraint but does not hold for the problem with constraint.

We refer to Figure 2 for a counterexample. Observe that any rectilinear tour that covers the 16 points under the constraint must contain either four horizontal segments on the four lines $y = j$, $0 \leq j \leq 3$, or four vertical segments on the four lines $x = i$, $0 \leq i \leq 3$. Assume the former without loss of generality. Then to connect the four horizontal segments into a rectilinear tour with 8 turns, we need two vertical segments with x-coordinates less than 0 and two vertical segments with x-coordinates more than 3; see Figure 2 for one such tour. To translate these four vertical segments until they cover points would violate the constraint. The algorithm for the rectilinear problem with constraint [8, Section 3.2.1] always outputs a tour in which every segment covers at least one point: each segment in the tour either is one of two candidate segments that cover a point p in the bounded search tree procedure as in [8, Figure 9], or is a

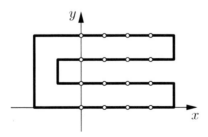

Fig. 2. A counterexample for the rectilinear problem with constraint. The point set consists of 16 points: (i, j) for $0 \leq i \leq 3$ and $0 \leq j \leq 3$. The number of turns is 8.

connecting segment incident to at least one extreme point of a candidate segment as in [8, Figure 10]. Consequently, the algorithm either fails to find a solution, or finds an invalid solution, to our counterexample illustrated in Figure 2.

3   Mistakes in Previous NP-Hardness Proofs

In this section, we show that the previous NP-hardness proofs for MINIMUM-LINK RECTILINEAR COVERING TOUR and MINIMUM-LINK RECTILINEAR SPANNING PATH in $\mathbb{R}^3$ [7, Theorem 15 and Theorem 16] are incorrect.

We briefly review the NP-hardness proof for MINIMUM-LINK RECTILINEAR COVERING TOUR in $\mathbb{R}^3$ [7, Theorem 15], which is based on a reduction from the NP-hard problem ONE-IN-THREE 3-SAT. The building block of the construction, illustrated in [7, Figure 5], consists of 22 line-segments arranged in two groups, with 11 segments in each group following a similar pattern, where each segment is represented by a large number of points on it. They claimed that each building block, when considered alone, admits only two minimum rectilinear tours with 34 turns. These two tours, illustrated in [7, Figures 6 and 7], can be represented schematically as

$$\bullet \ \oplus \ \circ \quad \text{and} \quad \circ \ \ominus \ \bullet$$

where $\oplus$ and $\ominus$ stand for the two different ways to traverse segments 2–21 depending on the directions $\bullet \ \circ$ or $\circ \ \bullet$ in which the two segments 1 and 22 are traversed.

Given a 3-SAT formula with n' variables and m clauses, the construction includes one variable gadget for each variable, and one clause gadget for each clause. Each variable gadget consists of m building blocks; see [7, Figure 9]. They claimed that each variable gadget, when considered alone, admits only two minimum rectilinear tours with $34m$ turns

$$\underset{1}{\bullet} \ \oplus \ \underset{2}{\circ} \ \underset{2}{\bullet} \ \oplus \ \circ \ \cdots \ \bullet \ \oplus \ \underset{m}{\circ} \ \underset{m}{\bullet} \ \oplus \ \underset{1}{\circ}$$

and

$$\underset{1}{\circ} \ \ominus \ \underset{2}{\bullet} \ \underset{2}{\circ} \ \ominus \ \bullet \ \cdots \ \circ \ \ominus \ \underset{m}{\bullet} \ \underset{m}{\circ} \ \ominus \ \underset{1}{\bullet}$$

where each pair

$$\underset{i}{\bullet} \quad \text{and} \quad \underset{i}{\circ}$$

of the same index i represent two "adjacent" segments parallel to the z-axis that are connected by one link parallel to either the x-axis or the y-axis. The building blocks of each variable gadget are then arranged in the space to pass through the clause gadgets of all clauses that contain the variable; see [7, Figure 10].

The authors of [7] observed that to connect the separate minimum tours for the variable gadgets into a single complete tour, it takes only two additional turns to merge any two tours into one. Their example [7, Section 4.4, Figure 11] corresponds to the following transformation:

$$\begin{array}{c} \underset{1}{\bullet} \ \oplus \ \underset{1}{\circ} \\[2mm] \underset{2}{\bullet} \ \oplus \ \underset{2}{\circ} \end{array} \quad \longrightarrow \quad \begin{array}{c} \underset{1'}{\bullet} \ \oplus \ \underset{2'}{\circ} \\[2mm] \underset{1'}{\bullet} \ \oplus \ \underset{2'}{\circ} \end{array}$$

where the two links 1 and 2 are replaced by two 2-segment chains $1'$ and $2'$. From this observation, they concluded in [7, Lemma 14] that the 3-SAT formula has a satisfying one-in-three assignment if and only if the point set thus constructed admits a rectilinear tour with $34mn' + 2(n' - 1)$ turns.

We note that the "if" direction of this lemma does not hold. Specifically, its proof overlooked the fact that there could be many different ways to merge the separate tours, which may disrupt the repeated pattern in each variable gadget. For example, when $n' = 3$ and $m = 2$, we can merge the $n' = 3$ separate tours into a complete tour with $2(n' - 1) = 4$ additional turns as follows

```
● ⊕ ○ ● ⊕ ○        ● ⊕ ○ ● ⊕ ○
1   2 2   1        1'   2 2   3'

● ⊕ ○ ● ⊕ ○   ⟶    ● ⊕ ○ ○ ⊖ ●
3   4 4   3        1'   4' 3'   5'

● ⊕ ○ ● ⊕ ○        ● ⊕ ○ ● ⊕ ○
5   6 6   5        5'   6 6   4'
```

where the four links 1, 3, 4, and 5 are replaced by four 2-segment chains $1'$, $3'$, $4'$, and $5'$, and the pattern in the second variable gadget is disrupted. Because of the disrupted patterns, it is possible that the point set admits a tour with $34mn' + 2(n' - 1)$ turns but the 3-SAT formula does not have a satisfying one-in-three assignment.

For a complete counterexample, consider the following 3-SAT formula with $n' = 4$ variables and $m = 4$ clauses:

$$(u_1 \vee u_2 \vee u_3) \wedge (u_1 \vee u_2 \vee u_4) \wedge (u_1 \vee u_3 \vee u_4) \wedge (u_2 \vee u_3 \vee u_4)$$

A satisfying one-in-three assignment for this formula must have exactly 4 true literals, one in each clause. But the number of true literals of any assignment is a multiple of 3, since each variable has exactly three positive literals. Thus this formula does not have any satisfying one-in-three assignment. If the boolean assignment is allowed to be *inconsistent*, for example, u_1 is true, u_3 and u_4 are false, but u_2 is false in the first two clauses and is true in the fourth clause, then the formula would have exactly one true literal in each clause. This corresponds to the following complete tour with a disrupted pattern in the second variable gadget (for simplicity, only indices of modified links are shown):

```
● ⊕ ○ ● ⊕ ○ ● ⊕ ○ ● ⊕ ○        ● ⊕ ○ ● ⊕ ○ ● ⊕ ○ ● ⊕ ○
a                       a        a'                      e'

○ ⊖ ● ○ ⊖ ● ○ ⊖ ● ○ ⊖ ●        ○ ⊖ ● ○ ⊖ ● ○ ⊖ ● ● ⊕ ○
b               e e     b   ⟶    b'                 a' e'  d'

○ ⊖ ● ○ ⊖ ● ○ ⊖ ● ○ ⊖ ●        ○ ⊖ ● ○ ⊖ ● ○ ⊖ ● ○ ⊖ ●
c                       c        b'                      c'

○ ⊖ ● ○ ⊖ ● ○ ⊖ ● ○ ⊖ ●        ○ ⊖ ● ○ ⊖ ● ○ ⊖ ● ○ ⊖ ●
d                       d        d'                      c'
```

Each of the four chains a', b', c', d' consists of two segments as in [7, Figure 11], one parallel to the x-axis and one parallel to the y-axis. The chain e' consists of three segments, with an additional middle segment parallel to the z-axis. These five chains together incur $1 + 1 + 1 + 1 + 2 = 6 = 2(n' - 1)$ more turns in addition to the $34mn'$ turns of the separate tours of the variable gadgets.

In our counterexample, the separate tours for the variable gadgets are cut open *between* building blocks. We did this for a clean presentation. Indeed these separate tours could be cut open at other places too, in particular, between any two parallel segments *within* a building block, such as segments 1 and 2, 2 and 3, 4 and 9, etc.; see [7, Figure 5]. A divided building block does not function as an atomic unit, and hence loses its boolean property to encode true or false. This would yield other, more complicated, counterexamples.

We have shown that the NP-hardness proof for MINIMUM-LINK RECTILINEAR COVERING TOUR in $\mathbb{R}^3$ [7, Theorem 15] does not hold. For the same reason, the NP-hardness proof for MINIMUM-LINK RECTILINEAR SPANNING PATH in $\mathbb{R}^3$ [7, Theorem 16] does not hold either.

4   New NP-Hardness Proofs

In this section, we prove the following theorem:

Theorem 1. MINIMUM-LINK RECTILINEAR COVERING TOUR *and* MINIMUM-LINK RECTILINEAR SPANNING PATH *in* $\mathbb{R}^{10}$ *are both NP-hard.*

We first prove the NP-hardness of MINIMUM-LINK RECTILINEAR COVERING TOUR in $\mathbb{R}^{10}$ by a reduction from the NP-hard problem HAMILTONIAN CIRCUIT in grid graphs [13]. A *grid graph* is a finite, vertex-induced, subgraph of the infinite graph with the points of integer coordinates in the plane as vertices, and with an edge between two vertices if and only the corresponding points have Euclidean distance exactly 1. Note that a grid graph is completely specified by its vertex set.

Let G be a grid graph specified by a set P of n grid points $p_i = (x_i, y_i) \in \mathbb{Z}^2$, $1 \le i \le n$, and let E be the set of m edges e_j, $1 \le j \le m$. We will construct a corresponding set Q of n grid points $q_i = (a_i, b_i, c_i, d_i, r_i, s_i, t_i, u_i, v_i, w_i) \in \mathbb{R}^{10}$, $1 \le i \le n$, such that G has a Hamiltonian circuit if and only if Q has a rectilinear tour of $8n$ turns.

For any two points p_k and p_l, we denote $p_k \prec_{xy} p_l$ if either $x_k < x_l$ or $x_k = x_l$ and $y_k < y_l$, and denote $p_k \prec_{x\bar{y}} p_l$ if either $x_k < x_l$ or $x_k = x_l$ and $y_k > y_l$. For each point p_i, we denote by $\mathrm{rank}_{xy}(p_i)$ and $\mathrm{rank}_{x\bar{y}}(p_i)$ the ranks of p_i in P ordered by $\prec_{xy}$ and $\prec_{x\bar{y}}$, respectively. For example, if G is the grid graph in Figure 3, then the values of $\mathrm{rank}_{xy}(p_i)$ and $\mathrm{rank}_{x\bar{y}}(p_i)$ are listed in Table 1.

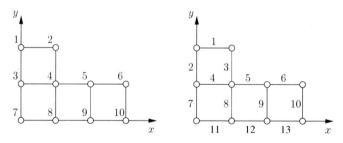

Fig. 3. A grid graph of 10 vertices and 13 edges. Left: vertex indices. Right: edge indices.

Table 1. Values of $\mathrm{rank}_{xy}(p_i)$ and $\mathrm{rank}_{x\bar{y}}(p_i)$ for the grid graph in Figure 3

i	1	2	3	4	5	6	7	8	9	10
$\mathrm{rank}_{xy}(p_i)$	3	6	2	5	8	10	1	4	7	9
$\mathrm{rank}_{x\bar{y}}(p_i)$	1	4	2	5	7	9	3	6	8	10

We now describe our construction. For each point q_i, $1 \le i \le n$, we set the four coordinates r_i, s_i, t_i, u_i to $\mathrm{rank}_{xy}(p_i)$ and set the two coordinates v_i, w_i to $\mathrm{rank}_{x\bar{y}}(p_i)$, then initialize the four coordinates a_i, b_i, c_i, d_i to the index i. Next, for each point p_i, and for each edge e_j that connects q_i to another point q_k, we update the four coordinates a_i, b_i, c_i, d_i of q_i according to the following cases:

1. $y_k = y_i$ (then $x_k = x_i - 1$ or $x_k = x_i + 1$)
   (a) If x_i is odd and $x_k = x_i - 1$, or if x_i is even and $x_k = x_i + 1$, set $a_i \leftarrow n + j$.
   (b) If x_i is odd and $x_k = x_i + 1$, or if x_i is even and $x_k = x_i - 1$, set $b_i \leftarrow n + j$.
2. $x_k = x_i$ (then $y_k = y_i - 1$ or $y_k = y_i + 1$)
   (a) If y_i is odd and $y_k = y_i - 1$, or if y_i is even and $y_k = y_i + 1$, set $c_i \leftarrow n + j$.
   (b) If y_i is odd and $y_k = y_i + 1$, or if y_i is even and $y_k = y_i - 1$, set $d_i \leftarrow n + j$.

This completes the construction. We refer to Figure 3 and Table 2 for an example.

Table 2. Coordinates of q_i for the grid graph in Figure 3

i	1	2	3	4	5	6	7	8	9	10
a_i	11	11	14	14	16	16	21	21	23	23
b_i	1	2	3	15	15	6	7	22	22	10
c_i	1	2	17	18	19	20	17	18	19	20
d_i	12	13	12	13	5	6	7	8	9	10
r_i	3	6	2	5	8	10	1	4	7	9
s_i	3	6	2	5	8	10	1	4	7	9
t_i	3	6	2	5	8	10	1	4	7	9
u_i	3	6	2	5	8	10	1	4	7	9
v_i	1	4	2	5	7	9	3	6	8	10
w_i	1	4	2	5	7	9	3	6	8	10

The following property of the construction can be easily verified by a case analysis:

Lemma 1. *For any two points q_i and q_j in Q, if the corresponding points p_i and p_j are adjacent in G, then q_i and q_j share one coordinate along one of the first 4 axes and have distinct coordinates along the other 9 axes, otherwise they have distinct coordinates along all 10 axes.*

Our reduction is clearly polynomial. To complete the proof of the NP-hardness of MINIMUM-LINK RECTILINEAR COVERING TOUR in $\mathbb{R}^{10}$, it remains to prove the following lemma:

Lemma 2. *G has a Hamiltonian circuit if and only if Q has a rectilinear tour of $8n$ turns.*

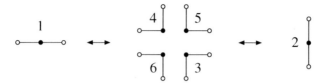

Fig. 4. Six types of turns

Proof. We first prove the direct implication. Suppose that G has a Hamiltonian circuit H. We will construct a rectilinear tour of $8n$ turns for Q. Refer to Figure 4. For each point p_i in P, select a line L_i through the corresponding point q_i in Q and parallel to one of the six axes r, s, t, u, v, w according to the following six cases about the two points p_k and p_l immediately before and after p_i in H:

1. $y_k = y_i = y_l$ (then $x_k < x_i < x_l$ or $x_l < x_i < x_k$): r for odd x_i, s for even x_i.
2. $x_k = x_i = x_l$ (then $y_k < y_i < y_l$ or $y_l < y_i < y_k$): r for odd y_i, s for even y_i.
3. $x_i = \min\{x_k, x_l\}$ and $y_i = \max\{y_k, y_l\}$: t
4. $x_i = \max\{x_k, x_l\}$ and $y_i = \min\{y_k, y_l\}$: u
5. $x_i = \min\{x_k, x_l\}$ and $y_i = \min\{y_k, y_l\}$: v
6. $x_i = \max\{x_k, x_l\}$ and $y_i = \max\{y_k, y_l\}$: w

Let $S_i \subset L_i$ be the rectilinear segment containing all points with coordinate between the coordinates of q_k and q_l (along the axis parallel to L_i). Recall that $r_i = s_i = t_i = u_i = \mathrm{rank}_{xy}(p_i)$ and $v_i = w_i = \mathrm{rank}_{x\bar{y}}(p_i)$. By our axis assignment, S_i must contain q_i in the interior.

Consider the six types of turns at p_i according to the six cases of p_k and p_l listed above. Observe that in any Hamiltonian circuit of G, (i) a turn of type 3, 4, 5, or 6 cannot be consecutive with another turn of the same type, (ii) a turn of type 1 or 2 is consecutive with another turn of the same type only if the middle vertices of the two turns are adjacent and hence have different parities, and (iii) a turn of type 1 cannot be consecutive with a turn of type 2. Thus by our choice of axis assignment, S_i is parallel to neither S_k nor S_l.

To connect the n segments $S_i, 1 \le i \le n$, into a rectilinear tour, we add $7n$ segments. Specifically, for each point p_i and its immediate successor p_l in the Hamiltonian circuit H, we add 7 more segments between the two segments S_i and S_l. Note that p_i and p_l are adjacent in G. So it follows by Lemma 1 that q_i and q_l share exactly one coordinate along one of the first four axes. Recall that the two segments S_i and S_l, which contain the two points q_i and q_l respectively, are not parallel to each other, but each of them is parallel to one of the last six axes. To connect the two segments S_i and S_l, we use one segment parallel to each axis except these three axes. For example, suppose that q_i and q_l share the same coordinate $a_i = a_l = \hat{a}$ along the a-axis, and that S_i and S_l are parallel to the r-axis and the w-axis, respectively. Then two segments S_i and S_l can be connected by 7 segments with 8 turns as follows:

$$q_i = (a_i, b_i, c_i, d_i, r_i, s_i, t_i, u_i, v_i, w_i) \to S_i \to$$
$$(\hat{a}, b_i, c_i, d_i, r_l, s_i, t_i, u_i, v_i, w_i)$$
$$(\hat{a}, b_l, c_i, d_i, r_l, s_i, t_i, u_i, v_i, w_i)$$
$$(\hat{a}, b_l, c_l, d_i, r_l, s_i, t_i, u_i, v_i, w_i)$$
$$(\hat{a}, b_l, c_l, d_l, r_l, s_i, t_i, u_i, v_i, w_i)$$
$$(\hat{a}, b_l, c_l, d_l, r_l, s_l, t_i, u_i, v_i, w_i)$$
$$(\hat{a}, b_l, c_l, d_l, r_l, s_l, t_l, u_i, v_i, w_i)$$
$$(\hat{a}, b_l, c_l, d_l, r_l, s_l, t_l, u_l, v_i, w_i)$$
$$(\hat{a}, b_l, c_l, d_l, r_l, s_l, t_l, u_l, v_l, w_i)$$
$$\to S_l \to (a_l, b_l, c_l, d_l, r_l, s_l, t_l, u_l, v_l, w_l) = q_l.$$

Altogether, there are exactly $8n$ turns in the resulting rectilinear tour for Q.

We next prove the reverse implication. Suppose that Q has a rectilinear tour R of $8n$ turns. We will find a Hamiltonian circuit in G. Consider any two points q_i and q_j in Q that are consecutively covered by R. By Lemma 1, the two points have distinct coordinates along at least 9 of the 10 axes. As a point moves in a rectilinear segment, only one of its coordinates changes. Therefore, besides the two segments S_i and S_j that contain q_i and q_j respectively, the chain of segments connecting q_i and q_j must include at least 7 other segments, and hence at least 8 turns. Since R has exactly $8n$ turns, it follows that each point in Q must be covered exactly once, and there must be exactly 8 turns between any two points that are covered consecutively. Any two consecutively covered points q_i and q_j in Q with 8 turns in between must have distinct coordinates along at most 9 axes. By our construction, they must correspond to two adjacent points p_i and p_j in P. Thus the rectilinear tour R for Q corresponds to a Hamiltonian circuit in the grid graph G specified by P.   □

We have proved the NP-hardness of MINIMUM-LINK RECTILINEAR COVERING TOUR in $\mathbb{R}^{10}$ by a reduction from HAMILTONIAN CIRCUIT in grid graphs. The same construction also gives a reduction to MINIMUM-LINK RECTILINEAR SPANNING PATH in $\mathbb{R}^{10}$ with specified starting and ending points from HAMILTONIAN PATH in grid graphs. Note that an instance of the problem HAMILTONIAN PATH in grid graphs consists of a grid graph G specified by a set P of n grid points as in the problem HAMILTONIAN CIRCUIT in grid graphs, and moreover has two special points p_i and p_j in P marked as the starting and ending vertices of the Hamiltonian path. For the reduction, we simply construct the point set Q corresponding to the point set P as before, then specify the two points q_i and q_j in Q as the first and last points to be covered by a rectilinear path. This leads to the following lemma analogous to Lemma 2:

Lemma 3. *G has a Hamiltonian path from p_i to p_j if and only if Q has a rectilinear spanning path of $8(n-1)$ turns from q_i to q_j.*

Proof. The proof is almost identical to that of Lemma 2 except that the spanning path from q_i to q_j need not be closed into a tour—this saves 7 segments and hence 8 turns. Also note the technicality that there is no point before the first point or after the last point in the Hamiltonian path. Without loss of generality, we categorize such degenerate cases into either case 1 or case 2 of the six cases illustrated in Figure 4.   □

We next prove the NP-hardness of MINIMUM-LINK RECTILINEAR SPANNING PATH in $\mathbb{R}^{10}$ (without specified starting and ending points). This is achieved by a Turing reduction from HAMILTONIAN PATH in grid graphs [13]. Given an instance (P, p_i, p_j) of HAMILTONIAN PATH in grid graphs, we first construct a corresponding instance (Q, q_i, q_j) of MINIMUM-LINK RECTILINEAR SPANNING PATH (with specified starting and ending points) in $\mathbb{R}^{10}$, then augment the point set Q to 400 instances of MINIMUM-LINK RECTILINEAR SPANNING PATH in $\mathbb{R}^{10}$ by enumerating all possible combinations of directions to extend the two end segments of a spanning path of Q from q_i to q_j. Since there are 10 axes and each axis has both a positive direction and a negative direction, there are $(10 \cdot 2)^2 = 400$ combinations of directions. For each combination of directions, we augment the point set Q to a point set Q' by adding two dummy points q_i' and q_j'. The dummy point q_i' is farther than q_i in the direction that extends the end segment at q_i; the dummy point q_j' is farther than q_j in the direction that extends the end segment at q_j. Moreover, q_i' (respectively, q_j') has the same coordinate as q_i (respectively, q_j) along one of the last 6 axes (recall that all points in Q have distinct coordinates along each of the last 6 axes), and has a distinct coordinate different from all others along each of the other 9 axes. We have the following lemma analogous to Lemma 3:

Lemma 4. *G has a Hamiltonian path from p_i to p_j if and only if at least one of the 400 instances of Q' has a rectilinear spanning path of $8(n + 1)$ turns.*

Proof. For the direct implication, suppose that G has a Hamiltonian path from p_i to p_j. Then, by Lemma 3, Q has a rectilinear spanning path of $8(n-1)$ turns from q_i to q_j. At least one of the 400 instances of Q' correctly guesses the directions to extend the two end segments of Q, and hence needs only 8 turns to connect q_i' to q_i and another 8 turns to connect q_j' to q_j. This yields an extended rectilinear spanning path of $8(n + 1)$ turns from q_i' to q_j'.

For the reverse implication, the crucial observation is that the two dummy points q_i' and q_j' only share coordinates with the two end points q_i and q_j, respectively. Thus under the constraint of at most 8 turns between any two consecutively covered points in a spanning path, we must have q_i' and q_j' at the two ends, and have q_i and q_j next to q_i' and q_j', respectively. Then the same argument as in the proof of Lemma 2 completes the proof.   □

Since HAMILTONIAN PATH in grid graphs is also NP-hard [13], this proves the NP-hardness of MINIMUM-LINK RECTILINEAR SPANNING PATH in $\mathbb{R}^{10}$.

References

1. Aggarwal, A., Coppersmith, D., Khanna, S., Motwani, R., Schieber, B.: The angular-metric traveling salesman problem. SIAM Journal on Computing 29, 697–711 (1999)
2. Arkin, E.M., Mitchell, J.S.B., Piatko, C.D.: Minimum-link watchman tours. Information Processing Letters 86, 203–207 (2003)

3. Arkin, E.M., Bender, M.A., Demaine, E.D., Fekete, S.P., Mitchell, J.S.B., Sethia, S.: Optimal covering tours with turn costs. SIAM Journal on Computing 35, 531–566 (2005)
4. Bereg, S., Bose, P., Dumitrescu, A., Hurtado, F., Valtr, P.: Traversing a set of points with a minimum number of turns. Discrete & Computational Geometry 41, 513–532 (2009)
5. Collins, M.J.: Covering a set of points with a minimum number of turns. International Journal of Computational Geometry and Applications 14, 105–114 (2004)
6. Collins, M.J., Moret, B.M.E.: Improved lower bounds for the link length of rectilinear spanning paths in grids. Information Processing Letters 68, 317–319 (1998)
7. Estivill-Castro, V., Heednacram, A., Suraweera, F.: NP-completeness and FPT results for rectilinear covering problems. Journal of Universal Computer Science 15, 622–652 (2010)
8. Estivill-Castro, V., Heednacram, A., Suraweera, F.: FPT-algorithms for minimum-bends tours. International Journal of Computational Geometry 21, 189–213 (2011)
9. Fekete, S.P., Woeginger, G.J.: Angle-restricted tours in the plane. Computational Geometry: Theory and Applications 8, 195–218 (1997)
10. Gaur, D.R., Bhattacharya, B.: Covering points by axis parallel lines. In: Proceedings of the 23rd European Workshop on Computational Geometry, pp. 42–45 (2007)
11. Grantson, M., Levcopoulos, C.: Covering a set of points with a minimum number of lines. In: Proceedings of the 22nd European Workshop on Computational Geometry, pp. 145–148 (2006)
12. Hassin, R., Megiddo, N.: Approximation algorithms for hitting objects with straight lines. Discrete Applied Mathematics 30, 29–42 (1991)
13. Itai, A., Papadimitriou, C.H., Szwarcfiter, J.L.: Hamiltonian paths in grid graphs. SIAM Journal on Computing 11, 676–686 (1982)
14. Langerman, S., Morin, P.: Covering things with things. Discrete & Computational Geometry 33, 717–729 (2005)
15. Kranakis, E., Krizanc, D., Meertens, L.: Link length of rectilinear Hamiltonian tours in grids. Ars Combinatoria 38, 177–192 (1994)
16. Megiddo, N., Tamir, A.: On the complexity of locating linear facilities in the plane. Operations Research Letters 1, 194–197 (1982)
17. Stein, C., Wagner, D.P.: Approximation Algorithms for the Minimum Bends Traveling Salesman Problem. In: Aardal, K., Gerards, B. (eds.) IPCO 2001. LNCS, vol. 2081, pp. 406–421. Springer, Heidelberg (2001)
18. Wagner, D.P.: Path Planning Algorithms under the Link-Distance Metric. Ph.D. thesis, Dartmouth College (2006)
19. Wang, J., Li, W., Chen, J.: A parameterized algorithm for the hyperplane-cover problem. Theoretical Computer Science 411, 4005–4009 (2010)

On Envy-Free Pareto Efficient Pricing

Xia Hua

School of Physical and Mathematical Sciences , Nanyang Technological University, Singapore
`huax0005@e.ntu.edu.sg`

Abstract. In a centralized combinatorial market, the market maker has a number of items for sale to potential consumers, who wish to purchase their preferred items. Different solution concepts (allocations of items to players) capture different perspectives in the market. Our focus is to balance three properties: revenue maximization from the market maker's perspective, fairness from consumers' perspective, and efficiency from the market's global perspective.

Most well-known solution concepts capture only one or two properties, e.g., Walrasian equilibrium requires fairness for consumers and uses market clearance to guarantee efficiency but ignores revenue for the market maker. Revenue maximizing envy-free pricing balances market maker's revenue and consumer's fairness, but ignores efficiency.

In this paper, we study a solution concept, envy-free Pareto efficient pricing, that lies between Walrasian equilibrium and envy-free pricing. It requires fairness for consumers and balances efficiency and revenue. We study envy-free Pareto efficient pricing in two domains, unit-demand and single-minded consumers, and analyze its existence, computation, and economic properties.

1 Introduction

In a centralized combinatorial market, a market maker sells a set of m items to n potential consumers, where each consumer i has a valuation function $v_i(\cdot)$ measuring the maximum amount that i is willing to pay for different combinations of items. As an outcome of the market, the market maker specifies a price vector $\mathbf{p} = (p_j)$ for all items and an allocation vector $X = (X_i)$ which indicates the subset of items that every consumer i obtains. That is, for the given outcome $(\mathbf{p}, X)$, consumer i obtains subset X_i with payment $p(X_i) = \sum_{j \in X_i} p_j$ to the market maker; therefore his *utility* is defined to be $v_i(X_i) - p(X_i)$.

The centralized combinatorial market is one of the most fundamental market models that has received a lot of attention in the literature [17,9]. It characterizes a number of applications, especially with the development of the Internet, e.g., Amazon's electronic market, Google and Yahoo's advertising markets, to name a few. A key question in such marketplaces is that for the given input information $v_i(\cdot)$, what kind of outcomes should the market output? In other words, what solution concepts should be selected in different applications with different focuses? Solution concepts play a critical role and finding them is the central question in economics and social choices. The focus of the current paper is to study different solution concepts in the centralized combinatorial marketplaces.

J. Snoeyink, P. Lu, K. Su, and L. Wang (Eds.): FAW-AAIM 2012, LNCS 7285, pp. 70–81, 2012.
© Springer-Verlag Berlin Heidelberg 2012

Before presenting the considered solution concepts, we will first examine a few critical properties that different parties would consider in the market.

- Envy-freeness (or fairness). It is natural to assume that all consumers are utility maximizers, i.e., for any given price vector, they would want to purchase a subset of items that gives them the maximum utilities. Envy-freeness captures such self-interested considerations of consumers and requires that no consumer envies any other allocation for the given market prices. In other words, every consumer's utility is maximized at his corresponding allocation, and therefore, is happy with the outcome of the market.
- Revenue. The market maker, on the other hand, would like to obtain as much revenue as possible, which is the total payment received from all consumers. Therefore, while consumers prefer to pay less to increase their utilities, the market maker's interest is revenue maximization.
- Social welfare. While consumers and the market maker have contrary interests in terms of payments, social welfare, defined to be the total utility of all participants (i.e., both consumers and the seller) in the market, unifies the interests of the whole system. Social welfare is one of the most important factors to evaluate an outcome of a system from a global point of view. Most seminal designs and benchmarks, e.g., VCG mechanism [22,7,13] and market equilibrium [23], are guaranteed to have solutions that maximize social welfare.
- Pareto efficiency. It is another condition that captures the overall performance of a system. Given a Pareto efficient solution, there is no way to improve someone's utility (either that of a consumer or of the seller in our combinatorial market) without hurting any other participants. Pareto efficiency is weaker than social welfare maximization since it only gives a locally optimal solution to maximize social welfare.

The above properties characterize different aspects of different solution concepts. Ideally, we would like to have one that satisfies all these properties. However, due to the contrary interests amongst different parties in the market, such a solution does not exist in general. Therefore, any feasible solution concept can only have a partial focus on each of these properties. We will consider the following solution concepts in the current paper.

Walrasian Equilibrium (WE). An outcome is called a *Walrasian equilibrium* if it satisfies, in addition to envy-freeness, a market clearance condition which says that every leftover item must be priced at zero. Walrasian equilibrium is one of the most important solution concepts in economics and has been studied extensively in the literature. It ensures individual fairness and overall market efficiency. In particular, the seminal first fundamental welfare theorem says that every Walrasian equilibrium maximizes social welfare; this implies that it is Pareto efficient as well.

However, it is well known that a Walrasian equilibrium may not exist in general; and even if one is guaranteed to exist (for example, when the valuation functions satisfy the gross substitutes condition [16]), the revenue it generates could be very small even for a revenue maximizing Walrasian equilibrium (RWE). These two issues largely limit the applicability of Walrasian equilibrium. For example, the market maker would like to

Table 1. Our results (last row) for unit-demand consumers

	Fairness	Social welfare	Pareto efficiency	Revenue	Computation	Existence
RWE	√	√	√	small	easy	yes
REF	√	×	×	large	hard	yes
REP	√	√	√	medium	hard	yes

Table 2. Our results (last row) for single-minded consumers

	Fairness	Social welfare	Pareto efficiency	Revenue	Computation	Existence
RWE	√	√	√	small	hard	no
REF	√	×	×	large	hard	yes
REP	√	×	√	medium	hard	no

seek to optimize his own objective, i.e., revenue maximization, which is one of the key business concerns in many applications. From this perspective, Example 1 shows that (revenue maximizing) Walrasian equilibrium is not a good solution concept in some applications.

Revenue Maximizing Envy-Freeness (REF). The limitation of Walrasian equilibrium relies on the requirement of the market clearance condition. In applications like Google and Yahoo's advertising markets, the market makers may hold some of their inventories unsold (i.e., priced at infinity) to achieve their own objectives. Removing the market clearance condition leads to the solution concept of envy-freeness (EF), which is guaranteed to always exist. Finding an envy-free pricing solution that maximizes revenue (REF) for the seller has attracted a lot of attention recently in computer science [15,1,5]: The optimization problem is NP-hard to solve even when all consumers only desire one item or a fixed subset of items; approximation algorithms have been considered for special cases.

A remarkable property of a REF pricing solution is that there are instances in which the REF revenue can be arbitrarily larger than that of the RWE.

Example 1. (REVENUE COMPARISON: RWE *vs* REF). There are two consumers i_1, i_2 and two items j_1, j_2. Both consumers want to get only one item, and their valuations are $v_{i_1 j_1} = k$, $v_{i_1 j_2} = k + 2$, and $v_{i_2 j_1} = 0$ and $v_{i_2 j_2} = 1$. In an optimal RWE solution, i_1 wins j_2 and i_2 wins j_1 at a price vector $(0, 2)$. In an optimal REF solution, however, we may charge a price vector $(k, k + 2)$ with the same allocation. Hence, the total revenue is increased from 2 to $k + 2$. (Note that in the REF solution, we are free to set any large price for item j_1, whereas in the RWE, its price has to be 0 by the market clearance condition.) Hence, the difference between an optimal RWE solution and REF solution can be arbitrarily large.

While REF does generate the largest possible revenue provided the fairness condition, it may not maximize social welfare; and further, it may not even be Pareto efficient, which is arguably the weakest solution concept to ensure overall performance of a system.

Example 2 (REF is not Pareto efficient). There are two consumers i_1, i_2 and two items j_1, j_2. Again both consumers only want to get one item, and their valuations are $v_{i_1 j_1} = 100$, $v_{i_1 j_2} = 10$, and $v_{i_2 j_1} = 180$ and $v_{i_2 j_2} = 100$, as the following figure (A) shows:

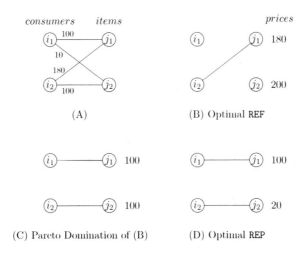

(A)

(B) Optimal REF

(C) Pareto Domination of (B)

(D) Optimal REP

In this example, an optimal REF solution is given by (B) with total revenue 180 (note that we can set j_2 with any large price to ensure that i_2 is happy with his allocation j_1 at price 180). However, this is Pareto dominated by the solution in (C) where both i_1 and i_2 obtain the same utility 0 and the market maker improves his revenue from 180 to 200. Note that the solution given by (C) is not envy-free. For this example, the optimal REP (defined below) is shown in (D).

Revenue Maximizing Envy-Free Pareto Efficiency (REP). Social welfare maximization, and the bottom line, Pareto efficiency, are critical solution conditions for a market to have healthy and long-term development. To capture these conditions, in particular, Pareto efficiency, we introduce a solution concept, envy-free Pareto efficiency (EP), which requires both envy-freeness and Pareto efficiency. A revenue maximizing envy-free Pareto efficient (REP) outcome is one that maximizes revenue among all envy-free Pareto efficient solutions. The figure (D) in Example 2 gives an optimal REP solution for the given instance, which is different from the optimal REF solution shown in figure (B); this implies that an algorithm for finding a REF solution will not suffice for finding a REP solution.

By the definitions of the three solution concepts and the first fundamental theorem, we know that EP lies between WE and EF, that is, the solution space has the following structure:

$$WE \subseteq EP \subseteq EF$$

This further implies that the total revenue generated by REP also sits between RWE and REF, i.e.,

$$RWE \leq REP \leq REF$$

where both inequalities can be strict.

Including Pareto efficiency strikes a balance between system efficiency and revenue: REF solutions yield good revenue, but, as mentioned earlier, the overall performance of the system is not addressed well. On the other hand, WE, while fair and efficient, is not practical because the revenue of the market maker can be arbitrarily small. Our solution concept EP, in addition to satisfying fairness (in the form of envy-freeness), addresses both overall performance (in the form of Pareto efficiency) and revenue (which is sandwiched between the revenues guaranteed by REF and WE solutions).

Our Results

We will study these solution concepts in two domains that are, arguably, the most applicable: *unit-demand* consumers (where each consumer desires exactly one item, as shown in Example 1 and 2) and *single-minded* consumers (where each consumer desires a fixed subset of items).

For a market with unit-demand consumers, the existence of an EP solution follows from the existence of a WE immediately. Further, in addition to satisfying fairness and efficiency, we show that EP (and therefore REP) has the property that it maximizes social welfare (Theorem 2). This property of EP solutions is quite remarkable because it does not hold for REF solutions. Hence, our solution concept possesses all the nice economic properties that WE has. On the other hand, while approximating the revenue of an REP solution is NP-hard (Theorem 4), we show that for a special case which includes most instances of advertising markets, an optimal REP can be computed in polynomial time (Theorem 5).

For a market with single-minded consumers, our solution concept, REP, aligns with WE and REF. Similar to the complexity of determining the existence of a WE [4], we show that determining the existence of an EP solution is NP-hard. Even if an EP solution is guaranteed to exist, computing one that maximizes revenue is NP-hard as well (Theorem 6); this has the same computational complexity as REF [15]. These hardness results do not rule out the applicability of the solution concepts, but rather, illustrate their computational features. This is similar to Nash equilibrium, which is extensively applied in game theory, but shown to be hard to compute [8,11,6].

Please note that in this paper, due to space limit, we do not include most of the proofs.

We summarize our results, and compare EP with WE and EF in Tables 1 and 2. Our results imply that EP captures fairness, and balances the tradeoff between efficiency and revenue. Therefore, it provides a good alternative for solution concepts in combinatorial markets.

Related Work

Envy-freeness and Pareto efficiency are among the most well-studied solution concepts in economics. For instance, Pazner and Schmeidler [18] gave examples in which no Pareto efficient allocation is envy-free. Varian [21] considered general utility functions for the buyers and showed that if preferences of consumers are monotonic, then a fair allocation exists. For Walrasian equilibrium, Shapley and Shubik [20] showed that a solution always exists in unit-demand settings. The result was later improved to a more general class of functions [10,12,19]. Kelso and Crawford [16], as well as Gul and

Stacchetti [14], proposed conditions (e.g., gross substitutes and single improvements) in which a Walrasian equilibrium always exists. More discussions are referred to the textbooks, e.g., [17,9]. We note the major difference between our work and the literature is that we consider the centralized seller as a participant of the market as well; thus, in our model the market is composed of $n + 1$ entities.

From a computational point of view, Bouveret and Lang [2] studied the complexity of deciding whether there exists a Pareto efficient and envy-free allocation in several contexts when preferences are represented compactly. Guruswami et al. [15] showed that a revenue maximizing envy-free solution is NP-hard to approximate, even for unit-demand and single-minded consumers. The revenue maximizing envy-free pricing problem was later shown by Briest [3] that it is even NP-hard to approximate within a ratio of $O(\log^\epsilon n)$ for some $\epsilon > 0$. Our results, on the other hand, suggest that REP, as an alternative solution concept, could be more applicable from computational perspective.

2   Model and Solution Concepts

In a marketplace, we have a set A of n consumers and a single seller (i.e., the market maker) with a set B of m items, where each item has unit supply. For every consumer i and subset of items $S \subseteq B$, there is a value $v_i(S) \geq 0$ denoting the valuation that i obtains from S. We assume that $v_i(\emptyset) = 0$. An output of the market is given by a *price* vector $\mathbf{p} = (p_j)_{j \in B}$ where p_j is the price of item j, and an *allocation* vector $X = (X_i)_{i \in A}$ where X_i is the subset of items that i obtains. For any $S \subseteq B$, its price is defined to be the sum of the prices of its elements, i.e., $p(S) = \sum_{j \in S} p_j$. Given an output $(\mathbf{p}, X)$, the *utility* of every consumer i is defined to be $u_i(\mathbf{p}, X) = v_i(X_i) - p(X_i)$. We assume all consumers are utility-maximizers, i.e., they would prefer outcomes with larger utilities.

In general, there could be many different outputs $(\mathbf{p}, X)$. A crucial study in economics and social choice is the selection of solution concepts that satisfy different criteria. All outputs considered in the current paper are assumed to be *feasible* (i.e., $X_i \cap X_{i'} = \emptyset$ for any $i \neq i'$) and *individually rational* (i.e., $u_i(\mathbf{p}, X) = v_i(X_i) - p(X_i) \geq 0$). One of the most notable solution concepts is that of Walrasian Equilibrium, which is defined formally as below.

Definition 1. (ENVY-FREENESS AND WALRASIAN EQUILIBRIUM). *An outcome $(\mathbf{p}, X)$ is called a* Walrasian Equilibrium *(WE) if it satisfies the following two conditions:*

- *(Market clearance) Every unsold item is priced at zero.*
- *(Envy-freeness) For any consumer i and any subset of items $S \subseteq B$, $v_i(X_i) - p(X_i) \geq v_i(S) - p(S)$.*

The first condition above is a market clearance condition, which requires that all unallocated items are priced at zero (or at any given reserve price). The second is a fairness condition, which says that at the given price vector, the utility of every consumer is maximized by the corresponding allocation, i.e., everyone is happy with his allocation.

While envy-freeness captures the interests of consumers, the seller, on the other hand, would like to maximize the total amount of payment received (i.e., *revenue*), defined by $r(\mathbf{p}, X) = \sum_{i \in A} p(X_i)$ for a given outcome $(\mathbf{p}, X)$. This motivates the following solution concepts:

- Revenue maximizing Walrasian equilibrium (RWE): As the name implies, it is a Walrasian equilibrium that offers the most revenue to the seller.
- Revenue maximizing envy-free solution (REF): These are envy-free solutions with maximum total revenue. Note that we do not require market clearance condition here, i.e., it is possible that an unallocated item has a positive price.

Although a (revenue maximizing) envy-free solution captures the interests of both the consumers and the seller to some extent, as Example 2 shows, it may not be Pareto efficient, which is defined formally as follows.

Definition 2 (Pareto efficiency). *An outcome* $(\mathbf{p}, X)$ *is said to Pareto dominate (or just dominate) another outcome* $(\mathbf{p}', X')$ *if (i) for all* $i \in A$, $u_i(\mathbf{p}, X) \geq u_i(\mathbf{p}', X')$; *(ii)* $r(\mathbf{p}, X) \geq r(\mathbf{p}', X')$, *and (iii) at least one of the above inequalities is strict. We say that an outcome* $(\mathbf{p}, X)$ *is* Pareto efficient *if it is not Pareto dominated by any other outcome.*

In the current paper, we will consider the following solution concept, combining revenue maximization (from the seller's perspective), envy-freeness (from the consumers' perspective), and Pareto efficiency (from the market's perspective).

Definition 3. (REVENUE MAXIMIZING ENVY-FREE PARETO EFFICIENCY). *An envy-free Pareto efficient (EP) solution is one that is both envy-free and Pareto efficient. A revenue maximizing envy-free Pareto efficient (REP) solution is one with maximum revenue among all EP solutions.*

Another important property considered in the current paper is social welfare, defined as below.

Definition 4 (Social welfare). *Given an outcome* $(\mathbf{p}, X)$, *the* social welfare *is defined to be the sum of utilities of all participants in the system, i.e.,* $\sum_{i \in A} u_i(\mathbf{p}, X) + r(\mathbf{p}, X)$.

Observe that the utility of every consumer i is $u_i(\mathbf{p}, X) = v_i(X_i) - p(X_i)$ and the utility of the seller is $\sum_{i \in A} p(X_i)$. Hence, in our setting social welfare is equal to $\sum_{i \in A} v_i(X_i)$, which is independent of the selected price vector. A feasible allocation that maximizes social welfare is called *optimal*, i.e., it has the maximum total valuations $\sum_{i \in A} v_i(X_i)$. Therefore, an outcome maximizes social welfare if and only if the corresponding allocation is optimal.

3   Unit Demand Consumers

We say a consumer i has *unit demand* if $v_i(S) = 0$ for any $S \subseteq B$ with $|S| \geq 2$. That is, i is interested in obtaining at most one item. In this section, we assume all consumers have unit demand and denote $v_{ij} = v_i(\{j\})$. Note that when all consumers have unit demand, any feasible allocation $X = (X_i)_{i \in A}$ corresponds to a matching between consumers and items.

3.1 Pareto Efficiency and Social Welfare

In this section, we will examine the relation between Pareto efficiency and social welfare maximization (hence, optimal allocation) in the market with unit-demand consumers.

We begin with a useful lemma.

Lemma 1. *If $(\mathbf{p}, X)$ is not Pareto efficient, then there is another solution $(\mathbf{p}', X')$ that dominates $(\mathbf{p}, X)$, but the strict improvement is restricted to the seller's revenue. That is, $u_i(\mathbf{p}', X') = u_i(\mathbf{p}, X)$ for all consumers $i \in A$ and $r(\mathbf{p}', X') > r(\mathbf{p}, X)$.*

Given the above characterization, we can show the following result.

Theorem 1. *Given an outcome $(\mathbf{p}, X)$, if X is an optimal allocation, then $(\mathbf{p}, X)$ is Pareto efficient.*

We comment that the reverse direction of the above claim does not hold, i.e., Pareto efficiency may not imply optimal allocation. For example, there are two consumers i_1, i_2, and one item j_1, with $v_{i_1 j_1} = 10$ and $v_{i_2 j_1} = 8$. Assigning the item to i_2 at price 0 is not an optimal allocation, but it is Pareto efficient since it is not dominated by any other outcome given that the utility of i_2 should be at least 8. That is, optimal allocation is only a sufficient condition to ensure Pareto efficiency.

However, as the next theorem implies, given envy-freeness, optimal allocation is also a necessary condition for Pareto efficiency. We note that the following result is known for WE, but we consider the solution concept EP with $n + 1$ participating entities.

Theorem 2. *For any envy-free solution $(\mathbf{p}, X)$, if X is not an optimal allocation, then $(\mathbf{p}, X)$ cannot be Pareto efficient. Therefore any EP solution has the maximum social welfare and the corresponding allocation is optimal.*

The above claim (with proof deferred to the full version) implies that if we enforce both envy-freeness and Pareto efficiency, the solution is guaranteed to have maximum social welfare. We have the following (main) conclusion.

Corollary 1. *In a market with unit demand consumers, given envy-freeness, an output is Pareto efficient if and only if it maximizes social welfare.*

3.2 Determining Pareto Efficiency

In this section, we consider the following question: Given a market with unit demand consumers, can we determine whether an output $(\mathbf{p}, X)$ is Pareto efficient or not? While optimality of an allocation ensures Pareto efficiency (Theorem 1), the converse is not necessarily true. Therefore, our approach does not entail Theorem 1.

Next we will give an algorithm to determine Pareto efficiency of any given market output $(\mathbf{p}, X)$. For simplicity, denote $r = r(\mathbf{p}, X)$ and $u_i = u_i(\mathbf{p}, X)$. Let N be the number of consumers with positive utility u_i, i.e., $N = |\{i \in A \mid u_i > 0\}|$. We construct a weighted bipartite graph $G(\mathbf{p}, X) = (A, B; E)$ as follows: For each consumer i and each item j, if $v_{ij} - u_i \geq 0$, then we add an edge connecting i and j with weight $w_{ij} = v_{ij} - u_i$. Let W be a sufficiently large number, e.g., $W = n^2 \cdot \max_{ij} v_{ij}$. Further, we define another graph $G^*(\mathbf{p}, X) = (A, B; E)$ derived from $G(\mathbf{p}, X)$ where the only difference is on the weights of edges: In graph G^*, the weight of every edge $(i, j) \in E$ is defined to be $w^*_{ij} = w_{ij} + W$ if $u_i > 0$ and $w^*_{ij} = w_{ij}$ if $u_i = 0$. If the total weight of the maximum matching in the graph $G^*(\mathbf{p}, X)$ exceeds $NW + r$, the algorithm reports that the market output $(\mathbf{p}, X)$ is not Pareto efficient; otherwise, it is Pareto efficient.

Theorem 3. *The total weight of the maximum matching in the graph $G^*(\mathbf{p}, X)$ exceeds $NW + r$ if and only if the market output $(\mathbf{p}, X)$ is not Pareto efficient. Therefore, the above algorithm determines whether $(\mathbf{p}, X)$ is Pareto efficient or not in polynomial time.*

3.3   Complexity of Computing REP

In this section, we consider the problem of computing a revenue maximization solution that is both envy-free and Pareto efficient (i.e., an REP solution) in a unit-demand market. The following claim says that the problem in general does not admit a polynomial time algorithm. (Indeed, it does not admit a $(1 + \epsilon)$ approximation algorithm for an arbitrarily small $\epsilon > 0$.)

Theorem 4. *For any constant $c > 1$, there is no polynomial time algorithm for computing an EP solution with revenue greater than $1/c$ times the revenue of an REP solution, unless P=NP.*

The proof is a gap preserving reduction from independent set problem.

Although the above hardness result shows that in general we cannot expect a polynomial time algorithm, the following claim says that for certain special cases, where all items have positive prices in a revenue maximizing WE solution, we can compute an optimal REP solution efficiently. Note that in most applications like advertising markets and housing markets, the items on sale are quite competitive and almost surely all items will be priced positively in a RWE solution.

Theorem 5. *Given a market with unit demand consumers, let $(\mathbf{p}, X)$ be an optimal RWE solution. If we have $p_j > 0$ for all items, then $(\mathbf{p}, X)$ is an optimal REP solution as well. Hence, REP can be computed in polynomial time.*

4   Single-Minded Consumers

We say a consumer i is *single-minded* if there is a subset of items $S_i \subseteq B$, called the demand set of i, such that $v_i(S) = v_i(S_i) > 0$ if $S_i \subseteq S$ and $v_i(S) = 0$ otherwise. That is, i is only interested in obtaining a subset containing S_i. In this section, we assume that all consumers are single-minded and denote $v_i(S_i) = v_i$. Note that when all consumers are single-minded, we can assume without loss of generality that every consumer i either wins subset S_i or wins nothing. Hence, we can encode an allocation $X = (x_i)_{i \in A}$ to be a $(0, 1)$-indicator vector where $x_i = 1$ implies that i wins subset S_i in X. Note that for any consumers $i \neq i'$ with $x_i = x_{i'} = 1$, we must have $S_i \cap S_{i'} = \emptyset$

in any feasible allocation since all items have unit supply. Given any solution $(\mathbf{p}, X)$, the revenue of the seller can be denoted by $r(\mathbf{p}, X) = \sum_{i \in A} x_i \cdot p(S_i)$.

Our goal is again to consider envy-free solutions which are Pareto efficient (EP) as well, and find one that maximizes the revenue among all such solutions (REP). Surprisingly, in a market with single-minded consumers, an EP solution may not exist at all; and even if one exists, an REP may not.

Example 3 (Non-existence of EP). There are three items j_1, j_2, and j_3 and six consumers i_k, $k = 1, \ldots, 6$. The first three consumers i_1, i_2, and i_3 are interested in $S_1 = \{j_1, j_2\}, S_2 = \{j_2, j_3\}$ and $S_3 = \{j_3, j_1\}$, each with valuation 2. The remaining consumers i_4, i_5, and i_6 are interested in items $j_1, j_2,$, and j_3, respectively, each with valuation 0.7. Suppose consumer i_1 is a winner in an envy free solution. Consumer i_6 is the only possible other winner. However, if i_6 is a winner, the solution will not be envy free as either i_2 or i_3 will have a valuation that exceeds the price of their demand set. Using similar argument for other cases, we can conclude that any envy free solution will have at most one winner.

Consider any solution that sells at most two items. Clearly, we can increase the revenue by selling the third item to the appropriate consumer i_k, where $k \in \{4, 5, 6\}$. Therefore, the solution that sells to at most two items cannot be Pareto efficient. This immediately implies that no Pareto efficient solution can have less than two winners. Therefore, in this example, there is no EP solution.

Example 4 (Non-existence of REP when EP exists). There are three items j_1, j_2, j_3 and four consumers $i_1, \ldots, i_4$ with demand subsets $S_1 = \{j_1, j_2\}, S_2 = \{j_2, j_3\}, S_3 = \{j_3, j_1\}$, and $S_4 = \{j_1, j_2, j_3\}$, and valuations $v_{i_1} = v_{i_2} = v_{i_3} = 2$ and $v_{i_4} = 2.5$. It is easy to see that at most one consumer can be the winner. Further, it can be seen that i_4 cannot be the winner given envy-freeness since there is no solution to the equation system:

$$\begin{aligned} p_{j_1} + p_{j_2} + p_{j_3} &\leq v_{i_4} = 2.5, \\ p_{j_1} + p_{j_2} &\geq v_{i_1} = 2, \\ p_{j_2} + p_{j_3} &\geq v_{i_2} = 2, \\ p_{j_3} + p_{j_1} &\geq v_{i_3} = 2. \end{aligned}$$

Since i_1, i_2, i_3 are symmetric, assume without loss of generality that i_1 is the winner and $p_{j_3} = \infty$. Then i_2, i_3, i_4 all get utility zero. Due to Pareto efficiency, we cannot charge i_1 at price $p(S_1) = p_{j_1} + p_{j_2} = v_{i_1} = 2$, since otherwise it is dominated by another solution where i_4 wins all three items at a total price of 2.5 (the seller obtains more revenue in this solution). Therefore, to guarantee envy-freeness and Pareto efficiency, we have to charge i_1 at a price strictly less than $v_{i_1} = 2$. Hence, any price vector $(p_{j_1}, p_{j_2}, \infty)$ where $p_{j_1} + p_{j_2} < 2$ gives an EP solution. However, in this example, there is no exact REP solution since the revenue of the seller can be arbitrarily close to 2. This example further implies that even if an EP solution exists, it may not maximize social welfare since i_4 should be the winner in the optimal allocation.

Our computational results, however, are quite negative. In particular, we show in Theorem 6 that, in addition to computing an optimal REP solution, even testing Pareto efficiency of a given solution and determining the existence of one EP solution is hard.

Theorem 6. *In a market with single-minded consumers, the following results hold:*

- *Determining whether a given solution* $(\mathbf{p}, X)$ *is Pareto efficient is NP-complete.*
- *Determining the existence of an* EP *solution is NP-complete.*
- *Given the existence of an* EP *solution, computing (a revenue maximizing) one is NP-hard.*

Despite the hardness results in Theorem 6, in a manner similar to Theorem 5, we present natural cases where optimal REP solutions can be easily computed.

Theorem 7. *Suppose a market with single minded consumers that admits an optimal* RWE *solution* $(\mathbf{p}, X)$ *in which* $p_j > 0$ *for all items* j. *Then,* $(\mathbf{p}, X)$ *is an optimal* REF *solution as well.*

Theorem 8. *Given a market with single minded consumers, let* $(\mathbf{p}, X)$ *be an optimal* REF *solution. If we have* $p_j > 0$ *for all items, then* $(\mathbf{p}, X)$ *is an optimal* REP *solution as well.*

Theorems 7 and 8 immediately lead to the following corollary.

Corollary 2. *Given a market with single minded consumers, let* $(\mathbf{p}, X)$ *be an optimal* RWE *solution. If we have* $p_j > 0$ *for all items, then* $(\mathbf{p}, X)$ *is an optimal* REP *solution as well.*

Finally, we note that the instances constructed in our reductions in Theorem 6 are somewhat artificial. Therefore, while the hardness results hold, we expect more natural instances that occur in real world scenarios to allow EP solutions.

Acknowledgments. I am grateful for John Augustine and Ning Chen for many helpful discussions and suggestions.

References

1. Balcan, M., Blum, A., Mansour, Y.: Item Pricing for Revenue Maximization. In: EC 2008, pp. 50–59 (2008)
2. Bouveret, S., Lang, J.: Efficiency and Envy-Freeness in Fair Division of Indivisible Goods: Logical Representation and Complexity. In: IJCAI 2005, pp. 935–940 (2005)
3. Briest, P.: Uniform Budgets and the Envy-Free Pricing Problem. In: Aceto, L., Damgård, I., Goldberg, L.A., Halldórsson, M.M., Ingólfsdóttir, A., Walukiewicz, I. (eds.) ICALP 2008, Part I. LNCS, vol. 5125, pp. 808–819. Springer, Heidelberg (2008)
4. Chen, N., Deng, X., Sun, X.: On Complexity of Single-Minded Auction. Journal of Computer and System Sciences 69(4), 675–687 (2004)
5. Chen, N., Ghosh, A., Vassilvitskii, S.: Optimal Envy-Free Pricing with Metric Substitutability. SIAM Journal on Computing 40(3), 623–645 (2011)

6. Chen, X., Deng, X., Teng, S.H.: Settling the Complexity of Computing Two-Player Nash Equilibria. Journal of the ACM 56(3) (2009)
7. Clarke, E.H.: Multipart Pricing of Public Goods. Public Choice 11, 17–33 (1971)
8. Conitzer, V., Sandholm, T.: Complexity Results about Nash Equilibria. In: IJCAI 2003, pp. 765–771 (2003)
9. Cramton, P., Shoham, Y., Steinberg, R.: Combinatorial Auctions. MIT Press (2006)
10. Crawford, V., Knoer, E.: Job Matching with Heterogeneous Firms and Workers. Econometrica 49(2), 437–450 (1981)
11. Daskalakis, C., Goldberg, P., Papadimitriou, C.: The complexity of computing a Nash equilibrium. SIAM Journal on Computing 39(1), 195–259 (2009)
12. Demange, G., Gale, D.: The Strategy of Two-Sided Matching Markets. Econometrica 53, 873–888 (1985)
13. Groves, T.: Incentives in Teams. Econometrica 41, 617–631 (1973)
14. Gul, F., Stacchetti, E.: Walrasian Equilibrium with Gross Substitutes. Journal of Economic Theory 87, 95–124 (1999)
15. Guruswami, V., Hartline, J., Karlin, A., Kempe, D., Kenyon, C., McSherry, F.: On Profit-Maximizing Envy-Free Pricing. In: SODA 2005, pp. 1164–1173 (2005)
16. Kelso, A., Crawford, V.: Job Matching, Coalition Formation, and Gross Substitutes. Econometrica 50, 1483–1504 (1982)
17. Mas-Colell, A., Whinston, M., Green, J.: Microeconomic Theory. Oxford University Press (1995)
18. Pazner, E., Schmeidler, D.: A Difficulty in the Concept of Fairness. Rev. Econ. Studies 41, 441–443 (1974)
19. Quinzii, M.: Core and Competitive Equilibria with Indivisibilities. International Journal of Game Theory 13, 41–60 (1984)
20. Shapley, L., Shubik, M.: The Assignment Game I: The Core. International Journal of Game Theory 1(1), 111–130 (1971)
21. Varian, H.: Equity, Envy, and Efficiency. Journal of Economic Theory 9, 63–91 (1974)
22. Vickrey, W.: Counterspeculation, Auctions and Competitive Sealed Tenders. Journal of Finance 16, 8–37 (1961)
23. Walras, L.: Elements of Pure Economics, 1877. Harvard University Press (1954)

Online Pricing for Multi-type of Items

Yong Zhang[1,2,*], Francis Y.L. Chin[2,**], and Hing-Fung Ting[2,***]

[1] Shenzhen Institutes of Advanced Technology, Chinese Academy of Sciences, China
[2] Department of Computer Science, The University of Hong Kong, Hong Kong
{yzhang,chin,hfting}@cs.hku.hk

Abstract. In this paper, we study the problem of online pricing for bundles of items. Given a seller with k types of items, m of each, a sequence of users $\{u_1, u_2, ...\}$ arrives one by one. Each user is single-minded, i.e., each user is interested only in a particular bundle of items. The seller must set the price and assign some amount of bundles to each user upon his/her arrival. Bundles can be sold fractionally. Each u_i has his/her value function $v_i(\cdot)$ such that $v_i(x)$ is the highest unit price u_i is willing to pay for x bundles. The objective is to maximize the revenue of the seller by setting the price and amount of bundles for each user. In this paper, we first show that the lower bound of the competitive ratio for this problem is $\Omega(\log h + \log k)$, where h is the highest unit price to be paid among all users. We then give a deterministic online algorithm, Pricing, whose competitive ratio is $O(\sqrt{k} \cdot \log h \log k)$. When $k = 1$ the lower and upper bounds asymptotically match the optimal result $O(\log h)$.

1 Introduction

Economy, a very important facet in the world, has received deep and wide study by scientists from economics, mathematics, and computer science for many years. In computer science, researchers often build theoretical models for some economic events, then solve the problems by using techniques derived from algorithm design, combinatorial optimization, randomness, etc.

In this paper, we study the problem of item pricing, which is one of the most important problems in computational economics. Item pricing contains two kinds of participators: the seller and the user. The seller has some items, which will be sold to the users at some designated prices; the user will buy the items at an acceptable price. The objective is to maximize the total revenue of the seller by assigning items to the users. To achieve this target, the prices of the items must be sold dynamically, i.e., the prices of items are different for different users, at different times, in different locations, with different amounts, ... If the designated price is higher than the expected price of a user, this user will reject the item; otherwise, this user will accept the item.

* Research supported by NSFC (No. 11171086) and Shenzhen Internet Industry Development Fund under grant No.JC201005270342A.
** Research supported by HK RGC grant HKU 7117/09E.
*** Research supported by HK RGC grant HKU-7171/08E.

J. Snoeyink, P. Lu, K. Su, and L. Wang (Eds.): FAW-AAIM 2012, LNCS 7285, pp. 82–92, 2012.
© Springer-Verlag Berlin Heidelberg 2012

Formally speaking, given a seller with k types of items, i_1, i_2, ..., i_k, the amount of each type is m, thus, the total amount of items is $m \cdot k$. A sequence of users $\{u_1, u_2, ...\}$ come one by one, each user is *single-minded*, i.e., each user is interested only in a particular bundle of items. For example, user u's bundle of interest is $I_u = \{i_1, i_2\}$ (or $I_u = \{1, 2\}$). The bundles can be sold fractionally, but the amount of each item in the sold bundle must be the same. Still considering the above example, the seller may sell a half bundle to u, i.e., half i_1 and half i_2. The seller must set the unit price and sell a certain number of bundles to each user on his/her arrival. In this paper, for ease of computation and comparison, the unit price is defined on items, not on bundles, even though one can convert to the other easily. For example, if the seller sells 1.5 bundles $I_u = \{1, 2\}$ to u at price 3, then the unit price is 1. Actually, if we define the unit price on a bundle, the results in this paper still hold, because the unit price on bundle I can be regarded as $|I|$ times the unit price on an item. However, defining the unit price on an item is more convenient since different bundles can be compared easily. Each u_i has his/her value function $v_i(\cdot)$ such that $v_i(x)$ is the highest unit price u_i is willing to pay for x bundles. Generally, the more bundles a user buys, the lower unit price he expects. Thus, in this paper, we assume that $v_i(x)$ is non-increasing. Let h be the highest value among all $v_i(x)$, i.e., $v_i(x) \leq h$ for all i and x. When user u comes with his bundle of interest I_u, suppose that the seller sets the item unit price p and assigns ℓ bundles to u. If $p > v_i(\ell)$, then user u cannot accept this price, and no bundle is bought by u. Otherwise, $p \leq v_i(\ell)$, and u accepts this price and pays $p \cdot \ell \cdot |I_u|$ to the seller.

To understand this model clearly, consider the example as shown in Figure 1. The seller has $k = 3$ types of items, and each type contains $m = 2$ items. There are three single-minded users who want to buy these items. User 1's bundle of interest is $I_1 = \{1, 2\}$; the unit prices at which user 1 is willing to buy his bundle are 5 and 5 for buying 1 and 2 bundles respectively, i.e., $v_1(1) = v_2(1) = 5$. User 2's bundle of interest is $I_2 = \{2, 3\}$; the unit prices at which user 2 is willing to buy his bundle are 6 and 4 for buying one and two bundles respectively, i.e., $v_2(1) = 6$ and $v_2(2) = 4$. User 3's bundle of interest is $I_3 = \{1, 3\}$; the unit prices at which user 3 is willing to buy his bundle are 7 and 4 for buying one and two bundles respectively, i.e., $v_3(1) = 7$ and $v_3(2) = 4$.

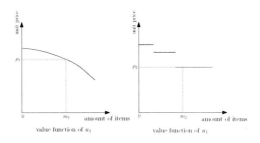

Fig. 1. An example of online pricing for multi items

When user 1 comes, to maximize the seller's revenue on this user, the seller will assign 2 bundles of $I_1 = \{1, 2\}$ at unit price 5 to him. When user 2 and user 3 come, there is no item 1 and item 2 left. In this case, user 2 and user 3 cannot buy anything and the total revenue of the seller is 20. However, the optimal strategy can achieve a total revenue of 36 by assigning one bundle of I_1 at unit price 5 to user 1, one bundle of $I_2 = \{2, 3\}$ at unit price 6 to user 2, and one bundle of $I_3 = \{1, 3\}$ at unit price 7 to user 3.

We consider the online version of this problem, i.e., before the i-th user comes, the seller has no information of the j-th user for $j \geq i$. To measure the performance of online algorithms, competitive analysis is generally used, i.e., to compare the outputs between the online algorithm and the optimal offline algorithm (which assumes all information is known in advance). Given the seller with item set B and a user sequence σ, let $A(B, \sigma)$ and $O(B, \sigma)$ denote the total revenue received by the seller according to the online algorithm A and the optimal offline algorithm O, respectively. The competitive ratio of the algorithm A is

$$R_A = \sup_{B, \sigma} \frac{O(B, \sigma)}{A(B, \sigma)}.$$

The pricing problem for items has been well studied in the past few years. Both multi-type and single-type items have been considered. Previous work has mainly focused on two supply models: the unlimited supply model [1, 2, 6, 10, 12] where the number of each type of item is unbounded and the limited supply model [1, 3–5, 7, 11, 13, 14] where the number of each type of item is bounded by some value. As for the users, there are several users' behaviors studied, including single-minded [7–10, 12, 14] (each user is interested only in a particular set of items), unit-demand [2–6, 12, 14] (each user will buy at most one item in total) and envy free [1, 5, 7, 10, 12] (after the assignment, no user would prefer to be assigned a different set of items with the designated prices, loosely speaking, each user is happy with his/her purchase). Most of the previous studies have considered a combination of the above scenarios (e.g. envy-free pricing for single-minded users when there is unlimited supply). In [15], Zhang et al. considered a more practical and realistic model in the sense that the seller has a finite number of items (one type of item with limited supply) and users can demand more than one items and arrive online. They proved that the lower bound of the competitive ratio is $O(\log h)$, moreover, they gave a deterministic online algorithm with competitive ratio $O(\log h)$, where h is the highest unit price.

In this paper, besides generalizing the problem to more than one type of items and bundles of items required by users, the idea used in our proposed online pricing algorithm is rooted by considering the amount of remaining items in additional to the user's value function in determining the price and amount of items to be sold. The proof in establishing the upper bound is more complicated by employing a fine price partition. The result in this paper can also match the optimal result $O(\log h)$ [15] asymptotically when there is only one type of item ($k = 1$) in the model.

This paper is organized as follows: Section 2 proves the lower bound of the competitive ratio for this variant to be $\Omega(\log h + \log k)$; in Section 3, a deterministic online algorithm whose competitive ratio is $O(\sqrt{k} \log h \log k)$ is given.

2   Lower Bound of the Competitive Ratio

In this part, our target is to show the lower bound of the competitive ratio of the online pricing problem for multi-type of items is $\Omega(\log h + \log k)$, where h is the highest unit price and k is the number of types.

To easily analyze the lower bound, we assume that $h = 2^\ell$ and $k = 2^j$, i.e., $\log h$ and $\log k$ are both integers. The lower bound is proved step by step. In each step, the adversary sends a user to the seller. In this proof, all value functions are flat, particularly, the value function $v(x)$ is some power of 2 for all x.

In the first $\log k$ steps, the value function $v(x) = 1$.

Step 1
  The adversary sends user u_1 to the seller, with bundle of interest $\{1\}$.
  If the seller assigns x_1 bundles to u_1 such that $x_1 \le m/(\log h + \log k)$, the adversary stops. In this case, the revenue of the seller is at most $m/(\log h + \log k)$, while the maximal revenue is m by assigning all m bundles to u_1. Thus, the ratio in this case is at least $\Omega(\log h + \log k)$.
  Otherwise, the seller assigns more than $m/(\log h + \log k)$ bundles to u_1. The adversary will send the next user to the seller.

Step 2
  The adversary sends user u_2 to the seller, with bundle of interest $\{1, 2\}$.
  If the seller assigns x_2 bundles to u_2 such that $x_1 + x_2 \le 2m/(\log h + \log k)$, the adversary stops. In this case, the revenue of the seller is at most $x_1 + 2x_2 \le 3m/(\log h + \log k)$, while the optimal revenue is $2m$ by assigning all i_1 and i_2 to user u_2. Thus, the ratio in this case is at least $\Omega(\log h + \log k)$.
  Otherwise, the seller assigns x_2 bundles to u_2 such that $x_1 + x_2 > 2m/(\log h + \log k)$.

...

Step ℓ: ($1 < \ell \le \log k$)
  The adversary sends user u_ℓ to the seller, with bundle of interest $\{1, 2^{\ell-2} + 1 - -2^{\ell-1}\}$.
  If the seller assigns x_ℓ bundles to u_ℓ such that $\sum_{t=1}^{\ell} x_t \le \ell \cdot m/(\log h + \log k)$, the adversary stops. In this case, the revenue achieved by the seller is

$$x_1 + 2x_2 + ... + (2^{\ell-2} + 1)x_\ell \qquad (1)$$

Lemma 1. *If the adversary stops at step ℓ, the total revenue is at most*

$$\frac{(2^{\ell-1} + \ell - 1)m}{\log h + \log k}.$$

Proof. From previous steps, we have

$$\sum_{p=1}^{t} x_p \geq \frac{tm}{\log h + \log k} \quad (1 \leq t < \ell),$$

thus,

$$\sum_{p=t+1}^{\ell} x_p \leq \frac{(\ell - t)m}{\log h + \log k} \quad (1 \leq t < \ell).$$

Therefore, Equation (1) achieves the maximal value $\frac{(2^{\ell-1}+\ell-1)m}{\log h+\log k}$ when each x_p equals to $m/(\log h + \log k)$. □

The optimal revenue is $(2^{\ell-2} + 1)m$ by assigning all m bundles to u_ℓ. Therefore, in this case, the competitive ratio is still bounded by $\Omega(\log h + \log k)$.

Otherwise, the adversary sends the next user to the seller.

...

In the following $\log h$ steps (step $\log k + 1 \leq \ell \leq \log k + \log h$), the bundles of interest are $\{1, k/2+1--k\}$, and the value functions are $v(x) = 2^{\ell-\log k}$ at step ℓ.

Step $\log k + 1$

The adversary sends user $u_{\log k+1}$ to the seller.

If the seller assigns $x_{\log k+1}$ bundles to $u_{\log k+1}$ such that $\sum_{\ell=1}^{\log k+1} x_\ell \leq (\log k + 1)m/(\log h + \log k)$, the adversary stops. In this case, the revenue achieved by the seller is

$$x_1 + 2x_2 + ... + (k/2 + 1)x_{\log k} + 2 \cdot (k/2 + 1)x_{\log k+1} \quad (2)$$

Similar to the proof in Lemma 1, we can find that the revenue achieved is at most

$$\frac{(3k/2 + \log k + 1)m}{\log k + \log h}.$$

The optimal revenue is $(k+2)m$ by assigning all m bundles to $u_{\log k+1}$. Therefore, in this case, the competitive ratio is still bounded by $\Omega(\log h + \log k)$.

Otherwise, the adversary sends the next user to the seller.

The analysis on Steps until Step $\log k + \log h - 1$ are similar to the above one.

Step $\log k + \log h$

The adversary sends user $u_{\log k+\log h}$ to the seller and the seller assigns $x_{\log k+\log h}$ bundles to the user. Since all bundles of interest include i_1, thus, the total number of all assigned bundles is no more than m. Thus, the adversary must stop at this step. Similar to the previous analysis, we can say that the ratio between the optimal solution and the revenue achieve by the online algorithm is at least $O(\log k + \log h)$.

Therefore, we have the following conclusion.

Theorem 1. *For the online pricing for multi-type of items, the lower bound of the competitive ratio is $\Omega(\log h + \log k)$.*

3   Online Algorithm

To maximize the revenue of the seller on a particular user u with bundle of interest I, a straightforward idea is finding unit price p and amount of bundles b such that p is acceptable when buying b bundles, and $b \cdot p$ is maximized. If we assign b bundles with unit price p to u, the revenue is $b \cdot p \cdot |I|$, which is maximal.

In our algorithm, we assign unit price 2^{ℓ} ($\ell \geq 0$) to each user. In this way, we have no need to consider all possible prices, and we will show that the performance doesn't be affected too much. Let (b, p) be the assignment such that $b \cdot p \cdot |I|$ is maximal. W.l.o.g., suppose $2^i \leq p < 2^{i+1}$, note that $v(x)$ is non-increasing, we have $v^{-1}(2^{i+1}) \leq b \leq v^{-1}(2^i)$. Thus,

$$b \cdot p \leq v^{-1}(2^i) \cdot p \leq v^{-1}(2^i) \cdot 2^{i+1} = 2 \cdot v^{-1}(2^i) \cdot 2^i.$$

If we choose the unit price equals to some power of 2, $(v^{-1}(2^i), 2^i)$ is a candidate of the assignment, which is at least half of the maximal value.

In our algorithm, we partition the amount of each type of item into $\lfloor \log h \rfloor$ stages, from stage 1 to stage $\lfloor \log h \rfloor$. The amount of items in stage i can be only assigned with unit price 2^{i-1}. Furthermore, partition the items in each stage into $\lfloor \log k \rfloor + 1$ levels, from level 0 to level $\lfloor \log k \rfloor$. For type i, items in level ℓ can be only assigned to users such that type i is in the user's bundle of interest and the size of the bundle is within $[2^{\ell-1} + 1, 2^{\ell}]$. For example, a user u's bundle of interest is $\{1, 2, 3, 4\}$, thus, in our algorithm, we choose items from level 2 in some stage to satisfy this user.

Let $\delta_i^{s,t}$ denote the available amount of items of type i in stage s level t. Initially, $\delta_i^{s,t} = m/(\lfloor \log h \rfloor (\lfloor \log k \rfloor + 1))$.

Next, we will formally describe the pricing algorithm for multi-type of items.

Algorithm 1. Pricing

1: Let I be the bundle of interest of the coming user u.
2: Let $\ell = \lceil \log |I| \rceil$ ▷ ℓ denotes the level which may assign items to user i.
3: Let x_j be the largest amount of bundles that user i is willing to buy given unit price 2^j and satisfying $x_j \leq m$.
4: Let $y_j = \min\{x_j, \min_{i \in I}\{\delta_i^{j+1,\ell}\}\}$.
5: Let $s = \arg\max_j y_j \cdot 2^j$ such that $y_j > 0$.
6: **if** no such s exists **then**
7:     Assign 0 bundles to user u.
8: **else**
9:     Set the unit price $p = 2^s$.
10:     Assign y_s bundles to user u.
11:     $\delta_i^{s+1,\ell} = \delta_i^{s+1,\ell} - y_s$ for all $i \in I$.
12: **end if**

According to the algorithm Pricing, if a user u with bundle of interest I cannot be satisfied, that means in each acceptable stage s, at least one type of item in I at level $\lceil \log |I| \rceil$ are all assigned to other users.

For a user sequence $\{u_1, u_2, ...\}$, let ALG denote the total revenue received from the algorithm Pricing, let OPT be the revenue achieved by the optimal algorithm. Next, we give the competitive ratio of the algorithm Pricing, i.e., prove the upper bound of the ratio between OPT and ALG.

After the processing of Pricing, for each type of item, some levels in some stages are full, the others still contain some available items. Classify all levels into two classes: L_i^f denotes the levels of type i which are full, i.e., if $\delta_i^{s,t} = 0$, level t in stage s of type i belongs to L_i^f; L_i^n denotes the levels of type i which contain available items, i.e., $\delta_i^{s,t} > 0$ for the corresponding levels.

Compare the assignments from the optimal algorithm and Pricing, we also partition the assignments from the optimal algorithm into two classes according to L_i^f and L_i^n. In the optimal solution, consider the assignment to a user with bundle of interest I, suppose the unit price $p \in [2^\ell, 2^{\ell+1})$. In the assignment from Pricing, if there exist $i \in I$ such that $\delta_i^{\ell+1, \lceil \log |I| \rceil} = 0$, i.e., this level belongs to L_i^f, we say the revenue of this assignment from the optimal algorithm belongs to O_f. Otherwise, if for any $i \in I$, $\delta_i^{\ell+1, \lceil \log |I| \rceil} > 0$ in the assignment w.r.t. Pricing, the revenue of this assignment belongs to O_n.

Moreover, we partition the assignment according to the size of the bundle. If $|I| \leq \sqrt{k}$, we say the size of the bundle is *small*, otherwise, the size is *large*. The revenue of the assignment from the optimal algorithm is further partitioned into four classes: O_f^s, O_f^l, O_n^s, and O_n^l, where O_f^s and O_n^s denote the revenue from small bundles, O_f^l and O_n^l denote the revenue from large bundles. Note that $O_f^s + O_f^l = O_f$ and $O_n^s + O_n^l = O_n$. Next, we compare these four classes with ALG respectively.

Lemma 2. $\frac{O_f^s}{ALG} \leq O(\sqrt{k} \cdot \log h \log k)$.

Proof. By the optimal algorithm, consider an assignment A of a bundle I with unit price $p \in [2^\ell, 2^{\ell+1})$. Assume that $|I| \leq \sqrt{k}$ and in the assignment from Pricing, at least one type $i \in I$ at level $\lceil \log |I| \rceil$ in stage $\ell+1$ is full. The optimal revenue of the assignment A is

$$p \cdot m \cdot |I| \tag{3}$$

From the algorithm Pricing, the revenue at level $\lceil \log |I| \rceil$ in stage $\ell+1$ of type i is

$$2^\ell \cdot m / (\lfloor \log h \rfloor (\lfloor \log k \rfloor + 1)) \tag{4}$$

The ratio between (3) and (4) is

$$O(\sqrt{k} \cdot \log h \log k).$$

Suppose by the optimal algorithm, more than one assigned small bundles, I_1, I_2, ..., I_j with unit prices $p_1, p_2, ..., p_j$ satisfy $p_{j'} \in [2^\ell, 2^{\ell+1})$ for $1 \leq j' \leq j$ and all $\lceil \log |I_{j'}| \rceil$ $(1 \leq j' \leq j)$ are equal, if these bundles share an item i such that in the assignment from Pricing, level $\lceil \log |I_1| \rceil$ in stage $\ell+1$ of item i is full, the total revenue for these bundles in the optimal solution is at most

$$\max\{p_{j'}\} \cdot m \cdot \max\{|I_{j'}|\} \quad 1 \leq j' \leq j \tag{5}$$

Compare the value in Equation (5) with the revenue from Pricing at level $\lceil \log |I| \rceil$ in stage $\ell + 1$ of type i (Equation (4)), the ratio is also

$$O(\sqrt{k} \cdot \log h \log k).$$

From the optimal pricing, all assignments in O_f^s can be partitioned into parts, each part contains the assignments mentioned above. Since the total revenues on such full levels is a lower bound of ALG, we have

$$\frac{O_f^s}{ALG} \leq O(\sqrt{k} \cdot \log h \log k).$$

□

Lemma 3. $\frac{O_f^l}{ALG} \leq O(\sqrt{k} \cdot \log h \log k).$

Proof. By the optimal algorithm, consider some amount of bundle I with unit price $p \in [2^\ell, 2^{\ell+1})$ is assigned to a user. Assume that $|I| > \sqrt{k}$ and there exist an item $i \in I$ such that in the assignment from Pricing, level $\lceil \log |I| \rceil$ in stage $\ell + 1$ of item i is full. Note that in Pricing, items in this level can be only assigned to bundles with size in between $(2^{\lceil \log |I| \rceil - 1}, 2^{\lceil \log |I| \rceil}]$, thus, the total revenue on level $\lceil \log |I| \rceil$ in stage $\ell + 1$ is at least

$$\frac{2^\ell \cdot m \cdot 2^{\lceil \log |I| \rceil - 1}}{\lfloor \log h \rfloor (\lfloor \log k \rfloor + 1)} \tag{6}$$

Note that the optimal revenue for bundles with unit price $p \in [2^\ell, 2^{\ell+1})$ is at most

$$2^{\ell+1} \cdot m \cdot k \tag{7}$$

The ratio between the above two equations is $O(\sqrt{k} \cdot \log h \log k)$. Combine all revenues in O_f^l, we can say that this lemma is true.   □

Lemma 4. $\frac{O_n^s}{ALG} \leq O(\sqrt{k} \cdot \log h \log k).$

Proof. Again, consider an assigned bundle I from the optimal algorithm, such that the revenue on I belongs to O_n^s and the unit price is $p \in [2^\ell, 2^{\ell+1})$. The algorithm Pricing chooses the unit price 2^j such that $2^j \cdot y_j$ is maximized. Note that $(2^\ell, y_\ell)$ is also a candidate for satisfying bundle I.

– If Pricing assigns $(2^\ell, y_\ell)$, since after the assignment, $\delta_i^{\ell+1, \lceil \log I \rceil} > 0$ for all $i \in I$, the revenue achieved on I by Pricing is at least half of the optimal revenue on this bundle.
– Otherwise, Pricing assigns $(2^j, y_j)$ such that $j \neq \ell$. From the choosing criteria, $2^j \cdot y_j \geq 2^\ell \cdot y_\ell$.

  • If $\delta_i^{\ell+1, \lceil \log I \rceil} > y_\ell$ for all $i \in I$, from above analysis, we can say the revenue on I by Pricing is at least half of the optimal revenue on this bundle.

- Otherwise, $\delta_i^{\ell+1,\lceil \log I\rceil} = y_\ell$ for some $i \in I$. Since $2^j \cdot y_j \geq 2^\ell \cdot y_\ell$, the revenue achieved on I by Pricing plus the current revenue on level $\lceil \log I\rceil$ in stage $\ell + 1$ of type i is at least

$$\frac{2^\ell \cdot m}{\lfloor \log h\rfloor(\lfloor \log k\rfloor + 1)} \tag{8}$$

This is because if we assign $(2^\ell, y_\ell)$ for this bundle, level $\lceil \log I\rceil$ in stage $\ell + 1$ of type i is full.

Similar to the analysis in Lemma 2, suppose more than one assigned small bundles I_1, I_2, ... with unit price p_1, p_2, ... within $[2^\ell, 2^{\ell+1})$ and the sizes of these bundles are all within $(2^{\lceil \log |I_1|\rceil - 1}, 2^{\lceil \log |I_1|\rceil}]$, if all these bundles share type i, the total revenue on such bundles by the optimal scheme is at most

$$2^{\ell+1} \cdot m \cdot 2^{\lceil \log |I_1|\rceil} \tag{9}$$

The ratio between the above two terms is $O(\sqrt{k} \cdot \log h \log k)$.

Mapping the assignments of O_n^s to the corresponding assignments of Pricing described above, each assignment by Pricing is counted at most TWICE. Combining the above analysis, we can say that

$$\frac{O_n^s}{ALG} \leq O(\sqrt{k} \cdot \log h \log k).$$

□

Lemma 5. $\frac{O_n^l}{ALG} \leq O(\sqrt{k} \cdot \log h \log k)$.

Proof. The proof of this lemma is similar to the proofs in Lemma 3 and Lemma 4. Consider an assigned bundle I from the optimal algorithm, such that the revenue on I belongs to O_n^l and the unit price is $p \in [2^\ell, 2^{\ell+1})$. The algorithm Pricing chooses the unit price 2^j such that $2^j \cdot y_j$ is maximized. Since any level $\lceil \log |I|\rceil$ in stage $\ell + 1$ of item $i \in I$ is not full, $(2^\ell, y_\ell)$ is a candidate.

- If Pricing assigns $(2^\ell, y_\ell)$, since after the assignment, $\delta_i^{\ell+1,\lceil \log I\rceil} > 0$ for all $i \in I$, the revenue achieved on I is at least half of the optimal revenue on this bundle.
- Otherwise, Pricing assigns $(2^j, y_j)$ such that $j \neq \ell$. From the choosing criteria, $2^j \cdot y_j \geq 2^\ell \cdot y_\ell$.
  - If $\delta_i^{\ell+1,\lceil \log I\rceil} > y_\ell$ for all $i \in I$, from above analysis, we can say the revenue on I by Pricing is at least half of the optimal revenue on this bundle.
  - Otherwise, $\delta_i^{\ell+1,\lceil \log I\rceil} = y_\ell$ for some $i \in I$. Since $2^j \cdot y_j \geq 2^\ell \cdot y_\ell$, the revenue achieved on I by Pricing plus the current revenue on level $\lceil \log I\rceil$ in stage $\ell + 1$ is at least

$$\frac{2^\ell \cdot m \cdot 2^{\lceil \log |I|\rceil - 1}}{\lfloor \log h\rfloor(\lfloor \log k\rfloor + 1)} \tag{10}$$

This is because if we assign $(2^\ell, y_\ell)$ for this bundle, level $\lceil \log I \rceil$ in stage $\ell + 1$ of type i is full, and the size of each bundle on this level is at least $2^{\lceil \log |I| \rceil - 1}$.

From the optimal scheme, the total revenue on assignments from unit price in between $[2^\ell, 2^{\ell+1})$ and bundle size in between $(2^{\lceil \log |I| \rceil - 1}, 2^{\lceil \log |I| \rceil}]$ is at most

$$2^{\ell+1} \cdot m \cdot k \tag{11}$$

The ratio between the above two terms is $O(\sqrt{k} \cdot \log h \log k)$

Mapping the revenue of assignments in O_n^l to the assignment by Pricing, from the above analysis, each assignment is counted at most twice. Thus,

$$\frac{O_n^l}{ALG} \leq O(\sqrt{k} \cdot \log h \log k).$$

$\square$

Now we give the main conclusion of this paper.

Theorem 2. *The competitive ratio of the algorithm Pricing is at most*

$$O(\sqrt{k} \cdot \log h \log k).$$

Proof. From the definition of O_f^s, O_f^l, O_n^s, and O_n^l, these four classes are disjoint. Note that if a requested bundle cannot be satisfied, it must belongs to O_f^s or O_f^l. Combining Lemma 2 until Lemma 5, we can say that the competitive ratio of the algorithm Pricing is $O(\sqrt{k} \cdot \log h \log k)$. $\square$

References

1. Balcan, N., Blum, A., Mansour, Y.: Item pricing for revenue maximization. In: Proc. of the 9th ACM Conference on Electronic Commerce (EC 2008), pp. 50–59 (2008)
2. Bansal, N., Chen, N., Cherniavsky, N., Rurda, A., Schieber, B., Sviridenko, M.: Dynamic pricing for impatient bidders. ACM Transactions on Algorithms 6(2) (March 2010)
3. Briest, P.: Uniform Budgets and the Envy-Free Pricing Problem. In: Aceto, L., Damgård, I., Goldberg, L.A., Halldórsson, M.M., Ingólfsdóttir, A., Walukiewicz, I. (eds.) ICALP 2008, Part I. LNCS, vol. 5125, pp. 808–819. Springer, Heidelberg (2008)
4. Briest, P., Krysta, P.: Buying cheap is expensive: hardness of non-parametric multi-product pricing. In: Proceedings of the Eighteenth Annual ACM-SIAM Symposium on Discrete Algorithms, New Orleans, Louisiana, January 07-09, pp. 716–725 (2007)
5. Chen, N., Deng, X.: Envy-Free Pricing in Multi-item Markets. In: Abramsky, S., Gavoille, C., Kirchner, C., Meyer auf der Heide, F., Spirakis, P.G. (eds.) ICALP 2010. LNCS, vol. 6199, pp. 418–429. Springer, Heidelberg (2010)
6. Chen, N., Ghosh, A., Vassilvitskii, S.: Optimal envy-free pricing with metric substitutability. In: Proc. of the 9th ACM Conference on Electronic Commerce (EC 2008), pp. 60–69 (2008)

7. Cheung, M., Swamy, C.: Approximation Algorithms for Single-minded Envy-free Profit-maximization Problems with Limited Supply. In: Proc. of 49th Annual IEEE Symposium on Foundations of Computer Science (FOCS 2008), pp. 35–44 (2008)
8. Elbassioni, K., Raman, R., Ray, S., Sitters, R.: On Profit-Maximizing Pricing for the Highway and Tollbooth Problems. In: Proceedings of the 2nd International Symposium on Algorithmic Game Theory, Paphos, Cyprus, October 18-20, pp. 275–286 (2009)
9. Elbassioni, K., Sitters, R., Zhang, Y.: A Quasi-PTAS for Profit-Maximizing Pricing on Line Graphs. In: Arge, L., Hoffmann, M., Welzl, E. (eds.) ESA 2007. LNCS, vol. 4698, pp. 451–462. Springer, Heidelberg (2007)
10. Fiat, A., Wingarten, A.: Envy, Multi Envy, and Revenue Maximization. In: Leonardi, S. (ed.) WINE 2009. LNCS, vol. 5929, pp. 498–504. Springer, Heidelberg (2009)
11. Grigoriev, A., van Loon, J., Sitters, R.A., Uetz, M.: How to Sell a Graph: Guidelines for Graph Retailers. In: Fomin, F.V. (ed.) WG 2006. LNCS, vol. 4271, pp. 125–136. Springer, Heidelberg (2006)
12. Guruswami, V., Hartline, J., Karlin, A., Kempe, D., Kenyon, C., McSherry, F.: On Profit-Maximizing Envy-Free Pricing. In: Proceedings of the 16th Annual ACM-SIAM Symposium on Discrete Algorithms (SODA 2005), pp. 1164–1173 (2005)
13. Im, S., Lu, P., Wang, Y.: Envy-Free Pricing with General Supply Constraints. In: Saberi, A. (ed.) WINE 2010. LNCS, vol. 6484, pp. 483–491. Springer, Heidelberg (2010)
14. Krauthgamer, R., Mehta, A., Rudra, A.: Pricing commodities. Theoretical Computer Science 412(7), 602–613 (2011)
15. Zhang, Y., Chin, F.Y.L., Ting, H.-F.: Competitive Algorithms for Online Pricing. In: Fu, B., Du, D.-Z. (eds.) COCOON 2011. LNCS, vol. 6842, pp. 391–401. Springer, Heidelberg (2011)

Algorithms with Limited Number of Preemptions for Scheduling on Parallel Machines⋆

Yiwei Jiang, Zewei Weng, and Jueliang Hu⋆⋆

Department of Mathematics, Zhejiang Sci-Tech University,
Hangzhou 310018, China
mathjyw@yahoo.com.cn, hujlhz@163.com

Abstract. In previous study on comparing the makespan of the schedule allowed to be preempted at most i times and that of the optimal schedule with unlimited number of preemptions, the worst case ratio was usually obtained by analyzing the structures of the optimal schedules. For m identical machines case, the worst case ratio was shown to be $2m/(m+i+1)$ for any $0 \leq i \leq m-1$[1], and they showed that LPT algorithm is an exact algorithm which can guarantee the worst case ratio for $i = 0$. In this paper, we propose a simpler method which is based on the design and analysis of the algorithm and finding an instance in the worst case. It can obtain the worst case ratio as well as the algorithm which can guarantee this ratio for any $0 \leq i \leq m-1$, and thus we generalize the previous results. We also make a discussion on the trade-off between the objective value and the number of preemptions. In addition, we consider the i-preemptive scheduling on two uniform machines. For both $i = 0$ and $i = 1$, we present the algorithms and give the worst-case ratios with respect to s, i.e., the ratio of the speeds of two machines.

Keywords: i-preemptive scheduling, approximation algorithm, worst case ratio, makespan.

1 Introduction

In the preemptive scheduling, we are given a sequence $\mathcal{J} = \{J_1, J_2, \ldots, J_n\}$ of n independent jobs with positive sizes $p_1, p_2, \ldots, p_n$, which must be scheduled onto m parallel machines $M_1, M_2, \cdots, M_m$. We identify the jobs with their sizes. At any time, each machine can handle at most one job and each job can be processed by at most one machine. Preemption is allowed, which means that each job may be preempted into a few pieces. These pieces are to be assigned to possibly different machines in non-overlapping time slots. Generally, the number of

---

⋆ Supported by the National Natural Science Foundation of China (11001242, 11071220) and Zhejiang Province Natural Science Foundation of China (Y6090175, Y6090554).

⋆⋆ Corresponding author.

J. Snoeyink, P. Lu, K. Su, and L. Wang (Eds.): FAW-AAIM 2012, LNCS 7285, pp. 93–104, 2012.

preemptions is not bounded and the costs of preempting a job may be neglected. However, in many practical systems, there arise costs with job preemptions such as the costs of moving a job off a machine before it is finished and the costs of saving information about the job while it is waiting to be resumed[1]. Hence, it is extremely important to consider the algorithms for the scheduling with limited number i of preemptions.

The optimal makespan for the preemptive scheduling on m identical machines is

$$\max\{\max_{1 \le i \le n} p_i, \frac{\sum_{i=1}^{n} p_i}{m}\} \qquad (1)$$

which can be obtained by McNaughton's *wrap around* rule[8] with no more than $m-1$ preemptions.

For the preemptive scheduling on m uniform machines, let s_i denote the speed of machine M_i and assume that $s_1 \ge s_2 \ge \cdots \ge s_m$ and $p_1 \ge p_2 \ge \cdots \ge p_m$. Let $S_j = \sum_{i=1}^{j} s_i$ and $P_j = \sum_{i=1}^{j} p_i$. Then the optimal makespan is

$$\max\{\max_{1 \le j < m} \{\frac{P_j}{S_j}\}, \frac{P_n}{S_m}\}. \qquad (2)$$

Liu and Yang[10] obtained the above result only for the case $s_1 \ge 1$ and $s_i = 1, 1 < i \le m$. For the general case, Horvath et al.[5] obtained the optimal schedule with as many as $n^2(m-1)$ preemptions. Gonzalez and Sahni[3] presented an algorithm which generates optimal schedule with at most $2(m-1)$ preemptions.

In the *i-preemptive scheduling*, it is allowed to preempt the jobs at most i times. For a job sequence $\mathcal{J}$, let $C^{i*}(\mathcal{J})$ (or in short C^{i*}) denote the optimal objective function value in the i-preemptive scheduling and let $C^*(\mathcal{J})$(or in short C^*) denote the optimal objective function value in preemptive scheduling where the number of preemptions is unbounded. In this paper, we consider the objective to minimize the makespan, i.e., the maximum completion time after scheduling all jobs. A natural question is to find the least upper bound R_i on the ratio of C^{i*} and C^*, that is,

$$R_i = \inf\{R | \frac{C^{i*}(\mathcal{I})}{C^*(\mathcal{I})} \le R, \forall \, \mathcal{I}\}.$$

For a job system with arbitrary precedence constraints, Liu[7] conjectured that $R_0 = \frac{2m}{m+1}$ when $i = 0$, i.e., the ratio the optimal nonpreemptive makespan versus the optimal preemptive makespan. Coffman and Garey[2] showed Liu's conjecture when $m = 2$, i.e, $R_0 = 4/3$. Hong and Leung[6] considered the cases for unit execution time (UET) and also for tree-structured job systems.

For the problem in which a set of n independent jobs has to be scheduled on m identical machines, we can obtain $R_i = 1$ when $i \ge m - 1$ from the result by McNaughton[8].

For $0 \le i \le m - 2$, Braun and Schmidt[1] showed

$$R_i = \frac{2m}{m+i+1},$$

based on the analysis of the structure of the optimal i-preemptive schedule. They further showed that the *Largest Processing Time* (LPT)[4] algorithm can guarantee the worst case ratio when $i = 0$, that is, the makepan of the schedule generated by LPT is not greater than $R_0 C^*$ for any instance. Klonowska et al.[9] generalized the results by Braun and Schmidt[1]. They considered the worst case ratio of the optimal makespans for i and j preemptions, $0 \le i < j \le m - 1$. In the case of $m \ge i + j + 1$, the worst case ratio is

$$2 \frac{\lfloor j/(i+1) \rfloor + 1}{\lfloor j/(i+1) \rfloor + 2}.$$

For the $m < i + j + 1$, they presented a formula based on the Stern-Brocot tree.

Most existing papers considered the ratio R_i by analyzing the structures of the optimal schedules. In this paper, we study the ratio R_i from a new perspective. That is, it can be obtained by analyzing the worst case ratio of an approximation algorithm and finding an instance in worst case. The *worst case ratio* of an approximation algorithm A is defined as

$$R_A = \inf\{R | \frac{C^A(\mathcal{I})}{C^*(\mathcal{I})} \le R, \forall \mathcal{I}\},$$

where $C^A(\mathcal{I})$ (or in short C^A) denotes the objective value produced by A, and $C^*(\mathcal{I})$ (or in short C^*) denotes the optimal preemptive objective value. More precisely, if there is a number α such that we can achieve the following two things:

a. *Designing an approximation algorithm A which is allowed to preempt at most i times and analyzing its worst case ratio $R_A \le \alpha$;*
b. *Finding an instance $\mathcal{I}^*$ such that the ratio $\frac{C^{i*}(\mathcal{I}^*)}{C^*(\mathcal{I}^*)} \ge \alpha$.*

Thus we can conclude that the ratio $R_i = \alpha$. This is our method to obtain the worst case ratio of the makespan of the optimal i-preemptive schedule and that of the optimal preemptive schedule with unlimited preemptions. It will be found later that this method is more tersely than the previous ones. What is more important is that we obtain the ratio R_i as well as an algorithm which can generate a schedule with the makespan no more than R_i times of the optimal preemptive schedule for any instance. Especially, it will generate an optimal schedule for the instance in the worst case.

In[1], Braun and Schmidt showed that LPT algorithm is an exact algorithm which can guarantee the worst case ratio R_i when $i = 0$. In this paper, we will present an algorithm for i-preemptive scheduling on m identical machines, which can guarantee the worst case ratio $R_i = \frac{2m}{m+i+1}$ for any $0 \le i \le m - 1$. In addition, we will discuss the trade-off between the objective of minimizing the makespan and the number of preemptions.

We will also consider the i-preemptive scheduling on two uniform machines. Without loss of generality, it is assumed that the speeds of two machines M_1 and M_2 are s and 1, respectively. From the results by Gonzalez and Sahni[3], we only need to consider two cases of $i = 0$ and $i = 1$. We will show that

$$R_0 = \begin{cases} \frac{2(s+1)}{3s}, & \text{for } 1 \leq s < 4/3, \\ \frac{s+1}{2}, & \text{for } 4/3 \leq s < 2, \\ \frac{s+1}{s}, & \text{for } s \geq 2, \end{cases}$$

and

$$R_1 = \frac{2s^2 + 2s - 1}{2s^2 + 1}.$$

The rest of the paper is organized as follows. Sections 2 and 3 consider the algorithms for i-preemptive scheduling on m identical machines and on two uniform machines, respectively. Finally, Section 4 presents some concluding remarks.

2 i-Preemptive Scheduling on m Identical Machines

In this section, we first present an algorithm which is allowed to preempt at most i number of preemptions and show its worst case ratio is $\frac{2m}{m+i+1}$. Together with the instance given by Braun and Schmidt[1], it is easy to obtain that the desired $R_i = \frac{2m}{m+i+1}$. Next, we will consider the problem that how many times of preemptions are needed at least in the worst case if the makespan should not be greater than α times of C^* for a given number α.

We now present our algorithm $LPT-iP$ based on McNaughton's *wrap around* rule and LPT rule. The algorithm splits at most i jobs and each of them are split into two pieces and scheduled on two distinct machines in the first i machines. The other $m-i$ machines are used to non-preemptively schedule jobs by the LPT rule. Formally, the algorithm $LPT-iP$ can be described as follows.

Algorithm $LPT-iP$:

1. **Sort** all the jobs such that $p_1 \geq p_2 \geq \cdots \geq p_n$ and compute the optimal makespan $C^* = \max\{p_1, \frac{\sum_{i=1}^{n} p_i}{m}\}$ by (1).
2. **Find** the job p_k such that $\sum_{i=1}^{k} p_i \geq iR_iC^*$ and $\sum_{i=1}^{k-1} p_i < iR_iC^*$. Split p_k into two pieces p_k^1 and p_k^2 with $p_k^1 = iR_iC^* - \sum_{i=1}^{k-1} p_i$.
3. **For** the job $p_1, p_2, \cdots, p_{k-1}, p_k^1$, schedule them in turn on the first i machines by the McNaughton's *wrap around* rule.
4. **For** the job $p_k^2, p_{k+1}, \cdots, p_n$, schedule them non-preemptively on the last $m-i$ machines by the LPT rule.
5. **Move** the job p_k^2 to be the first job on the machine that processes it.

Remark. We claim that the above algorithm $LPT-iP$ is well defined. That is, the time slots assigned to the different portions of each preempted job never overlap. Denoted by q_j the last job processed on M_j, $1 \leq j \leq i-1$. Noting that q_j is possibly preempted in the algorithm, see Figure 1. It is clear that the two portions (if any) of every q_j are not overlapping, since $q_j^1 + q_j^2 = q_j \leq C^* < R_iC^*$. Similarly, the assignment of the job p_k is feasible from the step 5 of the algorithm, see Figure 2.

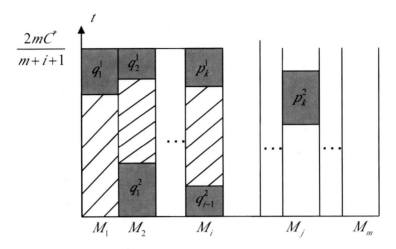

Fig. 1. The schedule π produced by step 4

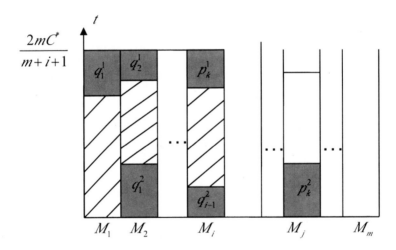

Fig. 2. The schedule π' produced by step 5

We now consider the worst case ratio of the algorithm $LPT - iP$.

Theorem 1. *The worst case ratio of the algorithm $LPT - iP$ is at most $\frac{2m}{m+i+1}$, i.e.,*

$$R_{LPT-iP} \leq \frac{2m}{m+i+1}.$$

Proof. Let π and π' denote the schedules produced by step 4 and step 5, respectively (see Figures 1 and 2). Note that the makespan is unchanged from the schedule π to π' in disregard of the feasibility of the schedule. Thus, for convenience, we only need to analyze the makespan of π.

Let p_l be the job which determines the makespan in the schedule π and let s_l be the start time of p_l. Then we have

$$C^{LPT-iP} = s_l + p_l. \tag{3}$$

If p_l is processed on the first i machines, we can conclude that $C^{LPT-iP} = \frac{2m}{m+i+1}C^*$ by step 3. We now turn to deal with the case that p_l is processed on the last $m - i$ machines below.

If $s_l = 0$, the desired result can be obtain trivially. Suppose $s_l > 0$, that is, there is at least one job processed on the same machine before the job p_l. We can claim that $p_l \leq s_l$, for all the jobs scheduled on these machines abide by the LPT rule. By (3), we have

$$s_l \geq \frac{C^{LPT-iP}}{2}. \tag{4}$$

We now focus on the completion time of the other machines except the machine to process the job p_l. Clearly, the completion time of the machine M_j for any $1 \leq j \leq i$ is $\frac{2m}{m+i+1}C^*$ and the total size is

$$\frac{2imC^*}{m+i+1}.$$

The completion time of each machine in the other $m - i - 1$ machines is at least s_l due to the LPT rule. Hence, we can conclude that

$$\frac{2imC^*}{m+i+1} + (m - i - 1)s_l + C^{LPT-iP} \leq \sum_{i=1}^{n} p_i,$$

which, together with the fact that the inequality (4) and $C^* \geq \sum_{i=1}^{n} p_i/m$ from (1), deduces that

$$\frac{2imC^*}{m+i+1} + \frac{C^{LPT-iP}(m - i - 1)}{2} + C^{LPT-iP} \leq mC^*.$$

By a simply computation, we can obtain

$$\frac{C^{LPT-iP}}{C^*} \leq \frac{2m}{m+i+1}.$$

The proof is complete.

By the above theorem, it is easy to obtain the result in[1] as follows.

Theorem 2. $R_i = \frac{2m}{m+i+1}$.

Proof. From Theorem 1, we can conclude that for any instance $\mathcal{I}$,

$$\frac{C^{i*}(\mathcal{I})}{C^*(\mathcal{I})} \leq \frac{C^{LPT-iP}(\mathcal{I})}{C^*(\mathcal{I})} \leq \frac{2m}{m+i+1}.$$

It follows that $R_i \leq \frac{2m}{m+i+1}$ by the definition of R_i.

On the other hand, let us consider the instance $\mathcal{I}^*$ of $m+i+1$ jobs with equal size of 1. Clearly, we have $C^* = \frac{m+i+1}{m}$.

Since at most i preemptions are allowed, there are at least $m+1$ jobs which are non-preemptively scheduled on the m machines. It implies that at least two of those $m+1$ jobs are scheduled on the same machine, resulting in $C^{i*} \geq 2$. Note that R_i is the worst case ratio of the optimal i−preemptive makespan C^{i*} and the optimal preemptive makespan C^*, thus we have

$$R_i \geq \frac{C^{i*}(\mathcal{I}^*)}{C^*(\mathcal{I}^*)} = \frac{2}{\frac{m+i+1}{m}} = \frac{2m}{m+i+1}.$$

Then the desired result is achieved.

To sum up the above arguments, it clearly shows that the result in [1] can be obtained simply by our method. At the same time we can give the algorithm to generate the schedule with the makespan not greater than $R_i C^*$ for all $0 \leq i \leq m-1$.

Next we will simply discuss the trade-off between the makespan and the number of preemption. That is, for a given $\alpha \geq 1$, how many preemptions are needed such that the makespan of produced schedule, denoted by C^{allow}, is not greater than αC^*.

It can be found that from the algorithm $LPT - iP$, we only need at most i preemptions for any $\alpha \in [\frac{2m}{m+i+1}, \frac{2m}{m+i})$ for some $1 \leq i \leq m-1$ and no preemption is needed if $\alpha \geq \frac{2m}{m+1}$.

Note that for the instance $\mathcal{I}^*$ of $m+i+1$ jobs with equal size of 1 mentioned in the proof of Theorem 2, the optimal non-preemptive makespan is identical to the optimal makespan with i preemptions, $1 \leq i \leq m-2$. Hence, we are much more concerned to know whether the i preemptions are required to ensure that $C^{allow} \leq \alpha C^*$.

In fact, suppose that there is an integer i such that $\alpha \in [\frac{2m}{m+i+1}, \frac{2m}{m+i})$. Let us consider the instance $\mathcal{J}^*$ of $m+i$ jobs with equal size of 1. Clearly, the makspan is at least 2 if we are allowed to preempt at most $i-1$ times, which follows that $\frac{C^{allow}}{C^*} \geq \frac{2}{(m+i)/m} > \alpha$. Thus we can conclude that the number of preemptions is at least i in this sense. Generally, we can draw a conclusion below.

Theorem 3. *For any given $1 \leq \alpha \leq \frac{2m}{m+1}$, there is an integer i such that $\alpha \in [\frac{2m}{m+i+1}, \frac{2m}{m+i})$. To ensure the desired makespan of the schedule is not greater than αC^* for any instance, we only need at most i preemptions. Moreover, it is a tight number of preemptions, i.e., there is an instance which needs exactly i preemptions.*

3 i-Preemptive Scheduling on Two Uniform Machines

In this section, we will consider the i-preemptive scheduling on two uniform machines. Without loss of generality, we assume that the speeds of two machines M_1 and M_2 are s and 1, $s \geq 1$. Note that from [3] the optimal scheduling on two uniform machines can be obtained by at most two preemptions. Therefore, we only need to consider the cases $i = 0$ and $i = 1$. In the following, we will give the worst case ratios R_0 and R_1 for the above cases. At the same time, we will present two algorithms to guarantee that the makespans produced by our algorithms are not greater than $R_0 C^*$ and $R_1 C^*$ for any instance, respectively.

Before going to present our algorithms, we first give the lower bound of R_i as follows, $i = 0, 1$. Let

$$
r_0 = \begin{cases} \frac{2(s+1)}{3s}, & \text{for } 1 \leq s < 4/3, \\ \frac{s+1}{2}, & \text{for } 4/3 \leq s < 2, \text{ and } r_1 = \frac{s^2 + 2s - 1}{s^2 + 1}. \\ \frac{s+1}{s}, & \text{for } s \geq 2. \end{cases}
$$

Lemma 1. $R_0 \geq r_0$ and $R_1 \geq r_1$.

Proof. To obtain the desired results, we only need to find some instances such that the ratio of the optimal makespan in the i-preemptive scheduling and the optimal makespan in the preemptive scheduling is exactly r_i, $i = 0, 1$.

In fact, it only needs to consider the instance of three identical jobs for $i = 0$ and $1 \leq s < \frac{4}{3}$, and the instance of two identical jobs for other cases. Here omits the details.

In the following two subsections, we will present the algorithm A_i with the worst case ratio of r_i, $i = 0, 1$.

3.1 Algorithm for Non-preemptive Scheduling

In this subsection, we will show the worst case ratio of the optimal non-preemptive makespan and the optimal preemptive makespan on two uniform machines is exactly r_0, i.e., $R_0 = r_0$. From Lemma 1, we only need to show $R_0 \leq r_0$ by presenting an algorithm without preemption, denoted by A_0, with the worst case ratio of $R_A \leq r_0$. The main idea of our algorithm is as follows: For the case of $s \geq 2$, it only needs to schedule all the jobs onto the fast machine. For the case of $1 \leq s < 2$, it firstly generates a schedule by the LPT rule, here the LPT rule schedules the largest one in all unscheduled jobs to a machine such that it can be completed as soon as possible. Then the algorithm will possibly make some adjustments according to the initial schedule generated by LPT. The detailed algorithm is below.

Algorithm A_0:

　　a. For $s \geq 2$. Schedule all jobs on the machine M_1.

　　b. For $1 \leq s < 2$.

1. **Sort** all the jobs such that $p_1 \geq p_2 \geq \cdots \geq p_n$. For any job p_j, $1 \leq j \leq n$, let c_i be the current completion time of M_i, $i = 1, 2$. If $c_1 + p_j/s \leq c_2 + p_j$, schedule p_j on the machine M_1, otherwise, on M_2. Denote this schedule by π_{LPT}.

2. **Suppose** p_l is the job that determines the makepan in the schedule π_{LPT}. Let $\mathcal{I}_i$ be the set of the jobs processed on the machine M_i except the job p_l, $i = 1, 2$. Denote $l_i = \sum\limits_{p_j \in \mathcal{I}_i} p_j$.

   If $l_1 \leq l_2 + p_l < 2l_1$, exchange all the jobs in $\mathcal{I}_1$ and $\mathcal{I}_2$ and schedule the job p_l on the machine M_1. Otherwise, keep the schedule π_{LPT} as the final schedule.

Before going to show the worst case ratio of the algorithm A_0 is r_0, we should illuminate that the desired ratio can not be obtained by simply using the LPT rule. Let us consider the instance of three jobs with $p_1 = s$ and $p_2 = p_3 = 1$, the LPT rule will schedule p_1 and p_3 on the fast machine M_1 and the other on M_2. Then we have $C^{LPT} = \frac{s+1}{s}$. With the optimal makespan $C^* = \frac{s+2}{s+1}$, it follows that for any $1 < s \leq \frac{4}{3}$

$$\frac{C^{LPT}}{C^*} \geq \frac{\frac{s+1}{s}}{\frac{s+2}{s+1}} = \frac{(s+1)^2}{s^2 + 2s} > \frac{2(s+1)}{3s} = r_0.$$

Therefore, it is essential to introduce the step 2 in the algorithm A_0. We next show the worst case ratio of the algorithm A_0 below.

Theorem 4. *The worst case ratio of the algorithm A_0 is at most r_0, i.e.,*

$$R_{A_0} \leq \begin{cases} \frac{2(s+1)}{3s}, & \text{for } 1 \leq s < 4/3, \\ \frac{s+1}{2}, & \text{for } 4/3 \leq s < 2, \\ \frac{s+1}{s}, & \text{for } s \geq 2. \end{cases}$$

Proof. By the notations in the algorithm and (2), we can obtain the optimal preemptive makespan is

$$C^* = \max\{\frac{p_1}{s}, \frac{l_1 + l_2 + p_l}{s+1}\}. \tag{5}$$

For the case of $s \geq 2$, it is clear that the makespan of our algorithm is $\frac{l_1 + l_2 + p_l}{s}$, which, together with (5), leads to $\frac{C^{A_0}}{C^*} \leq \frac{s+1}{s}$. We next focus on the case of $1 \leq s < 2$.

Without loss of generality, we assume that $\mathcal{I}_1 \neq \emptyset$, i.e., $l_1 > 0$. In fact, if $\mathcal{I}_1 = \emptyset$, from the algorithm, we can obtain that the job p_1 must be scheduled on the machine M_1. It implies that $p_l = p_1$ and thus we have $C^{A_0} = \frac{p_1}{s} = C^*$. By distinguishing three cases, we can show

$$\frac{C^{A_0}}{C^*} \leq \max\{\frac{s+1}{2}, \frac{2(s+1)}{3s}\} = \begin{cases} \frac{2(s+1)}{3s}, & \text{for } 1 \leq s < 4/3, \\ \frac{s+1}{2}, & \text{for } 4/3 \leq s < 2. \end{cases}$$

Case 1. $l_2 + p_l < l_1$. As the algorithm states that the final schedule is generated by the LPT rule, we can obtain that by the step 1 $C^{A_0} \leq \min\{l_2 + p_l, \frac{l_1 + p_l}{s}\} \leq l_2 + p_l$. Hence, by (5), we obtain that

$$\frac{C^{A_0}}{C^*} \leq \frac{l_2 + p_l}{\frac{l_1 + l_2 + p_l}{s+1}} = \frac{s+1}{1 + \frac{l_1}{l_2 + p_l}} \leq \frac{s+1}{2}. \tag{6}$$

For the **Case 2** of $l_1 \le l_2 + p_l < 2l_1$ and the **Case 3** of $l_2 + p_l \ge 2l_1$, the desired results can be obtained by similar discussion as Case 1.

It is clear that $R_0 \le R_{A_0} \le r_0$ with Theorem 4. By Lemma 1, we can obtain that the main result in this subsection.

Theorem 5. $R_0 = r_0$.

3.2   Algorithm for 1-Preemptive Scheduling

Similar to the above argument, we will show $R_1 = r_1$ in this subsection. Clearly, by Lemma 3.1, we only need to present an algorithm which is allowed to preempt at most one time, denoted by A_1, and show its worst case ratio is not greater than r_1. The algorithm is quite simple and can formally be described as follows.

Algorithm A_1:

1. **Sort** all the jobs such that $p_1 \ge p_2 \ge \cdots \ge p_n$ and compute the optimal makespan $C^* = \max\{\frac{p_1}{s}, \frac{\sum_{i=1}^{n} p_i}{s+1}\}$ by (2).
2. **Find** the job p_k, where $k = \min\{j \mid \sum_{i=1}^{j} p_i > sr_1 C^*\}$, and split it into two pieces $p_k^1 = sr_1 C^* - \sum_{i=1}^{k-1} p_i$ and $p_k^2 = p_k - p_k^1$.
3. **Schedule** $p_1, p_2, \cdots, p_{k-1}, p_k^1$ and $p_k^2, p_{k+1}, \cdots, p_n$ in turn on the machines M_1 and M_2 from zero time, respectively.

Noting that if k is not existing, we have $\sum_{i=1}^{n} p_i \le sr_1 C^*$. Therefore, the desired result can be achieved by simply scheduling all the jobs on the fast machine M_1. Hence, we assume that k is well defined and we must have $k > 1$ due to $C^* \ge \frac{p_1}{s}$.

Theorem 6. *The worst case ratio of the algorithm A_1 is at most r_1, i.e.,*

$$R_{A_1} \le \frac{2s^2 + s - 1}{2s^2}.$$

Proof. It is obvious that $C^{A_1} = \frac{\sum_{i=1}^{k-1} p_i + p_k^1}{s} = r_1 C^*$ and thus we only need to show that the assignment of the job p_k is feasible, that is, the time slots assigned to p_k^1 and p_k^2 are not overlapping. Hence, it is sufficient to prove that

$$p_k^2 + \frac{p_k^1}{s} \le r_1 C^*. \tag{7}$$

Note that $p_k^1 = sr_1 C^* - \sum_{i=1}^{k-1} p_i$ and $p_k \le \frac{\sum_{i=1}^{n} p_i}{2}$ with $k > 1$. Then

$$p_k^2 + \frac{p_k^1}{s} - r_1 C^* = p_k - p_k^1 + \frac{p_k^1}{s} - r_1 C^*$$

$$= p_k + \frac{s-1}{s} \sum_{i=1}^{k-1} p_i - sr_1 C^* = \frac{p_k}{s} + \frac{s-1}{s} \sum_{i=1}^{k} p_i - sr_1 C^*$$

$$\leq \frac{\sum_{i=1}^{n} p_i}{2s} + \frac{s-1}{s} \sum_{i=1}^{n} p_i - sr_1 C^*$$

$$\leq \frac{2s-1}{2s} \sum_{i=1}^{n} p_i - sr_1 \frac{\sum_{i=1}^{n} p_i}{s+1}$$

$$= (\frac{2s-1}{2s} \frac{s+1}{s} - r_1) \frac{s}{s+1} \sum_{i=1}^{n} p_i = 0.$$

Hence, the inequality (7) holds and the proof is complete.

By Lemma 1 and Theorem 6, we conclude that the worst case ratio of the optimal 1-preemptive makespan and the optimal preemptive makespan is exactly r_1. That is,

Theorem 7. $R_1 = r_1$.

4   Conclusions

In this paper, we studied the worst case ratio of the optimal makespan of i-preemptive schedule and the optimal preemptive schedule with unlimited preemptions. Compared with the technique used in the literature, our method is simpler and more effective. It can obtain the ratio $R_i = \frac{2m}{m+i+1}$ as well as an algorithm which can generate a schedule with the makespan no more than R_i times of the optimal preemptive schedule for any instance. We also discussed the trade-off between the objective of minimizing the makespan and the number of preemptions. In addition, we considered the i-preemptive scheduling on two uniform machines.

Further study is suggested to extend the result to m uniform machines, even for the special case of $s_1 = s$, $s_2 = \cdots = s_m = 1$. What's more, it is also worth studying the other objective functions.

References

1. Braun, O., Schmidt, G.: Parallel processor scheduling with limited number of preemptions. SIAM Journal on Computing 32(3), 671–680 (2003)
2. Coffman Jr., E.G., Garey, M.R.: Proof of the 4/3 conjecture for preemptive vs. nonpreemptive two-processor scheduling. Journal of the Association for Computing Machinery 20, 991–1018 (1993)
3. Gonzalez, T., Sahni, S.: Preemptive scheduling of uniform processor systems. Journal of the Association for Computing Machinery 25, 92–101 (1978)
4. Graham, R.L.: Bounds on multiprocessing timing anomalies. SIAM Journal on Applied Mathematics 17, 416–429 (1969)
5. Horvath, E.C., Lam, S., Sethi, R.: A level algorithm for preemptive scheduling. Journal of the Association for Computing Machinery 24, 32–43 (1977)
6. Hong, K.S., Leung, J.Y.-T.: Some results on Lius conjecture. SIAM Journal on Discrete Mathematics 5, 500–523 (1992)

7. Liu, C.L.: Optimal scheduling on multi-processor computing systems. In: Proceedings of the 13th Annual Symposium on Switching and Automata Theory, pp. 155–160. IEEE Computer Society, Los Alamitos (1972)
8. McNaughton, R.: Scheduling with deadlines and loss functions. Management Science 6, 1–12 (1959)
9. Klonowska, K., Lundberg, L., Lennerstad, H.: The maximum gain of increasing the number of preemptions in multiprocessor scheduling. Acta Informatica 46, 285–295 (2009)
10. Liu, J.W.S., Yang, A.: Optimal scheduing of independent tasks on heterogeneous computing systems. In: Proceedings of ACM Annual Conference, San Diego, Cahf, pp. 38–45 (1974)

Computing Maximum Non-crossing Matching in Convex Bipartite Graphs[*]

Danny Z. Chen, Xiaomin Liu, and Haitao Wang[**]

Department of Computer Science and Engineering,
University of Notre Dame, Notre Dame, IN 46556, USA
{dchen,xliu9,hwang6}@nd.edu

Abstract. We consider computing a maximum non-crossing matching in convex bipartite graphs. For a convex bipartite graph of n vertices and m edges, we present an $O(n \log n)$ time algorithm for finding a maximum non-crossing matching in the graph. The previous best algorithm for this problem takes $O(m + n \log n)$ time. Since $m = \Theta(n^2)$ in the worst case, our result improves the previous solution for large m.

1 Introduction

Matching problems are an important topic in combinatorics and operations research. In this paper, we study the problem of computing a maximum non-crossing matching in convex bipartite graphs and present an efficient algorithm for it. Roughly speaking, a matching is *non-crossing* if no two edges in its given embedding intersect each other. The formal problem definition is given below.

1.1 Notation and Problem Statement

A graph $G = (V, E)$ with vertex set V and edge set E is a *bipartite graph* if V can be partitioned into two subsets A and B (i.e., $V = A \cup B$ and $A \cap B = \emptyset$) such that every edge $e(a, b) \in E$ connects a vertex $a \in A$ and a vertex $b \in B$ (it is often also denoted by $G = (A, B, E)$). A bipartite graph $G = (A, B, E)$ is said to be *convex* on the vertex set B if there is a linear ordering on B, say $B = \{b_1, b_2, \ldots, b_{|B|}\}$, such that for each vertex $a \in A$ and any two vertices b_i and b_j in B with $i < j$, if both b_i and b_j are connected to a by two edges in E, then every vertex $b_t \in B$ with $i \leq t \leq j$ is connected to a by an edge in E. If G is convex on B, then G is called a *convex bipartite graph*. Figure 1 shows an example. In this paper, A, B, and E always refer to these sets in a convex bipartite graph $G = (A, B, E)$, and we assume that the vertices in B are ordered as discussed above.

We say that an edge $e(a, b) \in E$ is an *incident edge* of a and b, and a and b are *adjacent* to each other. For each vertex $a_k \in A$, suppose the adjacent vertices of a_k are $b_i, b_{i+1}, \ldots, b_j$ (i.e., all vertices in B from b_i to b_j); then we denote $begin(a_k) = i$ and $end(a_k) = j$.

---

[*] This research was supported in part by NSF under Grant CCF-0916606.
[**] Corresponding author.

J. Snoeyink, P. Lu, K. Su, and L. Wang (Eds.): FAW-AAIM 2012, LNCS 7285, pp. 105–116, 2012.
© Springer-Verlag Berlin Heidelberg 2012

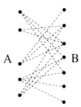

Fig. 1. An example of a convex bipartite graph

For simplicity, we assume $n = |A| = |B|$. Let $A = \{a_1, a_2, \ldots, a_n\}$. Let $m = |E|$. Note that although m may be $\Theta(n^2)$, the graph G can be represented *implicitly* in $O(n)$ time and $O(n)$ space by giving the two values $begin(a)$ and $end(a)$ for each vertex $a \in A$. A subset $M \subseteq E$ is a *matching* if no two distinct edges in M are connected to the same vertex. Two edges $e(a_i, b_j)$ and $e(a_h, b_l)$ in E are said to be *non-crossing* if either ($i < h$ and $j < l$) or ($i > h$ and $j > l$). A matching M is *non-crossing* if no two distinct edges in M intersect. A *maximum non-crossing matching* (MNCM for short) in G is a non-crossing matching M such that no other non-crossing matching in G has more edges than M.

1.2   Related Work

Finding a maximum matching in general graphs or bipartite graphs has been well studied [3,5,8,10,13,16]. Glover [9] considered computing a maximum matching in convex bipartite graphs with some industrial applications. Additional matching applications of convex bipartite graphs were given in [11]. A maximum matching in a convex bipartite graph can be obtained in $O(n)$ time [7,11,17]. Liang and Blum [14] gave a linear time algorithm for finding a maximum matching in *circular* convex bipartite graphs. Motivated by applications such as 3-side switch box routing in VLSI design, the problem of finding a maximum *non-crossing* matching (MNCM) in bipartite graphs was studied [12], which can be reduced to computing a longest increasing subsequence in a sequence of size m and thus is solvable in $O(m \log n)$ time [6,18]. An improved $O(m \log \log n)$ time algorithm was given by Malucelli *et al.* [15] for finding an MNCM in bipartite graphs; further, they showed that in a convex bipartite graph, an MNCM can be found in $O(m + (n - k) \log k)$ time where k is the size of the output MNCM [15], which is $O(m + n \log n)$ time in the worst case.

In this paper, we present a new algorithm for computing an MNCM in a convex bipartite graph in $O(n \log n)$ time. Since m can be $\Theta(n^2)$, our result improves the $O(m + n \log n)$ time solution by Malucelli *et al.* [15]. Our approach is based on the algorithm in [15]; the efficiency of our algorithm hinges on new observations on the problem as well as a data structure for efficiently processing certain frequent operations performed by the algorithm.

2   Preliminaries

In this section, we briefly review the algorithm by Malucelli *et al.* [15], called the *labeling algorithm* (for the full algorithmic and analysis details, see [15]). Our new algorithm given in Section 3 uses some ideas of this labeling algorithm.

For simplicity of discussion, we assume the vertices of each set of A and B are ordered on a vertical line in the plane from top to bottom by their indices and each edge in E is represented as a line segment connecting the two corresponding vertices. For any two non-crossing edges $e(a_i, b_j)$ and $e(a_h, b_l)$, we say $e(a_i, b_j)$ is *above* $e(a_h, b_l)$ if $i < h$ and $j < l$, and $e(a_i, b_j)$ is *below* $e(a_h, b_l)$ if $i > h$ and $j > l$.

The labeling algorithm [15] aims to compute a label $L(a, b)$ for each edge $e(a, b) \in E$, which is actually the cardinality of a "partial" MNCM if one considers only the edges of E above and including $e(a, b)$. After the labels for all edges of E are computed, an MNCM can be obtained in additional $O(m)$ time [15]. In order to compute the labels for all edges, the algorithm also computes a label $L(b)$ for each vertex $b \in B$, which is equal to the maximum label of all incident edges of b whose labels have been computed so far in the algorithm.

Initially, the label values for all edges of E and all vertices of B are zero. The algorithm considers the vertices in A one by one in their index order. For each vertex $a_i \in A$, there are two procedures for processing it. In the first procedure, for each incident edge $e(a_i, b_j)$ of a_i, the algorithm finds the vertex b_t with the maximum $L(b_t)$ such that $t < j$, and sets $L(a_i, b_j)$ as $L(b_t) + 1$, i.e., $L(a_i, b_j) = 1 + \max\{L(b_t) \mid t < j\}$. After the labels for all incident edges of a_i are computed, the second procedure does the following: For each incident edge $e(a_i, b_j)$ of a_i, set $L(b_j) = \max\{L(b_j), L(a_i, b_j)\}$. This finishes the processing of $a_i \in A$. The algorithm ends when all vertices in A have been processed. To analyze the running time, each edge in the second procedure takes constant time. The key is to implement the first procedure efficiently, i.e., computing the *key maximum value* $\max\{L(b_t) \mid t < j\}$ for each edge $e(a_i, b_j)$. As indicated in [15], by using a range query data structure [1], each key maximum value can be computed in $O(\log n)$ time. Consequently, the entire algorithm takes $O(m \log n)$ time. A further improvement is based on a new way of maintaining the label values for the vertices of B and using integer data structures (e.g., van Emde Boas tree [4]), as follows.

A map C is used in which each element $C(t)$ $(1 \le t \le n)$ stores the smallest index i such that $L(b_i) = t$, i.e., $L(b_{C(t)}) = t$ and there is no b_j with $j < C(t)$ and $L(b_j) = t$. For $1 \le i \le j \le n$, we use $C(i \ldots j)$ to denote the set of elements $C(i), C(i+1), \ldots, C(j)$. During the algorithm, the elements in $C(1 \ldots n)$ are computed in the index order. Let K be the number of elements in $C(1 \ldots n)$ that have been computed so far by the algorithm (K is also the index of the last element in $C(1 \ldots n)$ that has been computed). The value K is actually the cardinality of a partial MNCM that has been produced so far. Initially, $K = 0$, and for convenience let $C(0) = 0$. After the algorithm ends, the value K is the cardinality of the output MNCM (hence, $K \le n$). With the map C, in the first procedure for processing a vertex $a_i \in A$, for each edge $e(a_i, b_j)$, computing the value $\max\{L(b_t) \mid t < j\}$ becomes finding $\max\{k \mid C(k) < j, 0 \le k \le K\}$. Changes to the second procedure are also needed for updating the map C, as follows: If $L(a_i, b_j) \leq K$, then update $C(L(a_i, b_j))$ to be $\min\{j, C(L(a_i, b_j))\}$; otherwise, set $C(K+1) = j$ and $K = K+1$. For the running time, if the map C is implemented by an array, then each edge in the second procedure can still be processed in constant time. For the first procedure, a useful fact is that the values in $C(1 \ldots K)$ are monotonically increasing, and thus for each edge $e(a_i, b_j)$, the value $\max\{k \mid C(k) < j, 0 \leq k \leq K\}$ can be obtained in $O(\log K)$ time by a binary search on $C(1 \ldots K)$. If using an integer data structure [4] to implement the map C, each edge can be processed in $O(\log \log n)$ time since the universal values are $1, 2, \ldots, n$. Hence, the algorithm takes $O(m \log \log n)$ time.

When the graph is convex bipartite, there is a faster way to compute an MNCM. Here the map C is implemented by an array. Still, we focus on the first procedure for processing each vertex $a_i \in A$, which considers the incident edges of a_i in the inverse order of the indices of the vertices in B (i.e., the edge $e(a_i, b_{end(a_i)})$ is considered first). The algorithm first checks whether $end(a_i) > C(K)$. If $end(a_i) > C(K)$, then set $L(a_i, b_{end(a_i)}) = K+1$, $L(b_{end(a_i)}) = K+1$, $C(K+1) = end(a_i)$, and $K = K+1$; otherwise, the label $L(a_i, b_{end(a_i)})$ is computed by a binary search on C. After $L(a_i, b_{end(a_i)})$ is obtained, the labels of the other incident edges of a_i are computed by simultaneously scanning C from back to front and the incident edge list of a_i, and the labels of the vertices in B connecting to the above edges are updated accordingly. The array C is also updated accordingly. Note that the number of the scanned elements of C is no bigger than the number of incident edges of a_i. Thus, the above scanning procedure takes $O(m)$ time in the entire algorithm. For each vertex $a_i \in A$, we only perform at most one binary search on C. Therefore, the total running time of the algorithm is $O(m + n \log n)$.

3   Our Algorithm

In this section, we present our new algorithm for finding an MNCM in a convex bipartite graph $G = (A, B, E)$, whose running time is $O(n \log n)$. We follow the definitions and notation in Section 2.

Our algorithm still computes the labels for all edges of E and all vertices of B. However, a key difference is that our algorithm does so in an *implicit* manner, thus avoiding the $O(m)$ time overhead. In the previous labeling algorithm [15] sketched in Section 2, when a vertex $a_i \in A$ is being processed, the update of C is done after the labels of the incident edges (a_i, b_j) of a_i are explicitly computed. In contrast, our new algorithm updates C "directly" without explicitly computing the labels of the incident edges of a_i. This is possible due to our new observations. The efficiency of our algorithm also relies on a new way of implementing the map C and the operations performed by the algorithm.

We present our main algorithm in Section 3.1 but defer the implementation details and the time analysis to Section 3.2. However, this algorithm computes only the final map C without reporting an MNCM. In Section 3.3, we show that with the final map C and a little additional information, an MNCM can be produced in additional $O(n)$ time by a simple greedy approach.

3.1   The Description of the Main Algorithm

Our algorithm also considers the vertices in A from a_1 to a_n. For each $a_i \in A$, we compute the labels of all its incident edges and we also update the labels of all adjacent vertices of a_i. The labels of the edges and the vertices in B are defined the same as in Section 2. We also use the map C in the same way as before, i.e., each element $C(t)$ for $1 \le t \le n$ stores the smallest index i such that $L(b_i) = t$.

With the map C, for each vertex $a_i \in A$, the labels of a_i's incident edges will be computed implicitly and the labels of a_i's adjacent vertices in B will also be computed implicitly; in contrast, the map C will always be updated explicitly. The map C is implemented by a special balanced binary tree, which will be discussed in Section 3.2. Below, we give the algorithm details.

Again, as in Section 2, let K be the number of elements in $C(1 \ldots n)$ that have been computed so far by the algorithm and the value K is actually the cardinality of a partial MNCM that has been produced so far. Initially, we set $K = 0$, $C(0) = 0$, and $C(t) = +\infty$ for all $1 \le t \le n$. The labels of all edges and vertices in B are implicitly set to zero.

We discuss a general step where the vertex $a_i \in A$ is considered. Recall that its adjacent vertex set is $\{b_j \in B \mid begin(a_i) \le j \le end(a_i)\}$. Let x be the index with $0 \le x \le K + 1$ such that $C(x - 1) < begin(a_i) \le C(x)$, and y be the index with $0 \le y \le K + 1$ such that $C(y - 1) < end(a_i) \le C(y)$. Note that both x and y must exist. Also, note that $x \le y$ holds because $begin(a_i) \le end(a_i)$.

If $x > K$, then it is easy to see that x must be $K + 1$ since $C(K + 1) = +\infty$. By the definition of the index x, we have $C(K) < begin(a_i)$. Hence, the label of each edge $e(a_i, b_j)$ with $begin(a_i) \le j \le end(a_i)$ must be $K + 1$. Thus, the label of each vertex b_j with $begin(a_i) \le j \le end(a_i)$ must also be $K + 1$ (the previous values of these vertex labels must be less than $K + 1$ because the value of $C(K + 1)$ has not been set before, i.e., $C(K + 1) = +\infty$). It should be noted that we do not set these labels explicitly. According to the definition of the map C, the value of $C(K + 1)$ should be the smallest index t with $L(b_t) = K + 1$. In this case, such smallest index t is $begin(a_i)$. Thus, we simply set $C(K + 1) = begin(a_i)$ and $K = K + 1$. This finishes processing the vertex $a_i \in A$ for this case. Below, we discuss the case where $x \le K$.

Since $C(x - 1) < begin(a_i) \le C(x)$, by the definition of the map C, the label of the edge $e(a_i, b_{begin(a_i)})$ is $(x - 1) + 1 = x$. The label of the vertex $b_{begin(a_i)}$ should also be (implicitly) set to x since its previous value was at most x due to $begin(a_i) \le C(x)$. Thus, $C(x)$ should be updated to be $\min\{begin(a_i), C(x)\}$. Since $begin(a_i) \le C(x)$, we can simply set $C(x) = begin(a_i)$. If $x = y$, then the labels of all incident edges of a_i are x and the labels of all vertices in B adjacent to a_i are also x; thus we do not need to update other values in C and the processing of a_i is finished. Otherwise (i.e., $x \ne y$), we have $x < y$. In the following discussion, the value of $C(x)$ always refers to its previous value before we set $C(x) = begin(a_i)$. Since $x < y$, we have $C(x) < end(a_i)$ and in other words $C(x) + 1 \le end(a_i)$. Consider any edge $e(a_i, b_j)$ with $C(x) + 1 \le j \le \min\{end(a_i), C(x + 1)\}$. The label of the edge $e(a_i, b_j)$ must be $x + 1$. The label of the vertex b_j must also be $x + 1$ since its previous value must be at most $x + 1$

the map C

index	1	2	3	4	5	6	7
value	2	4	9	11	14	20	28

process a_i: begin(a_i)=8, end(a_i)=18, x=3, y=6

updated C

index	1	2	3	4	5	6	7
value	2	4	8	10	12	15	28

Fig. 2. Illustrating a range-shift operation in processing a vertex a_i with $begin(a_i) = 8$ and $end(a_i) = 18$. We have $x = 3$ and $y = 6$. The updated values of C are underlined.

due to $j \leq C(x + 1)$. Since $C(x) < C(x + 1)$ and both $C(x)$ and $C(x + 1)$ are integers, we have $C(x) + 1 \leq C(x + 1)$. Thus, the value of $C(x + 1)$ should be updated to $C(x) + 1$. If $y = x + 1$, we are done with updating C (after setting $C(x) = begin(a_i)$). If $y > x + 1$, we continue to consider the edges (a_i, b_j) with $C(x + 1) + 1 \leq j \leq \min\{end(a_i), C(x + 2)\}$ in a similar way (here, the value of $C(x + 1)$ is its previous value before we set $C(x + 1) = C(x) + 1$). Similarly, the labels of all these edges are $(x + 1) + 1 = x + 2$ and the labels of the vertices in B incident to these edges are also $x + 2$; thus the value of $C(x + 2)$ should be updated to $C(x + 1) + 1$ due to $C(x + 1) + 1 \leq C(x + 2)$.

In general, in the case of $x < y$, suppose C' is the new map after the update; then we have $C'(z) = C(z - 1) + 1$ for each $x + 1 \leq z \leq y$, and $C'(x) = begin(a_i)$. All other values of the map C do not need to be changed. Of course, in the implementation, it is not necessary to create a temporary map C' and we need to update C only from $C(y)$ back to $C(x + 1)$ and finally update $C(x)$, i.e., $C(y) = C(y - 1) + 1, C(y - 1) = C(y - 2) + 1, \ldots, C(x + 1) = C(x) + 1$, and $C(x) = begin(a_i)$. Intuitively, the values of C from $C(x)$ to $C(y - 1)$ are each increased by one and then *shift* to its right neighbor in the map C. We call the above process a *range-shift* operation on the index range from x to $y - 1$. Figure 2 gives such an example. Finally, if $y > K$, then y must be $K + 1$ and we update $K = K + 1$. This finishes the processing of the vertex $a_i \in A$.

3.2   The Algorithm Implementation and the Time Analysis

To implement our main algorithm efficiently, the key is to find a good way to store the map C in order to efficiently support the operations needed by the algorithm (e.g., the range-shift operation).

If using an array to store the map C, the algorithm runs in $O(m + n \log n)$ time. Specifically, for each $a_i \in A$, we can determine the two indices x and y by binary search in $O(\log n)$ time. A range-shift operation can be implemented in $O(y - x + 1)$ time. Since $y - x \leq end(a_i) - begin(a_i)$, the total time for performing all range-shift operations in the entire algorithm is bounded by $O(m)$. Thus, the overall running time of the algorithm is $O(m + n \log n)$. To speed up the algorithm, a clear target is to improve the range-shift operations. Below we present a new way to store the map C such that each range-shift operation is performed in $O(\log n)$ time (other operations needed in the algorithm can also be performed efficiently).

Note that this implies that the values of $C(x+1, x+2, \ldots, y)$ for each range-shift operation are set implicitly.

We use an augmented balanced binary search tree T (e.g., an AVL tree [2]) to store C. Each internal node of T has exactly two children except when T has at most one node. Each node v of T (a leaf or an internal node) stores a value, which we call the α *value* of v, denoted by $\alpha(v)$. For any two nodes v and u in T, the *path* from v to u always includes both v and u. Each leaf u is associated with a value that is the sum of the α values of all nodes in the path from u to the root of T, and we call it the β *value* of u, denoted by $\beta(u)$. Note that $\beta(u)$ is not explicitly stored in the leaf u. The β values of all leaves of T from left to right are monotonically increasing. For any node v of T, let $T(v)$ denote the subtree of T rooted at v. Each node v of T also explicitly stores two additional values $min(v)$ and $max(v)$, called the *min* and *max* values of v respectively, which are defined as follows. Denote by v_l the leftmost leaf of the subtree $T(v)$ and by v_r the rightmost leaf of $T(v)$. We define $min(v)$ as the sum of the α values of the nodes in the path of T from v_l to v, and $max(v)$ as the sum of the α values of the nodes in the path from v_r to v. In other words, if v is not the root, then $min(v)$ is equal to $\beta(v_l)$ minus the sum of the α values of the nodes in the path from the parent of v to the root, and similarly, $max(v)$ is equal to $\beta(v_r)$ minus the sum of the α values of the nodes in the path from the parent of v to the root. By the definitions of $min(v)$ and $max(v)$, if v is the root of T, $min(v) = \beta(v_l)$ and $max(v) = \beta(v_r)$; if v is a leaf, then $min(v) = max(v) = \alpha(v)$. As to be shown below, the *min* and *max* values are used to help search for a particular leaf. The α, *min*, and *max* values are also referred to as the *support values* of the nodes of T. The β values of the leaves of T are also referred to as the β values of T.

A key difference between the β values and the support values is that the support values are stored explicitly at the corresponding nodes while the β values are stored implicitly (they are inferred by the α values).

In our MNCM algorithm, we use T to store the map C. An element $C(i)$ of C is *valid* if $C(i)$ has already been computed, i.e., $1 \le i \le K$. Specifically, the map C is stored in T in a way that the β values of the leaves of T from left to right are exactly equal to the valid elements in C by the index order. Recall that the valid elements in C are monotone increasing by the index order. The details of doing so will be given later after we discuss some operations on T. Since all valid elements in C are distinct, in the following discussion, for ease of exposition, we also assume that the β values of T are distinct although the more general case can also be handled.

Suppose n is the number of leaves in T. Since T is balanced, the height of T is $O(\log n)$. Later, we will show that the following four operations on T can each be performed in $O(\log n)$ time, which are used by our MNCM algorithm. The first operation is the *search* operation: Given a value β', find the leaf whose β value is β'; if there is no such leaf, then report the *successor* leaf of β', i.e., the leaf whose β value is the smallest among those leaves whose β values are larger than β' (if β' is larger than all β values of T, then report "NONE"). The second operation is *insertion*: Given a value β', insert a new leaf to T such that its β

value is equal to β'. Here we assume the β values of other leaves of T are not equal to β'. The third operation is *deletion*: Given a value β' that is equal to the β value of a leaf in T, delete that leaf from T. The fourth operation is the *range-update* operation: Given a value β' and two other values β_1 and β_2 that are the β values of two leaves v_1 and v_2 of T with $\beta_1 \leq \beta_2$, update the tree T such that the β values of all the leaves of T between the two leaves v_1 and v_2 (including v_1 and v_2) are each increased by β'. Below, we describe the details of implementing these four operations, each in $O(\log n)$ time.

We begin with the search operation, which can be done similarly as searching in a binary search tree by making use of the support values of T.

Lemma 1. *Each search operation on T can be performed in $O(\log n)$ time.*

Proof. Given a value β', we first check whether β' is larger than the biggest β value of T. This can be done in $O(\log n)$ time by finding the rightmost leaf of T. Below, we focus on the case when β' is smaller than the biggest β value of T.

Denote by R the root of T. Initially, we are at R. Let v be the left child of R. By the definition of the *max* values, $max(v)$ is equal to the value $\beta(v_r)$ minus the value $\alpha(R)$, where v_r is the rightmost leaf of the subtree $T(v)$. In other words, the sum of $\alpha(R)$ and $max(v)$ is equal to $\beta(v_r)$. We compare the value β' with the sum of $\alpha(R)$ and $max(v)$. If β' is larger, then the sought leaf must be in the right subtree of R and we proceed to the right child of R; otherwise, we proceed to the left child of R. We continue this procedure recursively. Consider a general step when we are at a node u of T. If u is a leaf, then we stop and u is the leaf we seek. Otherwise, let $\sum$ be the sum of the α values of the nodes in the path from u to R. The value $\sum$ can be computed during our recursive procedure of moving from R down to the current node u. Let w be the left child of u. By the definition of the *max* values, $\beta(w_r)$ is the sum of $max(w)$ and $\sum$, where w_r is the rightmost leaf in the subtree $T(w)$. We compare the value β' with the sum of $max(w)$ and $\sum$. If β' is larger, then the sought leaf must be in the right subtree of u and we proceed to the right child of u (we also increase the value of $\sum$ by the α value of the right child of u); otherwise, we proceed to w (we also increase the value of $\sum$ by $\alpha(w)$). Since we only use constant time at each node, the sought leaf is found in $O(\log n)$ time. The lemma thus follows.

We next discuss the range-update operation.

Lemma 2. *Each range-update operation on T takes $O(\log n)$ time.*

Proof. Given β' and two other values β_1 and β_2 ($\beta_1 \leq \beta_2$), we first find the two leaves l and r in T with $\beta(l) = \beta_1$ and $\beta(r) = \beta_2$ in $O(\log n)$ time using the search operations in Lemma 1. Let $L_{[l,r]}$ denote the set of leaves of T from l to r (including l and r). From the two leaves l and r, we can find a set $N_{[l,r]}$ of $O(\log n)$ nodes of T such that the set of leaves in the subtrees rooted at the nodes of $N_{[l,r]}$ is exactly $L_{[l,r]}$ and each leaf of $L_{[l,r]}$ belongs to exactly one such subtree. The set $N_{[l,r]}$ can be obtained easily in $O(\log n)$ time by traversing the two paths from l and r to their lowest common ancestor in T. For each node

$v \in N_{[l,r]}$, we increase $\alpha(v)$ by β'. In this way, the β values of the leaves in $L_{[l,r]}$ are each (implicitly) increased by β'. However, we are not done yet. Due to our modifications on the α values of all nodes in $N_{[l,r]}$, we may also need to update the min and max values of some nodes, as discussed below.

By the definitions of the min and max values, these values at each node in $N_{[l,r]}$ need to be increased by β'. Further, proper ancestors of any node in $N_{[l,r]}$ may also need to have their min and max values increased. There are only $O(\log n)$ such ancestor nodes, since any such ancestor node v must be on either the path from l to the root or the path from r to the root. Let w be the lowest common ancestor of l and r. Each node $u \neq w$ in the two paths from l to w and from r to w has either a left child in $N_{[l,r]}$ or a right child in $N_{[l,r]}$; if left, we increase $min(u)$ by β', and if right we increase $max(u)$ by β'. For the path from w to the root, at each node u, if the min of the left child was increased, then we increase $min(u)$ by β', and if the max of the right child increased, then we increase $max(u)$ by β'. It is easy to see that all these min and max values are updated correctly by the above procedure, which takes $O(\log n)$ time.

The support values of all other nodes in T need not be updated. This completes the range-update operation on β', β_1, and β_2. The overall running time is $O(\log n)$. The lemma thus follows.

Now, we discuss the insertion operation, which is performed similarly as that in ordinary AVL trees plus we also need to update the support values accordingly.

Lemma 3. *Each insertion operation on T can be performed in $O(\log n)$ time.*

Proof. Due to the space limit, we only give a sketch. Given a value β', our goal is to insert a new leaf in T whose β value is β'. We assume that β' is smaller than the biggest β value of T (otherwise, the algorithm is very similar and simpler).

We first find the successor leaf u of β' in $O(\log n)$ time by a search operation in Lemma 1. At the position of u, we create a new internal node i, and let the parent of i be the parent of u and u be the right child of i. We create a new leaf a and let a be the left child of i. Next, we need to update the support values of T, after which the new leaf a should have β' as its β value.

First, set $\alpha(i) = 0$. Consider the new leaf a. Set $\alpha(a) = \beta' - \sum$, where $\sum$ is the sum of the α values of all nodes in the path from i to the root, and set $min(a) = max(a) = \alpha(a)$. For the node i, since $\alpha(i) = 0$, set $min(i) = \alpha(a)$ and $max(i) = \alpha(u)$. Further, we may also need to update the min values of some ancestors of a. Specifically, for any proper ancestor v of a, if a is the leftmost leaf in the subtree $T(v)$, then $min(v)$ needs to be set as the sum of the α values of all nodes in the path from a to v. If we check all a's ancestors in a bottom-up fashion, updating their min values can be easily done in totally $O(\log n)$ time.

At this moment, all support values of T after the new leaf a is inserted have been set correctly and $\beta(a) = \beta'$. However, due to the insertion of a, T may become unbalanced, in which case we need to perform rotations as in a normal AVL tree. After the rotation, we also need to update accordingly the support values of some nodes of T, whose total number is at most $O(\log n)$. Updating these support values takes totally $O(\log n)$ time. We omit the details.

The deletion operation can also be implemented in $O(\log n)$ time, which is similar to the deletion operation in the normal AVL trees except that some support values may need to be updated. This can be done similarly as the insertion operation and we omit the details.

Lemma 4. *Each deletion operation on T can be performed in $O(\log n)$ time.*

Below we show that our MNCM algorithm in Section 3.1 can be implemented in $O(n \log n)$ time by using T to store the map C.

The map C is stored in T such that the β values of the leaves of T from left to right are exactly equal to the valid elements in C by the index order. Initially, in the beginning of the algorithm, no element in C is valid, and thus $T = \emptyset$.

We consider the general step described in Section 3.1, i.e., the vertex $v_i \in A$ is processed. We assume that right before this step, all valid elements in the current map C have been stored in T in the way described above. First, we need to find the two indices x and y as well as the two values $C(x)$ and $C(y)$, which can be found in $O(\log n)$ time by a search operation on T, as follows. To find x, we perform a search operation on T with the value $\beta' = begin(a_i)$. If the operation returns "NONE", then we know that $begin(a_i)$ is larger than all β values of T, or equivalently, $begin(a_i)$ is larger than all valid elements of the current map C, implying that $x > K$. Otherwise, the search operation will return the leaf whose β value is equal to $C(x)$. However, this only gives us the value $C(x)$ and we also need to know the value x. An easy observation is that $x - 1$ is equal to the number of valid elements in C less than $C(x)$, or equivalently, $x - 1$ is exactly the number of leaves of T to the left of the leaf corresponding to $C(x)$. To find the number of leaves of T to the left of that leaf, we may need to modify the tree T such that each internal node v is also associated with the number of leaves in the subtree $T(v)$. We omit the details of this part. Hence, the value x can also be found in $O(\log n)$ time. Analogously, the values $C(y)$ and y can also be computed in $O(\log n)$ time. Below, we discuss other operations needed in our algorithm described in Section 3.1.

The operation of setting $C(K + 1) = begin(a_i)$ can be implemented by an insertion operation with the value $\beta' = begin(a_i)$. The operation of setting $C(x) = begin(a_i)$ for an $x \leq K$ can be implemented in $O(\log n)$ time by performing first a deletion operation with the value $\beta' = C(x)$ and then an insertion operation with the value $\beta' = begin(a_i)$. When $x < y$, the range-shift operation from x to $y - 1$ can be carried out in $O(\log n)$ time by a range-update operation with the values $\beta' = 1$, $\beta_1 = C(x)$, and $\beta_2 = C(y - 1)$. We also need to perform a deletion operation on the value $C(y)$ to delete the leaf with the β value $C(y)$. The subsequent operation of setting $C(x) = begin(a_i)$ can be implemented by an insertion operation with the value $begin(a_i)$ (note that no deletion is needed here because the original leaf with the β value $C(x)$ has been "shifted" to a new leaf with the β value $C(x) + 1$). In summary, processing each $a_i \in A$ takes $O(\log n)$ time. Thus, the final map C can be obtained in $O(n \log n)$ time.

We have not computed an actual matching yet. In Section 3.3, we show how to produce an actual MNCM.

3.3 Reporting an MNCM

In this section, we find an actual MNCM in additional $O(n)$ time based on the map C that has been computed by the main MNCM algorithm. After the main MNCM algorithm is over, the index of the last valid element of C (i.e., the value K) is the cardinality of an MNCM in G. To obtain an actual MNCM, we need to record some additional information in the main algorithm, as follows.

Suppose the main algorithm is processing a vertex $a_i \in A$. If the value K is increased (i.e., K is set to be $K + 1$) due to the processing of a_i, then this implies that there is a non-crossing matching of cardinality K (the value K here has already been set to $K + 1$) containing an incident edge of a_i. We associate this vertex a_i with K. For the implementation, we use a variable z to record it, i.e., set $z = a_i$ if the value K is increased due to the processing of a_i. At the end of the algorithm, z gives a vertex $a_i \in A$ such that an MNCM contains an incident edge of a_i but the MNCM does not contain any incident edge of $a_t \in A$ for any $t > i$. In Lemma 5 below, we show that knowing this particular vertex a_i is sufficient for us to find an MNCM in $O(n)$ time by a simple greedy algorithm.

Lemma 5. *Suppose an MNCM contains an incident edge of a_i but the MNCM does not contain any incident edge of $a_t \in A$ for any $t > i$. Then, an actual MNCM can be found in $O(n)$ time.*

Proof. Let $\mathcal{M}$ be an MNCM that contains an incident edge of a_i. Suppose $e(a_i, b_j)$ is the incident edge of a_i in $\mathcal{M}$. Let K be the cardinality of $\mathcal{M}$.

Since $\mathcal{M}$ does not contain any incident edge of $a_t \in A$ for any $t > i$, the other edges of $\mathcal{M}$ must be all above $e(a_i, b_j)$. We claim that there is an MNCM (denoted by $\mathcal{M}'$) that contains the edge $e(a_i, b_{end(a_i)})$. Indeed, if $j = end(a_i)$, then $\mathcal{M}' = \mathcal{M}$ and the claim holds. Otherwise, since $b_{end(a_i)}$ is below b_j (i.e., $j < end(a_i)$), by replacing the edge $e(a_i, b_j)$ in $\mathcal{M}$ with the edge $e(a_i, b_{end(a_i)})$, we obtain another feasible non-crossing matching $\mathcal{M}'$ whose cardinality is also K. Hence, $\mathcal{M}'$ is an MNCM and the claim holds. We pick the edge $e(a_i, b_{end(a_i)})$ as an edge in our sought MNCM.

Next, we consider the vertex a_{i-1}. If $begin(a_{i-1}) \geq end(a_i)$, then no incident edge of a_{i-1} can be in $\mathcal{M}'$ since every such edge intersects $e(a_i, b_{end(a_i)}) \in \mathcal{M}'$; thus, we continue to consider the vertex a_{i-2}. Otherwise (i.e., $begin(a_{i-1}) < end(a_i)$), there is an incident edge of a_{i-1} above the edge $e(a_i, b_{end(a_i)})$; we claim that there is an MNCM $\mathcal{M}''$ that contains both edges $e(a_i, b_{end(a_i)})$ and $e(a_{i-1}, b_t)$, where $t = \min\{end(a_i) - 1, end(a_{i-1})\}$. The argument is similar to the one above. Namely, if $e(a_{i-1}, b_t)$ is not contained in $\mathcal{M}'$, we can always find another MNCM $\mathcal{M}''$ by replacing an edge in $\mathcal{M}'$ by $e(a_{i-1}, b_t)$. We omit the details. We pick the edge $e(a_{i-1}, b_t)$ as an edge in our sought MNCM.

We then continue to consider a_{i-2} and the procedure is similar. After a_1 is considered, we obtain an MNCM. The total running time of the above greedy algorithm is $O(n)$. The lemma thus follows.

References

1. de Berg, M., Cheong, O., van Kreveld, M., Overmars, M.: Computational Geometry — Algorithms and Applications, 3rd edn. Springer, Berlin (2008)
2. Cormen, T., Leiserson, C., Rivest, R., Stein, C.: Introduction to Algorithms, 2nd edn. MIT Press (2001)
3. Edmonds, J.: Paths, trees and flowers. Canad J. Math. 17, 449–467 (1965)
4. van Emde Boas, P., Kaas, R., Zijlstra, E.: Design and implementation of an efficient priority queue. Theory of Computing Systems 10(1), 99–127 (1977)
5. Even, S., Tarjan, R.: Network flow and testing graph connectivity. SIAM J. Comput. 4, 507–518 (1975)
6. Fredman, M.: On computing the length of longest increasing subsequences. Discrete Mathematics 11(1), 29–35 (1975)
7. Gabow, H., Tarjan, R.: A linear-time algorithm for a special case of disjoint set union. Journal of Computer and System Sciences 30, 209–221 (1985)
8. Gabow, H.: An efficient implementation of Edmonds' algorithm for maximum matching on graphs. Journal of the ACM 23, 221–234 (1976)
9. Glover, F.: Maximum matching in a convex bipartite graph. Naval Res. Logist. Quart. 14, 313–316 (1967)
10. Hopcroft, J., Karp, R.: An $n^{5/2}$ algorithm for maximum matchings in bipartite graphs. SIAM J. on Comput. 2(4), 225–231 (1973)
11. Lipski Jr., W., Preparata, F.P.: Efficient algorithms for finding maximum matchings in convex bipartite graphs and related problems. Acta Informatica 15(4), 329–346 (1981)
12. Kajitami, Y., Takahashi, T.: The noncross matching and applications to the 3-side switch box routing in VLSI layout design. In: Proc. International Symposium on Circuits and Systems, pp. 776–779 (1986)
13. Kuhn, H.: The Hungarian method for the assignment problem. Naval Research Logistics Quarterly 2, 83–97 (1955)
14. Liang, Y., Blum, N.: Circular convex bipartite graphs: Maximum matching and Hamiltonian circuits. Information Processing Letters 56, 215–219 (1995)
15. Malucelli, F., Ottmann, T., Pretolani, D.: Efficient labelling algorithms for the maximum noncrossing matching problem. Discrete Applied Mathematics 47(2), 175–179 (1993)
16. Micali, S., Vazirani, V.: An $O(\sqrt{|V|}|E|)$ algorithm for finding maximum matching in general graphs. In: Proc. of the 21st Annual Symposium on Foundations of Computer Science, pp. 17–27 (1980)
17. Steiner, G., Yeomans, J.: A linear time algorithm for maximum matchings in convex, bipartite graphs. Computers and Mathematics with Applications 31(2), 91–96 (1996)
18. Widmayer, P., Wong, C.: An optimal algorithm for the maximum alignment of terminals. Information Processing Letters 10, 75–82 (1985)

Algorithms for Bandwidth Consecutive Multicolorings of Graphs
(Extended Abstract)

Kazuhide Nishikawa[1], Takao Nishizeki[1], and Xiao Zhou[2]

[1] School of Science and Engineering, Kwansei Gakuin University, Gakuen 2–1,
Sanda-shi, Hyogo, Japan
{nishikawa,nishi}@kwansei.ac.jp
[2] Graduate School of Information Sciences, Tohoku University,
Sendai-shi, Miyagi, Japan
zhou@ecei.tohoku.ac.jp

Abstract. Let G be a simple graph in which each vertex v has a positive integer weight $b(v)$ and each edge (v, w) has a nonnegative integer weight $b(v, w)$. A bandwidth consecutive multicoloring of G assigns each vertex v a specified number $b(v)$ of consecutive positive integers so that, for each edge (v, w), all integers assigned to vertex v differ from all integers assigned to vertex w by more than $b(v, w)$. The maximum integer assigned to a vertex is called the span of the coloring. In the paper, we first investigate fundamental properties of such a coloring. We then obtain a pseudo polynomial-time exact algorithm and a fully polynomial-time approximation scheme for the problem of finding such a coloring of a given series-parallel graph with the minimum span. We finally extend the results to the case where a given graph G is a partial k-tree, that is, G has a bounded tree-width.

Keywords: Bandwidth coloring, Channel assignment, Multicoloring, Series-parallel graph, Partial k-tree, Algorithm, Acyclic orientation, Approximation, FPTAS.

1 Introduction

An *ordinary coloring* of a graph G assigns each vertex a color so that, for each edge (v, w), the color assigned to v differ from the color assigned to w [7]. The problem of finding a coloring of a graph G with the minimum number $\chi(G)$ of colors often appears in the scheduling, task-allocation, etc. [7]. However, it is NP-hard, and difficult to find a good approximate solution. More precisely, for all $\varepsilon > 0$, approximating $\chi(G)$ within $n^{1-\varepsilon}$ is NP-hard [15], where n is the number of vertices in G.

The ordinary coloring has been extended in various ways [3–7, 9, 13]. A *multicoloring* assigns each vertex a specified number of colors so that, for each edge (v, w), the set of colors assigned to v is disjoint with the set of colors assigned to w [3–5, 14]. A *bandwidth coloring* assigns each vertex a positive integer as a

J. Snoeyink, P. Lu, K. Su, and L. Wang (Eds.): FAW-AAIM 2012, LNCS 7285, pp. 117–128, 2012.
© Springer-Verlag Berlin Heidelberg 2012

color so that the two integers assigned to the ends of each edge (v, w) differ by at least the specified weight $\omega(v, w)$ of (v, w) [9].

In this paper we deal with another generalized coloring, called a "bandwidth consecutive multicoloring." Let $G = (V, E)$ be a simple graph with vertex set V and edge set E. Each vertex $v \in V$ has a positive integer weight $b(v)$, while each edge $(v, w) \in E$ has a *non-negative* integer weight $b(v, w)$. A *bandwidth consecutive multicoloring* F of G is an assignment of positive integers to vertices such that

(a) each vertex $v \in V$ is assigned a set $F(v)$ of $b(v)$ consecutive positive integers; and

(b) for each edge $(v, w) \in E$, all integers assigned to v differ from all integers assigned to vertex w by *more than $b(v, w)$*.

We call such a bandwidth consecutive multicoloring F simply a *b-coloring* of G for a weight function b. The maximum integer assigned to a vetex is called the *span* of a b-coloring F, and is denoted by $span(F)$. We define the *b-chromatic number* $\chi_b(G)$ of a graph G to be the minimum span over all b-colorings F of G. A b-coloring F is called *optimal* if $span(F) = \chi_b(G)$. A *b-coloring problem* is to compute $\chi_b(G)$ for a given graph G.

Figure 1(a) depicts a weighted graph G together with an optimal b-coloring F of G, where a weight $b(e)$ is attached to an edge e, a weight $b(v)$ is written in a circle representing a vertex v, and a set $F(v)$ is attached to a vertex v. Since $span(F) = 11$, $\chi_b(G) = 11$.

The ordinary vertex-coloring is merely a b-coloring for the case $b(v) = 1$ for every vertex v and $b(v, w) = 0$ for every edge (v, w). The "bandwidth coloring" or "channel assignment" [9] is a b-coloring for the case $b(v) = 1$ for every vertex v and $b(v, w) = \omega(v, w) - 1$ for every edge (v, w). It should be noted that our edge weight $b(v, w)$ is one less than the ordinary edge weight $\omega(v, w)$ of a bandwidth coloring. (This convention will make the arguments and algorithms simple and transparent.)

A b-coloring arises in the assignment of radio channels in celluar communication systems [9] and in the non-preemptive task scheduling [10]. The $b(v)$ consecutive integers assigned to a vertex v correspond to the contiguous bandwidth of a channel v or the consecutive time periods of a task v. The weight $b(v, w)$ assigned to edge (v, w) represents the requirement that the frequency band or time period of v must differ from that of w by more than $b(v, w)$. The span of a b-coloring corresponds to the minimum total bandwidth or the minimum makespan.

One can find a multicoloring of a graph G with the minimum number of colors in time polynomial in the output size if G is a series-parallel graph or a partial k-tree, that is, a graph of bounded tree-width [5, 14]. The problem of finding a bandwidth coloring with the minimum number of colors is NP-hard even for partial 3-trees [9], and there is a fully polynomial-time approximation scheme (FPTAS) for the problem on partial k-trees [9]. Since our b-coloring problem is also NP-hard for partial 3-trees, it is desirable to obtain a good approximation algorithm. However, there are only heuristics for the b-coloring problem so far [8].

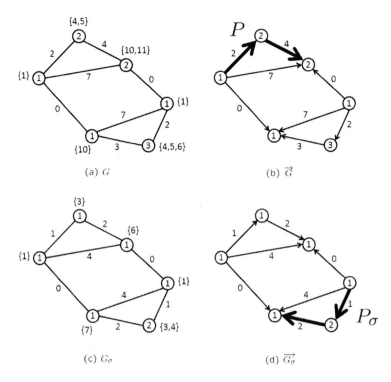

Fig. 1. (a) Series-parallel weighted graph G and its optimal b-coloring F, (b) acyclic orientation $\vec{G}$ and the longest path P, (c) graph G_σ with weights scaled by $\sigma = 2$ and its optimal b_σ-coloring F_σ, and (d) acyclic orientation $\vec{G_\sigma}$ and the longest path P_σ

In this paper, we first investigate fundamental properties of a b-coloring. In particular, we characterize the b-chromatic number $\chi_b(G)$ of a graph G in terms of the longest path in acyclic orientations of G. We then obtain a pseudo polynomial-time exact algorithm for the b-coloring problem on series-parallel graphs, which often appear in the task scheduling and electrical circuts [10, 11]. The algorithm takes time $O(B^3 n)$, where B is the maximum weight of G: $B = \max\limits_{x \in V \cup E} b(x)$. Using the algorithm, we then give a fully polynomial-time approximation scheme (FPTAS) for the problem. We finally extend these results to the case where G is a partial k-tree.

2   Preliminaries

In this section, we first give some definitions and then present three lemmas on a b-coloring.

Let $G = (V, E)$ be a simple graph without selfloops and multiple edges. We denote by n and m the number of vertices and edge in G, respectively.

The *chromatic number* $\chi(G)$ of G is the minimum number of colors required by an ordinary coloring of G.

Let $\mathbb{N}$ be the set of all positive integers, that are regarded as colors. A *b-coloring* $F : V \to 2^{\mathbb{N}}$ of G must satisfy the following (a) and (b):

(a) for every vertex $v \in V$, the set $F(v)$ consists of $b(v)$ consecutive positive integers, and hence

$$\min F(v) = \max F(v) - b(v) + 1; \text{ and}$$

(b) for every edge $(v, w) \in E$, all integers in $F(v)$ differ from those in $F(w)$ by more than $b(v, w)$.

A *b*-coloring F can be respresented by a function $f : V \to \mathbb{N}$ such that $f(v) = \max F(v)$ for every vertex $v \in V$. Clearly, for every vertex $v \in V$,

$$b(v) \leq f(v). \tag{1}$$

For every edge $(v, w) \in E$,

$$f(v) \neq f(w) \tag{2}$$

since $b(v, w) \geq 0$. For every edge $(v, w) \in E$ with $f(v) < f(w)$,

$$b(v, w) < (f(w) - b(w) + 1) - f(v).$$

and hence

$$f(v) + b(v, w) + b(w) \leq f(w). \tag{3}$$

Conversely, every function $f : V \to \mathbb{N}$ satisfying Eqs. (1), (2) and (3) represents a *b*-coloring F such that

$$F(v) = \{f(v) - b(v) + 1, f(v) - b(v) + 2, \cdots, f(v)\}.$$

Thus, such a function f is also called a *b-coloring* of G. Obviously, $span(F) = \max_{v \in V} f(v)$. We often denote $span(F)$ by $span(f)$. A *b*-coloring f is called *optimal* if $span(f) = \chi_b(G)$. The *b-coloring problem* is to compute $\chi_b(G)$ for a given graph G with weight $b(x)$ for each element $x \in V \cup E$.

The graph in Fig. 1(a) has the maximum weight $B = 7$. One can easily have the following lemmas.

Lemma 1. *For every weighted graph G, $B \leq \chi_b(G) \leq B(2\chi(G) - 1)$.*

Proof. Obviously $B \leq \chi_b(G)$. There is an ordinary coloring of G which uses a number $\chi(G)$ of colors c_i, $1 \leq i \leq \chi(G)$. Let $f : V \to \mathbb{N}$ be a function such that $f(v) = 2(i-1)B + b(v)$ if v is colored by c_i. Then f satisfies Eqs. (1), (2) and (3), and hence f is a *b*-coloring of G. Therefore, $\chi_b(G) \leq span(f) \leq B(2\chi(G) - 1)$.
$\square$

Lemma 2. *Let $G = (V, E)$ be a bipartite graph in which every vertex has degree one or more, and let*

$$\bar{B} = \max\{b(v) + b(v, w) + b(w) \mid (v, w) \in E\}. \tag{4}$$

Then $\chi_b(G) = \bar{B}$.

Lemma 2 implies that the b-coloring problem can be solved in linear time for bipartite graphs and hence for trees.

We then characterize $\chi_b(G)$ in terms of the longest path in acyclic orientations of G. Orient all edges of G so that the resulting directed graph $\overrightarrow{G}$ is acyclic. The directed graph is called an *acyclic orientation* of G. Figure 1(b) depicts an acyclic orientation of the graph G in Fig. 1(a).

The length $\ell(P, D)$ of a directed path P in an acyclic graph D is the sum of the weights of all vertices and edges in P. We denote by $\ell_{\max}(D)$ the length of the longest directed path in D. For the acyclic graph $\overrightarrow{G}$ in Fig. 1(b) $\ell_{\max}(\overrightarrow{G}) = 11$, and the longest directed path P in $\overrightarrow{G}$ is drawn by thick lines.

Extending the Gallai-Roy theorem on the ordinary coloring (see for example [12]), we have the following lemma on the b-coloring.

Lemma 3. *For every graph G with weight function b*

$$\chi_b(G) = \min_{\overrightarrow{G}} \ell_{\max}(\overrightarrow{G}),$$

where the minimum is taken over all acyclic orientations $\overrightarrow{G}$ of G.

Proof. We first prove that $\chi_b(G) \geq \min_{\overrightarrow{G}} \ell_{\max}(\overrightarrow{G})$. Let f be an optimal b-coloring of G. Then $span(f) = \chi_b(G)$. Orient each edge $(v, w) \in E$ from v to w if and only if $f(v) < f(w)$. Then clearly the resulting directed graph D is acyclic. Let $P = v_1, e_1, v_2, e_2, \cdots, v_{p-1}, e_{p-1}, v_p$ be the longest directed path in D, where edge e_i, $1 \leq i \leq p - 1$, goes from vertex v_i to v_{i+1}. Then

$$\ell_{\max}(D) = \sum_{i=1}^{p} b(v_i) + \sum_{i=1}^{p-1} b(v_i, v_{i+1}).$$

Since f is a b-coloring of G, by Eqs. (1) and (3) we have

$$b(v_1) \leq f(v_1) \tag{5}$$

and

$$b(v_i, v_{i+1}) + b(v_{i+1}) \leq f(v_{i+1}) - f(v_i) \tag{6}$$

for every i, $1 \leq i \leq p - 1$. Taking the sum of Eq. (5) and Eq. (6) for all i, $1 \leq i \leq p - 1$, we have

$$\ell_{\max}(D) \leq f(v_p) \leq span(f) = \chi_b(G).$$

Since $\min_{\overrightarrow{G}} \ell_{\max}(\overrightarrow{G}) \leq \ell_{\max}(D)$, we have $\chi_b(G) \geq \min_{\overrightarrow{G}} \ell_{\max}(\overrightarrow{G})$.

We then prove that $\chi_b(G) \leq \min_{\overrightarrow{G}} \ell_{\max}(\overrightarrow{G})$. Let D be an acyclic orientation of G such that

$$\ell_{\max}(D) = \min_{\overrightarrow{G}} \ell_{\max}(\overrightarrow{G}).$$

Let $f : V \to \mathbb{N}$ be a mapping such that $f(v)$ is the length of the longest directed path in D ending at v for each vertex v of D. Then, for every directed edge (v, w) of D, $f(v) + b(v, w) + b(w) \leq f(w)$ and hence $f(v) \neq f(w)$. The definition of f implies that $b(v) \leq f(v)$ for every vertex $v \in V$. Thus f is a b-coloring of G and $span(f) = \ell_{\max}(D)$. Hence $\chi_b(G) \leq span(f) = \min_{\overrightarrow{G}} \ell_{\max}(\overrightarrow{G})$.   □

3   Exact Algorithm for Series-Parallel Graphs

Many problems can be solved for series-parallel graphs in polynomial time or mostly in linear time [11]. In this section we show that the b-coloring problem can be solved for series-parallel graphs in pseudo polynomial-time $O(B^3 n)$. It should be noted that $B^3 n$ is polynomial in n and B.

A *series-parallel graph* is recursively defined as follows [11]:

1. A graph G of a single edge is a series-parallel graph, and has the ends of the edge as terminals s and t of G. (See Fig. 2(a).)
2. Let G_1 be a series-parallel graph with terminals s_1 and t_1, and let G_2 be a series-parallel graph with terminals s_2 and t_2. (See Fig. 2(b).)
   (a) A graph G obtained from G_1 and G_2 by identifying t_1 with s_2 is a series-parallel graph, whose terminals are s_1 and t_2. Such a connection is called a *series connection*. (See Fig. 2(c).)
   (b) A graph obtained from G_1 and G_2 by identifying s_1 with s_2 and identifying t_1 with t_2 is a series-parallel graph, whose terminals are $s_1 = s_2$ and $t_1 = t_2$. Such a connection is called a *parallel connection*. (See Fig. 2(d).)

Every series-parallel graph G can be represented by a "*binary decomposition tree*." Figure 3 illustrates a decomposition tree T of the series-parallel graph G in Fig. 1(a). Labels **s** and **p** attached to internal nodes in T indicate series and parallel connections, respectively. Every leaf of T represents a subgraph of G induced by a single edge. A node u of T corresponds to a subgraph G_u of G induced by all edges represented by the leaves that are descendants of u in T. Thus $G = G_r$ for the root r of T. One can find a decomposition tree of a given series-parallel graph in linear time [11].

The definition immediately implies that every series-parallel graph G has an ordinary coloring with at most three colors, that is, $\chi(G) \leq 3$. Therefore, by Lemma 1, we have $\chi_b(G) \leq 5B$. For a series-parallel graph G with terminals s and t and integers i and j, $1 \leq i, j \leq 5B$, we define

$$\chi_{ij}(G) = \min_f span(f)$$

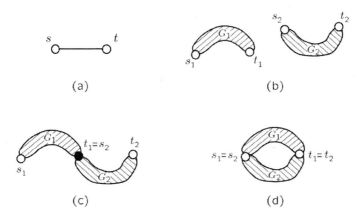

Fig. 2. Definition of series-parallel graphs

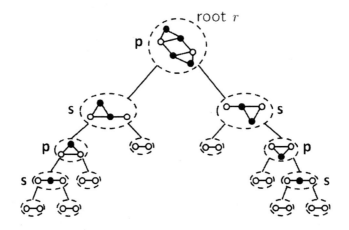

Fig. 3. Decomposition tree T of the series-parallel graph in Fig. 1(a)

where the minimum is taken over all b-colorings f of G such that $f(s) = i$ and $f(t) = j$. Let $\chi_{ij}(G) = \infty$ if there is no such b-coloring.

One can recursively compute $\chi_{ij}(G)$, $1 \leq i, j \leq 5B$, as follows. Consider first the case where G consists of a single edge $e = (s,t)$ as illustrated in Fig. 2(a). Then $\chi_{ij}(G) = \max\{i, j\}$ if the following (a)–(c) hold:

(a) $i \neq j$, $b(s) \leq i$, and $b(t) \leq j$;
(b) $i < j$ implies $i + b(s,t) + b(t) \leq j$; and
(c) $j < i$ implies $j + b(s,t) + b(s) \leq i$.

Otherwise, $\chi_{ij}(G) = \infty$. Consider next the case where G is obtained from G_1 and G_2 by a series connection as illustrated in Fig. 2(c). Then

$$\chi_{ij}(G) = \min_{1 \leq k \leq 5B} \max\{\chi_{ik}(G_1), \chi_{kj}(G_2)\}. \tag{7}$$

Consider finally the case where G is obtained from G_1 and G_2 by a parallel connection as illustrated in Fig. 2(d). Then

$$\chi_{ij}(G) = \max\{\chi_{ij}(G_1), \chi_{ij}(G_2)\}. \tag{8}$$

One may assume that a series-parallel graph G has no multiple edges. Then one can easily prove by induction that $m \leq 2n - 3$. Since the binary decomposition tree T of G has m leaves, T has exactly $m - 1(\leq 2n - 4)$ internal nodes. We compute $\chi_{ij}(G_u)$, $1 \leq i, j \leq 5B$, for all nodes u of T from leaves to the root r. It takes time $O(B^3 n)$. Since $G = G_r$, we compute $\chi_b(G)$ from $\chi_{ij}(G_r)$ in time $O(B^2)$ as follows:

$$\chi_b(G) = \min_{1 \leq i, j \leq 5B} \chi_{ij}(G_r)$$

Thus we have the following theorem.

Theorem 1. *The b-coloring problem can be solved in time $O(B^3 n)$ for a series-parallel graph G, where n is the number of vertices in G and B is the maximum weight of G.*

Clearly, $B^3 n$ is polynomial in n if B is bounded above by a polynomial in n.

4   FPTAS

In this section we give a fully polynomial-time approximation scheme (FPTAS) for the b-coloring problem on series-parallel graphs.

Let G be a graph with a weight function b, and let σ be a scaling factor which is a positive integer. Then we denote by G_σ a graph which is isomorphic with G but has a weight function b_σ such that

$$b_\sigma(x) = \lceil b(x)/\sigma \rceil \tag{9}$$

for every element $x \in V \cup E$. Figure 1(c) depicts G_σ with $\sigma = 2$ for the graph G in Fig. 1(a). An optimal b_σ-coloring F_σ of G_σ is also depicted in Fig. 1(c). We now have the following lemma.

Lemma 4. *Let G be a graph with weight function b, let σ be a positive integer, and let f_σ be an optimal b_σ-coloring of G_σ. Then, a function f such that $f(v) = \sigma f_\sigma(v)$ for every vertex v is a b-coloring of G, and hence $\chi_b(G) \leq \sigma \chi_{b_\sigma}(G_\sigma)$.*

Proof. Since f_σ is an optimal b_σ-coloring of G_σ, we have $span(f_\sigma) = \chi_{b_\sigma}(G_\sigma)$, $b_\sigma(v) \leq f_\sigma(v)$ for every vertex $v \in V$, $f_\sigma(v) \neq f_\sigma(w)$ for every edge (v, w), and $f_\sigma(v) + b_\sigma(v, w) + b_\sigma(w) \leq f_\sigma(w)$ for every edge (v, w) with $f_\sigma(v) < f_\sigma(w)$. Therefore, we have $b(v) \leq \sigma b_\sigma(v) \leq \sigma f_\sigma(v) = f(v)$ for every vertex v. Similarly, we have $f(v) \neq f(w)$ for every edge (v, w), and $f(v) + b(v, w) + b(w) \leq f(w)$ for every edge (v, w) with $f(v) < f(w)$. Thus f is a b-coloring of G, and hence $\chi_b(G) \leq span(f) = \sigma \cdot span(f_\sigma) = \sigma \chi_{b_\sigma}(G_\sigma)$. $\square$

Consider the following approximation scheme.

Approximation Scheme

1. Choose a scaling factor σ appropriately. (We will later choose $\sigma = \lfloor \varepsilon B/4n \rfloor$ for a desired approximation error rate ε.)
2. Find an optimal b_σ-coloring f_σ of G_σ (by a pseudo polynomial-time exact algorithm, say the algorithm in Section 3).
3. Output, as an approximate solution, a b-coloring f of G such that $f(v) = \sigma f_\sigma(v)$ for every vertex v.

We now have the following lemma on the longest paths in acyclic orientations $\overrightarrow{G}$ and $\overrightarrow{G_\sigma}$.

Lemma 5. *Let G be a weighted graph of n vertices, let $\overrightarrow{G}$ be an acyclic orientation of G, and let P be the longest directed path in $\overrightarrow{G}$. Let σ be a positive integer, let $\overrightarrow{G_\sigma}$ be the acyclic graph obtained from G_σ by orienting each edge in the same direction as in $\overrightarrow{G}$, and let P_σ be the longest directed path in $\overrightarrow{G_\sigma}$. (See Fig. 1.) Then*

$$\ell(P_\sigma, \overrightarrow{G_\sigma}) < \ell(P, \overrightarrow{G_\sigma}) + 2n \tag{10}$$

Using Lemmas 3 and 5, we then have the following lemma on the error of the approximation scheme above.

Lemma 6. *For a positive integer σ and a graph G with weight function b*

$$\sigma \chi_{b_\sigma}(G_\sigma) < \chi_b(G) + 4\sigma n.$$

Proof. Lemma 3 implies that there are an acyclic orientation $\overrightarrow{G}$ of G and the longest path P in $\overrightarrow{G}$ such that

$$\chi_b(G) = \ell_{\max}(\overrightarrow{G}) = \ell(P, \overrightarrow{G}). \tag{11}$$

Let $\overrightarrow{G_\sigma}$ be the acyclic orientation obtained from G_σ by orienting each edge in the same direction as in $\overrightarrow{G}$. The path P contains at most $2n - 1$ elements (vertices and edges). Therefore, we have

$$
\begin{aligned}
\chi_b(G) + 2\sigma n &= \sum_{x \in P} b(x) + 2\sigma n \\
&> \sigma \sum_{x \in P} \left(\frac{b(x)}{\sigma} + 1 \right) \\
&> \sigma \sum_{x \in P} b_\sigma(x) \\
&= \sigma \ell(P, \overrightarrow{G_\sigma}).
\end{aligned}
\tag{12}
$$

Let P_σ be the longest path in $\overrightarrow{G_\sigma}$, then by Lemma 5 we have

$$\ell(P, \overrightarrow{G_\sigma}) + 2n > \ell(P_\sigma, \overrightarrow{G_\sigma}). \tag{13}$$

By Lemma 3 we have

$$\ell(P_\sigma, \overrightarrow{G_\sigma}) \geq \chi_{b_\sigma}(G_\sigma). \tag{14}$$

By Eqs. (12)−(14) we have

$$\begin{aligned}
\chi_b(G) + 4\sigma n &> \sigma\ell(P, \overrightarrow{G_\sigma}) + 2\sigma n \\
&> \sigma\ell(P_\sigma, \overrightarrow{G_\sigma}) \\
&\geq \sigma\chi_{b_\sigma}(G_\sigma).
\end{aligned}$$

$\square$

Let $\varepsilon(> 0)$ be a desired approximation error rate. If $\varepsilon B/4n \leq 1$, then we compute $\chi_b(G)$ by a pseudo polynomial-time exact algorithm, say the algorithm in Section 3; the computation time is bounded by a polynomial in n and $1/\varepsilon$ since $B \leq 4n/\varepsilon$. One may thus assume that $\varepsilon B/4n > 1$. We then choose $\sigma = \lfloor \varepsilon B/4n \rfloor (\geq 1)$, and find an approximately optimal b-coloring $f(= \sigma f_\sigma)$ of G by the approximation scheme above. By Lemmas 1 and 6 one can bound the error as follows:

$$\begin{aligned}
span(f) - \chi_b(G) &= \sigma\chi_{b_\sigma}(G_\sigma) - \chi_b(G) \\
&< 4\sigma n \\
&\leq \varepsilon B \\
&\leq \varepsilon\chi_b(G). \tag{15}
\end{aligned}$$

We thus have the following theorem.

Theorem 2. *If there is an exact algorithm to solve the b-coloring problem for a class of graphs in time polynomial in n and B, then there is a fully polynomial-time approximation scheme for the class.*

Proof. Suppose that the algorithm finds an optimal b-coloring of a graph G in the class in time $p(n, B)$, where $p(n, B)$ is a polynomial in n and B. Find an optimal b_σ-coloring f_σ of G_σ in time $p(n, B_\sigma)$ by the algorithm, and output a b-coloring $f = \sigma f_\sigma$ of G by the approximation scheme above, where $\sigma = \lfloor \varepsilon B/4n \rfloor$ and $B_\sigma = \lceil B/\sigma \rceil$ is the maximum weight of G_σ. By Eq. (15) the error is less than $\varepsilon\chi_b(G)$. Since $B_\sigma = O(n/\varepsilon)$, the computation time $p(n, B_\sigma)$ of the scheme is bounded by a polynomial in n and $1/\varepsilon$.

From Theorems 1 and 2 we thus have the following corollary.

Corollary 1. *There is a fully polynomial-time approximation scheme for the b-coloring problem on series-parallel graphs, and the computation time is $O(B_\sigma{}^3 n) = O(n^4/\varepsilon^3)$.*

5   Partial k-Trees

The class of partial k-trees, that is, graphs with bounded tree-width, contains trees, outerplanar graphs, series-parallel graphs, etc. In this section we show that the results in Sections 3 and 4 can be extended to partial k-trees.

For a bounded positive integer k, a k-tree is recursively defined as follows [1, 2]:

(1) A complete graphs with k vertices is a k-tree.
(2) If $G = (V, E)$ is a k-tree and k vertices $v_1, v_2, \cdots, v_k$ induce a complete subgraph of G, then $G' = (V \cup \{w\}, E \cup \{(v_i, w) : 1 \le i \le k\})$ is a k-tree where w is a new vertex not contained in G.

Any subgraph of a k-tree is called a *partial k-tree*.

A binary tree $T = (V_T, E_T)$ is called a *tree decomposition* of a partial k-tree $G = (V, E)$ if T satisfies the following conditions (a)−(e) [2]:

(a) every node $X \in V_T$ is a subset of V and $|X| = k + 1$;
(b) $\bigcup_{X \in V_T} X = V$;
(c) for each edge $e = (u, v) \in E$, T has a leaf $X \in V_T$ such that $u, v \in X$;
(d) if node X_p lies on the path in T from node X_q to node X_r, then $X_q \cap X_r \subseteq X_p$; and
(e) each internal node X_i of T has exactly two children, say X_ℓ and X_r, such that $|X_\ell - X_r| = 1$ and either $X_i = X_\ell$ or $X_i = X_r$.

One can easily observe from the definitions above that $\chi(G) \le k + 1$ for every partial k-tree G. Therefore, by Lemma 1 we have $\chi_b(G) \le (2k + 1)B$. Similarly as in Section 3, we compute the counterparts of χ_{ij} from leaves to the root of a tree decomposition T of G and compute $\chi_b(G)$ from the counterparts for the root, which corresponds to G. Thus $\chi_b(G)$ can be computed in time

$$((2k + 1)B)^{k+1} \times (2k + 1)B \times n = O(B^{k+2}n),$$

because $\chi_b(G) \le (2k + 1)B$ and $|X| = k + 1$ for every node X of T and hence there are a number $((2k + 1)B)^{k+1}$ of counterparts of χ_{ij}, $|X_\ell - X_r| = 1$ for every internal node X_i of T, and T has $O(n)$ nodes. The time is bounded by a polynomial in n and B. Therefore, by Theorem 2, the scheme in Section 4 is an FPTAS and takes time

$$B_\sigma^{k+2}n = O\left(\left(\frac{n}{\varepsilon}\right)^{k+2} n\right).$$

We thus have the following corollary.

Corollary 2. *There is a fully polynomial-time approximation scheme for the b-coloring problem on partial k-trees.*

References

1. Arnborg, S., Proskurowski, A.: Linear time algorithms for NP-hard problems restricted to partial k-trees. Disc. Appl. Math. 23, 11–24 (1989)
2. Bodlaender, H.L.: Treewidth: Algorithmic Techniques and Results. In: Privara, I., Ružička, P. (eds.) MFCS 1997. LNCS, vol. 1295, pp. 19–36. Springer, Heidelberg (1997)
3. Halldórsson, M.M., Kortsarz, G.: Tools for multicoloring with applications to planar graphs and partial k-trees. J. of Algorithms 42(2), 334–366 (2002)
4. Halldórsson, M.M., Kortsarz, G., Proskurowski, A.: Multicoloring trees. Information and Computation 180(2), 113–129 (2003)
5. Ito, T., Nishizeki, T., Zhou, X.: Algorithms for multicolorings of partial k-trees. IEICE Trans. Inf. & Syst. E86-D(2), 191–200 (2003)
6. Jansen, K., Scheffler, P.: Generalized coloring for tree-like graphs. Disc. Appl. Math. 75(2), 135–155 (1997)
7. Jensen, T.R., Toft, B.: Graph Coloring Problems. Wiley, New York (1994)
8. Malaguti, E., Toth, P.: An evolutionary approach for bandwidth multicoloring problems. Europian Journal of Operation Reseach 189, 638–651 (2008)
9. McDiamid, C., Reed, B.: Channel assignment on graphs of bounded treewidth. Discrete Mathematics 273, 183–192 (2003)
10. Pinedo, M.L.: Scheduling: Theory, Algorithms and Systems. Springer Science, New York (2008)
11. Takamizawa, K., Nishizeki, T., Saito, N.: Linear-time computability of combinatorial problems on series-parallel graphs. J. Assoc. Comput. Mach. 29, 623–641 (1982)
12. West, D.B.: Introduction to Graph Theory. Prentice-Hall, Englewood Cliffs (1996)
13. Zhou, X., Kanari, Y., Nishizeki, T.: Generalized vertex-colorings of partial k-trees. IEICE Trans. Fundamentals E83-A(4), 671–678 (2000)
14. Zhou, X., Nishizeki, T.: Multicolorings of series-parallel graphs. Algorithmica 38, 271–297 (2004)
15. Zuckerman, D.: Linear degree extractors and the inapproximability of max clique and chromatic number. Theory of Computing 3, 103–128 (2007)

Independent Domination on Tree Convex Bipartite Graphs

Yu Song[1], Tian Liu[1,*], and Ke Xu[2,*]

[1] Key Laboratory of High Confidence Software Technologies, Ministry of Education,
Institute of Software, School of Electronic Engineering and Computer Science,
Peking University, Beijing 100871, China
{songyufish,lt}@pku.edu.cn
[2] National Lab. of Software Development Environment,
Beihang University, Beijing 100191, China
kexu@nlsde.buaa.edu.cn

Abstract. An independent dominating set in a graph is a subset of
vertices, such that every vertex outside this subset has a neighbor in this
subset (dominating), and the induced subgraph of this subset contains no
edge (independent). It was known that finding the minimum independent
dominating set (Independent Domination) is $\mathcal{NP}$-complete on bipartite
graphs, but tractable on convex bipartite graphs. A bipartite graph is
called tree convex, if there is a tree defined on one part of the vertices,
such that for every vertex in another part, the neighborhood of this vertex
is a connected subtree. A convex bipartite graph is just a tree convex one
where the tree is a path. We find that the sum of larger-than-two degrees
of the tree is a key quantity to classify the computational complexity of
independent domination on tree convex bipartite graphs. That is, when
the sum is bounded by a constant, the problem is tractable, but when
the sum is unbounded, and even when the maximum degree of the tree
is bounded, the problem is $\mathcal{NP}$-complete.

1 Introduction

A *dominating set* in a graph $G = (V, E)$ is a subset D of vertices, such that
every vertex in $V \setminus D$ has a neighbor in D. An *independent dominating set* D
is a special kind of dominating set which is also independent, that is, there
is no edge whose both ends are in D. The problem of finding the minimum
independent dominating set (IDS, in short) is $\mathcal{NP}$-complete on general graphs
[3], chordal graphs [2], bipartite graphs [4], chordal bipartite graphs [7], etc.

A bipartite graph $G = (A, B; E)$ is called *tree convex*, if there is a tree $T = (A, F)$, such that for all vertex b in B, the neighborhood of b is a connected
subtree in T [5,6]. When T is a path, G is called *convex*. It was known that
IDS is $\mathcal{NP}$-complete on bipartite graphs [4], but becomes tractable on convex
bipartite graphs [1]. A natural question is

---

* Corresponding authors.

J. Snoeyink, P. Lu, K. Su, and L. Wang (Eds.): FAW-AAIM 2012, LNCS 7285, pp. 129–138, 2012.
© Springer-Verlag Berlin Heidelberg 2012

- *What is the boundary between intractability and tractability of IDS on bipartite graphs?*

In this paper, we answer this question. We first explore IDS on some simple cases, showing an intractability on star convex bipartite graphs, where T is a star, i.e. a bipartite complete graph $K_{1,|A|-1}$, and a tractability on triad convex bipartite graphs, where T is a triad, i.e. three paths with a common end. These results have already extended the known results of [4] and [1], respectively. Finally, we find the exact condition to differentiate $\mathcal{NP}$-completeness and $\mathcal{P}$: whether

$$t = \sum_{v_i : deg_T(v_i) > 2} deg_T(v_i)$$

is bounded by a constant or not, where $deg_T(v)$ is the degree of vertex v in tree T. The results of this paper are pictured in Figure 1.

Fig. 1. The results of this paper

The remaining part of this paper is organized as follows. The $\mathcal{NP}$-completeness of IDS is shown on star convex bipartite graphs in Section 2, and then on more general graph classes in Section 3. Tractability of IDS is shown in Section 4. The conclusion and discussion are in Section 5.

2   Intractability of IDS on Star Convex Bipartite Graphs

IDS is $\mathcal{NP}$-complete in bipartite graphs [4]. We can refine this intractability into star convex bipartite graphs by a similar reduction.

Theorem 1. *IDS is $\mathcal{NP}$-complete on star convex bipartite graphs.*

Proof. We reduce from SAT to IDS on star convex bipartite graphs. Given an instance I of SAT, which has m variables $x_1, ..., x_m$ and n clauses $C_1, ..., C_n$, we construct a star convex bipartite graph $G = (A, B; E)$, such that I is satisfiable if and only if G has an IDS of size $2m$, as follows.

1. For each variable x_i in I ($1 \le i \le m$), there is a small gadget involving six vertices $\{x_i, \bar{x}_i, u_i, v_i, y_i, z_i\}$ and six edges in G, as shown in Figure 2.

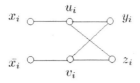

Fig. 2. The gadget for the literals $x_i, \bar{x}_i$

2. For each clause C_j in I $(1 \leq i \leq n)$, there is a vertex C_j in G.
3. For all literals $x_i, \bar{x}_i$ $(1 \leq i \leq m)$, and for all clauses C_j $(1 \leq i \leq n)$, we connect x_i and C_j if x_i is in C_j, connect $\bar{x}_i$ and C_j if $\bar{x}_i$ is in C_j.
4. Add a vertex v_0, which is connected to every u_i, v_i $(1 \leq i \leq m)$ and C_j $(1 < j < n)$.

Clearly, the construction is in polynomial time. An example of this construction is in Figure 3 for a SAT instance with two clauses $C_1 = x_1 \vee x_2$ and $C_2 = \bar{x}_1 \vee x_2$.

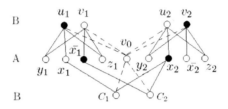

Fig. 3. An example of the construction of G

Lemma 1. *Graph G is star convex bipartite.*

Proof. The graph G is bipartite with respect to the following partition of vertices

$$A = \{x_i, \bar{x}_i, y_i, z_i | 1 \leq i \leq m\} \cup \{v_0\}, B = \{u_i, v_i | 1 \leq i \leq m\} \cup \{C_j | 1 \leq j \leq n\}.$$

Also G is star convex with the tree T on A be a star with central vertex v_0 and $4m$ leaves, since every vertex b in B is connected to v_0, the neighborhood of b is a subtree in T.   □

Lemma 2. *If I is satisfiable, G has an IDS of size no more than $2m$.*

Proof. If there is a satisfying assignment to I, then the set

$$D = \bigcup_i \{x_i, v_i | x_i = true\} \cup \bigcup_i \{\bar{x}_i, u_i | x_i = false\}$$

is an independent set of size $2m$. The set D is also a dominating set, since every gadget is dominated by $\{x_i, v_i\}$ or $\{\bar{x}_i, u_i\}$, the vertex C_j is dominated by one of $\{x_i\}$ or $\{\bar{x}_i\}$ by the satisfying property, and v_0 is dominated by $\{u_i\}$ or $\{v_i\}$. In the example shown in Figure 3, a satisfying assignment of I is $x_1 = false$, $x_2 = true$, and $D = \{\bar{x}_1, u_1, x_2, v_2\}$.   □

Lemma 3. *If G has an IDS of size no more than 2m, I is satisfiable.*

Proof. Suppose there is an IDS of size no more than $2m$. Since there are m gadgets, and we can not use only one vertex to dominate all six vertices in one gadget, no matter whether we choose v_0 and C_j or not, we must choose exactly two vertices from each gadget. The limitation of the size $2m$ makes v_0 and C_j outside the IDS. For each i, the pair x_i and $\bar{x}_i$ can not be both in IDS, for otherwise, neither y_i nor z_i will be dominated. If we assign variable x_i to be true whenever IDS contains x_i, and false otherwise, we get an assignment that satisfies I, since each vertex C_j must be dominated, which implies that every clause C_j contains a true literal and thus is satisfied.   □

The proof of Theorem 1 is finished.   □

3   Intractability of IDS on Tree Convex Bipartite Graphs

We can transform the star T constructed in last section into a new tree whose maximum degree is bounded by a constant d_{max} as follows. We split the single central vertex v_0 with $4m$ leaves into a set of central vertices, whose cardinality is $\lceil 4m/(d_{max} - 2) \rceil$, to from a path in T. Every $(d_{max} - 2)$ of the original $4m$ leaves of v_0 form a group to connect to one of the central vertices. Figure 4 is an example of this transformation with $m = 5$ and $d_{max} = 6$. Amazingly, after this transformation, the whole reduction still works with some slight modification.

Theorem 2. *IDS is $\mathcal{NP}$-complete on tree convex bipartite graphs where the maximum degree of tree T is bounded by a constant.*

Proof. We still reduce from SAT. The modified whole reduction is as follows.

1. For each variable x_i in I ($1 \le i \le m$), there is a small gadget in G involving six vertices $\{x_i, \bar{x}_i, u_i, v_i, y_i, z_i\}$ and six edges in G, as shown in Figure 2.
2. For each clause C_j in I ($1 \le j \le n$), there is a vertex C_j in G.
3. If variable x_i is in clause C_j, we connect vertices x_i and C_j in G. If negated variable $\bar{x}_i$ is in C_j, we connect vertices $\bar{x}_i$ and C_j in G.
4. There are $p = \lceil 4m/(d_{max} - 2) \rceil$ central vertices $v_{01}, v_{02}, ..., v_{0p}$ to form a path in T (not in G). The vertices $y_i, x_i, \bar{x}_i, z_i$ ($1 \le i \le m$) are ordered by $y_1, x_1, \bar{x}_1, z_1, y_2, x_2, \bar{x}_2, z_2$, and so on. Every $(d_{max} - 2)$ consecutive vertices in this order form a group, and one by one each group are leaves of one of the p central vertices in T (again not in G).

Fig. 4. Transforming a star into a tree of bounded maximum degree

5. For each C_j, find the minimum subtree T_j in T containing $\{x_i | x_i \in C_j\} \cup \{\bar{x}_i | \bar{x}_i \in C_j\}$, and connect C_j to the central vertices on T_j.
6. For x_i or $\bar{x}_i$ in C_j, connect u_i and v_i to the central vertices on above T_j.

Clearly, the construction is in polynomial time. Figure 5 shows an example of the construction with $d_{max} = 4$ and $p = 4$, for a SAT instance with two clauses $C_1 = x_1 \vee x_2$ and $C_2 = \bar{x}_1 \vee x_2$.

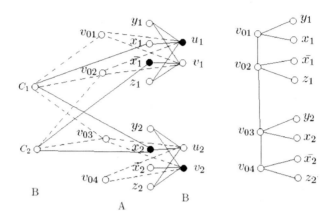

Fig. 5. An example of the constructions of G and T

Lemma 4. *Graph G is tree convex bipartite where the maximum degree of tree T is bounded by constant d_{max}.*

Proof. The graph G is bipartite with respect to the partition of vertices

$$A = \{x_i, \bar{x}_i, y_i, z_i | 1 \le i \le m\} \cup \{v_{0k} | 1 \le k \le p\},$$

$$B = \{u_i, v_i | 1 \le i \le m\} \cup \{C_j | 1 \le j \le n\}.$$

That the G is tree convex with maximum degree d_{max} in T is ensured by the construction, especially by the fifth and the sixth steps above.   □

Lemma 5. *If I is satisfiable, G has an IDS of size no more than $2m$.*

Proof. If there is a satisfying assignment of I, the set

$$D = \bigcup_i \{x_i, v_i : x_i = true\} \cup \bigcup_i \{\bar{x}_i, u_i : x_i = false\}$$

is an independent set of size $2m$. The set D is also a dominating set, since every gadget is dominated by $\{x_i, v_i\}$ or $\{\bar{x}_i, u_i\}$, and C_j is dominated by x_i or $\bar{x}_i$ by the satisfying property, and the sixth step of construction ensures that each v_{0k} is connected to a u_i and a v_i. In the example in Figure 5, the satisfying assignment is $x_1 = false$ and $x_2 = true$, and the IDS is $D = \{\bar{x}_1, u_1, x_2, v_2\}$.   □

Lemma 6. *If G has an IDS of size no more than $2m$, I is satisfiable.*

Proof. Suppose there is an IDS of size no more than $2m$. Since there are m gadgets, and we can not use a single vertex to dominate all six vertices in one gadget, no matter whether we choose v_{0k} and C_j ($1 \leq k \leq p$, $1 \leq j \leq m$) or not, so we must choose exactly two vertices from each gadget. The limitation of the size $2m$ kills both v_{0k} and C_j from the IDS. For each i, the pair x_i and $\bar{x}_i$ can not be both in IDS, for otherwise neither y_i nor z_i will be dominated. If we assign variable x_i to be true whenever IDS contains x_i, and false otherwise, we get an satisfying assignment for I, since each C_j must be dominated, which implies that every clause contains a true variable. □

This finishes the proof of Theorem 2. □

Note that in above two reductions, the sum of larger-than-two degrees in tree T is unbounded. Formally, let $t = \sum_{v_i:deg_T(v_i)>2} deg_T(v_i)$. Then $t = 4m$ for the T constructed in last section, and $t = 4m + p - 1$ for the T constructed in this section. In the former case, the number of larger-than-two degrees is bounded, but the maximum degree is unbounded, while in the later case, the number of larger-than-two degrees is unbounded, but the maximum degree is bounded, all in T. In both cases, the sum of larger-than-two degrees in tree T is unbounded. Thus we have the following intractability result.

Theorem 3. *IDS is $\mathcal{NP}$-complete on tree convex bipartite graphs whose t, the sum of larger-than-two degrees in tree T, is unbounded.*

4   Tractability of IDS on Tree Convex Bipartite Graphs

IDS is polynomial time on convex bipartite graphs [1]. We can extend this tractability onto more general tree convex bipartite graphs by a similar dynamic programming. We start with a simple situation as follows. Recall that a triad is just three paths with a common end.

Theorem 4. *IDS is in polynomial time on triad convex bipartite graphs.*

Proof. Our algorithm is an extension of the dynamic programming in [1]. Let D be a minimum subset of vertices with a desired property $\mathcal{Q}$. D is constructed as an increasing sequence $D_1 \subseteq D_2 \subseteq \cdots \subseteq D_m$ as follows.

- Step (a) (initialization): generate all possible versions of D_1.
- Step (b) (branching): extend all versions of present D_{i-1} to all possible versions of D_i with $D_{i-1} \subseteq D_i$. (Possible means that the versions satisfy $\mathcal{Q}$.)
- Step (c) (classification): classify the versions of D_i, such that if versions V and V' belong to the same class, then $V \subseteq D'$ and D' satisfies $\mathcal{Q}$ imply that $(D' \setminus V) \cup V'$ satisfies $\mathcal{Q}$.
- Step (d) (deletion): delete all versions except the one of minimum cardinality in each class.

The algorithm proceeds at steps (a), $(c)_1$, $(d)_1$, $(b)_2$, $(c)_2$, $(d)_2$, $(b)_3$, $(c)_3$, $\cdots$, $(b)_m$, $(c)_m$, $(d)_m$. We take some version of D_m with the minimum cardinality as the output minimum IDS.

Then we define some notations on triad convex bipartite graphs. Suppose $G = (A, B; E)$ and there is a triad T on A. The triad T consists of three paths $p = 1, 2, 3$ with a common vertex, as shown in the leftmost in Figure 6. For

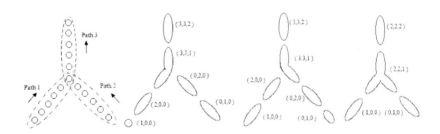

Fig. 6. Three paths in triad convex tree, and the label of A_i

each vertex x, $N(x)$ denotes the neighborhood of x in G. We partition A into nonempty sets A_1, A_2, $\cdots$, A_m, such that the following conditions hold:

- Each A_i consists of consecutive vertices in T.
- All vertices $x \in A_i$ have the same $N(x)$.
- Each A_i is maximal with respect to the above two conditions.
- Each A_i has a three-dimension label (x_i, y_i, z_i), where the p-th bit represents the order of A_i in path p of T in a bottom-up manner. Figure 6 shows some examples on how to label A_i.
- In the following text, we will also call A_i: $A_{(x_i, y_i, z_i)}$, and define $A_{(x_i, y_i, z_i)}$ $[1] := x_i$, $A_{(x_i, y_i, z_i)}[2] := y_i$, $A_{(x_i, y_i, z_i)}[3] := z_i$. Further defined three-dimension variable holds the same definition.

For each vertex y in B, we define $l(y)$ and $r(y)$ as a three-dimension variable, to record the range where y covers in each path. They are calculated as follows. If $N(y)$ include the common vertex of three paths, then

$$l(y) := \min\{A_{(x_i,y_i,z_i)}[p] | A_{(x_i,y_i,z_i)} \subseteq N(y) \cap \{v|v \text{ is on path } i \text{ in } T\}\},$$

$$r(y) := \max\{A_{(x_i,y_i,z_i)}[p] | A_{(x_i,y_i,z_i)} \subseteq N(y) \cap \{v|v \text{ is on path } i \text{ in } T\}\}.$$

Otherwise, $N(y)$ must intersect with only one path. If $N(y) \subseteq$ path 1, then $l(y)[2] = l(y)[3] = r(y)[2] = r(y)[3] = 0$. If $N(y) \subseteq$ path 2, then $l(y)[1] = l(y)[3] = r(y)[1] = r(y)[3] = 0$. If $N(y) \subseteq$ path 3, then $l(y)[1] = r(y)[1] = A_{i'}[1]$, $l(y)[2] = r(y)[2] = A_{i'}[2]$, where $A_{i'}$ contains the common vertex of tree paths. Figure 7 shows an example of triad convex bipartite graph and the label of vertices of A and B.

There are some properties about A_i. Suppose that $N(A_i)$ are vertices connected to A_i in B, and D is the minimum IDS. If $N(A_i) \cap D \neq \emptyset$, from the

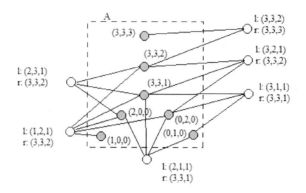

Fig. 7. The label of A_i, $l(y)$ and $r(y)$

independence of D it follows $A_i \cap D = \emptyset$. In the other case $N(A_i) \cap D = \emptyset$, to dominate A_i there must be $A_i \subseteq D$. For each A_i, we denote b_i with $b_i = 1$ iff $A_i \subseteq D$, and $b_i = 0$ iff $A_i \cap D = \emptyset$. Since there are only two possible b_i, we execute the dynamic programming process by dealing with one A_i per branching step, with a reverse order of breadth-first search, and extend D_{i-1} to D_i by enumerating the value of b_i.

Further, in each classification part of step i in the algorithm, we define s_i to be a three-dimension variable as follows. The p-th bit of s_i is

$$s_i[p] := max\{A_k[p] | A_k \cap \text{ path } p \neq \emptyset, A_k[p] < A_i[p], b_k = 1\}.$$

Actually, s_i represents the latest position where IDS has already dominated in each path of the convex tree. We define this variable because if we know s_i and whether A_i is chosen or not, i.e. b_i, we know how to choose vertices in B into IDS. So we can categorize versions with the same s_i into a class, and do the deletion procedure.

Recall the definition of D_i to be the minimum IDS found in step i. Then D_i has the following properties:

- $D_i = \bigcup \{A_k - k \leq i, b_k = 1 \} \cup \{y \in B \mid r(y)[1] \leq A_k[1], r(y)[2] \leq A_k[2], r(y)[3] \leq A_k[3], b_k = 0 \text{ for all } A_k \subseteq N(y)\}$;
- (opt) If $b_k = 0$, and every bit of $A_k \leq s_i$, then D_i must contain some vertex of $N(A_k)$.

Now we can represent our IDS algorithm as follows.

```
(a) version b1=0:
       D1:={y is vertex in B | r(y):=(D1[1],D1[2],D1[3])};
       s1:=(0,0,0).
    version b1=1:
       D1:=A1;
       s1:=(A1[1],A1[2],A1[3]).
(b) version bi=0:
```

```
    Di:=D(i-1) + {y is vertex in B|l(y)[1] > s(i-1)[1],
        l(y)[2] > s(i-1)[2], l(y)[3] > s(i-1)[3],
        r(y) = (Ai[1],Ai[2],Ai[3])};
    si:=s(i-1).
    If i=m, and Ai is not dominated, then delete the version.
    version bi=1:
        check (opt), if it fails, then delete the version;
        If (opt) hols, then Di:=D(i-1)+Ai;
        If Ai have common vertices with path p, si[p]:=Ai[p];
        else si[p]:=s(i-1)[p].
(c) The version with equal sk belong to the same class.
(d) Delete all versions except one of minimum cardinality in
    each class.
```

We can briefly analyze the running time of this algorithm as follows.

- Labeling each A_i and $l(y)$, $r(y)$ for each y in B cost $O(|A| + |A| \cdot |B|)$ time.
- Step (a) costs $O(|B|)$ time.
- Steps (b)(c)(d) will repeat $O(|A|)$ times. In each loop, there are at most $O(|A|^3)$ versions, we should calculate in $O(|A| + |B|)$ time for each version. Steps (c) and (d) cost $O(|A|^3)$ time. The total cost is $O(|A|^5 + |A|^4 \cdot |B|)$.

So the total running time is $O(|A|^5 + |A|^4 \cdot |B|)$, which is a polynomial.   □

This algorithm is easily extended to more general situations. For tree convex bipartite graphs with the sum of larger-than-two degrees in the tree is bounded by t (this property will be call $\mathcal{Q}$), the tree is split into m paths, where $m \leq t$. Figure 8 briefly presents an example.

So, as long as we redefine the label of A_i, $l(y)$ and $r(y)$ $(y \in B)$ to be m-dimension variables, in which each bit records the position and covering range of A_i and y in each path of the convex tree, the algorithm can operate as the same way:

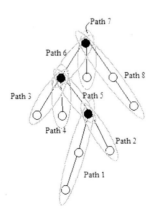

Fig. 8. An example of graphs satisfies $\mathcal{Q}$

- Initializing is as the same.
- Each A_i will be all or none chosen, which consist the two possibilities of branching in each step.
- We should record s_i, which is also a m-dimension variable, presenting the latest position where IDS has already dominated in each path. Versions with same s_i are classified into the same class.
- We only keep the version with minimum cardinality in each class.

With the same analysis, the above algorithm runs in time $O(|A|^{t+2} + |A|^{t+1} \cdot |B|)$, which proves the following theorem.

Theorem 5. *IDS is in polynomial time on tree convex bipartite graphs where the sum of larger-than-two degrees of the tree is bounded by a constant.*

5   Conclusion and Open Problems

We have shown a dichotomy of complexity of IDS on tree convex bipartite graphs: the problem is intractable when the sum of larger-than-two degrees in the tree is unbounded, and tractable when the sum is bounded by a constant. In our intractability reductions, the sum increases linearly. Can we make a reduction with an arbitrarily slow increasing of the sum? In our tractable algorithm, the running time is exponential in the sum. Can we get a better running time?

Acknowledgments. This research was partially supported by the National 973 Program of China (Grant No. 2010CB328103) and the National Natural Science Foundation of China (Grant No. 60973033).

References

1. Damaschke, P., Muller, H., Kratsch, D.: Domination in Convex and Chordal Bipartite Graphs Information Processing Letters 36, 231–236 (1990)
2. Farber, M.: Independent Domination in Chordal Graphs. Operations Research Letters 1, 134–138 (1982)
3. Garey, M.R., Johnson, D.S.: Computers and Intractability. A Guide to the Theory of NP-Completeness (1979)
4. Irving, W.: On approximating the minimum independent dominating set. Information Processing Letters 37, 197–200 (1991)
5. Jiang, W., Liu, T., Ren, T.N., Xu, K.: Two Hardness Results on Feedback Vertex Sets. In: Atallah, M., Li, X.-Y., Zhu, B. (eds.) FAW-AAIM 2011. LNCS, vol. 6681, pp. 233–243. Springer, Heidelberg (2011)
6. Jiang, W., Liu, T., Xu, K.: Tractable Feedback Vertex Sets in Restricted Bipartite Graphs. In: Wang, W., Zhu, X., Du, D.-Z. (eds.) COCOA 2011. LNCS, vol. 6831, pp. 424–434. Springer, Heidelberg (2011)
7. Muller, H., Brandstadt, A.: The NP-completeness of Steiner Tree and Dominating Set for Chordal Bipartite Graphs. Theoretical Computer Science 53, 257–265 (1987)

On-Line Scheduling of Parallel Jobs in Heterogeneous Multiple Clusters*

Deshi Ye and Lili Mei

College of Computer Science, Zhejiang University, Hangzhou 310027, China
yedeshi@zju.edu.cn

Abstract. We consider the on-line scheduling of parallel jobs in heterogeneous multiple clusters, in which a set of clusters is given and the parallel jobs arrive one by one, and the goal is to schedule all the jobs while minimizing the makespan. A cluster consists of many identical processors. A parallel job may require several processors in one cluster to execute it simultaneously. In this paper, we investigate two variants of the heterogeneous clusters. First, for the clusters of different widths (number of processors) but identical processor speeds, we provide an on-line algorithm with a competitive ratio at most of 14.2915. Second, for the clusters of different speeds but identical widths, we provide an on-line algorithm with a competitive ratio at most of 18.2788.

Keywords: Multiple cluster scheduling, Parallel jobs scheduling, Strip packing, On-line algorithms.

1 Introduction

A computer cluster is a set of identical processors connected by a local interconnection network. We consider a grid computing environment, in which several clusters share their computing resources to reach a common goal. A parallel job may require several processors in a cluster for processing. Parallel jobs arrive in an on-line manner and the objective is to minimize the time when all the jobs are completed.

Formally, our problem can be described as multiple cluster scheduling (MCS). We are given m heterogeneous clusters $Cl_1, Cl_2, \ldots, Cl_m$. A set of jobs $J = \{T_1, T_2, \ldots\}$ arrives one by one. Each job T_j is described by a processing time $h(T_j)$ and a width $w(T_j)$ (the number of required processors). The work (or area) of a job T_j is $w(T_j) \cdot h(T_j)$. A cluster Cl_i has w_i identical processors and identical speed v_i. A job T_j is allowed to be scheduled in only one cluster and without preemption, and its processing time in cluster Cl_i is $p_j^i = h(T_j)/v_i$ if $w(T_j) \leq w_i$ else $p_j^i = \infty$. The objective is to minimize the makespan, which is the maximum completion time of a job.

This problem is closer to the multiple strip packing problem (MSP), in which a set of rectangles is packed into multiple two-dimensional strips of certain widths

---

* Research was supported by NSFC(11071215).

J. Snoeyink, P. Lu, K. Su, and L. Wang (Eds.): FAW-AAIM 2012, LNCS 7285, pp. 139–148, 2012.
© Springer-Verlag Berlin Heidelberg 2012

such that the sides of the rectangles are parallel to the strip sides and no rectangle interiors intersect. The objective is to minimize the maximum height of the strip used to pack a given sequence of rectangles without rotation. These two problems are equivalent if we require in addition that a job must use consecutive processors, i.e., we regard a cluster as a strip and a job as a rectangle.

In this paper, we study on-line algorithms for the MCS problem. In the *on-line* variation, jobs arrive one by one and the allocated position for the current job must be immediately and irrevocably made without any information of next jobs. The provided algorithms in this paper always assign a job on consecutive processors, and thus the algorithms are valid for the MCS problem.

Performance Measures. We adopt the classical competitive analysis [4] to measure the performance of on-line algorithms. For any instance I and an on-line algorithm A, we denote by $OPT(I)$ and $A(I)$, respectively, the makespans given by an optimal off-line algorithm and by the on-line algorithm A to schedule all the jobs in I. An on-line algorithm is ρ-competitive if $A(I) \leq \rho OPT(I)$. The *competitive ratio* of the algorithm A is the supremum value of ρ, i.e., $R_A = \sup_I \{A(I)/OPT(I)\}$.

Related Work. The multiple strips packing (MSP) problem was first considered by Zhuk [21]. The main result of their paper is a 10-approximation algorithm that is semi on-line. The jobs arrive on-line and must be distributed to the strips in the on-line manner, but the packing of items in a strip is off-line. The packing is performed after all the items have been deployed to the strips. The author also showed that the problem cannot be approximated within a factor of 2 even if there are only two identical strips.

For the off-line MSP with identical widths, Ye et al. [20] provided a $(2 + \varepsilon)$-approximation algorithm. Bougeret et al. [7] improved the result to 2-approximation, which therefore is tight. Also, they provided an asymptotic FPTAS for the multiple strip packing problem. If we remove the constraint of using consecutive processors, 5/2-approximation algorithms were given in [6,5].

For the off-line MSP with different widths, Bougeret et al. [8] provided 5/2-approximation algorithms. They also presented an off-line 5/2-approximation algorithm for the MSP problem with different speeds.

The on-line MSP with identical speeds have been studied in [17,20]. In [20] the authors studied the on-line model, as in our paper, and investigated both deterministic and randomized algorithms. However, their results are limited to the MSP problem with identical widths and identical speeds. The maximum competitive ratio is 6.6623 for single strip, and the competitive ratio decreases when the number of strips increases. On the other hand, in [17] the authors studied the model in which jobs arrive over time, and designed a 3-competitive algorithm for the problem without release times and a 5-competitive algorithm with release times. Further notice that their model is non-clairvoyant, i.e. the running time of every job is only known after the completion of this job. Their model holds for the variants with different widths.

As far as we know, there is no previous work on MSP with different speeds. The existing work focuses on the sequential jobs scheduling in related machines.

Aspnes et al. [1] provided an 8-competitive deterministic algorithm and a 5.436-competitive randomized algorithm for the on-line related machine scheduling. Berman et al. [3] improved the competitive ratios to 5.828 and 4.311, respectively. Besides, they gave lower bounds of 2.4380 and 1.8372 on the competitive ratios of deterministic algorithms and randomized algorithms, respectively. Currently, Ebenlendr and Sgall [11] improved the deterministic lower bound to 2.564.

The classical one strip packing problem has been extensively studied. Coffman et al. [10] studied NFDH (Next Fit Decreasing Height) and FFDH (First Fit Decreasing Height) algorithms and showed that their asymptotic approximation ratios are 2 and 1.7, respectively. The current best asymptotic approximation is an asymptotic FPTAS due to Kenyon and Rémila [15], Jansen and Stee [14] for the model if rotation of rectangles is allowed. For the absolute approximation ratio, 2-approximation algorithms have been provided in [16,18]. The current best algorithm is due to Harren et al. [12] with an approximation ratio of $5/3 + \varepsilon$, where ε is an arbitrarily small positive number.

For the on-line single strip packing problem, Baker and Schwarz [2] developed shelf algorithms. The next fit shelf algorithm (NFS) and the first fit shelf algorithm (FFS) achieve competitive ratios 7.46 and 6.99, respectively. Revised shelf algorithms were proposed with competitive ratio of 6.66 [19,13].

Our Contributions. In this paper, we study the on-line algorithms in heterogeneous multiple clusters. On-line means that jobs not only arrive on-line to be distributed to clusters, but also that each job shall be immediately and irrevocably assigned. We study two variants of heterogeneous multiple clusters. Heterogeneous means the clusters might have different widths or different speeds. First, in Section 2, for clusters with different widths but the same speeds, we provide an on-line algorithm with a competitive ratio of 14.2915. Second, in Section 3, for clusters with different speeds but the same widths, we provide an on-line algorithm with a competitive ratio of 18.2788. In the remaining paper, the terms *strip* and *rectangle* can be used interchangeably with terms *cluster* and *job*, respectively.

2   Different Widths

In this section, we investigate the MCS problem with identical speeds but different widths. Without loss of generality, we assume that the clusters are arranged in non-decreasing order of widths, i.e., $w_1 \leq w_2 \leq \ldots \leq w_m$.

A semi on-line algorithm [21] was provided for the MCS problem for identical speeds but different widths, in which the items arrive one by one and are assigned to one cluster immediately, but one can pack the items in one cluster by an off-line algorithm after all the items have been distributed.

Our on-line algorithm consists of two steps. In the first step, we extend the algorithm [21] to distribute items to clusters. Next, we apply revised first fit shelf RS_r [19] algorithm to pack the items in that strip in the second step. To be self-content of the paper, the revised first fit shelf algorithm RS_r is given in Algorithm 1.

Algorithm 1. RS_r [19] (assign jobs in the cluster i)

1: Given any $r > 1$, and for any incoming job T_j, the assignment is illustrated as below.
2: **if** Job T_j is *big*, i.e. $w(T_j) > 1/2 * w_i$ **then**
3:    Open a new shelf with height $h(T_j)$ and use it.
4: **end if**
5: **if** Job T_j is *small*, i.e., $w(T_j) \leq 1/2 * w_i$ **then**
6:    Choose a value $k \in \mathbb{Z}$ such that $r^k < h(T_j) \leq r^{k+1}$, then pack it into a shelf of height r^{k+1} by the FF(First Fit) algorithm. That is to say, the job (or rectangle) T_j is packed left most to a lowest shelf with height r^{k+1} which has enough room for it. Otherwise, create a new shelf with height r^{k+1} at the current top of the strip, and then place this rectangle into it.
7: **end if**

Algorithm 2.

1: Specify a constant parameter $0 < \beta \leq 1$. For each incoming job T_j, we select the cluster i among the admissible clusters such that the ratio S_i/w_i is minimized.
2: Then T_j is packed in the ith cluster by Algorithm 1 (the Revised First Fit (RS_r) [19] shelf algorithm).

Let $0 < \beta \leq 1$. For each incoming job T_j, let first(T_j) be the minimum i such that $w_i \geq w(T_j)$ and last(T_j) be the minimum r such that $\sum_{i=first(T_j)}^{r} w_i \geq \beta \sum_{i=first(T_j)}^{m} w_i$. Let S_i be the area of jobs packed in the i-th cluster. The clusters between first(T_j) and last(T_j) are called admissible clusters for job T_j. The detailed on-line algorithm for our problem is given in Algorithm 2.

Theorem 2.1. *Let* $\beta = 1/2$. *For any* $r > 4/3$, *the competitive ratio of Algorithm 2 for on-line MCS problem with different widths is at most of* $1 + 6r + r^2/(r-1)$. *The competitive ratio is at most of* 14.2915 *if we let* $r = 1 + \sqrt{7}/7 \approx 1.378$.

Proof. We extend the criterion for distributing jobs to clusters for the semi online algorithm in [21] to our fully on-line algorithm. In [21], the β was setting to be $1/2$. Denote OPT to be the optimal off-line makespan. Suppose T_a is the last job that was added to the k-th cluster when the makespan happens. In [21], for single strip, the following inequality is obtained.

$$\frac{S_k}{w_k} \leq 4 \cdot OPT. \tag{2.1}$$

By specifying the parameter β, we can obtain that

$$\frac{S_k}{w_k} \geq \beta(1 - \beta)OPT.$$

Then the best possible value for β is $1/2$, which therefore the inequality 2.1 still holds.

Denote ALG to be the makespan of our on-line algorithm. Let us consider the schedule generated by Algorithm 1, which consists of shelves. Denote by H_F the length of shelves that contain big jobs. For small jobs, the shelves with the same length r^k forms class k, in which the *sparse* shelf is define to be the last shelf if the number of shelves used is odd, and all the other shelves are *dense*. Denote H_S and H_D to be the height of sparse shelves and dense shelves before packing T_a, respectively. Then $ALG \leq H_S + H_D + H_F + h(T_a)$. Denote μ to be the longest item in the k-th strip.

Lemma 2.2. *[19] The total work assigned in a cluster by the Algorithm 1 is at least $\frac{H_F}{2} + \frac{2}{3} \cdot \frac{1}{r} \cdot H_D$.*

Proof. The detailed proof is given in [19], we simply restate this fact here. For full shelves, the width of each job is larger than $1/2$ which implies the total work of full shelves is at least $H_F/2$. For dense shelves, the length of each job is at least $1/r$ the height of each shelf. Moreover, except the last shelf of each class, at most one shelf of which the total width of the assigned jobs is less than $2/3$. If it happens, we add the work of last shelf to this shelf, and the last shelf is added to the set of sparse shelves. □

Let $r \geq 4/3$, and from Lemma 2.2, we have the work of a cluster is at least $\frac{H_F + H_D}{1.5r}$, i.e., $S_k \geq \frac{H_F + H_D}{1.5r} w_k$. Combine the equation (2.1), we have

$$ALG = H_S + H_D + H_F + h(T_a)$$

$$\leq \frac{r^2}{r-1}\mu + 1.5r\frac{S_k}{w_k} + \mu$$

$$\leq (\frac{r^2}{r-1} + 6r + 1)OPT.$$

By setting $r = 1 + \sqrt{7}/7 \approx 1.378$, we obtain that the minimum value of the competitive ratio is 14.2915. □

3 Different Speeds

In this section, we deal with the variant of the MCS problem for clusters with the same widths but different speeds, which are assumed to be $v_1 \leq v_2 \leq \ldots \leq v_m$. Without loss of generality, we assume the widths of all clusters to be 1. Our main idea to design the on-line algorithm is to guess an optimal value for the off-line schedule. If the value is correctly guessed, all the jobs must be scheduled with some factor of the guessed value. This idea is used widely in scheduling theory, such as on-line scheduling algorithms for related machine or unrelated machine [1,3].

Suppose that Λ is a guess of an optimal off-line solution. Let $L(v, \Lambda, J)$ be the work of those jobs with $h(J_i) > \Lambda \cdot v$, i.e., the total work of jobs that cannot be processed with speed of at most v if the makespan is bounded by Λ, since the width of a cluster is 1. Let $Cap(v)$ be the sum of speeds larger than v.

Lemma 3.1. *If $L(v, \Lambda, J) > \Lambda \cdot Cap(v)$, then the set of job J cannot be executed with makespan bounded by Λ.*

Proof. $L(v, \Lambda, J)$ means that the set of jobs J must be processed in the clusters with speeds larger than v with bounded makespan Λ. Since all clusters have a width of 1 and there is at most $Cap(v)$ clusters, the makespan must be larger than Λ. □

Algorithm 3.

1: The clusters are indexed such that $v_1 \leq v_2 \leq \ldots \leq v_m$. Specify a parameter r_1 as the parameter r chosen in Algorithm 1. Choose a constant $r_2 > 1$, which will be specified later.
2: Let $\Lambda = h(T_1)/v_m$, $c_0 = \Lambda$, $x_0 = 0$, and $i = 0$.
3: **for** each incoming job T_j **do**
4:     We assign this job to a cluster as below.
5:     **while** $L(v, \Lambda, J) > \Lambda \cdot Cap(v)$ for any $v \in V = \{0, v_1, v_2, \ldots, v_m\}$ **do**
6:         $i = i + 1$;
7:         $\Lambda = r_2 \cdot \Lambda$;
8:         $x_i = \frac{r_1^2}{r_1 - 1} \Lambda + \Lambda$;
9:         $c_i = c_{i-1} + \frac{3r_1}{2} \Lambda$.
10:     **end while**
11:     Job T_j is assigned to the cluster p such that $p = \min\{q | S(q) \leq c_i + x_i, h(T_j)/v_q \leq \Lambda\}$, where $S(q)$ is the makespan of the cluster q if the job T_j is assigned to this cluster by the Algorithm 1.
12:     Assign Job T_j in cluster p using Algorithm 1, where $r = r_1$.
13: **end for**

Our algorithm consists of two levels. One is to distribute the incoming job to a specified cluster and then the second level is to assign this job in the cluster. The second level is the classical on-line strip packing algorithm, we adopt the existing algorithm revised first fit shelf (RS_r) [19], which is described in Algorithm 1. Thus, the essential part is the first level which is described in line 1 to line 9 in Algorithm 3.

The technique we used in Algorithm 3 is inspired from the on-line related machine scheduling [3]. The on-line related machine scheduling is a special case of our problem if we let the size of any job is exactly the width of a cluster. One typical technique for on-line algorithms is doubling [9], the 8-competitive algorithm for the related machine scheduling uses the doubling algorithm [1], in which the algorithm guesses an optimal off-line makespan, if the guess is not correct and then doubling the guess, otherwise the algorithm is shown to be correct for assigning all the jobs within twice the guess. This method is extended in [3], which increases the guess by a parameter r instead of doubling. To efficiently use the clusters and reduce the makespan, we cannot simply use the classical on-line algorithms for related machine scheduling to our problem. Instead, we need to carefully design an algorithm while considering both levels.

Theorem 3.2. *Algorithm 3 for on-line MCS problem with different speeds achieves the competitive ratio at most of* 18.2788, *if we choose* $r_1 = 1.43126$ *in Algorithm 1 and* $r_2 = 1.5214$ *in Algorithm 3.*

Proof. Without loss of generality, we assume that the optimal off-line makespan is 1. The initial value of Λ is defined to be Λ_0. Suppose that the Algorithm 3 stops at the l-th *iteration* and in each iteration i, the value of Λ is defined to be $\Lambda_i = r_1^i \Lambda_0$ for any $1 \leq i \leq l$. Since the optimal makespan is 1, we have $\Lambda_l \leq r_2$ and $\Lambda_{l-1} < 1$ from Lemma 3.1 and Line 5 in Algorithm 3. Then the competitive ratio of our algorithm is computed as below. Let ALG be the makespan of our algorithm. Thus,

$$ALG \leq c_l + x_l = \frac{3r_1}{2}\sum_{i=0}^{l}\Lambda_i + \frac{r_1^2}{r_1-1}\Lambda_l + \Lambda_l$$

$$\leq \frac{3r_1}{2}\sum_{i=-\infty}^{1}r_2^i\Lambda_{l-1}$$

$$+(\frac{r_1^2}{r_1-1}r_2 + r_2)\Lambda_{l-1}$$

$$\leq \frac{3r_1}{2}\sum_{i=-\infty}^{1}r_2^i + \frac{r_1^2}{r_1-1}r_2 + r_2$$

$$\leq \frac{3r_1}{2}\cdot\frac{r_2^2}{r_2-1} + \frac{r_1^2}{r_1-1}r_2 + r_2.$$

In the following, we will show our algorithm is correct if $r_1 \geq 4/3$. Then the best value of ALG is 18.2788, if we choose $r_1 = 1.43126$ and $r_2 = 1.5214$.

Now we need to show that the correctness of our algorithm if we let $r_1 \geq 4/3$. Denote J^k to be the set of jobs scheduled in the iteration of k, where $0 \leq k \leq l$ and the definition of iteration is given at the beginning of this proof. By Algorithm 3, the incoming job is preferred to be assigned to the slowest cluster. Let J_i^k be the set of jobs that in phase k cluster $i+1$ received or cluster i ignored. To prove the correctness of this theorem, we only need to show that the set J_m^k is empty for every phase k, i.e., $L(0, \Lambda_t, J_m^t) = 0$ for any $t \leq k$.

In each iteration k, we have

$$L(v, \Lambda_k, J^k) \leq \Lambda_k \cdot Cap(v) \tag{3.1}$$

for every speed v and phase k.

Lemma 3.3. *For every $i = 0,\ldots,m$ and phase k, if $r_1 \geq 4/3$ in Algorithm 1, then we have*

$$\sum_{t=0}^{k}L(0, \Lambda_t, J_i^t) \leq (\sum_{t=0}^{k}\Lambda_t)(\sum_{j=i+1}^{m} v_j).$$

Proof. The proof is done by the induction on i. In initial step, we show the lemma holds for $i = 0$ and $l = 0$. From the inequality (3.1), for any t we have

$$L(0, \Lambda_t, J_0^t) = L(0, \Lambda_t, J^t) \leq \Lambda_t Cap(0) = \Lambda_t(\sum_{j=1}^{m} v_j).$$

Thus, the claim is true for $i = 0$. For $l = 0$, since J^0 is empty and then J_i^0 is empty for all i, which therefore the left side of the inequality is zero.

Suppose that now the claim is true for $(i, l-1)$ and $(i-1, l)$. We will prove the claim for (i, l). Let us consider the schedule generated in Algorithm 1 in cluster i, the schedule consists of shelves. Again we let H_F be the length of shelves that contain big jobs. For small jobs, the shelves with the same length r^y consists of class y. The last shelf in any class y is called *sparse* if the number of shelves used is odd, while all the other shelves are *dense*. Let H_S and H_D be the total length of sparse shelves and dense shelves over all y, respectively.

Case 1: $(H_F + H_D)/v_i \leq c_l = 1.5r_1 \sum_{t=0}^{l} \Lambda_t$. By the hypothesis of $(i, l-1)$, it is sufficient to show that

$$L(0, \Lambda_l, J_i^l) \leq \Lambda_l(\sum_{j=i+1}^{m} v_j).$$

Since all the jobs assigned in cluster i are bounded by $v_i \cdot \Lambda_l$, we have $H_S/v_i \leq \Lambda_l(1 + r_1^{-1} + r_1^{-2} + \ldots) \leq \frac{r_1^2}{r_1-1}\Lambda_l$. Note that in iteration l, $c_l + x_l = c_l + \frac{r_1^2}{r_1-1}\Lambda_l + \Lambda_l$. Consequently, in this case cluster i accepted all jobs with length at most $\Lambda_l \cdot v_i$ from the set J_{i-1}^l. Therefore, the set J_i^l consists only of the jobs that must be executed on clusters faster than v_i, i.e. $L(0, \Lambda_l, J_i^l) = L(v_i, \Lambda_l, J_i^l)$. From the inequality (3.1), we have $L(0, \Lambda_l, J_i^l) = L(v_i, \Lambda_l, J_i^l) \leq \Lambda_l Cap(v_i) \leq \Lambda_l(\sum_{j=i+1}^{m} v_j)$.

Case 2: $(H_F + H_D)/v_i > c_l$. From the hypothesis of $(i-1, l)$, it is sufficient to show that

$$\sum_{t=0}^{l}(L(0, \Lambda_t, J_{i-1}^t) - L(0, \Lambda_t, J_i^t)) \geq (\sum_{t=0}^{l} \Lambda_t)v_i. \tag{3.2}$$

The left-hand side of inequality (3.2) is the total work of the jobs accepted by cluster i during these phases. Since $(H_F + H_D)/v_i > c_l = 1.5r_1 \sum_{t=0}^{l} \Lambda_t$, and it remains to show that the total work accepted in cluster i is at least $(H_F + H_D)/(1.5r_1)$. This is done by Lemma 2.2 and let $r_1 \geq 4/3$.   □

For $i = m$, the right-hand of the inequality in the Lemma 3.3 is zero, which implies that $\sum_{t=0}^{k} L(0, \Lambda_t, J_i^t) \leq 0$, and then $L(0, \Lambda_t, J_m^t) = 0$ for any $t \leq k$.   □

4   Conclusions

In this paper, we have investigated the on-line scheduling of parallel jobs in multiple clusters. First, for clusters with different widths but identical speeds, we designed an on-line algorithm with a competitive ratio of no more than 14.2915. Second, for clusters with different speeds but identical speeds, we provided an on-line algorithm with a competitive ratio at most of 18.2788.

Though we did not consider the case where clusters have different widths and different speeds, we believe that the technique used in this paper can be extended to this general case. It is an interesting question whether we could improve the competitive ratios for all variants of the problem in this work.

Acknowledgment. We thank anonymous referees for helpful comments to improve the presentation of this paper. We also thank Jack Snoeyink for his comments and suggestions regarding the presentation of this paper.

References

1. Aspnes, J., Azar, Y., Fiat, A., Plotkin, S., Waarts, O.: On-line load balancing with applications to machine scheduling and virtual circuit routing. Journal of the ACM 44, 486–504 (1997)
2. Baker, B.S., Schwartz, J.S.: Shelf algorithms for two-dimensional packing problems. SIAM Journal on Computing 12, 508–525 (1983)
3. Berman, P., Charikar, M., Karpinski, M.: On-line load balancing for related machines. Journal of Algorithms 35, 108–121 (2000)
4. Borodin, A., El-Yaniv, R.: Online Computation and Competitive Analysis. Cambridge University Press (1998)
5. Bougeret, M., Dutot, P.-F., Jansen, K., Otte, C., Trystram, D.: A Fast 5/2-Approximation Algorithm for Hierarchical Scheduling. In: D'Ambra, P., Guarracino, M., Talia, D. (eds.) Euro-Par 2010. LNCS, vol. 6271, pp. 157–167. Springer, Heidelberg (2010)
6. Bougeret, M., Dutot, P.F., Jansen, K., Otte, C., Trystram, D.: Approximating the non-contiguous multiple organization packing problem. In: Proceedings of Theoretical Computer Science: the 6th IFIP WG 2.2 International Conference (TCS), pp. 316–327 (2010)
7. Bougeret, M., Dutot, P.F., Jansen, K., Otte, C., Trystram, D.: Approximation Algorithms for Multiple Strip Packing. In: Bampis, E., Jansen, K. (eds.) WAOA 2009. LNCS, vol. 5893, pp. 37–48. Springer, Heidelberg (2010)
8. Bougeret, M., Dutot, P.F., Trystram, D.: An extention of the 5/2-approximation algorithm using oracle. Research Report (2010)
9. Chrobak, M., Kenyon, C.: Competitiveness via doubling. In: SIGACT News, pp. 115–126 (2006)
10. Coffman, E.G., Garey, M.R., Johnson, D.S., Tarjan, R.E.: Performance bounds for level oriented two-dimensional packing algorithms. SIAM Journal on Computing 9, 808–826 (1980)
11. Ebenlendr, T., Sgall, J.: A lower bound on deterministic online algorithms for scheduling on related machines without preemption. In: Proc. of the 9th Workshop on Approximation and Online Algorithms, WAOA 2011 (2012)
12. Harren, R., Jansen, K., Prädel, L., van Stee, R.: A $(5/3 + \epsilon)$-Approximation for Strip Packing. In: Dehne, F., Iacono, J., Sack, J.-R. (eds.) WADS 2011. LNCS, vol. 6844, pp. 475–487. Springer, Heidelberg (2011)

13. Hurink, J.L., Paulus, J.J.: Online Algorithm for Parallel Job Scheduling and Strip Packing. In: Kaklamanis, C., Skutella, M. (eds.) WAOA 2007. LNCS, vol. 4927, pp. 67–74. Springer, Heidelberg (2008)
14. Jansen, K., van Stee, R.: On strip packing With rotations. In: Proc. 37th Symp. Theory of Computing (STOC), pp. 755–761 (2005)
15. Kenyon, C., Remila, E.: Approximate Strip Packing. In: Proc. 37th Symp. Foundations of Computer Science (FOCS), vol. 37, pp. 31–37 (1996)
16. Schiermeyer, I.: Reverse-Fit: A 2-Optimal Algorithm for Packing Rectangles. In: van Leeuwen, J. (ed.) ESA 1994. LNCS, vol. 855, pp. 290–299. Springer, Heidelberg (1994)
17. Schwiegelshohn, U., Tchernykh, A., Yahyapour, R.: Online scheduling in grids. In: IEEE International Symposium on Parallel and Distributed Processing (IPDPS), pp. 1–10 (2008)
18. Steinberg, A.: A strip-packing algorithm with absolute performance bound 2. SIAM Journal on Computing 26, 401–409 (1997)
19. Ye, D., Han, X., Zhang, G.: A note on online strip packing. Journal of Combinatorial Optimization 17(4), 417–423 (2009)
20. Ye, D., Han, X., Zhang, G.: Online multiple-strip packing. Theoretical Computer Science 412(3), 233–239 (2011)
21. Zhuk, S.: Approximate algorithms to pack rectangles into several strips. Discrete Mathematics and Applications 16(1), 73–85 (2006)

On Multiprocessor Temperature-Aware Scheduling Problems

Evripidis Bampis[1,*], Dimitrios Letsios[1,2,*], Giorgio Lucarelli[1,2,*],
Evangelos Markakis[3], and Ioannis Milis[3]

[1] LIP6, Université Pierre et Marie Curie, France
{Evripidis.Bampis,Giorgio.Lucarelli}@lip6.fr
[2] IBISC, Université d' Évry, France
dimitris.letsios@ibisc.univ-evry.fr
[3] Dept. of Informatics, Athens University of Economics and Business, Greece
{markakis,milis}@aueb.gr

Abstract. We study temperature-aware scheduling problems under the
model introduced by Chrobak et al. in [9], where unit-length jobs of given
heat contributions are to be scheduled on a set of parallel identical processors. We consider three optimization criteria: makespan, maximum
temperature and (weighted) average temperature. On the positive side,
we present polynomial time approximation algorithms for the minimization of the makespan and the maximum temperature, as well as, optimal
polynomial time algorithms for minimizing the average temperature and
the weighted average temperature. On the negative side, we prove that
there is no $(\frac{4}{3}-\epsilon)$-approximation algorithm for the problem of minimizing
the makespan for any $\epsilon > 0$, unless $\mathcal{P} = \mathcal{NP}$.

1 Introduction

The exponential increase in the processing power of recent (micro)processors has
led to an analogous increase in the energy consumption of computing systems
of any kind, from compact mobile devices to large scale data centers. This has
also led to vast heat emissions and high temperatures affecting the processors'
performance and reliability. Moreover, high temperatures reduce the lifetime of
chips and may permanently damage the processors. For this reason, manufactur-
ers have set appropriate thresholds in processors' temperature and use cooling
systems. However, the energy consumption and heat emission of these cooling
systems have to be added to that of the whole system.

The issues of the energy and thermal management, in the (micro)processor and
system design levels, date back to the first computer systems. During the last few
years these issues have been also addressed at the operating system's level, generating new interesting questions. In this context the operating system has to decide
the order in which the jobs should be scheduled so that the system's temperature

* Research supported by the French Agency for Research under the DEFIS program
TODO, ANR-09-EMER-010, and by GDR du CNRS, RO.

J. Snoeyink, P. Lu, K. Su, and L. Wang (Eds.): FAW-AAIM 2012, LNCS 7285, pp. 149–160, 2012.
© Springer-Verlag Berlin Heidelberg 2012

(and/or energy consumption) remains as low as possible, while at the same time some standard user or system oriented criterion (e.g. makespan, response time, throughput, etc) is optimized. Clearly, the minimization of the temperature and the optimization of the scheduling criteria are typically in conflict, and several models have been proposed in the literature in order to analyze such conflicts and trade-offs. A first model is based on the speed-scaling technique for energy saving and the Newton's law of cooling; see for example [5,4] as well as recent reviews on speed scaling in [12,1,2]. In another model proposed in [15], a thermal RC circuit is utilized to capture the temperature profile of a processor.

Apart from the above models, in this paper we adopt the simplified model for cooling and thermal management introduced by Chrobak et al. [9] who were motivated by [14]. We consider a set of unit-length jobs (corresponding to slices of the processes to be scheduled), each one of a given heat contribution, and model the thermal behavior of the system as follows: If a job of heat contribution h is executed on a processor in a time interval $[t-1, t)$, $t \in \mathbb{N}$, and the temperature of the processor at time $t-1$ is Θ, then the processor's temperature at time t is $\frac{\Theta+h}{2}$. Although in practise [16] the heat contribution of the executed jobs and the cooling effect are spread over time, we consider a simplified discrete process in which we first add to the temperature the heat contribution of the current job and then we multiply by one half, in order to take into account the cooling effect. We consider two natural variants of the model:

- the *threshold thermal model* in which a given threshold on the temperature of the processors cannot be violated. This makes necessary the introduction of idle times in a schedule.
- the *optimization thermal model* in which there is no explicit upper bound on the temperature of the processors. The lack of such an explicit bound is counterbalanced by the fact that the minimization of the (maximum or average) temperature becomes the goal of the scheduler.

The constraints that are introduced by such temperature management models give rise to interesting and technically challenging scheduling problems, which is the focus of our work. In particular, our goal is to schedule a set of jobs on a set of m parallel identical processors so as to minimize (i) the makespan in the threshold thermal model and (ii) the maximum or average temperature in the optimization thermal model.

Related Results and Our Contribution. In [9], Chrobak et al. consider the threshold thermal model with a given temperature threshold θ. They study the problem of scheduling a set of unit-length jobs with release dates and deadlines on a single processor so as to maximize the throughput, i.e. the number of jobs that meet their deadlines, without exceeding the temperature threshold θ at any time $t \in \mathbb{N}$. Extending the well-known three-field notation for scheduling problems, this problem is denoted as $1|r_i, p_i = 1, h_i, \theta| \sum U_i$. They prove that this problem is NP-hard even for the special case when all jobs are released at time 0 and their deadlines are equal, i.e. $1|p_i = 1, h_i, \theta| \sum U_i$. Furthermore, they study the on-line version of the throughput maximization problem in the presence of release dates and deadlines. They prove that a family of reasonable list scheduling algorithms, including *coolest first* and *earliest deadline first* algorithms, have a competitive ratio of at most two. In the negative side, they also give an instance that shows that there is no deterministic on-line algorithm with competitive ratio less than two. This result implies also an approximation factor of two for the off-line problem.

The same discrete thermal model has been also adopted by Birks et al. in [7,6,8] where online algorithms for several generalizations of the throughput maximization problem have been studied. In fact, in [7] the cooling effect is taken into account by multiplying the temperature by $1/c$, where $c > 1$, instead of one half. In [6] the weighted throughput objective is considered, while in [8] the jobs have equal (non-unit) processing times.

We initiate the study of three additional optimization criteria under the threshold thermal model of [9] as well as under the optimization thermal model. Furthermore, we study all criteria for the case of multiple processors, unlike [9]. In Section 3 we address the problem of minimizing the schedule length (makespan) in the threshold thermal model ($P|p_i = 1, h_i, \theta|C_{max}$). We prove that this problem cannot be approximated within a factor less than $4/3$ and we present a generic algorithm of approximation ratio 2ρ, where ρ is the approximation ratio of an algorithm $\mathcal{A}$ for the classical makespan problem on parallel machines, used as a subroutine in our algorithm. This leads to a $(2 + \epsilon)$-approximation ratio, within a running time that is polynomial in n but exponential in $1/\epsilon$, for m processors (by using the known PTAS's for minimizing makespan) and a 2-approximation ratio for a single processor, within $O(n \log n)$ time. If instead the standard LPT $(\frac{4}{3} - \frac{1}{3m})$-approximation algorithm is used in the generic algorithm, we are able to give a tighter analysis, improving the 2ρ-approximation ratio to $\frac{7}{3} - \frac{1}{3m}$, while the overall running time is in $O(n \log n)$. In Sections 4 and 5 we move to the optimization thermal model. In Section 4, we study the problem of minimizing the maximum temperature of a schedule ($P|p_i = 1, h_i|\Theta_{max}$), and we give a $4/3$ approximation algorithm. In Section 5, we prove that the problem of minimizing the average temperature of a schedule ($P|p_i = 1, h_i|\sum \Theta_i$), as well as a time-dependent weighted version of this problem are both solvable in polynomial time. We conclude in Section 6.

2   Notation and Preliminaries

We consider a set $J = \{J_1, J_2, \ldots, J_n\}$ of n jobs to be executed on a system of m identical processors. All jobs have unit processing times and for each one of them we are given a heat contribution h_i, $1 \leq i \leq n$. We consider each job J_i executed in a time interval $[t - 1, t)$, $t \in \mathbb{N}$, which we call slot t, on some processor. By Θ_t^j we denote the temperature of processor j at time t. As in [9] if we start executing job J_i at time $t - 1$, then $\Theta_t^j = \frac{\Theta_{t-1}^j + h_i}{2}$. The initial temperature of each processor (the ambient temperature) is considered to be zero, i.e., $\Theta_0^j = 0$. In what follows, we simplify the notation by using Θ_t instead of Θ_t^j, when the processor is specified by the context.

The threshold thermal model. In this model, the temperature is not allowed to exceed a threshold θ at any time $t \in \mathbb{N}$. It is clear that, for a given instance in this model, a feasible schedule may exist only if $h_i \leq 2 \cdot \theta$ for each job J_i. By normalizing the values of h_i's and θ we can assume w.l.o.g. that $0 < h_i \leq 2$ and $\theta = 1$. Moreover, if a processor at time $t - 1$ has temperature Θ_{t-1} and it holds that $\frac{\Theta_{t-1} + h_i}{2} > 1$, for every job J_i that has not yet been scheduled, then this processor will remain idle for the slot $[t - 1, t)$ and its temperature at time t will be reduced by half, i.e., $\frac{\Theta_{t-1}}{2}$. Note also that once a processor has executed some job(s) its temperature will never become exactly zero. Therefore, in this model, a feasible instance can not contain more than m jobs of heat contributions equal to 2, as there are m slots with $\Theta_0 = 0$ (the first slots in each one of the m available processors).

The optimization thermal model. In this model, no explicit temperature threshold is given and the problems we will study are the minimization of maximum and average temperature. For any instance in this model, any schedule of length at least $\lceil \frac{n}{m} \rceil$ is feasible, independently of the range of the jobs' heat contributions. However, the optimum value of our objectives depends on the time available to execute the given set of jobs: the maximum or average temperature of a schedule of length $\lceil \frac{n}{m} \rceil$ is, clearly, greater than that of a schedule of bigger length, where we are allowed to introduce idle slots. Hence, we are interested in minimizing these two objective functions with respect to a given schedule length $d \geq \lceil \frac{n}{m} \rceil$. In what follows, we only consider instances with $n = m \cdot d$, since for instances with $n < md$ we can simply add $md - n$ fictive jobs of heat contribution equal to zero. Thus, the number of jobs equals the total number of available slots of the m processors.

We close this section by elaborating on the complexity of the problems studied in the rest of the paper. It is already mentioned in [9] that even for a single processor, the NP-hardness of the maximum throughput problem for the case where jobs are released at time 0 and all deadlines are equal ($1|p_i = 1, h_i, \theta| \sum U_i$) implies the NP-hardness of the makespan minimization problem ($1|p_i = 1, h_i, \theta|C_{max}$). In fact, the decision version of the latter problem asks for the existence of a feasible schedule where all jobs complete their execution by some given deadline d. Moreover, the decision version of the maximum temperature problem for a single processor ($1|p_i = 1, h_i|\Theta_{max}$) asks for the existence of a schedule where all jobs complete their execution by some given deadline d without exceeding a given temperature threshold θ. Therefore, the same reduction gives NP-hardness for both makespan and maximum temperature minimization problems. The NP-hardness for our problems on an arbitrary number of parallel processors follows trivially.

3   Makespan Minimization

In this section we study the approximability of makespan minimization under the threshold thermal model, that is $P|p_i = 1, h_i, \theta|C_{max}$.

We start with a negative result on the approximability of our problem. The proof of the next theorem is along the same lines with the NP-hardness reduction for the throughput maximization problem under the same model [9] and it is omitted.

Theorem 1. *It is NP-hard to approximate the minimum makespan problem* $(P|p_i = 1, h_i, \theta|C_{\max})$ *within a factor better than* $4/3$.

In what follows in this section, we present an approximation algorithm for the minimum makespan problem. Note that, in order to respect the temperature threshold, a schedule may have to contain idle slots. To argue about the number of idle slots that are needed before the execution of each job, we will introduce first an appropriate partition of the set of jobs according to their heat contribution. In particular, for each integer $k \geq 0$, we can argue separately for jobs whose heat contribution belongs to the interval $(\frac{2^k-1}{2^{k-1}}, \frac{2^{k+1}-1}{2^k}]$; recall that $h_i \leq 2$, $1 \leq i \leq n$. Moreover, the interval to which a job of heat contribution h_i belongs to is indexed by k_i, that is

$$k_i = \max\{k \in \mathbb{N} \mid h_i > \tfrac{2^k-1}{2^{k-1}}\}.$$

Our algorithm and its analysis are based on the following proposition for the structure of any feasible schedule.

Proposition 1
(i) Let J' be the set of jobs of heat contribution $h_i > 1$; $|J'| = n'$. Any feasible schedule can be transformed into another feasible one of at most the same length where exactly $\max\{m - n', 0\}$ jobs in $J \setminus J'$ are executed in the first slot of the processors.
(ii) Any schedule where every J_i is executed after at least k_i idle slots is feasible.
(iii) In an optimal schedule, between the execution on the same processor of jobs J_j and J_i of heat contributions $h_j, h_i > 1$, there are at least $k_i - 1$ slots, which are either idle or execute jobs of heat contribution at most one.

In what follows we consider instances with $n > m$ for otherwise the problem becomes trivial. By Proposition 1(i) we assume also that the number of jobs of heat contribution $h_i > 1$ is greater than m, for otherwise all jobs are executed without any idle before them and the length of an optimal schedule is exactly $\lceil \frac{n}{m} \rceil$. We consider the jobs in non-increasing order of their heat contributions, i.e., $h_1 \geq h_2 \geq \ldots \geq h_n$, and we define $A = \{J_1, J_2, \ldots, J_m\}$ and $B = \{J_{m+1}, J_{m+2}, \ldots, J_n\}$. Our algorithm schedules first the jobs in A to the first slot of each processor. Each one of the jobs in B is scheduled by leaving before its execution *exactly* k_i idle slots, according to the Proposition 1(ii). In this way, our problem, for the jobs in B, is transformed to an instance of the classical makespan problem on parallel machines, $P||C_{\max}$, where the processing time of each job is $p_i = k_i + 1$, that is, k_i idle slots plus its original unit processing time. Then, these jobs are scheduled using an algorithm $\mathcal{A}$ for $P||C_{\max}$.

We denote by SOL the length of the schedule S provided by Algorithm MAX_C and by OPT the length of an optimal schedule S^* for our problem.

Algorithm MAX_C

1: Sort the jobs in non-increasing order of their heat contributions: $h_1 \geq h_2 \geq \ldots \geq h_n$;
2: Let $A = \{J_1, J_2, \ldots, J_m\}$, and $B = \{J_{m+1}, J_{m+2}, \ldots, J_n\}$;
3: Schedule each job $J_i \in A$ to the first slot of processor i;
4: For each job $J_i \in B$, let $p_i = k_i + 1$;
5: Schedule the jobs in B by running an algorithm $\mathcal{A}$ for $P||C_{\max}$;

By $\mathcal{S}(k_i + 1)$ we denote the schedule found by an algorithm $\mathcal{A}$ for the $P||C_{\max}$ problem for the jobs in B with processing times $p_i = k_i + 1$; let $\mathcal{C}(k_i + 1)$ be the length of this schedule. Clearly, $SOL = 1 + \mathcal{C}(k_i + 1)$. Similarly, we denote an optimal schedule for the same instance of $P||C_{\max}$ by $\mathcal{S}^*(k_i + 1)$ and its length by $\mathcal{C}^*(k_i + 1)$.

To analyze our Algorithm MAX_C, we first need a lower bound on the optimal makespan. To derive this bound we will also utilize an optimal schedule $\mathcal{S}^*(k_i)$ for the jobs of B assuming that they have processing times $p_i = k_i$. Note that for jobs with $h_i \in (0, 1]$, $k_i = 0$, hence the schedule $\mathcal{S}^*(k_i)$ involves only jobs for which $h_i > 1$. Let $\mathcal{C}^*(k_i)$ denote the length of such a schedule.

Lemma 1. *For the optimal makespan it holds that $OPT \geq \max\{\frac{n}{m}, 1 + \mathcal{C}^*(k_i)\}$.*

Proof. The first bound on the optimal makespan follows trivially by considering all jobs requiring a single slot for their execution.

For the second bound, let A^*, $|A^*| = m$, be the set of jobs executed in the first slot of the m processors in an optimal solution and $B^* = J \setminus A^*$.

Consider, first, an auxiliary schedule of length OPT^-, identical to the optimal apart from the fact that each job in $B^* \cap A$ has been replaced by a different job in $A^* \cap B$. Observe that in this schedule, the jobs executed in the first slot of the processors remain A^* while the jobs executed in the remaining slots are the jobs in B. Since each job in B has smaller or equal heat contribution than any job in A, it follows that $OPT \geq OPT^-$.

Consider, next, the schedule $\mathcal{S}^*(k_i)$. For this schedule it holds that, $OPT^- \geq 1 + \mathcal{C}^*(k_i)$, since by Proposition 1(i),(iii) each job in B requires at least k_i slots to be executed; recall that we consider instances where the number of jobs of heat contribution $h_i > 1$ is greater than m and that jobs in B with $h_i \leq 1$, and hence $k_i = 0$, do not appear in the schedule $\mathcal{S}^*(k_i)$. $\qquad\square$

It is well-known that the $P||C_{\max}$ problem is strongly NP-hard and a series of constant approximation algorithms and PTASs have been proposed. Our main result in this Section is that in step 5 of Algorithm MAX_C we can use any algorithm $\mathcal{A}$ for $P||C_{\max}$ to obtain twice the approximation ratio of $\mathcal{A}$ for our problem.

Theorem 2. *Algorithm MAX_C achieves an approximation ratio $\frac{SOL}{OPT} \leq 2\rho$, where ρ is the approximation ratio of the algorithm $\mathcal{A}$ for $P||C_{\max}$.*

Proof. A ρ-approximation algorithm $\mathcal{A}$ implies that $\frac{\mathcal{C}(k_i+1)}{\mathcal{C}^*(k_i+1)} \leq \rho$. Hence, $SOL = 1 + \mathcal{C}(k_i + 1) \leq 1 + \rho \cdot \mathcal{C}^*(k_i + 1)$.

To obtain an upper bound to $\mathcal{C}^*(k_i+1)$ we start from the schedule $\mathcal{S}^*(k_i)$. The processing times of jobs in the latter schedule are reduced by one with respect to the former one, and the jobs in B with $h \leq 1$ do not appear in schedule $\mathcal{S}^*(k_i)$. Let $B' \subseteq B$ be this set of jobs.

We transform the schedule $\mathcal{S}^*(k_i)$ to a new schedule $\mathcal{S}'(k_i+1)$ in two successive steps: (i) we increase the processing time of jobs in $B \setminus B'$ from k_i to $k_i + 1$, and (ii) we introduce the jobs in B' with unit processing time, at the end of the resulting schedule in a first-fit manner. Clearly, for the length, $\mathcal{C}'(k_i + 1)$, of this new schedule it holds that $\mathcal{C}^*(k_i + 1) \leq \mathcal{C}'(k_i + 1)$ as both of them refer to the same instance. Let us now bound $\mathcal{C}'(k_i + 1)$ in terms of $\mathcal{C}^*(k_i)$.

If $\mathcal{C}'(k_i + 1) \leq 2\mathcal{C}^*(k_i)$, then $\frac{SOL}{OPT} \leq \frac{1+2\rho\mathcal{C}^*(k_i)}{1+\mathcal{C}^*(k_i)} \leq 2\rho$, since $\rho \geq 1$.

If $\mathcal{C}'(k_i + 1) > 2\mathcal{C}^*(k_i)$, then we consider the construction of $\mathcal{S}'(k_i + 1)$ and we argue about the completion time of a critical processor in $\mathcal{S}^*(k_i)$, i.e., the processor that finishes last. By step (i), the length of schedule $\mathcal{S}^*(k_i)$ increases at most twice, since each job in $B \setminus B'$ has processing time at least one and this is increased by 1. As $\mathcal{C}'(k_i+1) > 2\mathcal{C}^*(k_i)$, in the last slot of $\mathcal{S}'(k_i+1)$ all non-idle processors execute jobs of B'. By step (ii), all but the last time slots of $\mathcal{S}'(k_i+1)$ are busy. Hence, the critical processor in $\mathcal{S}^*(k_i)$ finishes in $\mathcal{S}'(k_i + 1)$ the earliest at time $\mathcal{C}'(k_i + 1) - 1$. Moreover, this processor is assigned the minimum total increase at the end of the transformation, since it finishes last in $\mathcal{S}^*(k_i)$. As the total increase of the processing times from $\mathcal{S}^*(k_i)$ to $\mathcal{S}'(k_i + 1)$ is $n - m$, it follows that the length of the critical processor increases at most by $\frac{n-m}{m}$. Hence, $\mathcal{C}'(k_i + 1) - 1 \leq \mathcal{C}^*(k_i) + \frac{n-m}{m}$, that is $\mathcal{C}'(k_i + 1) \leq \mathcal{C}^*(k_i) + \frac{n}{m}$. Thus, by Lemma 1 we get $\frac{SOL}{OPT} \leq \frac{1+\rho(\mathcal{C}^*(k_i)+\frac{n}{m})}{\max\{\frac{n}{m},1+\mathcal{C}^*(k_i)\}} \leq \frac{1+\rho\mathcal{C}^*(k_i)}{1+\mathcal{C}^*(k_i)} + \frac{\rho\frac{n}{m}}{\frac{n}{m}} \leq 2\rho.$ □

For the case of a single processor the $1||C_{max}$ problem is trivially polynomial, whereas for multiple processors there are well known PTAS's, e.g., [11,3]. Hence the main implication of Theorem 2 is:

Corollary 1. *For any $\epsilon > 0$, Algorithm MAX_C achieves a $2+\epsilon$-approximation ratio for $P|p_i = 1, h_i, \theta|C_{\max}$. For a single processor, it achieves an approximation ratio of 2.*

To obtain the ratio of $2 + \epsilon$, as stated above, one needs to use a PTAS for the classical makespan problem in step 5 of Algorithm MAX_C, resulting in a running time that is exponential in $1/\epsilon$. To achieve more practical running times, we can investigate the use of other algorithms for step 5. In particular, if the standard Longest Processing Time (LPT) algorithm is used, then Theorem 2 leads to a $2(\frac{4}{3} - \frac{1}{3m})$ approximation ratio within $O(n \log n)$ time. Recall that the LPT algorithm simply assigns the next job (in the non-increasing order of their processing times) to the first available processor [10]. In the next theorem we are able to improve this ratio to $7/3$, based on an LPT oriented analysis of Algorithm MAX_C.

Theorem 3. *Algorithm* MAX_C *using the LPT rule in step* 5 *achieves an approximation ratio of* $\frac{7}{3} - \frac{1}{3m}$ *within* $O(n \log n)$ *time.*

4 Maximum Temperature Minimization

Now, we turn our attention to the optimization thermal model and to the problem of minimizing the maximum temperature, i.e., $P|p_i = 1, h_i|\Theta_{max}$. Recall that as we discussed in Section 2, we consider a schedule length $d \geq \lceil \frac{n}{m} \rceil$ and that $n = m \cdot d$, by adding the appropriate number of fictive jobs. By Θ^*_{max} we denote the maximum temperature of an optimal schedule.

We start with the observation that any algorithm for this problem achieves a 2 approximation ratio. Indeed, it holds that $\Theta^*_{\max} \geq h_{max}/2$, no matter how we schedule the job of maximum heat contribution. It also holds that for any algorithm, $\Theta_{\max} \leq h_{max}$, with $\Theta_{\max}$ being the maximum temperature of the algorithm's schedule. Therefore, $\Theta_{\max} \leq 2 \cdot \Theta^*_{\max}$.

To improve this trivial ratio we propose the following algorithm which is based on the intuitive idea of alternating the execution of hot and cool jobs.

Algorithm MAX_T

1: Sort the jobs in non-increasing order of their heat contributions: $h_1 \geq h_2 \geq \ldots \geq h_n$;
2: Using the order of Step 1, schedule the $\lceil \frac{d}{2} \rceil m$ hottest jobs to the *odd* slots of the processors using Round-Robin;
3: Using the reverse order of Step 1, schedule the $\lfloor \frac{d}{2} \rfloor m$ coolest jobs to the *even* slots of the processors using Round-Robin;

To elaborate a little more on how the algorithm works, note that processor 1 will be assigned the job J_1, followed by J_n, then followed by J_{m+1}, and then by J_{n-m} and this alternation of hot and cool jobs will continue till the end of the schedule. Similarly processor 2 will de assigned the jobs J_2, J_{n-1}, J_{m+2}, J_{n-m-1}, and so on. The schedule is illustrated further in Table 1.

To analyze the Algorithm MAX_T, we start with the proposition below, which is implied by the Round-Robin scheduling of jobs in its Steps 2 and 3.

Proposition 2. *In the schedule returned by Algorithm* MAX_T:
(i) A job J_i, $i \geq (\lfloor \frac{d}{2} \rfloor + 1)m + 1$, *is succeeded by the job* $J_{n-i+m+1}$.
(ii) A job J_i, $m + 1 \leq i \leq \lceil \frac{d}{2} \rceil m$, *is preceded by the job* $J_{n-i+m+1}$.

Table 1. The schedule produced by Algorithm MAX_T

1	J_1	J_n	J_{m+1}	J_{n-m}	J_{2m+1}	...
2	J_2	J_{n-1}	J_{m+2}	J_{n-m-1}	J_{2m+2}	...
...	...	...	...	...	...	...
m	J_m	J_{n-m+1}	J_{2m}	J_{n-2m+1}	J_{3m}	...

The maximum temperature may appear at various points of the schedule of Algorithm MAX_T. The next lemma states that one of these points satisfies a certain property regarding the heat contribution of the job executed right before.

Lemma 2. *In the schedule returned by Algorithm* MAX_T, *the maximum temperature is achieved after the execution of a job* J_i, *with* $i \leq (\lfloor \frac{d}{2} \rfloor + 1)m$.

Proof. Assume that all the points where the maximum temperature $\Theta_{\max}$ occurs are after the execution of a job J_i, with $i \geq (\lfloor \frac{d}{2} \rfloor + 1)m + 1$. By Proposition 2, such a job is succeeded by a job $J_{i'}$, $i' = n - i + m + 1$, in the schedule returned by Algorithm MAX_T. It is easy to check that $i > i'$, hence $h_{i'} \geq h_i$. Let $\Theta, \Theta' \leq \Theta_{\max}$ be the temperatures before the execution of J_i and after the execution of $J_{i'}$, respectively. Then, $\Theta_{\max} = \frac{\Theta + h_i}{2}$ and $h_i \geq \Theta_{\max}$, since $\Theta_{\max} \geq \Theta$. Moreover, $\Theta' = \frac{\Theta_{\max} + h_{i'}}{2} \geq \Theta_{\max}$, since $h_{i'} \geq h_i$. This implies that $\Theta' = \Theta_{\max}$, since $\Theta' \leq \Theta_{\max}$. But this means that the maximum temperature is also achieved after the execution of job $J_{i'}$, which is a contradiction because
$$i' = n - i + m + 1 \leq m(d - \lfloor \tfrac{d}{2} \rfloor) \leq m(\lfloor \tfrac{d}{2} \rfloor + 1)$$
contrary to what we assumed in the beginning of the proof. $\square$

Lemma 3. *For the maximum temperature of an optimal schedule it holds that* $\Theta^*_{max} \geq \frac{h_{n-i+m+1}}{4} + \frac{h_i}{2}$, *for any* i, $m + 1 \leq i \leq \lceil \frac{d}{2} \rceil m$.

Proof. Consider a job J_i and let $J_{i'}$ be its previous job in the same processor in an optimal schedule S^*. The jobs executed in the first slot of each processor in S^* do not have a previous one. To simplify the presentation of our proof, we assume that they are preceded by hypothetical jobs J_{n+j}, $1 \leq j \leq m$.

If $i' \leq n - i + m + 1$, then $\Theta^*_{max} \geq \frac{h_{i'}}{4} + \frac{h_i}{2} \geq \frac{h_{n-i+m+1}}{4} + \frac{h_i}{2}$, since $h_{i'} \geq h_{n-i+m+1}$.

If $i' > n - i + m + 1$, then let $B = \{J_{n-i+m+2}, J_{n-i+m+3}, \ldots, J_n, J_{n+1}, \ldots, J_{n+m}\}$ and let A be the set of jobs that precede the jobs $J_1, J_2, \ldots, J_{i-1}$ in the optimal schedule. Clearly, $|B| = |A| = i - 1$, $J_{i'} \in B$ and $J_{i'} \notin A$ since $J_{i'}$ precedes J_i in S^*.

Therefore, there is a job $J_{k'} \in A$ such that $J_{k'} \notin B$, that is $k' < n - i + m + 2$. For any $i \leq \lceil \frac{d}{2} \rceil m$, the job $J_{k'}$ precedes a job J_k in S^* and since $J_{k'} \in A$ it follows, by the definition of the set A, that $k < i$. Hence, $\Theta^*_{max} \geq \frac{h_{k'}}{4} + \frac{h_k}{2} \geq \frac{h_{n-i+m+1}}{4} + \frac{h_i}{2}$, since $h_k \geq h_i$ and $h_{k'} \geq h_{n-i+m+1}$. $\square$

Theorem 4. *Algorithm* MAX_T *achieves a* $\frac{4}{3}$ *approximation ratio.*

Proof. By Lemma 2 the maximum temperature in the schedule, S, obtained by Algorithm MAX_T occurs after the execution of a job J_i, $i \leq (\lfloor \frac{d}{2} \rfloor + 1)m$ (the maximum may be achieved in other timeslots as well).

If $1 \leq i \leq m$, then the maximum occurs at the first processor and $\Theta_{\max} = \frac{h_1}{2} \leq \Theta^*_{\max}$ and, hence, the algorithm returns an optimal schedule.

If $m + 1 \leq i \leq \lceil \frac{d}{2} \rceil m$ then by Proposition 2, the job J_i is preceded in the schedule S by the job $J_{n-i+m+1}$. Let Θ be the temperature before the execution of the job $J_{n-i+m+1}$. By Lemma 3, and since $\Theta \leq \Theta_{\max}$, $\Theta_{\max} = \frac{\Theta}{4} + \frac{h_{n-i+m+1}}{4} + \frac{h_i}{2} \leq \frac{\Theta_{\max}}{4} + \Theta^*_{\max}$. Hence, $\Theta_{\max} \leq \frac{4}{3} \cdot \Theta^*_{\max}$.

Note that if d is odd, then $\lceil \frac{d}{2} \rceil m = (\lfloor \frac{d}{2} \rfloor + 1)m$ and the analysis of the previous case holds. Hence the only remaining case is that d is even and $\lceil \frac{d}{2} \rceil m + 1 \leq i \leq (\lfloor \frac{d}{2} \rfloor + 1)m$. For this case, let $\Theta' \leq \Theta_{\max}$ be the temperature before the execution of J_i. Then, $h_i \geq \Theta_{\max}$, since $\Theta_{\max} = \frac{\Theta' + h_i}{2}$ and $\Theta_{\max} \geq \Theta'$. Thus, there are at least $\lceil \frac{d}{2} \rceil m + 1$ jobs of heat contribution at least $\Theta_{\max}$. Note that, in any schedule, each processor can execute at most $\lceil \frac{d}{2} \rceil$ jobs without any pair of them scheduled in two consecutive slots. Hence, in an optimal schedule, there are at least two jobs J_p and J_q, $p, q \leq i$, executed in consecutive slots in the same processor. Therefore, $\Theta^*_{\max} \geq \frac{h_p}{4} + \frac{h_q}{2} \geq \frac{\Theta_{\max}}{4} + \frac{\Theta_{\max}}{2} = \frac{3}{4} \cdot \Theta_{\max}$, that is $\Theta_{\max} \leq \frac{4}{3} \cdot \Theta^*_{\max}$. $\qquad \square$

For the tightness of the analysis of Algorithm MAX_T consider an instance of m processors, mn^2 jobs and $d = n^2$; mn hot jobs of heat contribution $h = 2$ and $mn(n - 1)$ cool jobs of heat contribution $h = \epsilon$. We consider n to be sufficiently large and that ϵ tends to 0. The algorithm in each processor alternates n hot jobs with $n - 1$ cool jobs and schedules $n(n - 2) + 1$ cool jobs at the end. The maximum temperature of the algorithm's schedule is attained exactly after the execution of the last hot job on each processor. This job is executed at slot $2n - 1$, and thus $\Theta_{max} = \frac{2}{2^{2n-1}} + \frac{\epsilon}{2^{2n-2}} + \frac{2}{2^{2n-3}} + \frac{\epsilon}{2^{2n-4}} + \ldots + \frac{\epsilon}{2^2} + \frac{2}{2^1} \simeq 2\frac{\frac{1}{2}}{1 - \frac{1}{4}} = \frac{4}{3}$.

On the other hand, the optimal solution alternates in each processor a hot job with $n - 1$ cool jobs. The temperature before the execution of any hot job tends to zero and the maximum temperature is one.

5   Average Temperature Minimization

In this section, we look at the problem of minimizing the average temperature, that is $P|p_i = 1, h_i| \sum \Theta_i$, instead of the maximum temperature. We will again consider a schedule length d and assume that the number of jobs is $n = md$. Contrary to the maximum temperature, we show that minimizing the average temperature of a schedule is solvable in polynomial time. Our algorithm is based on the following lemma whose proof is omitted.

Lemma 4. *In any optimal solution for the average temperature, jobs are scheduled in a coolest first order, i.e., for any pair of jobs J_i, J_j such that $h_i > h_j$ scheduled at slots t and t', respectively, it holds that $t' \leq t$, regardless of the processor they are assigned to.*

The previous lemma leads directly to the next simple algorithm.

Algorithm AVR_T finds a schedule in $O(n \log n)$ time. The optimality of this schedule follows directly by the Round-Robin scheduling of the jobs in nondecreasing order of their heat contributions and Lemma 4.

Theorem 5. *An optimal schedule for the problem of minimizing the average temperature $(P|p_i = 1, h_i| \sum \Theta_i)$ can be found in polynomial time.*

Algorithm AVR_T

1: Sort the jobs in non-decreasing order of their heat contributions: $h_1 \leq h_2 \leq ... \leq h_n$;

2: According to this order schedule the jobs to processors using Round-Robin;

5.1   Weighted Average Temperature Minimization

In what follows, we consider a time-dependent weighted version of average temperature minimization. In particular, we consider each slot of every processor to be associated with a given positive weight w_i, $1 \leq i \leq d$, and our problem is denoted as $P|p_i = 1, h_i| \sum w_i \cdot \Theta_i$. The weights w_i could represent the interest of the system manager to keep its processors/computers cool during specific time periods of peak loads. This leads to special, but more practical cases, of our formulation where the weights of some slots (e.g. the same or consecutive slots in all processors) could be considered equal.

To be more precise with our presentation we denote the weight of the t-th slot of processor p by w_t^p, $1 \leq t \leq d$, $1 \leq p \leq m$. Similarly with the un-weighted case, we consider a job J_i of heat contribution h_i scheduled in the t-th slot of processor p in a schedule S. The contribution of this job to the weighted temperature of the s-th slot of processor p, with $t \leq s \leq d$, is $w_s^p \cdot \frac{h_i}{2^{s-t+1}}$, and this job does not affect the temperature of any other slot in any processor. Hence, the contribution of job J_i to the total weighted temperature of the schedule S is $\sum_{s=t}^{d} w_s^p \cdot \frac{h_i}{2^{s-t+1}} = h_i \cdot \sum_s^d \frac{w_s^p}{2^{s-t+1}}$. Clearly, the quantity $c_t^p = \sum_s^d \frac{w_s^p}{2^{s-t+1}}$ is a constant that depends only on the slot t of processor p and not on the job executed in this slot.

Based on this, our problem can be transformed into a weighted bipartite matching problem and the next theorem holds.

Theorem 6. *The problem of minimizing the weighted average temperature ($P| p_i = 1, h_i| \sum w_i \cdot \Theta_i$) is polynomially solvable.*

6   Conclusions

The most important open question is to improve the approximation ratio for the problems of minimizing the makespan and minimizing the maximum temperature. Also it would be interesting to generalize our results in the case where the cooling effect is different than one half, as in [7,6,8], and to consider other classical scheduling objectives under this thermal model. Resolving these questions seems technically more challenging than the classic scheduling problems due to the different nature of the constraints that are introduced by temperature management models. Note that scheduling problems under the threshold thermal model can be seen as scheduling problems with sequence-dependent setup times; such a setup time for a job corresponds to the idle slots required to respect the temperature threshold. In this context (see for example [13]), the set-up time of a job usually depends only on the job itself and the previous job in the schedule. However, in our case, the number of idle slots, required before executing a job, depends on all the jobs scheduled before as well as on their order, hence existing results from the literature cannot be applied.

References

1. Albers, S.: Energy-efficient algorithms. Commun. ACM 53, 86–96 (2010)
2. Albers, S.: Algorithms for dynamic speed scaling. In: STACS 2011. LIPIcs, vol. 9. Schloss Dagstuhl - Leibniz-Zentrum fuer Informatik (2011)
3. Alon, N., Azar, Y., Woeginger, G., Yadid, T.: Approximation schemes for scheduling on parallel machines. Journal of Scheduling 1, 55–66 (1998)
4. Atkins, L., Aupy, G., Cole, D., Pruhs, K.: Speed Scaling to Manage Temperature. In: Marchetti-Spaccamela, A., Segal, M. (eds.) TAPAS 2011. LNCS, vol. 6595, pp. 9–20. Springer, Heidelberg (2011)
5. Bansal, N., Kimbrel, T., Pruhs, K.: Speed scaling to manage energy and temperature. J. ACM 54(1), Article 3 (2007)
6. Birks, M., Cole, D., Fung, S.P.Y., Xue, H.: Online Algorithms for Maximizing Weighted Throughput of Unit Jobs with Temperature Constraints. In: Atallah, M., Li, X.-Y., Zhu, B. (eds.) FAW-AAIM 2011. LNCS, vol. 6681, pp. 319–329. Springer, Heidelberg (2011)
7. Birks, M., Fung, S.P.Y.: Temperature Aware Online Scheduling with a Low Cooling Factor. In: Kratochvíl, J., Li, A., Fiala, J., Kolman, P. (eds.) TAMC 2010. LNCS, vol. 6108, pp. 105–116. Springer, Heidelberg (2010)
8. Birks, M., Fung, S.P.Y.: Temperature Aware Online Algorithms for Scheduling Equal Length Jobs. In: Atallah, M., Li, X.-Y., Zhu, B. (eds.) FAW-AAIM 2011. LNCS, vol. 6681, pp. 330–342. Springer, Heidelberg (2011)
9. Chrobak, M., Dürr, C., Hurand, M., Robert, J.: Algorithms for Temperature-Aware Task Scheduling in Microprocessor Systems. In: Fleischer, R., Xu, J. (eds.) AAIM 2008. LNCS, vol. 5034, pp. 120–130. Springer, Heidelberg (2008)
10. Graham, R.L.: Bounds on multiprocessing timing anomalies. SIAM J. Appl. Math. 17, 416–426 (1969)
11. Hochbaum, D.S., Shmoys, D.B.: Using dual approximation algorithms for scheduling problems: theoretical and practical results. J. ACM 34, 144–162 (1987)
12. Irani, S., Pruhs, K.R.: Algorithmic problems in power management. ACM SIGACT News 36, 63–76 (2005)
13. Pinedo, M.: Scheduling: Theory, Algorithms and Systems. Prentice-Hall (1995)
14. Yang, J., Zhou, X., Chrobak, M., Zhang, Y., Jin, L.: Dynamic thermal management through task scheduling. In: ISPASS 2008, pp. 191–201. IEEE Computer Society (2008)
15. Zhang, S., Chatha, K.S.: Approximation algorithm for the temperature-aware scheduling problem. In: ICCAD 2007, pp. 281–288. IEEE Press (2007)
16. Zhou, X., Yang, J., Chrobak, M., Zhang, Y.: Performance-aware thermal management via task scheduling. ACM Trans. Archit. Code Optimizat. 7, 1–31 (2010)

Online Minimum Makespan Scheduling with a Buffer

Yan Lan[1], Xin Chen[2], Ning Ding[2], György Dósa[3], and Xin Han[2,*]

[1] Dalian Neusoft Institute of Information
`lanyan@neusoft.edu.cn`
[2] Software School, Dalian University of Technology
`chenx_dlut@163.com`, {`dingning,hanxin`}`@dlut.edu.cn`
[3] Department of Mathematics, University of Pannonia, Veszprém, Hungary
`dosagy@almos.vein.hu`

Abstract. In this paper we study an online minimum makespan scheduling problem with a reordering buffer. We obtain the following results, which improve on work from FOCS 2008: i) for m identical machines, we give a 1.5-competitive online algorithm with a buffer of size $1.5m$, which is better than the previous best result : 1.5-competitive algorithm with a buffer of size $1.6197m$; ii) for three identical machines, to give an optimal online algorithm we reduce the size of the buffer from nine to six; iii) for m uniform machines, using a buffer of size m, we improve the competitive ratio from $2 + \epsilon$ to $2 - 1/m + \epsilon$, where $\epsilon > 0$ is sufficiently small.

1 Introduction

In the classic minimum makespan scheduling problem, we are asked to allocate a set of jobs with processing times (here, we also call it size) to m parallel machines without preemption. The target is to minimize the makespan, i.e., the time when all the jobs are precessed. This problem is NP-hard in the strong sense [10]. Normally before we assign all the jobs to machines we know the information of all the jobs, this version of the problem is called offline version. Another version of this problem is called online version, where the information of all the jobs is given gradually, after the current incoming job is handled, we know the next job. And once the decision is made, we cannot change it. This condition is strict in some sense. To relax this strict condition, we consider an online non-preemptive scheduling with a reordering buffer, at each time step, we have two choices: assign the job to some machine or store it the buffer temporally, once the job is assigned on some machine, it cannot be stored in the buffer. In the final step when the input ends, all the jobs in the buffer must be allocated to some machines. Since the buffer has a size limitation, the key point is how to select jobs and put them into the buffer. This problem has been studied in [15,7,5,4,19]. To evaluate online algorithms, we use one of the standard measures: competitive ratio. Given any

---

* Partially supported by NSFC(11101065).

sequence of jobs if the makespan by an online algorithm is at most c times the one by an optimal algorithm, we call the online algorithm c-competitive.

Previous Results: For the online minimum makespan scheduling problem on m identical machines, Graham first gave a $(2 - \frac{1}{m})$-competitive algorithm, called *list scheduling*. The upper bound has been improved in [2,14,1]. The best known one is 1.9201 due to Fleischer and Wahl [8] . The best lower bound known to date is 1.88 due to Rudin and Chandrasekaran [13]. Kellerer et al [15], Zhang [19] studied the online minimum makespan scheduling problem on two identical machines with a buffer of size one and obtained a $\frac{4}{3}$-competitive algorithm, which is the best possible we can do. Recently, a seminal paper is by Englert et al [7], they gave several results on identical machines and uniform machines. For m parallel machines with a reordering buffer of size $\Theta(m)$, they gave a lower bound $4/3 \le r_m \le 1.4659$ (for example, $r_2 = 4/3$ and $r_3 = \frac{15}{11} \approx 1.3636$), and proposed an optimal online algorithm with a buffer of size $\lceil 2.5m \rceil + 1$. For m related machines, a $(2 + \epsilon)$-competitive algorithm was obtained. In that paper several lower and upper bounds are given, for example a 1.5-competitive online algorithm with a buffer of size $1.6197m$ is given. The *preemptive* online minimum makespan scheduling problem on m identical machines with a buffer of size k was studied by Dósa and Epstein [5]. The *non-preemptive* online minimum makespan scheduling problem on two related machines with a buffer of size k was studied by Dósa and Epstein [4].

Related Models: Some similar models have been investigated in the last years such as i) online scheduling problems with *bounded migration* [17]; when a new job comes some already scheduled jobs can be reassigned. Here the bound migration means that the total size of rescheduled jobs is bounded or the total cost of the rescheduled jobs is bounded. ii) Online scheduling problems with *bounded rearrangement*, where we are allowed to reschedule a bounded number of jobs in order to allocate a new job and there are several variants of this model [6,18,16,3].

Our Contribution: i) For m identical machines, we give a 1.5-competitive online algorithm with a buffer of size $1.5m$, which is better than the previous best result : 1.5-competitive algorithm with a buffer of size $1.6197m$ in [7]; ii) for three identical machines, we propose an optimal online algorithm with a buffer of size six, which is better than the previous result of a buffer with size nine in [7]; iii) for m uniform machines, using a buffer of size m, we improve the competitive ratio from $2 + \epsilon$ in [7] to $2 - 1/m + \epsilon$, where $\epsilon > 0$ is arbitrarily small.

2   Preliminaries

Our problem is the same with the one presented in the paper [7].

Input: Given a job sequence $J = \{j_1, j_2, ..., j_n\}$ composed of n jobs, each job is associated with processing time (also called as size) $p_i (1 \le i \le n)$, and a set of machines $M_1, M_2, ..., M_m$, and a buffer of size k.

Output: Online allocate all the jobs in J on m machines without preemption such that the makespan(the maximal completion time) is minimum. During the assignment, we are allowed to store some jobs in the buffer, but once some job is assigned on some machine, that job cannot be stored in the buffer and we cannot reschedule that job.

3 Two Algorithms for Identical Machines

In this section we consider online algorithms for identical machines. We propose a $\frac{3}{2}$-competitive online algorithm for m machine with a buffer of size $1.5m$, which is better than the previous result $1.6197m$. We also give an optimal online algorithm with a buffer of size six for three identical machines.

3.1 m Machines with a Buffer of Size $\frac{3m}{2}$

In this subsection, we give a $\frac{3}{2}$-competitive online algorithm for m identical machines with a buffer of size $1.5m$, which is better than the previous result [7], a $\frac{3}{2}$-competitive online algorithm with a buffer of size $1.6197m$. Here assume $m = 6n$ for simplicity, where $n \geq 1$ is an integer. Our result can be generalized without this condition. The main ideas are below: i) we keep the largest $\frac{3m}{2}$ jobs in the buffer, ii) allocate all the other jobs on m machines in such a way that in some machine there is a room for one of the largest m jobs in the buffer without violating the condition: the makespan is at most 1.5 times the optimal value, iii) when the input ends, we assign the largest m jobs to their corresponding machines, and allocate all the other jobs by a greedy way.

There are two phases in our algorithm: iteration phase and final phase. In the iteration phase, we handle the incoming job j_t with size p_t as below. Let T be the total size of all the jobs assigned on m machines so far, initially it is zero. Let $\pi_{t,i}$ be the i-th largest job in the buffer after time t. We also use $\pi_{t,i}$ to denote its size.

In the iteration phase:

---

1. Store in the buffer the larger job between jobs j_t and $\pi_{t-1,\frac{3m}{2}}$, i.e., if $p_t > \pi_{t-1,\frac{3m}{2}}$ then remove job $\pi_{t-1,\frac{3m}{2}}$ from the buffer and put j_t in the buffer, else do nothing.
2. For $i = 1$ to $\frac{3m}{2}$, update $\pi_{t,i}$.
3. Assign the smaller job of j_t and $\pi_{t-1,\frac{3m}{2}}$ into machine M_k by the following way(we will prove that such machine must exist). Define $p = \min\{p_t, \pi_{t-1,\frac{3m}{2}}\}$.
   (a) If there is machine M_k with load at most

   $$w_k(T + \delta_k p) - p,$$

   assign the job to M_k, where $w_k = \frac{3}{2m}$ and $\delta_k = \frac{5m}{6}$ for $1 \leq k \leq m/3$, for other k, we have $w_k = \frac{1}{2(k-1)}$ and $\delta_k = \frac{m}{2} + k$.

In the final phase: we assign the jobs in the buffer as below:

---

1. For $1 \le i \le m$, allocate job $\pi_{t,i}$ to machine M_{m+1-i}.
2. Allocate all the other jobs in the buffer one by one to a machine with minimum load.

Observation 1. *For any $t \ge 0$ and $i \ge 1$ we have $\pi_{t+1,i} \ge \pi_{t,i}$ and $\pi_{t,i} \ge \pi_{t,i+1}$.*

Observation 2. $\sum_{i=\frac{m}{3}+1}^{m} w_i = \sum_{i=\frac{m}{3}+1}^{m} \frac{1}{2(i-1)} \ge \frac{\ln 3}{2}$.

Proof. $\sum_{i=\frac{m}{3}+1}^{m} \frac{1}{2(i-1)} \ge \frac{1}{2} \cdot \int_{\frac{m}{3}}^{m} \frac{1}{x} dx \ge \frac{\ln 3}{2}$. □

Observation 3. $\sum_{i=\frac{m}{3}+1}^{m} w_i \cdot \delta_i = \sum_{i=\frac{m}{3}+1}^{m} \frac{1}{2(i-1)} \cdot (i + \frac{m}{2}) \ge \frac{m}{3} + \frac{m \ln 3}{4} > \frac{7m}{12}$.

Lemma 1. *In the iteration phase there is always a machine M_k with load at most*

$$w_k(T + \delta_k p) - p$$

for $1 \le k \le m$, where $p = \min\{p_t, \pi_{t-1,\frac{3m}{2}}\}$.

Proof. Assume the lemma does not hold. Let L_k be the load of machine M_k. Then for $1 \le k \le m/3$, we have

$$L_k > \frac{3}{2m}(T + \frac{5m}{6}\min\{p_t, \pi_{t-1,\frac{3m}{2}}\}) - \min\{p_t, \pi_{t-1,\frac{3m}{2}}\},$$

for $m/3 < k \le m$, we have

$$L_k > \frac{1}{2(k-1)}(T + (\frac{m}{2} + k)\min\{p_t, \pi_{t-1,\frac{3m}{2}}\}) - \min\{p_t, \pi_{t-1,\frac{3m}{2}}\}.$$

Then we have

$$T = \sum_{k=1}^{m} L_k > (\frac{3}{2m} \times \frac{m}{3} + \sum_{i=\frac{m}{3}+1}^{m} \frac{1}{2(i-1)}) \cdot T$$

$$+ (\sum_{i=\frac{m}{3}+1}^{m} \frac{1}{2(i-1)} \cdot (i + \frac{m}{2}) + \frac{3}{2m} \times \frac{5m}{6} \times \frac{m}{3} - m) \cdot \min\{p_t, \pi_{t-1,\frac{3m}{2}}\}$$

$$> \frac{1 + \ln 3}{2} \cdot T + (\frac{7m}{12} + \frac{5m}{12} - m) \cdot \min\{p_t, \pi_{t-1,\frac{3m}{2}}\}$$

$$> T,$$

where the last second inequality holds by Observations 2 and 3. Hence the assumption is wrong and the lemma holds. □

Theorem 1. *Our online algorithm is 1.5-competitive.*

Proof. For a given input L, let $A(L)$ be the makespan by our online algorithm, let $OPT(L)$ be the makespan by an optimal offline algorithm. Next we will prove that $A(L) \le 1.5OPT(L)$.

By Lemma 1, we know our online algorithm cannot fail at the iteration phase. Assume the iteration phase ends at time t. Consider machine M_k. Let job j with size p be the last job assigned on M_k during the iteration phase. Let T' be the total size of jobs assigned on m machines just before assigning job j. Let T be the total size of jobs assigned on m machines at the end of the iteration phase. Then for $1 \le k \le \frac{m}{3}$ we have

$$L_k \le \frac{3}{2m}\left(T' + \frac{5m}{6} \cdot p\right) = \frac{3}{2m}\left(T + \left(\frac{5m}{6} - 1\right) \cdot p\right),$$

for $\frac{m}{3} + 1 \le k \le m$, we have

$$L_k \le \frac{1}{2(k-1)}\left(T' + \left(k + \frac{m}{2}\right) \cdot p\right) = \frac{1}{2(k-1)}\left(T + \left(k + \frac{m}{2} - 1\right) \cdot p\right).$$

It is not difficult to see

$$\frac{T + \sum_{i=1}^{3m/2} \pi_{t,i}}{m} \le OPT(L). \tag{1}$$

After the first step in the final phase, for $1 \le k \le \frac{m}{3}$, by (1) and Observation 1 we have

$$T + \left(\frac{5m}{6} - 1\right) \cdot p < m \cdot OPT(L) - \frac{2m}{3} \cdot \pi_{t,\frac{2m}{3}},$$

hence after assigning job $\pi_{t,m+1-k}$, the load of machine M_k is

$$L'_k \le \frac{3}{2m}\left(T + \left(\frac{5m}{6} - 1\right) \cdot p\right) + \pi_{t,m+1-k}$$
$$< \frac{3}{2m}\left(m \cdot OPT(L) - \frac{2m}{3} \cdot \pi_{t,\frac{2m}{3}}\right) + \pi_{t,\frac{2m}{3}} = \frac{3}{2}OPT(L),$$

for $\frac{m}{3} + 1 \le k \le m$, by (1) and Observation 1 we have

$$T + \left(k + \frac{m}{2} - 1\right) \cdot p \le m \cdot OPT(L) - (m + 1 - k) \cdot \pi_{t,m+1-k},$$

hence after assigning job $\pi_{t,m+1-k}$, the load of machine M_k is

$$L'_k \le \frac{1}{2(k-1)}\left(T + \left(k + \frac{m}{2} - 1\right) \cdot p\right) + \pi_{t,m+1-k}$$
$$\le \frac{1}{2(k-1)}\left(m \cdot OPT(L) - (m + 1 - k) \cdot \pi_{t,m+1-k}\right) + \pi_{t,m+1-k}$$
$$= \frac{m}{2(k-1)} \cdot OPT(L) + \frac{3(k-1) - m}{2(k-1)}\pi_{t,m+1-k} \le \frac{3}{2}OPT(L),$$

where the last inequality holds from $\pi_{t,m+1-k} \le OPT(L)$ and $k \ge \frac{m}{3} + 1$. So, after the first step in the final phase the makespan of our algorithm is at most $1.5OPT(L)$. We know the average load is always at most $OPT(L)$. It is also not difficult to see $\pi_{t,k} \le \frac{OPT(L)}{2}$, where $k > m$. Since we assign $\pi_{t,k}$ to a machine with minimal load for all $m+1 \le k \le 3m/2$, after the assignment the load of that machine is always at most $1.5OPT(L)$. Hence we have that $A(L) \le 1.5OPT(L)$.

$\square$

3.2   Three Machines with a Buffer of Size Six

In this subsection we give an online algorithm with a competitive ratio $\frac{15}{11} \approx$ 1.3636, which matches the lower bound, for three identical machines with size six, i.e., we reduce the size from nine [7] to six. The main ideas are similar with the one in [7], but we use a different threshold for each machine when we assign jobs in the iteration phase. Because of the different threshold, the size of the buffer is reduced.

There are two phases: iteration phase and final phase. In the iteration phase, we handle the incoming job j_t with size p_t as below. Let T be the total size of all the jobs assigned on three machines just before time t, initially it is zero. Let $\pi_{t,i}$ be the i-th largest job in the buffer at time t. We also use $\pi_{t,i}$ to denote its size. Define $w_1 = \frac{5}{11}$, $w_2 = \frac{4}{11}$, $w_3 = \frac{2}{11}$, and $\delta_1 = \frac{13}{5}$, $\delta_2 = 3$ and $\delta_3 = 4$.

The iteration phase

---

1. Store the larger job of j_t and $\pi_{t-1,6}$ in the buffer, i.e., if $p_t > \pi_{t-1,6}$ then remove job $\pi_{t-1,6}$ from the buffer and put j_t in the buffer, else do nothing.
2. Assign the smaller job of j_t and $\pi_{t-1,6}$ to machine M_k with load at most $w_i(T + \delta_i \cdot p) - p$, where $p = \min\{p_t, \pi_{t-1,6}\}$ (We will prove such machine must exist).

In the final phase: we assign the jobs in the buffer as below:

---

1. Let OPT' be the optimal value for the jobs in the buffer.
2. Put all the jobs in the buffer with size larger than $OPT'/3$ into a subset G_b, put all the others in the buffer into a subset G_s.
3. Call LPT(Largest Processing Time) [11,9] algorithm to allocate the jobs in G_b to three virtual machines M_i', where $1 \le i \le 3$ (note that LPT is an optimal algorithm for at most six jobs on three machines). Assume that $L(M_1') \le L(M_2') \le L(M_3')$, where $L(M_i')$ is the load of machine M_i'. Then assign all the jobs in M_i' to M_i for all $1 \le i \le 3$.
4. schedule all the jobs in G_s one by one on a machine with minimum load.

Lemma 2. *In the iteration phase there is always a machine M_k with load at most*

$$w_k\big(T + \delta_k \min\{p_t, \pi_{t-1,6}\}\big) - \min\{p_t, \pi_{t-1,6}\}$$

for $1 \le k \le 3$, where p_t is the size of the incoming job j_t.

Proof. Assume the lemma does not holds. Let L_k be the load of machine M_k. Then for $1 \le k \le 3$, we have

$$L_k > w_k\big(T + \delta_k \min\{p_t, \pi_{t-1,6}\}\big) - \min\{p_t, \pi_{t-1,6}\}.$$

Then we have

$$T = \sum_{k=1}^{3} L_k > \left(\sum_{k=1}^{3} w_k\right) \cdot T + \left(\sum_{k=1}^{3} w_k \cdot \delta_k - 3\right) \cdot \min\{p_t, \pi_{t-1,6}\}$$
$$= T,$$

where the last second inequality holds by the definitions of w_k and δ_k. Hence the assumption is wrong and the lemma holds. □

Lemma 3. *[7] For three identical machines, any online algorithm cannot have a competitive ratio strictly less than $\frac{15}{11}$.*

Theorem 2. *The above online algorithm is $\frac{15}{11} \approx 1.3636$-competitive for three machines with a buffer of size 6.*

Proof. For a given input L, let $A(L)$ be the makespan by our online algorithm, let $OPT(L)$ be the makespan by an optimal offline algorithm. Next we will prove that $A(L) \leq \frac{15}{11}OPT(L)$.

Consider machine M_k for $1 \leq k \leq 3$. Let job j with size p be the last job assigned on M_k during the iteration phase. We have

$$L_k \leq w_k(T + \delta_k \cdot \min\{p, \pi_{t-1,6}\}),$$

since T and $\pi_{t-1,6}$ are increasing functions over t. Observe that

$$T + p + \sum_{i=1}^{6} \pi_{t,i} \leq 3OPT(L). \tag{2}$$

We will prove for each $k \in [1,3]$, after the the final stage, the makespan by our algorithm is at most $\frac{15}{11}OPT(L)$. After the third step in the final stage, consider M_1. If $L(M_1') \leq 2p$, we have

$$L_1 \leq \frac{5}{11}(T + \frac{13}{5}p) + L(M_1') \leq \frac{5}{11}(3OPT(L) - 6p + \frac{8}{5}p) + 2p = \frac{15}{11}OPT(L),$$

else $L(M_1') > 2p$, we have

$$L_1 \leq \frac{5}{11}(T + \frac{13}{5}p) + L(M_1') \leq \frac{5}{11}(3OPT(L) - 3L(M_1') + \frac{8}{5}p) + L(M_1')$$
$$\leq \frac{5}{11}(3OPT(L) - 3L(M_1') + \frac{4}{5}L(M_1')) + L(M_1') = \frac{15}{11}OPT(L).$$

After the third step in the final stage, consider M_2. If $L(M_1') = 0$, then machines M_2' and M_3' have at most one job by LPT algorithm, then there are at least four jobs in G_s. Else if there is one job on M_1' then there are at least one job in G_s. The reason is that: there are at most two jobs in each virtual machine since each job in G_b has size larger than $OPT'/3$. Hence we have

$$L(G_s) + L(M_1') \geq 2p,$$

by (2) we have $T + 3p + 2L(M_2') \leq T + p + L(G_s) + \sum_{i=1}^{3} L(M_i') \leq 3OPT(L)$
and

$$L_2 \leq \frac{4}{11}(T + 3p) + L(M_2') \leq \frac{4}{11}(3OPT(L) - 2L(M_2')) + L(M_2')$$
$$= \frac{12}{11}OPT(L) + \frac{3}{11}L(M_2') \leq \frac{15}{11}OPT(L),$$

since $L(M_2') \leq OPT(L)$.

After the third step in the final stage, consider M_3. If $L(M_2') = 0$, then we have machine M_3' has at most one job by LPT algorithm, then there are at least five jobs in G_s. Else if there is one job in M_2' then by the above analysis, we have $L(G_s) + L(M_1') \geq 2p$. Hence we have

$$L(G_s) + L(M_1') + L(M_2') \geq 3p$$

by (2), we have $T + 4p + L(M_3') \leq T + p + L(G_s) + \sum_{i=1}^{3} L(M_i') \leq 3OPT(L)$
and

$$L_3 \leq \frac{2}{11}(T + 4p) + L(M_3') \leq \frac{2}{11}(3OPT(L) - L(M_3')) + L(M_3')$$
$$= \frac{6}{11}OPT(L) + \frac{9}{11}L(M_3') \leq \frac{15}{11}OPT(L),$$

since $L(M_3') \leq OPT(L)$.

Now we we consider the last step in the final stage, i.e., all the jobs in G_s with size at most $OPT(L)/3$. We know the average load is always at most $OPT(L)$. It is also not difficult to see all the jobs in G_s are assigned on machines with load at most $OPT(L)$ and after the assignment the load of the machine is at most $4OPT(L)/3$. Hence this theorem holds.   □

4   A Simple Algorithm for Uniform Machines

Our algorithm is similar with the one in [7]. There are two phases: the iteration phase and the final phase. We use the iteration phase to deal a new job, when there is no new job given, we execute the final phase. The iteration phase is the same as the one in [7], however we fully use the buffer of size m while the algorithm [7] uses a buffer with size $m - 1$. The final phase of our algorithm is different from the one in [7]: if the largest job in the buffer is very large, then we use the same approach as the one in [7], otherwise we use the same approach as the one in the iteration phase (this part is different from the one in [7]).

Let $C_i(t)$ be the completion time on machine i for all $1 \leq i \leq m$ just after time step t. Let s_i be the speed of machine i, where $1 \leq i \leq m$. Let j_{min} be the smallest job in the buffer. Let j_t be the job arrived at time step t. The iteration phase of our online algorithm performs as below:

The iteration phase

1. If job j_t is larger than job j_{min} then remove job j_{min} from the buffer and store job j_t in the buffer. Let j be the smaller job between jobs j_t and j_{min}. Update job j_{min} if possible.
2. Assign job j into machine i such that

$$C_i(t-1) \le \frac{T(t-1) + m \cdot p(j)}{\sum_{k=1}^{m} s_i} - \frac{p(j)}{s_i},$$

where $T(t-1)$ is the total load on all the machines after time $t-1$, i.e., the total processing time of all the jobs so far. (Such machine always exists).

In the final phase: we assign the jobs in the buffer as below:

1. Let j_{max} be the largest job in the buffer. If $p(j_{max}) > \frac{T}{m}$ then allocate all the jobs in the buffer by the PTAS due to Hochbaum and Shmoys [12] on m machines without considering the scheduled jobs, where T is the total load of all the jobs given.
2. Else assign all the jobs in the buffer one by one by the previous approach, i.e., at time step t assign job j to a machine i such that

$$C_i(t-1) \le \frac{T(t-1) + m \cdot p(j)}{\sum_{k=1}^{m} s_i} - \frac{p(j)}{s_i}.$$

Lemma 4. *[7] In Step 2 of the iteration phase, for job j there always exists machine M_i such that*

$$C_i(t-1) \le \frac{T(t-1) + m \cdot p(j)}{\sum_{k=1}^{m} s_i} - \frac{p(j)}{s_i}.$$

Lemma 5. *After time step $t \ge 1$, for all $1 \le i \le m$ on machine i,*

$$C_i(t) \le \frac{T(t) + (m-1)p(j)}{\sum_{k=1}^{m} s_i}.$$

Theorem 3. *Our algorithm is $\left(\frac{2m-1}{m} + \epsilon\right)$-competitive on m uniform machines with a buffer of size m.*

Proof. Again let T be the total load of all the jobs given. Let OPT be the optimal solution of an offline algorithm. So, we have $OPT \ge \frac{T}{\sum_{k=1}^{m} s_i}$. Consider the time when there are no new jobs given. For $\forall i$, let job j be the last job assigned on machine i at time step t in the iteration phase. There are two cases.

Case 1: $j_{max} \le \frac{T}{m}$. Then by Lemma 5,

$$C_i(t) \le \frac{T(t) + (m-1)p(j)}{\sum_{k=1}^{m} s_i} \le \frac{T + (m-1) \cdot \frac{T}{m}}{\sum_{k=1}^{m} s_i} = \frac{(2m-1)T}{m \cdot \sum_{k=1}^{m} s_i} \le \frac{(2m-1)OPT}{m}.$$

Case 2: $j_{max} > \frac{T}{m}$. Then we have

$$T(t) + (m-1)p(j) \le T - p(j_{max}) < \frac{(m-1)T}{m}$$

since any job in the buffer is not smaller than job j. At the end of the iteration phase, we have

$$C_i(t) \le \frac{T(t) + (m-1)p(j)}{\sum_{k=1}^{m} s_i} \le \frac{m-1}{m} \frac{T}{\sum_{k=1}^{m} s_i} \le \frac{m-1}{m} OPT.$$

After the final phase we have $ALG \le (1 + \epsilon + \frac{m-1}{m})OPT$.   □

Remarks: Some open questions are left. For example, i) for m identical machines, can we reduce the size of the buffer smaller than $2.5m + 2$ to achieve an optimal online algorithm? ii) for m identical machines, using a buffer of size m, can we have an online algorithm with a competitive ratio less than $\frac{2+r_m}{2}$?

References

1. Albers, S.: Better bounds for online scheduling. SIAM J. Comput. 29(2), 459–473 (1999)
2. Bartal, Y., Fiat, A., Karloff, H., Vohra, R.: New algorithms for an ancient scheduling problem. J. Comput. Syst. Sci. 51(3), 359–366 (1995)
3. Chen, X., Lan, Y., Benko, A., Dósa, G., Han, X.: Optimal algorithms for online scheduling with bounded rearrangement at the end. Theor. Comput. Sci. 412(45), 6269–6278 (2011)
4. Dósa, G., Epstein, L.: Online scheduling with a buffer on related machines. J. Comb. Optim. 20(2), 161–179 (2010)
5. Dósa, G., Epstein, L.: Preemptive online scheduling with reordering. SIAM J. Discrete Math. 25(1), 21–49 (2011)
6. Dósa, G., Wang, Y., Han, X., Guo, H.: Online scheduling with rearrangement on two related machines. Theor. Comput. Sci. 412(8-10), 642–653 (2011)
7. Englert, M., Özmen, D., Westermann, M.: The power of reordering for online minimum makespan scheduling. In: Proc. 48th Symp. Foundations of Computer Science (FOCS), pp. 603–612 (2008)
8. Fleischer, R., Wahl, M.: Online scheduling revisited. Journal of Scheduling 3, 343–353 (2000)
9. Friesen, D.K.: Tighter bounds for lpt scheduling on uniform processors. SIAM J. Comput. 16(3), 554–560 (1987)
10. Garey, M.R., Johnson, D.S.: Computers and Intractability: A Guide to the Theory of NP-Completeness. W.H. Freeman (1979)
11. Graham, R.L.: Bounds on multiprocessing timing anomalies. SIAM Journal of Applied Mathematics 17(2), 416–429 (1969)
12. Hochbaum, D.S., Shmoys, D.B.: A polynomial approximation scheme for scheduling on uniform processors: Using the dual approximation approach. SIAM Journal on Computing 17(3), 539–551 (1988)
13. Rudin III, J.F., Chandrasekaran, R.: Improved bounds for the online scheduling problem. SIAM J. Comput. 32(3), 717–735 (2003)

14. Karger, D.R., Phillips, S.J., Torng, E.: A better algorithm for an ancient scheduling problem. J. Algorithms 20(2), 400–430 (1996)
15. Kellerer, H., Kotov, V., Speranza, M.G., Tuza, Z.: Semi on-line algorithms for the partition problem. Oper. Res. Lett. 21(5), 235–242 (1997)
16. Liu, M., Xu, Y., Chu, C., Zheng, F.: Online scheduling on two uniform machines to minimize the makespan. Theor. Comput. Sci. 410(21-23), 2099–2109 (2009)
17. Sanders, P., Sivadasan, N., Skutella, M.: Online scheduling with bounded migration. Math. Oper. Res. 34(2), 481–498 (2009)
18. Tan, Z., Yu, S.: Online scheduling with reassignment. Oper. Res. Lett. 36(2), 250–254 (2008)
19. Zhang, G.: A simple semi on-line algorithm for p2//c_{max} with a buffer. Information Processing Letters 61, 145–148 (1997)

A Dense Hierarchy of Sublinear Time Approximation Schemes for Bin Packing

Richard Beigel[1] and Bin Fu[2]

[1] CIS Department, Temple University,
Philadelphia, PA 19122-6094, USA
beigel@cis.temple.edu

[2] Department of Computer Science, University of Texas-Pan American,
Edinburg, TX 78539, USA
bfu@utpa.edu

Abstract. The bin packing problem is to find the minimum number of bins of size one to pack a list of items with sizes $a_1, \ldots, a_n$ in $(0, 1]$. Using uniform sampling, which selects a random element from the input list each time, we develop a randomized $O(\frac{n(\log \log n)}{\sum_{i=1}^{n} a_i} + (\frac{1}{\epsilon})^{O(\frac{1}{\epsilon})})$ time $(1+\epsilon)$-approximation scheme for the bin packing problem. We show that every randomized algorithm with uniform random sampling needs $\Omega(\frac{n}{\sum_{i=1}^{n} a_i})$ time to give an $(1 + \epsilon)$-approximation. For each function $s(n) : N \to N$, define $\sum(s(n))$ to be the set of all bin packing problems with the sum of item sizes equal to $s(n)$. We show that $\sum(n^b)$ is NP-hard for every $b \in (0, 1]$. This implies a dense sublinear time hierarchy of approximation schemes for a class of NP-hard problems, which are derived from the bin packing problem. We also show a randomized streaming approximation scheme for the bin packing problem such that it needs only constant updating time and constant space, and outputs an $(1 + \epsilon)$-approximation in $(\frac{1}{\epsilon})^{O(\frac{1}{\epsilon})}$ time. Let $S(\delta)$-bin packing be the class of bin packing problems with each input item of size at least δ. This research also gives a natural example of NP-hard problem ($S(\delta)$-bin packing) that has a constant time approximation scheme, and a constant time and space sliding window streaming approximation scheme, where δ is a positive constant.

1 Introduction

The bin packing problem is to find the minimum number of bins of size one to pack a list of items with sizes $a_1, \ldots, a_n$ in $(0, 1]$. It is a classical NP-hard problem and has been widely studied. The bin packing problem has many applications in the engineering and information sciences. Some approximation algorithm has been developed for bin packing problem: for examples, the first fit, best fit, sum-of-squares, or Gilmore-Gomory cuts [2,8,7,16,15]. The first linear time approximation scheme is shown in [11]. Recently, a sublinear time $O(\sqrt{n})$ with weighted sampling and a sublinear time $O(n^{1/3})$ with a combination of weighted and uniform samplings were shown for bin packing problem [3].

J. Snoeyink, P. Lu, K. Su, and L. Wang (Eds.): FAW-AAIM 2012, LNCS 7285, pp. 172–181, 2012.

We study the bin packing problem in randomized offline sublinear time model, randomized streaming model, and randomized sliding window streaming model. We also study the bin packing problem that has input item sizes to be random numbers in $[0, 1]$. Sublinear time algorithms have been found for many computational problems, such as checking polygon intersections [5], estimating the cost of a minimum spanning tree [6,9,10], finding geometric separators [13], and property testing [22,17], etc. Early research on streaming algorithms dealt with simple statistics of the input data streams, such as the median [21], the number of distinct elements [12], or frequency moments [1]. Streaming algorithm is becoming more and more important due to the development of internet, which brings a lot of applications. There are many streaming algorithms that have been proposed from the areas of computational theory, database, and networking, etc.

Due to the important role of bin packing problem in the development of algorithm design and its application in many other fields, it is essential to study the bin packing problem in these natural models. Our offline approximation scheme is based on the uniform sampling, which selects a random element from the input list each time. Our first approach is to approximate the bin packing problem with a small number of samples under uniform sampling. We identify that the complexity of approximation for the bin packing problem inversely depends on the sum of the sizes of input items.

Using uniform sampling, we develop a randomized $O(\frac{n(\log \log n)}{\sum_{i=1}^{n} a_i} + (\frac{1}{\epsilon})^{O(\frac{1}{\epsilon})})$ time $(1 + \epsilon)$-approximation scheme for the bin packing problem. We show that every randomized algorithm with uniform random sampling needs $\Omega(\frac{n}{\sum_{i=1}^{n} a_i})$ time to give an $(1 + \epsilon)$-approximation. Based on an adaptive random sampling method developed in this paper, our algorithm automatically detects an approximation to the weights of summation of the input items in time $O(\frac{n(\log \log n)}{\sum_{i=1}^{n} a_i})$ time, and then yields an $(1 + \epsilon)$-approximation.

For each function $s(n) : N \to N$, define $\sum(s(n))$ to be the set of all bin packing problems with the sum of item sizes equal to $s(n)$. For a constant $b \in (0, 1)$, every problem in $\sum(n^b)$ has an $O(n^{1-b}(\log \log n) + (\frac{1}{\epsilon})^{O(\frac{1}{\epsilon})})$ time $(1 + \epsilon)$-approximation for an arbitrary constant ϵ. On the other hand, there is no $o(n^{1-b})$ time $(1 + \epsilon)$-approximation scheme for the bin packing problems in $\sum(n^b)$ for some constant $\epsilon > 0$. We show that $\sum(n^b)$ is NP-hard for every $b \in (0, 1]$. This implies a dense sublinear time hierarchy of approximation schemes for a class of NP-hard problems that are derived from bin packing problem. We also show a randomized single pass streaming approximation scheme for the bin packing problem such that it needs only constant updating time and constant space, and outputs an $(1 + \epsilon)$-approximation in $(\frac{1}{\epsilon})^{O(\frac{1}{\epsilon})}$ time. This research also gives an natural example of NP-hard problem that has a constant time approximation scheme, and a constant time and space sliding window single pass streaming approximation scheme.

The streaming algorithms in this paper for bin packing problem only approximate the minimum number of bins to pack those input items. It also gives a packing plan that allows an item position to be changed at different moment. This has no contradiction with the existing lower bound [4,19] that no approximation scheme exists for online algorithm that does not change bins of already packed items.

2 Models of Computation and Overview of Methods

Algorithms for bin packing problem in this paper are under four models, which are deterministic, randomized, streaming, and sliding windows streaming models.

Definition 1. – *A bin packing is an allocation of the input items of sizes $a_1, \ldots, a_n$ in $(0, 1]$ to bins of size 1. We want to minimize the total number of bins. We often use $Opt(L)$ to denote the least number bins for packing items in L.*
  – *Assume that c and η are constants in $(0, 1)$, and k is a constant integer. There are k kinds of bins of different sizes. If $c \le s_i \le 1$, and $\eta \le w_i \le 1$ for all $i = 1, 2, \ldots, k$, then we call the k kinds of bins to be (c, η, k)-related, where w_i and s_i are the cost and size of the i-th kind of bin, respectively.*
  – *A bin packing with (c, η, k)-related bins is to allocate the input items $a_1, \ldots, a_n$ in $(0, 1]$ to (c, η, k)-related bins. We want to minimize the total costs $\sum_{i=1}^{k} u_i w_i$, where u_i is the number of bins of cost w_i. We often use $Opt_{c, \eta, k}(L)$ to denote the least cost for packing items in L with (c, η, k)-related bins. It is easy to see $Opt(L) = Opt_{1,1,1}(L)$.*
  – *For a positive constant δ, a $S(\delta)$-bin packing problem is the bin packing problem with all input items at least δ.*
  – *For a nondecreasing function $f(n) : N \to N$, a $\sum(f(n))$-bin packing problem is the bin packing problem with all input items $a_1, \ldots, a_n$ satisfying $\sum_{i=1}^{n} a_i = f(n)$.*

Deterministic Model: The bin packing problem under the deterministic model has been well studied. We give a generalized version of bin packing problem that allows multiple sizes of bins to pack them. It is called as bin packing with (c, η, k) related bins in Definition 1.

Randomized Models: Our main model of computation is based on the uniform random sampling. We give the definitions for both uniform and weighted random samplings below.

Definition 2. *Assume that $a_1, \ldots, a_n$ is an input list of items in $(0, 1]$ for a bin packing problem.*

  – *A uniform sampling selects an element a from the input list with $\Pr[a = a_i] = \frac{1}{n}$ for $i = 1, \ldots, n$.*
  – *A weighted sampling selects an element a from the input list with $\Pr[a = a_i] = \frac{a_i}{\sum_{i=1}^{n} a_i}$ for $i = 1, \ldots, n$.*

We feel that the uniform sampling is more practical to implement than weighted sampling. In this paper, our offline randomized algorithms are based on uniform sampling. The weighted sampling was used in [3].

Streaming Computation: A data stream is an ordered sequence of data items $p_1, p_2, \ldots, p_n$. Here, n denotes the number of data points in the stream. A *streaming algorithm* is an algorithm that computes some function over a data stream and has the following properties: 1. The input data are accessed in the sequential order of the data stream. 2. The order of the data items in the stream is not controlled by the algorithm.

Sliding Window Model: In the sliding window streaming model, there is a window size n for the most recent n items. The bin packing problem for the sliding window streaming algorithm is to pack the most recent n items.

Bin Packing with Random Inputs: We study the bin packing problem such that the input is a series of sizes that are random numbers in $[0, 1]$. It has a constant time approximation scheme.

2.1   Overview of Our Method

We develop algorithms for the bin packing problem under offline uniform random sampling model, the streaming computation model, and sliding window streaming model (only for $S(\delta)$-bin packing with a positive constant δ). The brief ideas are given below.

Sublinear Time Algorithm for Offline Bin Packing. Since the sum of input item sizes is not a part of input, it needs $O(n)$ time to compute its exact value, and it's unlikely to be approximated via one round random sampling in a sublinear time. We first approximate the sum of sizes of items through a multiphase adaptive random sampling. Select a constant φ to be the threshold for large items. Select a small constant $\gamma = O(\epsilon)$. All the items from the input are partitioned into intervals $[\pi_1, \pi_0], [\pi_2, \pi_1] \ldots, [\pi_{i+1}, \pi_i), \ldots$ such that $\pi_0 = 1, \pi_1 = \varphi$, and $\pi_{i+1} = \pi_i/(1 + \gamma)$ for $i = 2, \ldots$. We approximate the number of items in each interval $(\pi_{i+1}, \pi_i]$ via uniform random sampling. Those intervals with very a small number of items will be dropped. This does not affect much of the ratio of approximation. One of worst cases is that all small items are of size $\frac{1}{n^2}$ and all large size items are of size 1. In this case, we need to sample $\Omega(\frac{n}{\sum_{a_i=1} 1})$ number of items to approximate the number of 1s. This makes the total time to be $\Omega(\frac{n}{\sum_{i=1}^{n} a_i})$. Packing the items of large size is adapted the method in [11], which uses a linear programming method to pack the set of all large items, and fills small items into those bins with large items to waste only a small piece of space for each bin. Then the small items are put into bins that still have space left after packing large items. When the sum of all item sizes is $O(1)$, we need $O(n)$ time. Thus, the $O(n)$ time algorithm is a part of our algorithm for the case $\sum_{i=1}^{n} a_i = O(1)$.

Streaming Algorithm for Bin Packing. We apply the above approximation scheme to construct a single pass streaming algorithm for bin packing problem. A crucial step is to sample some random elements among those input items of size at least δ, which is set according to ϵ. The weights of small items are added to a variable s_1. After packing large items of size at least δ, we pack small items into those bins so that each bin does not waste more than δ space while there is small items unpacked.

Sliding Window Streaming Algorithm for $S(\delta)$-Bin Packing. Our sliding window single pass streaming algorithm deals with the bin packing problem that all input items are of size at least a constant δ. Let n be the size of sliding window instead of the total number of input items. Select a sufficiently large constant k. There are k sessions to approximate the bin packing. After receiving every $\frac{n}{k}$ items, a new session is started to approximate the bin packing. The approximation ratio is guaranteed via ignoring at most $\frac{n}{k}$ items. As each item is of large size at least δ, ignoring $\frac{n}{k}$ items only affect a small ratio of approximation.

3   Adaptive Random Sampling for Bin Packing

In this section, we develop an adaptive random sampling method to get the rough information for a list of items for the bin packing problem. We show a randomized algorithm to approximate the sum of the sizes of input items in $O((\frac{n}{\sum_{i=1}^{n} a_i}) \log \log n))$ time. This is the core step of our randomized algorithm, and is also or main technical contribution.

Definition 3.   – *For each interval I and a list of items S, define $C(I,S)$ to be the number of items of S in I.*
   – *For $\varphi, \delta,$ and γ in $(0,1)$, a $(\varphi, \delta, \gamma)$-partition for $(0,1]$ divides the interval $(0,1]$ into intervals $I_1 = [\pi_1, \pi_0], I_2 = [\pi_2, \pi_1), I_3 = [\pi_3, \pi_2), \ldots, I_k = (0, \pi_{k-1})$ such that $\pi_0 = 1, \pi_1 = \varphi, \pi_i = \pi_{i-1}(1-\delta)$ for $i = 2, \ldots, k-1$, and π_{k-1} is the first element $\pi_{k-1} \leq \frac{\gamma}{n^2}$.*
   – *For a set A, $|A|$ is the number of elements in A. For a list S of items, $|S|$ is the number of items in S.*

Lemma 1. *For parameters $\varphi, \delta,$ and γ in $(0,1)$, a $(\varphi, \delta, \gamma)$-partition for $(0,1]$ has the number of intervals $k = O(\frac{\log n}{\gamma \theta})$.*

We need to approximate the number of large items, the total sum of the sizes of items, and the total sum of the sizes of small items. For a $(\varphi, \delta, \gamma)$-partition $I_1 \cup I_2 \ldots \cup I_k$ for $(0,1]$, Algorithm Approximate-Intervals(.) below gives the estimation for the number of items in each I_j if interval I_j has a number items to be large enough. Otherwise, those items in I_j can be ignored without affecting much of the approximation ratio. We have an adaptive way to do random samplings in a series of phases. Phase $t+1$ doubles the number of random samples of phase t ($m_{t+1} = 2m_t$). For each phase, if an interval I_j shows sufficient number of items from the random samples, the number of items $C(I_j, S)$ in I_j can be sufficiently approximated by $\hat{C}(I_j, S)$. Thus, $\hat{C}(I_j, S)\pi_j$ also gives an approximation for the sum of the sizes of items in I_j. The sum $app_w = \sum_{I_j} \hat{C}(I_j, S)\pi_j$ for those intervals I_j with large number of samples gives an approximation for the total sum

$\sum_{i=1}^{n} a_i$ of items in the input list. Let m_t denote the number of random samples in phase t. In the early stages, app_w is much smaller than $\frac{n}{m_t}$. Eventually, app_w will surpass $\frac{n}{m_t}$. This happens when m_t is more than $\frac{n}{\sum_{i=1}^{n} a_i}$ and app_w is close to the sum $\sum_{i=1}^{n} a_i$ of all items from the input list. This indicates that the number of random samples is sufficient for approximation algorithm. For those intervals with small number of samples, their items only need small fraction of bins to be packed. This process is terminated when ignoring all those intervals with none or small number of samples does not affect much of the accuracy of approximation. The algorithm gives up the process of random sampling when m_t surpasses n, and switches to use a deterministic way to access the input list, which happens when the total sum of the sizes of input items is $O(1)$. The lengthy analysis is caused by the multi-phases adaptive random samplings. We show two examples below.

Algorithm Approximate-Intervals$(\varphi, \delta, \gamma, \theta, \alpha, P, n, S)$

Input: a parameter $\varphi \in (0,1)$, a small parameter $\theta \in (0,1)$, a failure probability upper bound α, a $(\varphi, \delta, \gamma)$ partition $P = I_1 \cup \ldots \cup I_k$ for $(0,1]$ with $\delta, \gamma \in (0,1)$, an integer n, a list S of n items $a_1, \ldots, a_n$ in $(0,1]$. Parameters $\varphi, \delta, \gamma, \theta$, and α do not depend on the number of items n.

Steps:

1. Phase 0:
2.   Let $z := \xi_0 \log \log n$, where ξ_0 is a parameter such that $8(k+1)(\log n)$ $g(\theta)^{z/2} < \alpha$ for all large n.
3.   Let parameters $c_0 := \frac{1}{100}, c_2 := \frac{1}{3(1+\delta)c_0}, c_3 := \frac{\delta^4}{2(1+\delta)}, c_4 := \frac{8}{(1-\theta)(1-\delta)\varphi c_0}$, and $c_5 := \frac{12\xi_0}{(1-\theta)c_2 c_3}$.
4.   Let $m_0 := z$.
5. End of Phase 0.
6. Phase t:
7.   Let $m_t := 2m_{t-1}$.
8.   Sample m_t random items $a_{i_1}, \ldots, a_{i_{m_t}}$ from the input list S.
9.   Let $d_j := |\{j : a_{i_j} \in I_j \text{ and } 1 \leq j \leq m_t\}|$ for $j = 1, 2, \ldots, k$.
10.   For each I_j,
11.   if $d_j \geq z$,
12.   then let $\hat{C}(I_j, S) := \frac{n}{m_t} d_j$ to approximate $C(I_j, S)$.
13.   else let $\hat{C}(I_j, S) := 0$.
14.   Let $app_w := \sum_{d_j \geq z} \hat{C}(I_j, S)\pi_j$ to approximate $\sum_{i=1}^{n} a_n$.
15.   If $app_w \leq \frac{c_5 n \log \log n}{c_0 m_t}$ and $m_t < n$ then enter Phase $t+1$.
16.   else
17.   If $m_t < n$
18.   then let $app'_w := \sum_{d_j \geq z \text{ and } j > 1} \hat{C}(I_j, S)\pi_j$ to approximate $\sum_{a_i < \delta, 1 \leq i \leq n} a_i$.
19.   else let $app_w := \sum_{i=1}^{n} a_i$ and $app'_w := \sum_{a_i < \varphi} a_i$.

20.   Output app_w, app'_w and $\hat{C}(I_1, S)$ (the approximate number of items of size at least φ).
21. End of Phase t.

End of Algorithm

Lemma 2 uses several parameters $\varphi, \delta, \gamma, \alpha$ and θ that will be determined by the approximation ratio for the the bin packing problem. If the approximation ratio is fixed, they all become constants.

Lemma 2. *Assume that $\varphi, \delta, \gamma, \alpha$ and θ are parameters in $(0,1)$, and those parameters do not depend on the number of items n.. Then there exists a randomized algorithm described in Approximate-Intervals(.) such that given a list S of items of size $a_1, \ldots, a_n$ in the range $(0, 1]$ and a $(\varphi, \delta, \gamma)$-partition for $(0, 1]$, with probability at most α, at least one of the following statements is false after executing the algorithm:*

1. *For each I_j with $\hat{C}(I_j, S) > 0$, $C(I_j, S)(1 - \theta) \leq \hat{C}(I_j, S) \leq C(I_j, S)(1 + \theta)$;*
2. *$\sum_{a_i \in I_j \text{ and } \hat{C}(I_j,S)=0} a_i \leq \frac{\delta^3}{2}(\sum_{i=1}^{n} a_i) + \frac{\gamma}{n}$;*
3. *$(1 - \theta)(1 - \delta)\varphi(\frac{\sum_{i=1}^{n} a_i}{2} - \frac{2\gamma}{n}) \leq app_w \leq (1 + \theta)(\sum_{i=1}^{n} a_i)$;*
4. *If $\sum_{i=1}^{n} a_i \geq 4$, then $\frac{1}{4}(1 - \theta)(1 - \delta)\varphi(\sum_{i=1}^{n} a_i) \leq app_w \leq (1 + \theta)(\sum_{i=1}^{n} a_i)$; and*
5. *It runs in $O(\frac{1}{(1-\theta)\delta^4 \log g(\theta)} \min(\frac{n}{\sum_{i=1}^{n} a_i}, n) \log \log n)$ time. In particular, the complexity of the algorithm is $O(\min(\frac{n}{\sum_{i=1}^{n} a_i}, n) \log \log n)$ if $\varphi, \delta, \gamma, \alpha$ and θ are constants in $(0, 1)$.*

Lemma 2 implies that with probability at least $1 - \alpha$, all statements 1 to 5 are true.

4   Main Results

We list the main results that we achieve in this paper. The proof of Theorem 1 is omitted in the conference version of this paper. The full version can be downloaded at http://arxiv.org/abs/1007.1260.

Theorem 1 (Main). *Approximate-Bin-Packing(.) is a randomized approximation scheme for the bin packing problem such that given an arbitrary $\tau \in (0, 1)$ and a list of items $S = a_1, \ldots, a_n$ in $(0, 1]$ for the bin packing problem, it gives an approximation app with $Opt(S) \leq app \leq (1 + \tau)Opt(S) + 1$ in $O(\frac{n(\log \log n)}{\sum_{i=1} a_i} + (\frac{1}{\tau})^{O(\frac{1}{\tau})})$ time with probability at least $\frac{3}{4}$.*

We show a lower bound for those bin packing problems with bounded sum of sizes $\sum_{i=1}^{n} a_i$. The lower bound always matches the upper bound.

Theorem 2. *Assume $f(n)$ is a nondecreasing unbounded function from N to N with $f(n) = o(n)$. Every randomized $(2 - \epsilon)$ approximation algorithm for bin packing problems in $\sum(f(n))$ needs $\Omega(\frac{n}{f(n)})$ time, where ϵ is an arbitrary small constant in $(0, 1)$.*

Proof. Since $f(n)$ is unbounded, assume n is large enough such that

$$(f(n) + 2)(2 - \epsilon) < 2(f(n) - 2). \tag{1}$$

We design two input list of items.

The first list contains $m = 2(f(n) - 2))$ elements of size $\frac{1}{2} + \delta$, where $\delta = \frac{1}{2(f(n)-2)}$. The rest $n - m$ items are of the same size $\gamma = \frac{1}{n-m} = o(1)$. We have $m(\frac{1}{2} + \delta) + (n - m)\gamma = 2(f(n) - 2)(\frac{1}{2} + \frac{1}{2(f(n)-2)}) + 1 = f(n)$. Therefore, the first list is a bin packing problem is in $\sum(f(n))$.

The second list contains $n - f(n)$ elements of size γ and the rest $f(n)$ items are of size equal to $1 - \tau$, where $\tau = \frac{(n-f(n))\gamma}{f(n)} = o(1)$. We have $f(n)(1 - \tau) + (n - f(n))\gamma = f(n)$. The second list is also a bin packing problem is in $\sum(f(n))$.

Both γ and τ are small. Packing the first list needs at least $2(f(n) - 2)$ bins. Packing the second list only needs at most $f(n) + 2$ bins since two bins of size one is enough to pack those items of size τ.

Assume that an algorithm only has computational time $o(\frac{n}{f(n)})$ for computing $(2 - \epsilon)$-approximation for bin packing problems in $\sum(f(n))$. The algorithm has an $o(1)$ probability to access at least one item of size at least $\frac{1}{2}$ in both lists. Therefore, the two lists have the same output for approximation by the same randomized algorithm. For the second list, the output for the number of bins should be at most $(f(n) + 2)(2 - \epsilon)$. By inequality (1), it is impossible to pack the first list items. This brings a contradiction. ☐

Corollary 1. *There is no $o(\frac{n}{\sum_{i=1}^{n} a_i})$ time randomized approximation scheme algorithm for the bin packing problem.*

Theorem 3. *For each constant $b \in (0, 1)$, the bin packing problem in $\sum(n^b)$ is NP-hard.*

5   Streaming Approximation Scheme

In this section, we show a constant time and constant space streaming algorithm for the bin packing problem. For the streaming model of the bin packing problem, we output a plan to pack the items that have come from the input list, and the number of bins to approximate the optimal number of bins. Our algorithm only holds a constant number of items. Therefore, it has a constant updating time and constant space complexity.

Lemma 3. *There is an $O(u)$ updating time algorithm to select u random elements from a stream of input elements.*

Proof. We set up u positions to put the u elements. There is a counter n to count the total number of elements arrived. For each new arrived element a_n, the i-th position uses probability $\frac{1}{n}$ to replace the old element at the i-th position with the new element. For each element a_i, with probability $\frac{1}{i}\frac{i}{i+1} \ldots \frac{n-1}{n} = \frac{1}{n}$, it is kept at each of the u positions after processing n elements. Therefore, we keep u-random elements from the input list. ☐

Theorem 4. *There is a single pass streaming randomized approximation scheme algorithm for the bin packing problem such that the algorithm has $O(1)$ updating time and $O(1)$ space, and computes an approximate packing solution $Apx(n)$ with $\mathrm{Sopt}(n) \leq App(n) \leq (1 + \epsilon)\mathrm{Sopt}(n) + 1$ in $(\frac{1}{\epsilon})^{O(\frac{1}{\epsilon})}$ time, where $\mathrm{Sopt}(n)$ is the optimal solution for the first n items in the input stream, and $App(n)$ is an approximate solution for the first n items in the input stream.*

Definition 4. *The 3-partition problem is to decide whether a given multiset of integers in the range $(\frac{B}{4}, \frac{B}{2})$ can be partitioned into triples that all have the same sum B, where B is an integer. More precisely, given a multiset S of $n = 3t$ positive integers, can S be partitioned into m subsets $S_1, S_2, \ldots, S_t$ such that the sum of the numbers in each subset is equal?*

It is well known that 3-partition problem is NP-complete [14]. It is used in proving the following NP-hard problems (Theorem 3 and Theorem 5). We show that the $S(\delta)$-bin packing problem is NP-hard if δ is at least $\frac{1}{4}$.

Theorem 5. *For each δ at most $\frac{1}{4}$, the $S(\delta)$-bin packing problem is NP-hard.*

Proof. We reduce the 3-partition problem to $S(\delta)$-bin packing problem. Assume that $S = \{a_1, \ldots, a_{3m}\}$ is an input of 3-partition. We design that a $S(\delta)$-bin packing problem as below: the bin size is 1 and the items are $\frac{a_1}{B}, \ldots, \frac{a_{3m}}{B}$. The size of each item is at least $\frac{1}{4}$ since each $a_i > \frac{B}{4}$. It is easy to see that there is a solution for the 3-partition problem if and only if those items for the bin packing problem can be packed into m bins. $\qquad\square$

Theorem 6. *Assume that c, η, and k are constants. Let δ be an arbitrary constant. Then there is a single pass sliding window streaming randomized approximation algorithm for the $S(\delta)$-bin packing problem with (c, η, k)-related bins that has $O(1)$ updating time and $O(1)$ space, and computes an approximate packing solution $App(.)$ with $\mathrm{Sopt}_{c,\eta,k}(n) \leq App(n) \leq (1 + \gamma)\mathrm{Sopt}_{c,\eta,k}(n)$ in $(\frac{1}{\gamma})^{O(\frac{1}{\gamma})}$ time, where $\mathrm{Sopt}_{c,\eta,k}(n)$ is the optimal solution for the last n items in the input stream, and $App(n)$ is an approximate solution for the most recent n items in the input stream.*

Acknowledgements. We would like to thank Xin Han for his helpful suggestions which improves the presentation of this paper. We would like to thank the helpful comments from the reviewers at FAW-AAIM 2012. This research is supported in part by National Science Foundation Early Career Award 0845376. An earlier version of this paper is posted at http://arxiv.org/abs/1007.1260.

References

1. Alon, N., Matias, Y., Szegedy, M.: The space complexity of approximating the frequency moments. In: Proceedings of the Symposium on Theory of Computing, pp. 20–29 (1996)

2. Applegate, D., Buriol, L., Dillard, B., Johnson, D., Shore, P.: The cutting-stock approach to bin packing: Theory and experiments. In: Proceedings of Algorithm Engineering and Experimentation (ALENEX), pp. 1–15 (2003)
3. Batu, T., Berenbrink, P., Sohler, C.: A sublinear-time approximation scheme for bin packing. Theoretical Computer Science 410, 5082–5092 (2009)
4. Brown, D.: A lower bound for on-line one-dimensional bin packing problem. Technical Report 864, University of Illinois, Urbana, IL (1979)
5. Chazelle, B., Liu, D., Magen, A.: Sublinear geometric algorithms. SIAM Journal on Computing 35, 627–646 (2005)
6. Chazelle, B., Rubinfeld, R., Trevisan, L.: Approximating the minimum spanning tree weight in sublinear time. SIAM Journal on Computing 34, 1370–1379 (2005)
7. Csirik, J.A., Johnson, D.S., Kenyon, C., Shor, P.W., Weber, R.R.: A Self Organizing Bin Packing Heuristic. In: Goodrich, M.T., McGeoch, C.C. (eds.) ALENEX 1999. LNCS, vol. 1619, pp. 246–265. Springer, Heidelberg (1999)
8. Csirik, J., Johnson, D., Kenyon, C., Orlin, J., Shore, P., Weber, R.: On the sum-of-squares algorithm for bin-packing. In: Proceedings of the 22nd Annual ACM Symposium on Theory of Computing (STOC), pp. 208–217 (2000)
9. Czumaj, A., Ergun, F., Fortnow, L., Magen, I.N.A., Rubinfeld, R., Sohler, C.: Sublinear approximation of euclidean minimum spanning tree. SIAM Journal on Computing 35, 91–109 (2005)
10. Czumaj, A., Sohler, C.: Estimating the weight of metric minimum spanning trees in sublinear-time. In: Proceedings of the 36th Annual ACM Symposium on Theory of Computing, pp. 175–183 (2004)
11. Fernandez de la Vega, W., Lueker, G.S.: Bin packing can be solved within 1+epsilon in linear time. Combinatorica 1(4), 349–355 (1981)
12. Flajolet, P., Martin, G.: Probabilistic counting algorithms for data base application. Journal of Computer and System Sciences 31, 182–209 (1985)
13. Fu, B., Chen, Z.: Sublinear-time algorithms for width-bounded geometric separators and their applications to protein side-chain packing problems. Journal of Combinatorial Optimization 15, 387–407 (2008)
14. Garey, M.R., Johnson, D.S.: Computers and Intractability. W. H. Freeman and Company, New York (1979)
15. Gilmore, M., Gomory, R.: A linear programming approach to the cutting-stock problem - part ii. Operations Research
16. Gilmore, M., Johnson, D.: A linear programming approach to the cutting-stock problem. Operations Research
17. Goldreich, O., Ron, D.: On testing expansion in bounded-degree graphs. Technical Report 00-20, Electronic Colloquium on Computational Complexity (2000), http://www.eccc.uni-trier.de/eccc/
18. Li, M., Ma, B., Wang, L.: On the closest string and substring problems. Journal of the ACM 49(2), 157–171 (2002)
19. Liang, F.: A lower bound for on-line bin packing. Information Processing Letters 10, 76–79 (1980)
20. Motwani, R., Raghavan, P.: Randomized Algorithms. Cambridge University Press (2000)
21. Munro, J.I., Paterson, M.S.: Selection and sorting with limited storage. Theoretical Computer Science 12, 315–323 (1980)
22. Goldreich, S.G.O., Ron, D.: Property testing and its connection to learning and approximation. J. ACM 45, 653–750 (1998)

Multivariate Polynomial Integration and Differentiation Are Polynomial Time Inapproximable Unless P=NP

Bin Fu

Department of Computer Science, University of Texas-Pan American,
Edinburg, TX 78539, USA
bfu@utpa.edu

Abstract. We investigate the complexity of approximate integration and differentiation for multivariate polynomials in the standard computation model. For a functor $F(\cdot)$ that maps a multivariate polynomial to a real number, we say that an approximation $A(\cdot)$ is a *factor $\alpha\colon N \to N^+$ approximation* iff for every multivariate polynomial f with $A(f) \geq 0$, $\frac{F(f)}{\alpha(n)} \leq A(f) \leq \alpha(n)F(f)$, and for every multivariate polynomial f with $F(f) < 0$, $\alpha(n)F(f) \leq A(f) \leq \frac{F(f)}{\alpha(n)}$, where n is the length of f, $len(f)$.

For integration over the unit hypercube, $[0,1]^d$, we represent a multivariate polynomial as a product of sums of quadratic monomials: $f(x_1,\ldots,x_d) = \prod_{1 \leq i \leq k} p_i(x_1,\ldots,x_d)$, where $p_i(x_1,\ldots,x_d) = \sum_{1 \leq j \leq d} q_{i,j}(x_j)$, and each $q_{i,j}(x_j)$ is a single variable polynomial of degree at most two and constant coefficients. We show that unless P = NP there is no $\alpha\colon N \to N^+$ and $A(\cdot)$ that is a factor α polynomial-time approximation for the integral $I_d(f) = \int_{[0,1]^d} f(x_1,\ldots,x_d) d\,x_1,\ldots,d\,x_d$.

For differentiation, we represent a multivariate polynomial as a product quadratics with $0, 1$ coefficients. We also show that unless P = NP there is no $\alpha\colon N \to N^+$ and $A(\cdot)$ that is a factor α polynomial-time approximation for the derivative $\frac{\partial f(x_1,\ldots,x_d)}{\partial x_1,\ldots,\partial x_d}$ at the origin $(x_1,\ldots,x_d) = (0,\ldots,0)$. We also give some tractable cases of high dimensional integration and differentiation.

1 Introduction

Integration and differentiation are basic operations in classical mathematics. Integrations with a large number of variables have been found applications in many areas such as finance, nuclear physics, and quantum system, etc. The complexity for approximating multivariate integration has been studied by measuring the number of function evaluations. For example, Sloan and Wozniakowski proved an exponential lower bounds 2^s of function evaluations in order to obtain an approximation with error less than the integration itself that has s variables [9]. The integration $\int_{[0,1]^d} f(x_1,\cdots,x_d)d x_1 \cdots d x_d$ is over the cubic $[0,1]^d$ for some function $f(x_1,\cdots,x_d)$. In the quasi-Monte Carlo method for computing $\int_{[0,1]^d} f(x)d x$, it is approximated by $\frac{1}{n}\sum_{i=1}^{n} f(p_i)$, where $p_1,\cdots,p_n$ are n random points in $[0,1]^d$.

J. Snoeyink, P. Lu, K. Su, and L. Wang (Eds.): FAW-AAIM 2012, LNCS 7285, pp. 182–191, 2012.
© Springer-Verlag Berlin Heidelberg 2012

This approximation has an error $\Theta(\frac{(\ln n)^{d-1}}{n})$ that grows exponentially on the dimension number d (see e.x., [7,6] and the reference papers there).

We study the polynomial-time approximation limitation for the high dimensional integration for some easily defined functions. Valiant showed that computing the high dimensional integration and differentiation are both $\#P$-hard via reducing permanent problem to them [12]. Valiant's results do not imply any inapproximability of the two problems since permanent has a polynomial-time approximation scheme [5].

In this paper, we consider the high dimensional integration for multivariate polynomials. For a functor $F(\cdot)$ that maps a multivariate polynomial to a real number, we say that an approximation $A(\cdot)$ is a *factor* $\alpha\colon N \to N^+$ *approximation* iff for every multivariate polynomial f with $A(f) \geq 0$, $\frac{F(f)}{\alpha(n)} \leq A(f) \leq \alpha(n)F(f)$, and for every multivariate polynomial f with $F(f) < 0$, $\alpha(n)F(f) \leq A(f) \leq \frac{F(f)}{\alpha(n)}$, where n is the length of f, $len(f)$. For integration over the unit hypercube, $[0,1]^d$, we represent a multivariate polynomial as a product of sums of quadratic monomials: $f(x_1, \ldots, x_d) = \prod_{1 \leq i \leq k} p_i(x_1, \ldots, x_d)$, where $p_i(x_1, \ldots, x_d) = \sum_{1 \leq j \leq d} q_{i,j}(x_j)$, and each $q_{i,j}(x_j)$ is a single variable polynomial of degree at most two and constant coefficients. Its integration can be computed in polynomial space. It is easy to deal with the sum of monomials. Studying the product of multiple polynomial of multiple variables is of essential importance in complexity theory. A $\prod \sum \prod$ polynomial is a multiplication of several polynomial $\prod_{i=1}^{m} p_i(x_1, \cdots, x_d)$ and each $p_i(x_1, \cdots, x_d)$ is a sum of monomials. The sum of product expansion of a $\prod \sum \prod$ polynomial may have exponential number of terms. Testing some properties such as the multilinear monomial in the sum of product expansion of a $\prod \sum \prod$ polynomial is NP-hard [2]. We show that there is not any-factor polynomial-time approximation to the integration problem unless P = NP.

A similar hardness of approximation result is also derived for the differentiation of the polynomial function. The recent development of monomial testing theory [1,3,2] can be used to explain the hardness for computing the differentiation for a $\prod \sum \prod$ polynomial. For differentiation, we represent a multivariate polynomial as a product quadratics with $0, 1$ coefficients. We also show that unless P = NP, there is not any-factor polynomial-time approximation to its differentiation $\frac{\partial f^{(d)}(x_1, \cdots, x_d)}{\partial x_1 \cdots \partial x_d}$ at the origin point $(x_1, \cdots, x_d) = (0, \cdots, 0)$. Since both integration and differentiation are widely used, this approach may help understand the complexity of approximation of some mathematics systems that involve high dimension integration or differentiation.

Partial derivatives were used in developing deterministic algorithms for the polynomial identity problem (for example, see [8]), a fundamental problem in the computational complexity theory. Our intractability result for the high dimension differentiation over multivariate polynomial points out a barrier of this approach.

We also give some tractable cases of high dimension integration and differentiation.

2   Notations

Let $N = \{0, 1, 2, \cdots\}$ be the set of all natural numbers. Let $N^+ = \{1, 2, \cdots\}$ be the set of all positive natural numbers.

Assume that function $r(n)$ is from N to N^+. For a functor $F(.)$ that converts a multivariate polynomial into a real number, an algorithm $A(.)$ gives an $r(n)$-factor approximation to $F(f)$ if it satisfies the following conditions: if $F(f) \geq 0$, then $\frac{F(f)}{r(n)} \leq A(f) \leq r(n)F(f)$; and if $F(f) < 0$, then $r(n)F(f) \leq A(f) \leq \frac{F(f)}{r(n)}$, where n is the length of f.

Assume that functions $r(n)$ and $s(n)$ are from N to N^+. For a functor $F(.)$, an algorithm $A(.)$ gives an $(r(n), s(n))$-factor approximation to $F(f)$ if $F(f) \geq 0$, then $\frac{F(f)}{r(n)} - s(n) \leq A(f) \leq r(n)F(f) + s(n)$; and if $F(f) < 0$, then $r(n)F(f) - s(n) \leq A(f) \leq \frac{F(f)}{r(n)} + s(n)$, where n is the length of f.

In this paper, we consider two kinds of functors. The first one is the integration in the unit cube for a multivariate polynomial: $\int_{[0,1]^d} f(x_1, \cdots, x_d)dx_1 \cdots dx_d$. The second is the differentiation $\frac{\partial f^{(d)}(x_1, \cdots, x_d)}{\partial x_1 \cdots \partial x_d}$ at the origin point $(x_1, \cdots, x_d) = (0, \cdots, 0)$.

For the complexity of multivariate integration, we consider the functions with the format $f(x_1, \cdots, x_d) = p_1(x_1, \cdots, x_d)p_2(x_1, \cdots, x_d) \cdots p_k(x_1, \cdots, x_d)$, where each $p_i(x_1, \cdots, x_d) = \sum_{j=1}^{d} q_{i,j}(x_j)$ with each single variable polynomial $q_{i,j}(x_j)$ of constant degree. This kind multivariate polynomial is called $\prod \sum S_c$ if the degree of each $q_{i,j}(x_j)$ is at most c.

For the complexity of multivariate differentiation, we consider the functions with the format $f(x_1, \cdots, x_d) = p_1(x_1, \cdots, x_d)p_2(x_1, \cdots, x_d) \cdots p_k(x_1, \cdots, x_d)$, where each polynomial $p_i(x_1, \cdots, x_d)$ is of a constant degree. The polynomial $f(x_1, \cdots, x_d)$ is called a $\prod \sum \prod_k$ *polynomial* if the degree of each $p_i(x_1, \cdots, x_d)$ is at most k.

3   Overview of Our Methods

In this section, we show the brief idea to derive the main result of this paper (Theorem 1). 3SAT is an NP-complete problem proved by Cook [4]. We show that approximating the integration of a $\prod \sum S_2$ polynomial is NP-hard by a reduction from 3SAT problem to it. It is still NP-hard to decide a conjunctive normal form that each variable appears at most three times with at most one negative time. We assume that each variable has its negation appears at most one time (Otherwise, we replace it by its negation).

We show (see Lemma 2) that there exist integer coefficients polynomial functions $g_1(x) = ax^2 + bx + c$, $g_2(x) = ux + v$, and $f(x) = 2x$ satisfy that $\int_0^1 g_1(x)dx = 1$, $\int_0^1 g_2(x)dx = 1$, $\int_0^1 f(x)dx = 1$, $\int_0^1 g_1(x)g_2(x)dx = 4$, $\int_0^1 g_1(x)f(x)dx = 0$, $\int_0^1 g_2(x)f(x)dx = 0$, and $\int_0^1 g_1(x)g_2(x)f(x)dx = 0$.

Example 1. Consider the logical formula $F = (x_1 + x_2)(x_1 + \overline{x}_2)(\overline{x}_1 + x_2)$, which has the sum of product expansion $x_1 x_1 \overline{x}_1 + x_1 x_1 x_2 + x_1 \overline{x}_2 \overline{x}_1 + x_1 \overline{x}_2 x_2 + x_2 x_1 \overline{x}_1 + x_2 x_1 x_2 + x_2 \overline{x}_2 \overline{x}_1 + x_2 \overline{x}_2 x_2$. The term $x_1 x_1 x_2$ can bring a truth assignment

$x_1 = true$ and $x_2 = true$ to make F true. As each variable appears at most 3 times with at most one negative appearance, the first positive x_i is replaced by $g_1(y_i)$, the second positive x_i is replaced by $g_2(y_i)$, and the negative $\bar{x}_i$ is replaced by $f(y_i)$. It is converted into the polynomial

$$p(y_1, y_2) = (g_1(y_1) + g_1(y_2))(g_2(y_1) + f(y_2))(f(y_1) + g_2(y_2)).$$

The polynomial $p(y_1, y_2)$ has the sum of product expansion

$$g_1(y_1)g_2(y_1)f(y_1) + g_1(y_1)g_2(y_1)g_2(y_2) + g_1(y_1)f(y_2)f(y_1) +$$
$$g_1(y_1)f(y_2)g_2(y_2) + g_1(y_2)g_2(y_1)f(y_1) + g_1(y_2)g_2(y_1)g_2(y_2) +$$
$$g_1(y_2)f(y_2)f(y_1) + g_1(y_2)f(y_2)g_2(y_2).$$

Consider the integration $\int_{[0,1]^2} p(y_1, y_2) dy_1 dy_2$. The integration can be distributed into those product terms. $\int_{[0,1]^2} g_1(y_1)g_2(y_1)g_2(y_2) dy_1 dy_2$ is one of them. We have $\int_{[0,1]^2} g_1(y_1)g_2(y_1)g_2(y_2) dy_1 dy_2 = (\int_{[0,1]} g_1(y_1)g_2(y_1) dy_1)(\int_{[0,1]} g_2(y_2) dy_2)$ $= 4 \cdot 1 = 4$.

The integrations for other terms are all non-negative integers. Thus, $\int_{[0,1]^2} p(y_1, y_2) dy_1 dy_2$ is a positive integer due to the satisfiability of F.

Example 2. Consider the logical formula $G = (x_1 + x_2)\bar{x}_1\bar{x}_2$, which has the sum of product expansion $x_1\bar{x}_1\bar{x}_2 + x_1\bar{x}_2\bar{x}_2$. Neither $x_1\bar{x}_1\bar{x}_2$ nor $x_1\bar{x}_2\bar{x}_2$ can be satisfied. As each variable appears at most 3 times with at most one negative appearance, the first positive x_i is replaced by $g_1(y_i)$, the second positive x_i is replaced by $g_2(y_i)$, and the negation case $\bar{x}_i$ is replaced by $f(y_i)$. It is converted into the polynomial $q(y_1, y_2) = (g_1(y_1) + g_1(y_2))f(y_1)f(y_2)$. The polynomial $q(y_1, y_2)$ has the sum of product expansion $g_1(y_1)f(y_1)f(y_2) + g_1(y_2)f(y_1)f(y_2)$.

Consider the integration $\int_{[0,1]^2} q(y_1, y_2) dy_1 dy_2$ that is identical to $\int_{[0,1]^2} g_1(y_1)f(y_1)f(y_2) dy_1 dy_2 + \int_{[0,1]^2} g_1(y_2)f(y_1)f(y_2) dy_1 dy_2$. We have

$$\int_{[0,1]^2} g_1(y_1)f(y_1)f(y_2) dy_1 dy_2 = (\int_{[0,1]} g_1(y_1)f(y_1) dy_1)(\int_{[0,1]} f(y_2) dy_2) = 0 \cdot 1 = 0.$$

We also have

$$\int_{[0,1]^2} g_1(y_2)f(y_1)f(y_2) dy_1 dy_2 = (\int_{[0,1]} f(y_1) dy_1)(\int_{[0,1]} g_1(y_2)f(y_2) dy_2) = 1 \cdot 0 = 0.$$

Therefore, $\int_{[0,1]^2} q(y_1, y_2) dy_1 dy_2 = 0$ due to the unsatisfiability of G. Therefore, for any-factor $a(n) > 0$, a polynomial-time factor $a(n)$-approximation to the integration of a $\prod \sum S_2$ polynomial implies a polynomial-time decision for the satisfiability of the corresponding boolean formula.

4   Intractability of High Dimensional Integration

In this section, we show that the integration in high dimensional cube $[0, 1]^d$ does not have any-factor approximation. We will reduce an existing NP-complete problem to the integration problem. Our main technical contribution is in converting a logical formula into a polynomial. We often use a basic property of integration, which can be found in some standard text books of calculus (for example [11]). Assume function $f(x_1, \cdots, x_d) = f_1(x_{i_1}, \cdots, x_{i_{d_1}}) f_2(x_{j_1}, \cdots, x_{j_{d_2}})$, where $\{x_1, \cdots, x_d\}$ is the disjoint union of $\{x_{i_1}, \cdots, x_{i_{d_1}}\}$ and $\{x_{j_1}, \cdots, x_{j_{d_2}}\}$. Then we have $\int_{[0,1]^d} f(x_1, \cdots, x_d) dx_1 \cdots dx_d =$

$$\left(\int_{[0,1]^{d_1}} f_1(x_{i_1}, \cdots, x_{i_{d_1}}) dx_{i_1} \cdots dx_{i_{d_1}} \right) \cdot \left(\int_{[0,1]^{d_2}} f(x_{j_1}, \cdots, x_{j_{d_2}}) dx_{j_1} \cdots dx_{j_{d_2}} \right).$$

In order to make the conversion from logical operation to algebraic operation, we represent conjunctive normal form with the following format. For example, the formula $(x_1 + x_2)(x_1 + \overline{x}_2)(\overline{x}_1 + x_2)$ is a conjunctive normal form with two boolean variables x_1 and x_2, where $+$ represents the logical $\bigvee$, and . represent the logical $\bigwedge$.

Definition 1

- A 3SAT instance is a conjunctive form $C_1 \cdot C_2 \cdots C_m$ such each C_i is a disjunction of at most three literals.
- 3SAT is the language of those 3SAT instances that have satisfiable assignments.
- A $(3, 3)$-SAT instance is an instance G for 3SAT such that for each variable x, the total number of times of x and $\overline{x}$ in G is at most 3, and the total number of times of $\overline{x}$ in G is at most 1.
- $(3, 3)$-SAT is the language of those $(3, 3)$-SAT instances that have satisfiable assignments.

It is well known 3SAT is NP-complete [4]. For examples, $(x_1 + x_2 + x_3)(x_1 + \overline{x}_2)(\overline{x}_1 + x_2)$ is both 3SAT and $(3, 3)$-SAT instance, and also belongs to both 3SAT and $(3, 3)$-SAT. On the other hand, $(x_1 + x_2 + x_3)(\overline{x}_1 + \overline{x}_2)(\overline{x}_1 + x_2)$ is not a $(3, 3)$-SAT instance since $\overline{x}_1$ appears twice in the formula. The following lemma is similar to a result derived by Tovey [10].

Lemma 1. *There is a polynomial-time reduction from 3SAT to $(3, 3)$-SAT.*

Proof. Let F be an instance for 3SAT. Let's focus on one variable x_i that appears m times in F. Introduce a series of new variables $y_{i,1}, \cdots, y_{i,m}$ for x_i. Convert F to F' by changing the j-th occurrence of x_i in F to $y_{i,j}$ for $j = 1, \cdots, m$. Define

$$G_{x_i} = (x_i \rightarrow y_{i,1}) \cdot (y_{i,1} \rightarrow y_{i,2}) \cdot (y_{i,2} \rightarrow y_{i,3}) \cdot (y_{i,3} \rightarrow y_{i,4}) \cdots (y_{i,m-1} \rightarrow y_{i,m}) \cdot$$
$$(y_{i,m} \rightarrow x_i)$$
$$= (\overline{x}_i + y_{i,1}) \cdot (\overline{y}_{i,1} + y_{i,2}) \cdot (\overline{y}_{i,2} + y_{i,3}) \cdot (\overline{y}_{i,3} + y_{i,4}) \cdots (\overline{y}_{i,m-1} + y_{i,m}) \cdot$$
$$(\overline{y}_{i,m} + x_i).$$

Each logical formula $(x \rightarrow y)$ is equivalent to $(\overline{x} + y)$. If G_{x_i} is true, then $x_i, y_{i,1}, \cdots, y_{i,m}$ are equivalent.

Convert F' into F'' such that $F'' = F' G_{x_1} \cdots G_{x_k}$, where $x_1, \cdots, x_k$ are all variables in F.

For each variable x in F'' with more than one $\overline{x}$, create a new variable y_x, replace each positive x of F by $\overline{y_x}$, and each negative $\overline{x}$ by y_x. Thus, F'' becomes F'''. It is easy to see that $F \in$ 3SAT iff F'' is satisfiable iff $F''' \in (3,3)$-SAT. $\square$

4.1   Integration of $\prod \sum S_2$ Polynomial

Lemma 2 is our main technical lemma. It is used to convert a $(3,3)$-SAT instance into a $\prod \sum S_2$ polynomial.

Lemma 2. *Let* $g_1(x) = 30x^2 - 36x + 9$, $g_2(x) = -6x + 4$, *and* $f(x) = 2x$. *They satisfy the following conditions*

1. $\int_0^1 g_1(x)dx$, $\int_0^1 g_2(x)dx$, $\int_0^1 f(x)dx$, *and* $\int_0^1 g_1(x)g_2(x)dx$ *are all positive integers, and*
2. $\int_0^1 g_1(x)f(x)dx$, $\int_0^1 g_2(x)f(x)dx$, *and* $\int_0^1 g_1(x)g_2(x)f(x)dx$ *are all equal to 0.*

Proof. For a polynomial $h(x) = a_n x^n + a_{n-1}x^{n-1} + \cdots + a_0$, we can compute its integration as $\int_0^1 h(x)dx = \frac{a_n}{n+1} + \frac{a_{n-1}}{n} + \cdots + a_0$. It is straightforward to verify the lemma with the concrete expressions for the three functions $g_1(x), g_2(x)$ and $f(x)$.   □

Lemma 3. *There is a polynomial-time algorithm h such that given a $(3,3)$-SAT instance $s(x_1, \cdots, x_d)$, it produces a $\prod \sum S_2$ polynomial $h(s(x_1, \cdots, x_d)) = q(y_1, \cdots, y_d)$ to satisfy the following two conditions:*

1. *if* $s(x_1, \cdots, x_d)$ *is satisfiable, then* $\int_{[0,1]^d} q(y_1, \cdots, y_d)dy_1 \cdots dy_d$ *is a positive integer; and*
2. *if* $s(x_1, \cdots, x_d)$ *is not satisfiable, then* $\int_{[0,1]^d} q(y_1, \cdots, y_d)dy_1 \cdots dy_d$ *is zero.*

Proof. We give two examples to show how a logical formula is converted into a multivariate polynomial in section 3. Let polynomials $g_1(y), g_2(y)$, and $f(y)$ be defined according to those in Lemma 2.

For a $(3,3)$-SAT problem $s(x_1, \cdots, x_d)$, let $q(y_1, \cdots, y_d)$ be defined a follows.

- For the first positive literal x_i in $s(x_1, \cdots, x_d)$, replace it with $g_1(y_i)$.
- For the second positive literal x_i in $s(x_1, \cdots, x_d)$, replace it with $g_2(y_i)$.
- For the negative literal $\overline{x}_i$ in $s(x_1, \cdots, x_d)$, replace it with $f(y_i)$.

The formula $s(x_1, \cdots, x_d)$ has a sum of product form. It is satisfiable if and only if one term does not contain both positive and negative literals for the same variable. If a term contains both x_i and $\overline{x}_i$, the corresponding term in the sum of product for $q(.)$ contains both $g_j(y_i)$ and $f(y_i)$ for some $j \in \{1, 2\}$. This makes it zero after integration by Lemma 2. Therefore, $s(x_1, \cdots, x_d)$ is satisfiable if and only if $\int_{[0,1]^d} q(y_1, \cdots, y_d)dy_1 \cdots dy_d$ is not zero. Furthermore, it is satisfiable, the integration is a positive integer by Lemma 2. See the two examples in section 3. The computating time of h is clearly polynomial since we convert s to $h(s)$ by replacing each literal by a single variable function of degree at most 2.   □

Theorem 1. *Let $a(n)$ be an arbitrary function from N to N^+. Then there is no polynomial-time $a(n)$-factor approximation for the integration of a $\prod \sum S_2$ polynomial $p(x_1, \cdots, x_d)$ in the region $[0, 1]^d$ unless $P = NP$.*

Proof. Assume that $A(.)$ is a polynomial-time $a(n)$-factor approximation for the integration $\int_{[0,1]^d} p(y_1, \cdots, y_d) dy_1 \cdots dy_d$ with $\prod \sum S_2$ polynomial $p(y_1, \cdots, y_d)$. For a $(3,3)$-SAT instance $s(x_1, \cdots, x_d)$, let $p(y_1, \cdots, y_d) = h(s(x_1, \cdots, x_d))$ according to Lemma 3. By Lemma 3, a $(3,3)$-SAT instance $s(x_1, \cdots, x_d)$ is satisfiable if and only if the integration $J = \int_{[0,1]^d} p(y_1, \cdots, y_d) dy_1 \cdots dy_d$ is not zero. Assume that $s(x_1, \cdots, x_d)$ is not satisfiable, then we have $A(J) \in [J/a(n), J \cdot a(n)] = [0, 0]$ that implies $A(J) = 0$. Assume that $s(x_1, \cdots, x_d)$ is satisfiable, then we have $A(J) \in [J/a(n), J \cdot a(n)] \subseteq (0, +\infty)$ that implies $A(J) > 0$. Thus, $s(x_1, \cdots, x_d)$ is satisfiable if and only if $A(J) > 0$.

Therefore, there is a polynomial-time algorithm for solving $(3,3)$-SAT that is NP-complete by Lemma 1. So, P $=$ NP. □

As n is often used as the length of input, the time complexity is measured by a function on n. We use a function $a(1^n)$ instead of function $a(n)$ to control the accuracy of approximation since n has only $\log n$ bits in binary expression, and 1^n has n bits.

Lemma 4. *Assume that $a(1^n)$ is a polynomial-time computable function from $\{1\}^*$ to N^+ with $a(1^n) > 0$ for n. There is a polynomial-time algorithm such that given a $(3,3)$-SAT instance $s(x_1, \cdots, x_d)$, it generates a $\prod \sum S_2$ polynomial $p(y_1, \cdots, y_d)$ such that if $s(x_1, \cdots, x_d)$ is satisfiable, then $\int_{[0,1]^d} p(y_1, \cdots, y_d) dy_1 \cdots y_d$ is a positive integer at least $3a(1^n)^2$; and if $s(x_1, \cdots, x_d)$ is not satisfiable, $\int_{[0,1]^d} p(y_1, \cdots, y_d) dy_1 \cdots y_d$ is zero.*

Proof. For a $(3,3)$-SAT problem $s(x_1, \cdots, x_d)$, let $q(y_1, \cdots, y_d) = h(s(x_1, \cdots, x_d))$ be constructed as Lemma 3.

Since $a(1^n)$ is polynomial-time computable, let $p(y_1, \cdots, y_d) = 3a(1^n)^2 q(y_1, \cdots, y_d)$ that can be computed in a polynomial-time . □

Theorem 2. *Let $a(1^n)$ be a polynomial-time computable function from $\{1\}^*$ to N^+. Then there is no polynomial-time $(a(1^n), a(1^n))$-approximation for the integration problem $\int_{[0,1]^d} f(x_1, \cdots, x_d) dx_1 \cdots dx_d$ for a $\prod \sum S_2$ polynomial $f(.)$ unless P $=$ NP.*

Proof. Assume that there is a polynomial-time $(a(1^n), a(1^n))$-approximation $App(.)$ for the integration problem $\int_{[0,1]^d} f(x_1, \cdots, x_d) dx_1 \cdots dx_d$ for a $\prod \sum S_2$ polynomial $f(.)$.

Let $s(x_1, \cdots, x_d)$ be an arbitrary $(3,3)$-SAT instance. Let $p(y_1, \cdots, y_d)$ be the polynomial according to Lemma 4.

Let $J = \int_{[0,1]^d} p(y_1, \cdots, y_d) dy_1 \cdots y_d$. If $s(x_1, \cdots, x_d)$ is not satisfiable, then $J = 0$. Otherwise, $J \geq 3a(1^n)^2$.

Assume that $s(x_1, \cdots, x_d)$ is not satisfiable. Since $App(J)$ is an $(a(1^n), a(1^n))$-approximation, we have $App(J) \leq J \cdot a(1^n) + a(1^n) = a(1^n)$ by the definition in section 2.

Assume that $s(x_1, \cdots, x_d)$ is satisfiable. Since $App(J)$ is an $(a(1^n), a(1^n))$-approximation, we have $App(J) \geq \frac{J}{a(1^n)} - a(1^n) \geq \frac{3a(1^n)^2}{a(1^n)} - a(1^n) = 2a(1^n)$ by the definition in section 2.

Therefore, $s(x_1, \cdots, x_d)$ is satisfiable if and only if $App(J) \geq 2a(1^n)$. Thus, if there is a polynomial-time $(a(1^n), a(1^n))$–approximation, then there is a polynomial-time algorithm for solving $(3,3)$-SAT. By Lemma 1, P = NP. □

5   Inapproximation of Differentiation

In this section, we study the hardness of high dimensional differentiation. We derive the inapproximation results under both NP $\neq$ P.

Definition 2. *A monomial is an expression* $x_1^{a_1} \cdots x_d^{a_d}$ *and its degree is* $a_1 + \cdots + a_d$. *A monomial* $x_1^{a_1} \cdots x_d^{a_d}$, *in which* $x_1, \cdots, x_d$ *are different variables, is a multilinear if* $a_1 = a_2 = \cdots = a_d = 1$.

For example, $(x_1 x_3 + x_2^2)(x_2 x_4 + x_3^2)$ is a $\prod \sum \prod_2$ polynomial. It has a multilinear monomial $x_1 x_2 x_3 x_4$ in its sum of products expansion.

We give Lemma 5 to convert an instance f for $(3,3)$-SAT into a $\prod \sum \prod_2$ polynomial. The technology developed in [1,2] will be applied in the construction.

Lemma 5. *Let* $a(1^n)$ *be a polynomial-time computable function from* N *to* N^+. *Then there is a polynomial-time algorithm* A *such that given a* $(3,3)$-SAT *instance* $F(y_1, \cdots, y_d)$, *the algorithm returns a* $\prod \sum \prod_2$ *polynomial* $G(x_1, \cdots, x_d)$ *such that*

1. *If* F *is not satisfiable, then* G *does not have a multilinear monomial with an nonzero coefficient in its sum of product expansion.*
2. *If* F *is satisfiable, then* G *has the multilinear monomial* $x_1 \cdots x_d$ *with a positive integer coefficient at least* $3a(1^n)^2$ *in its sum of product expansion.*

Theorem 3. *Assume that* $r(n)$ *is a function from* N *to* N^+. *If there is a polynomial-time algorithm* A *such that given a* $\prod \sum \prod_2$ *polynomial* $g(x_1, \cdots, x_d)$, *it gives an* $r(n)$-*factor approximation to* $\frac{\partial g^{(n)}(x_1, \cdots, x_d)}{\partial x_1 \cdots \partial x_d}$ *at the origin point* $(x_1, \cdots, x_d) = (0, \cdots, 0)$, *then* P = NP.

Theorem 4. *Let* $a(1^n)$ *be a polynomial-time computable function from* $\{1\}^*$ *to* N^+. *Then there is no polynomial-time* $(a(1^n), a(1^n))$-*approximation for* $\frac{\partial g^{(n)}(x_1, \cdots, x_d)}{\partial x_1 \cdots \partial x_d}$ *with* $g(x_1, \cdots, x_d)$ *as a* $\prod \sum \prod_2$ *polynomial at the origin point* $(x_1, \cdots, x_d) = (0, \cdots, 0)$, *unless* P = NP.

6   Some Tractable Integrations and Derivatives

In this section, we present some polynomial-time algorithms for integration with some restrictions. We also show a case that the differentiation can fully polynomial-time approximation scheme.

6.1 Bounded Width Product

Definition 3. *A formula $f_1 \cdot f_2 \cdots f_m$ is c-wide if for each variables x_i, there is an index j such that x_i only appears in $f_j, f_{j+1}, \cdots f_{j+c-1}$, where each f_i is a sum of monomials.*

Theorem 5. *There is an $O(mn^{3c})$ time algorithm to compute the integration $\int_{[0,1]^d} F(x_1, \cdots, x_d)$ for a c-wide formula $F(x_1, \cdots, x_d) = f_1 \cdots f_m$, where n is the total length of F.*

Proof. Apply the divide and conquer method. Convert F into $F_1 G F_2$ such that G is a product of at most c sub-formulas $f_i \cdots f_j$ with $j - i = c$ in the middle region of F (we can let $i = \lceil \frac{m-c}{2} \rceil + 1$, and $j = \lceil \frac{m-c}{2} \rceil + c$).

Let S_1 be the set of variables that are only in F_1, S_2 be the set of variables that are only in F_2, and S be the set of variables that appear in G. The set of variables in F is partitioned into S_1, S, and S_2.

As $F_1 = f_1 \cdots f_{i-1}$, we convert F_1 into $F_1^* = f_1 \cdots f_{i-c} f_1^*$, where f_1^* is the product of the last c sub-formulas: $f_1^* = f_{i-c} \cdots f_{i-1}$. Similarly, as $F_2 = f_{j+1} \cdots f_m$, we convert F_2 into $F_2^* = f_2^* f_{j+c+1} \cdots f_m$, where f_2^* is the product of the first c sub-formulas: $f_2^* = f_{j+1} \cdots f_{j+c}$. Convert G into the sum of products.

We have

$$\int_{[0,1]^d} F(x_1, \cdots, x_d) dx_1 \cdots dx_d = \int_{[0,1]^{|S|}} G \cdot \left(\int_{[0,1]^{|S_1|}} F_1 dS_1 \right) \cdot \left(\int_{[0,1]^{|S_2|}} F_2 dS_2 \right) dS$$

$$= \int_{[0,1]^{|S|}} G \cdot \left(\int_{[0,1]^{|S_1|}} F_1^* dS_1 \right) \cdot \left(\int_{[0,1]^{|S_2|}} F_2^* dS_2 \right) dS.$$

The integration $\int_{[0,1]^{|S_1|}} F_1 dS_1$ can be expressed as a polynomial of variables in S. The integration $\int_{[0,1]^{|S_2|}} F_2 dS_2$ can be expressed as a polynomial of variables in S.

We have the recursive equation for the computational time $T(m) = 2T(m/2) + O(n^{3c})$. This gives $T(m) = O(mn^{3c})$. □

6.2 Tractable Differentiation

In this section, we give a polynomial-time randomized approximation scheme by using the theory of testing monomials developed by Chen and Fu [2,1].

Definition 4. *Let $f(x_1, \cdots, x_d) = p_1(x_1, \cdots, x_d) \cdots p_k(x_1, \cdots, x_d)$ be a $\prod \sum$ polynomial. If for each $p_i(x_1, \cdots, x_d)$, each variable's coefficient is either 0 or 1, then f is called a $\prod \sum^*$ polynomial.*

We show that the differentiation for a $\prod \sum^*$ polynomial has a polynomial-time approximation scheme. Chen and Fu derived the following theorem by a reduction from the number of perfect matchings in a bipartite.

Theorem 6 (Chen and Fu [1]). *There is a polynomial-time randomized algorithm to approximate the coefficient of a $\prod \sum^*$ polynomial.*

Theorem 7. *Let ϵ be an arbitrary constant in $(0,1)$. Then there is a polynomial-time randomized algorithm that given a $\prod \sum^*$ polynomial f, it returns a $(1+\epsilon)$-approximation for $\frac{\partial f(x_1,\cdots,x_d)^{(d)}}{\partial x_1 \cdots \partial x_d}$ at the point $(0,\cdots,0)$.*

Proof. For a $\prod \sum^*$ polynomial $f(x_1,\cdots,x_d)$, its $\frac{\partial f(x_1,\cdots,x_d)^{(d)}}{\partial x_1 \cdots \partial x_d}$ at the point $(0,\cdots,0)$ is identical to the coefficient of the monomial $x_1 \cdots x_d$ in the sum of products in the expansion of $f(x_1,\cdots,x_d)$. The theorem follows from Theorem 6.
$\square$

Acknowledgements. The author is grateful to Jack Snoeyink for his suggestions that improves the presentation of this paper. The author would also like to thank the referees at FAW-AAIM 2012 for their helpful comments. This research is supported in part by the National Science Foundation Early Career Award CCF-0845376.

References

1. Chen, Z., Fu, B.: Approximating Multilinear Monomial Coefficients and Maximum Multilinear Monomials in Multivariate Polynomials. In: Wu, W., Daescu, O. (eds.) COCOA 2010, Part I. LNCS, vol. 6508, pp. 309–323. Springer, Heidelberg (2010)
2. Chen, Z., Fu, B.: The Complexity of Testing Monomials in Multivariate Polynomials. In: Wang, W., Zhu, X., Du, D.-Z. (eds.) COCOA 2011. LNCS, vol. 6831, pp. 1–15. Springer, Heidelberg (2011)
3. Chen, Z., Fu, B., Liu, Y., Schweller, R.: Algorithms for Testing Monomials in Multivariate Polynomials. In: Wang, W., Zhu, X., Du, D.-Z. (eds.) COCOA 2011. LNCS, vol. 6831, pp. 16–30. Springer, Heidelberg (2011)
4. Cook, S.A.: The complexity of theorem-proving procedures. In: Proceedings of the 2nd Annual ACM Symposium on Theory of Computing, pp. 151–158 (1971)
5. Jerrum, M., Sinclaire, A., Vigoda, E.: A polynomial-time ap- priximation algorithm for the permanent of a matrix with nonnegative entries. Journal of the ACM 51(4), 671–697 (2004)
6. Niederreiter, H.: Quasi-monte carlo methods and pseudo-random numbers. Bulletin of the American Mathematical Society 84(6), 957–1041 (1978)
7. Niederreiter, H.: Random Number Generation and quasi-Monte Carlo Methods, vol. 63. SIAM, Philadelphia (1992)
8. Shpilka, A., Volkovich, I.: Read-once polynomial identity testing. In: Proceedings of the 40th Annual ACM Symposium on Theory of Computing, pp. 507–516 (2008)
9. Sloan, I.H., Wozniakowski, H.: An intractability result for multiple integration. Math. Comput. 66(219), 1119–1124 (1997)
10. Tovey, C.A.: A simplified satisfiability problem. Discrete Applied Mathematics 8, 85–89 (1984)
11. Trench, W.F.: Advanced Calculus. Harper & Row, New York (1978)
12. Valiant, L.: Completeness classes in algebra. In: Proceedings of the Eleventh Annual ACM Symposium on Theory of Computing, pp. 249–261 (1979)

Some Remarks on the Incompressibility of Width-Parameterized SAT Instances

Bangsheng Tang

Institute for Interdisciplinary Information Sciences,
Tsinghua University, 100084, Beijing
bangsheng.tang@gmail.com

Abstract. Compressibility of a formula regards reducing the length of the input, or some other parameter, while preserving the solution. Any 3-**SAT** instance on N variables can be represented by $O(N^3)$ bits; [4] proved that the instance length in general cannot be compressed to $O(N^{3-\epsilon})$ bits under the assumption **NP** $\not\subseteq$ **coNP/poly**, which implies that the polynomial hierarchy does not collapse. This note initiates research on compressibility of **SAT** instances parameterized by width parameters, such as tree-width or path-width. Let $\mathbf{SAT}_{\mathsf{tw}}(w(n))$ be the satisfiability instances of length n that are given together with a tree-decomposition of width $O(w(n))$, and similarly let $\mathbf{SAT}_{\mathsf{pw}}(w(n))$ be instances with a path-decomposition of width $O(w(n))$. Applying simple techniques and observations, we prove conditional incompressibility for both instance length and width parameters: (i) under the exponential time hypothesis, given an instance ϕ of $\mathbf{SAT}_{\mathsf{tw}}(w(n))$ it is impossible to find within polynomial time a ϕ' that is satisfiable if and only if ϕ is satisfiable and tree-width of ϕ' is half of ϕ; and (ii) assuming a scaled version of **NP** $\not\subseteq$ **coNP/poly**, any 3-$\mathbf{SAT}_{\mathsf{pw}}(w(n))$ instance of N variables cannot be compressed to $O(N^{1-\epsilon})$ bits.

1 Introduction

Satisfiability(**SAT**) is the problem of deciding whether a given conjunctive normal form(**CNF**) formula is satisfiable. Denote by n the input length of the formula, and by N the number of variables. **SAT** has been playing a central role in both theoretical and applied aspects of computer science. It is the prototypical **NP**-complete problem in complexity theory, and has found enormous applications in various practical areas, e.g. artificial intelligence, machine learning, decision making, automated theorem proving.

Although it is not possible to solve **SAT** in its most general form within polynomial time unless **NP** $\neq$ **P**, there is an apparent need for practically efficient algorithms. Real world instances often come with structure. Parameterization is a general way of quantifying the structure. The quest for efficient algorithms for fixed parameters (*fixed-parameter tractable*) has received significant attention in the past few years. One way of parameterizing **SAT** is using width parameters (tree-width, path-width, branch-width, etc.), which are graph-theoretic parameters of a graph associated with the instance. It is shown in a series of moves, e.g. [1, 13, 12, 5, 6, 2] that width-parameterized **SAT** with parameter k can be solved time-efficiently in simultaneously $2^{O(k)}n^{O(1)}$ time and $2^{O(k)}n^{O(1)}$ space, or space-efficiently in simultaneously $2^{O(k \log n)}n^{O(1)}$ time

J. Snoeyink, P. Lu, K. Su, and L. Wang (Eds.): FAW-AAIM 2012, LNCS 7285, pp. 192–198, 2012.
© Springer-Verlag Berlin Heidelberg 2012

and $n^{O(1)}$ space. Rather involved time-space trade-off algorithms achieving better time and space are also given in [2].

We use the term *compressibility of* **SAT** *instances* to refer the fact that the length of input instance or parameters can be reduced by an efficient algorithm, in a way that the compressed instance preserves satisfiability. Compressing input instances has been considered in [4] under the term *sparsification*. The following two-player communication game between an efficient algorithm and an oracle was considered: the first player is a verifier who has an d-**CNF** formula, and she wants to decide its satisfiability within deterministic polynomial time; the second player is an oracle with unbounded computational power but without knowing the formula beforehand. The goal of this communication game is to minimize the number of bits the verifier must communicate with the oracle, such that the verifier can decide the satisfiability of the formula. N^d bits will suffice, because this is the total number of possible clauses and the first player can send an N^d bit string, where the ith bit indicates whether the ith clause is present. In [4], it is proved that the trivial way of communication is essentially optimal in the following sense: if the satisfiability of d-**CNF** formulas can be decided by communication within $O(N^{d-\epsilon})$ bits for any $\epsilon > 0$, then **NP** $\subseteq$ **coNP/poly**. How about **SAT** instance of bounded width parameters? At one extreme, when the width is $O(\log n)$, the verifier can compute satisfiability by herself without any communication; at another end, when the width is $\Omega(n)$, the formula is in its most general form, previous result can be applied. The technically interesting case is for the intermediate range of this width parameter.

In this note, we focus on the compressibility of **SAT** instances parameterized by two types of width parameters, i.e. tree-width and path-width. In the setting of parameterized problems, we can discuss compression of input instances as above, or compression of parameters. Note that algorithms for parameterized problems have running time and space super-polynomial in the chosen parameter times a polynomial of the input instance length. Efficient compression of parameters would indicate significant improvement in the time and space resource requirements of the state-of-the-art algorithms. We show that compressing the width-parameter by a constant factor within polynomial time is not possible under the assumption that unless exponential time hypothesis (**ETH**) fails. **ETH** states that solving **SAT** on N variables requires $2^{\Omega(N)}$ time, which is stronger than **P** $\neq$ **NP** and has important implications in computational complexity (see e.g. [7, 8, 10]). Let **NL**$[r(n)]$ be the class of problems decidable by a logspace machine equipped with a read-only non-deterministic, polynomially long witness tape, where the machine can have $O(r(n))$ passes (as the head reverses) over the witness tape. We strengthen the assumption that **NP** $\nsubseteq$ **coNP** to **NL**$[\omega(1)]$ $\nsubseteq$ **coNP**. In fact, we assume further that

$$\mathbf{NL}[\omega(1)] \nsubseteq \mathbf{coNP/poly}$$

to obtain an incompressibility result of input length for width-parameterized **SAT**.

A complexity theoretic study of this assumption is interesting on its own right, and it is left for future work. Here are some indications on its validity: (i) the belief that **NP** $\nsubseteq$ **coNP** is because usually people think that in fact the required certificate size blows up to exponential (not merely super-polynomial) - i.e. some kind of exhaustive enumeration is required - and (ii) given a non-uniform advice of polynomial size will not help either (in particular, an easy extension of Karp-Lipton shows that if **NP** $\subseteq$ **coNP/poly** the polynomial hierarchy collapses).

Results in this note are based on preliminary observations and simple techniques. Nevertheless, this work initiates the study of compressibility or sparsification of width-parameterized **SAT** instances, and makes conceptual contributions to better understanding the complexity of width-parameterized **SAT**. Techniques that work for general instances fail dramatically for width-parameterized instances. Intuition and speculation on why previous techniques fail are also included in this note. We believe that improving our results is a very interesting research direction for both theory and practice.

2 Preliminary

Notation and terminology used in this note basically follow [2].

Definition 1. *Let $G = (V, E)$ be an undirected graph. A* tree decomposition *of G is a tuple (T, X), where $T = (W, F)$ is a tree, and $X = \{X_1, \cdots, X_{|W|}\}$ where $X_i \subseteq V$ s.t. (1) $\cup_{i=1}^{|W|} X_i = V$, (2) $\forall (i, j) \in E$, $\exists t \in W$, s.t. $i, j \in X_t$, and (3) $\forall i$, the set $\{t : i \in X_t\}$ forms a subtree of T.*

Each of X_i is called a bag, *the* width *of (T, X) is defined as $max_{t \in W} |X_t| - 1$, and the* tree-width $\mathcal{TW}(G)$ *of graph G is defined as the minimum width over all possible tree decompositions.*

When the tree decomposition $T = (W, F)$ is restricted to a path, the decomposition is called *path decomposition*, and the specific tree-width is called path-width $\mathcal{PW}(G)$.

Definition 2. *The* incidence graph G_ϕ *of a* **SAT** *instance ϕ is a bipartite graph, where in one side of the bipartization each node is associated with a distinct unsigned variable, and in the other each node is associated with a clause. There is an edge between a clause-node and a variable-node if and only if the variable appears in a literal of the clause.*

The tree-width of a formula ϕ is the tree-width of its incidence graph, $\mathcal{TW}(\phi) = \mathcal{TW}(G_\phi)$. When it is clear from the context we may abuse notation and write $\mathcal{TW}(\phi)$ to denote the width of a given decomposition of G_ϕ. Note that any tree-decomposition (path-decomposition) of an instance graph will have tree-width (path-width) at most N, because one can always construct a path of the number of the clauses, and put each clause into a different bag arbitrarily, and copy all the variables into all the bags. This is a valid path-decomposition, and therefore a valid tree-decomposition of width N. Without loss of generality we assume that in number of bags is upper bounded by a polynomial of the number of the clauses and variables. This is assured by a property of decompositions called *nice*. A nice one can be constructed from any decomposition efficiently (see e.g. [9, 3]).

Denote by $\mathbf{SAT}_{tw}(w(n))$ the problem of deciding satisfiability of a **CNF** formula, which is given together with a tree-decomposition of width $O(w(n))$ as input, where n

is input length. $\mathbf{SAT}_{pw}(w(n))$ is the path-decomposition version. $3\text{-}\mathbf{SAT}_{tw}(w(n))$ and $3\text{-}\mathbf{SAT}_{pw}(w(n))$ are the variants where the input formulas are 3-**CNF** formulas. The following lemma shows that there is no essential difference between $3\text{-}\mathbf{SAT}_{pw}(w(n))$ and $\mathbf{SAT}_{pw}(w(n))$.

Lemma 1 ([11]). $\mathbf{SAT}_{pw}(w(n))$ *is reducible to* $3\text{-}\mathbf{SAT}_{pw}(w(n))$, *under logspace many-to-one reductions.*

Although the width parameter remains asymptotically unchanged, the number of variables is increased in the reduction. For each clause of k literals, at most k new variables need to be introduced. Recall that we have defined the complexity class $\mathbf{NL}[r(n)]$, which is of interest because it characterizes width-parameterized $\mathbf{SAT}$.

Lemma 2 ([11]). $\mathbf{SAT}_{pw}(w(n))$ *is complete for* $\mathbf{NL}\left[\frac{w(n)}{\log n}\right]$, *under logspace many-to-one reductions.*

As in the literature, denote by $\mathbf{NP/poly}$ the class of languages accepted by a non-deterministic polynomial time Turing machine with a polynomial length advice, and $\mathbf{coNP/poly}$ its complement.

3   Incompressibility

3.1   Incompressibility of Width Parameters

We start with some preliminary observations stating that no non-trivial compression can be done to reduce the width parameter. Suppose we have an $\mathbf{SAT}_{tw}$ instance ϕ together with an optimal tree decomposition of width $\mathcal{TW}(\phi) = \omega(\log n)$. A *width-compression algorithm* A with *compression ratio* α: $0 < \alpha < 1$, is an algorithm satisfying the following property (*):

> A takes ϕ and the tree decomposition as input, runs in polynomial time and then outputs another instance ϕ' along with a new tree decomposition, such that ϕ is satisfiable if and only if ϕ' is satisfiable, and $\mathcal{TW}(\phi') = \alpha \mathcal{TW}(\phi)$.

Theorem 1. *No width-compression algorithm with* $\alpha = n^{-\frac{1}{n^c}}$ *(c > 1 is a constant) for* $\mathbf{SAT}_{tw}$ *instances with tree-width* $\omega(\log n)$, *satisfying (*) can exist, under* $\mathbf{ETH}$.

Proof. By $\mathbf{ETH}$, deciding satisfiability of the sub-formula by picking the clauses included in a specific bag in the decomposition in general requires $2^{\Omega(\mathcal{TW}(\phi))} = 2^{\omega(\log n)} = n^{\omega(1)}$ time. This also lower bounds the running time for any algorithm deciding satisfiability of ϕ under $\mathbf{ETH}$.

Suppose for the sake of contradiction, such an algorithm A exists. If we repeatedly run A for $\log_\alpha \mathcal{TW}(\phi) = O(\log n / \log(n^{-\frac{1}{n^c}})) = O(n^c)$ times upon ϕ, we will obtain an instance $\overline{\phi}$, where $\mathcal{TW}(\overline{\phi})$ is a constant and has the same satisfiability as ϕ. Satisfiability of $\overline{\phi}$ and the transformation from ϕ to $\overline{\phi}$ can be computed in polynomial time, which in turn implies that satisfiability of ϕ can be decided in polynomial time. However, this is impossible due to the super-polynomial lower bound for running time under $\mathbf{ETH}$ given in the previous paragraph.   □

If we allow the tree-width of the instances be up to linear in n, namely general **SAT** instances, the same incompressibility result holds, while $\mathbf{P} \neq \mathbf{NP}$ is sufficient for contradiction. Namely,

Proposition 1. *No width-compression algorithm with* $\alpha = n^{-\frac{1}{n^c}}$ *(c > 1 is a constant) for* $\mathbf{SAT_{tw}}$ *instances with tree-width* $\Omega(n)$, *satisfying* (*) *can exist, assuming* $\mathbf{P} \neq \mathbf{NP}$.

The compression ratio $\alpha = n^{-\frac{1}{n^c}}$ is a slowly increasing function as n increases with upper bound 1. When n is large enough, we can actually replace α with any constant, and the following corollary holds.

Corollary 1. *No width-compression algorithm with* $\alpha = \alpha_0$ *(a constant,* $0 < \alpha_0 < 1$), *for* $\mathbf{SAT_{tw}}$ *instances with tree-width* $\omega(\log n)$, *satisfying* (*) *exists, under* **ETH**.

3.2   Incompressibility of Instance Length

Next we turn to the question of interactively "compressing" the instance length à la [4]. Let L be a language, denote $\mathbf{OR}(L)$ with k instances is the problem: given a k-tuple $(x_1, x_2, \cdots, x_k)$, deciding whether there is an x_i, s.t. $x_i \in L$. The following lemma is crucial for the proof.

Lemma 3 ([4]). *Let L be a language, with instance length n and $t : \mathbb{Z}^+ \to \mathbb{Z}^+$ be polynomially bounded s.t. the problem of $\mathbf{OR}(L)$ with $t(n)$ instances can be decided by sending $O(t(n) \log t(n))$ bits, then $L \in \mathbf{coNP/poly}$.*

Note that for 3-**CNF** formulas, input length n is $O(N^3)$, therefore $\log n = \Theta(\log N)$. Now we are ready to state the incompressibility result for 3-$\mathbf{SAT_{pw}}(w(n))$ instances, where $w(n) = \Omega(\log n)$.

Theorem 2. *If satisfiability of every* 3-**CNF** *formula on N variables, with a path-decomposition of width $O(w(n))$, where n is the input length, can be decided by the verifier through communicating $O(N^{1-\epsilon})$ bits with the oracle, then* $\mathbf{NL}[\frac{w(n)}{\log n}] \subseteq \mathbf{coNP/poly}$.

Proof. Consider an $\mathbf{OR}(3\text{-}\mathbf{SAT_{pw}}(w(n)))$ instance, with $t(n)$ 3-$\mathbf{SAT_{pw}}(w(n))$ instances each with N variables can be represented by a 3-$\mathbf{SAT_{pw}}(w(n))$ instance with $t(n)N$ variables, and all the instances use different variables. We choose $t(n)$ to be polynomially bounded.

Suppose the instances are ϕ_i, $\forall i$, and each has a corresponding path-decomposition $\mathcal{P}_i$, variables $v_{i,j}$, clauses $C_{i,j}$. Merely joining all the path-decompositions sequentially will impose an AND-relation. To impose an OR-relation, additional operations are required after joining. Let a be a group of variables of length $O(\log n)$ acting as a selector, namely, for a fixed i, $(a = i)$ denotes the clause with semantic meaning "a representing the binary expansion of i". Since $t(n)$ is a polynomial in n, $O(\log n)$ bits are sufficient. For each $\mathcal{P}_i$, replace each clause $C_{i,j}$ by a clause representing $(a = i) \to C_{i,j}$, or equivalently $\overline{(a = i)} \vee C_{i,j}$. To preserve the connectivity requirement of a path-decomposition, the variables of a need to be added to each bag of the joined path.

One last problem is that each newly created clause is of $O(\log n)$ variables. To obtain a 3-$\textbf{SAT}_{\textsf{pw}}(w(n))$ instance, we apply the reduction by Lemma 1, blowing up the number of variables by a factor of $O(\log n)$.

In the end, a 3-$\textbf{SAT}$ instance of $O(t(n)N \log n)$ variables with path-width $O(w(n))$ is constructed. Now by hypothesis, when $t(n)$ is a large enough polynomial this instance can be decided by the verifier through communicating $O((t(n)N \log n)^{1-\epsilon}) = O(t(n) \log t(n))$ bits with the oracle. By Lemma 3, this means 3-$\textbf{SAT}_{\textsf{pw}}(w(n))$ is in coNP/poly. Combining this and the characterization in Lemma 2 concludes the proof.

$\square$

The incompressibility result for width-parameterized $\textbf{SAT}$ seems much weaker than that for general $\textbf{SAT}$ as in [4]. There is a crucial step called *packing lemma*, failed to be applied in width-parameterized setting. The lemma describes a procedure which combines OR of $t(n)$ $\textbf{SAT}$ instances (each of length n) into a semantically equivalent $\textbf{SAT}$ instance, without requiring large number of variables (only $(t(n)n)^{\frac{1}{3}}$) by allowing the clauses corresponding to different original instances to share variables. However, the same technique does not work in the width-parameterized setting since the same procedure did not take width into consideration and actually will blow up the width of resulting instance to n. Therefore, a straightforward way of combining was used in the proof of Theorem 2 which in turn requires $t(n)n$ variables. One direction of improving the result will be finding a new packing technique for width-parameterized settings.

4   Conclusion

In this note, we proved two incompressibility results, one for width parameters, the other for instance lengths. Our techniques are elementary, and future improvements with new techniques tailored for width-parameterized $\textbf{SAT}$ are left for future work.

Acknowledgments. This work was supported in part by the National Basic Research Program of China Grant 2011CBA00300, 2011CBA00301, the National Natural Science Foundation of China Grant 61033001, 61061130540, 61073174. The author would like to thank Periklis Papakonstantinou for supervising this research.

References

[1] Alekhnovich, M., Razborov, A.A.: Satisfiability, branch-width and Tseitin tautologies. In: Foundations of Computer Science (FOCS), pp. 593–603. IEEE (2002)

[2] Allender, E., Chen, S., Lou, T., Papakonstantinou, P., Tang, B.: Width-parameterized sat: Time-space tradeoffs (2012) (manuscript)

[3] Bodlaender, H.L.: A partial k-arboretum of graphs with bounded treewidth. Theoretical Computer Science 209(1-2), 1–45 (1998)

[4] Dell, H., van Melkebeek, D.: Satisfiability allows no nontrivial sparsification unless the polynomial-time hierarchy collapses. In: Symposium on Theory of Computing (STOC), pp. 251–260. ACM (2010)

[5] Fischer, E., Makowsky, J.A., Ravve, E.V.: Counting truth assignments of formulas of bounded tree-width or clique-width. Discrete Applied Mathematics 156(4), 511–529 (2008)

[6] Georgiou, K., Papakonstantinou, P.A.: Complexity and Algorithms for Well-Structured k-SAT Instances. In: Kleine Büning, H., Zhao, X. (eds.) SAT 2008. LNCS, vol. 4996, pp. 105–118. Springer, Heidelberg (2008)

[7] Impagliazzo, R., Paturi, R.: Complexity of k-sat. In: Proceedings of Fourteenth Annual IEEE Conference on Computational Complexity, pp. 237–240. IEEE (1999)

[8] Impagliazzo, R., Paturi, R., Zane, F.: Which problems have strongly exponential complexity? Journal of Computer and System Sciences (JCSS) 63(4), 512–530 (2001); (also FOCS 1998)

[9] Kloks, T.: Treewidth: computations and approximations, vol. 842. Springer (1994)

[10] Lokshtanov, D., Marx, D., Saurabh, S.: Known algorithms on graphs of bounded treewidth are probably optimal. In: 22nd ACM/SIAM Symposium on Discrete Algorithms (SODA 2011), pp. 777–789 (2011)

[11] Papakonstantinou, P.A.: A note on width-parameterized sat: An exact machine-model characterization. Information Processing Letters (IPL) 110(1), 8–12 (2009)

[12] Samer, M., Szeider, S.: A fixed-parameter algorithm for# sat with parameter incidence treewidth. Arxiv preprint cs/0610174 (2006)

[13] Szeider, S.: On Fixed-Parameter Tractable Parameterizations of SAT. In: Giunchiglia, E., Tacchella, A. (eds.) SAT 2003. LNCS, vol. 2919, pp. 188–202. Springer, Heidelberg (2004)

Kernels for Packing and Covering Problems
(Extended Abstract)

Jianer Chen[1,2], Henning Fernau[3], Peter Shaw[4],
Jianxin Wang[1], and Zhibiao Yang[1]

[1] School of Information Science and Engineering, Central South University,
Changsha 410083, P.R. China
{jianer,jxwang}@mail.csu.edu.cn, csu_yzb@126.com
[2] Department of Computer Science and Engineering, Texas A&M University,
College Station, Texas 77843-3112, USA
chen@cs.tamu.edu
[3] FB IV—Abteilung Informatik, Universität Trier, D-54286 Trier, Germany
fernau@uni-trier.de
[4] School of Engineering and Information Technology, Charles Darwin University,
Darwin, Northern Territory, Australia, 0909
peter.shaw@cdu.edu.au

Abstract. We show how the notion of combinatorial duality, related to
the well-known notion of duality from linear programming, may be used
for translating kernel results obtained for packing problems into kernel
results for covering problems. We exemplify this approach by having a
closer look at the problems of packing a graph with vertex-disjoint trees
with r edges. We also improve on the best known kernel size for packing
graphs with trees containing two edges, which has been well studied.

1 Introduction

There are several lines of motivation for our study:[1]

(1) Parameterized complexity has gained much interest over the last decade. An
essential tool for proving a problem to be fixed-parameter tractable is the notion of kernelization, which also characterizes the lowest complexity class $\mathcal{FPT}$.
Fellows [11], one of the pioneers of this area, sees the quest for smaller kernel
sizes as one of the "races" in that area. Recently, the first meta-results for obtaining small kernels appeared, see [4,15,17,18,21]. By way of contrast, at about
the same time, several lower bound techniques for kernel sizes were established:
[3,5,6,9,12,16,20,19]. Hence, it would be good to see more results of parameterized problems having small problem kernels. Also in recent years, several rather
general scenarios for developing kernelization results have been obtained, see
[4,15,17,18,21]. Our paper adds to the latter list several packing and covering
problems, focussing on vertex-linear kernels for graph problems.

---

[1] Precise definitions of all necessary concepts will be given at the end of this section,
to which we refer readers not familiar with parameterized complexity.

J. Snoeyink, P. Lu, K. Su, and L. Wang (Eds.): FAW-AAIM 2012, LNCS 7285, pp. 199–211, 2012.

(2) More specifically, one of the many open questions in the area of kerneliza-tion is the question whether or not r-HITTING SET has a kernel with less than $\Theta(k^{r-1})$ vertices; this bound can be attained due to [2]. Similar questions can be raised for the corresponding packing problems [1]. Since hitting set problems can be used to model various problems in a natural way, it is natural to look into such specialized problems. A natural restriction are vertex-deletion prob-lems in graphs as studied in [21]. We will focus here on the problem(s) formally described below.

COVERING-H'S	H-PACKING
Given: a graph $G = (V, E)$, $k \in \mathbb{N}$	**Given:** a graph $G = (V, E)$, $k \in \mathbb{N}$
Parameter: k	**Parameter:** k
Question: Is there a set C with $\leq$ k vertices in G such that $G[V \setminus C]$ contains no subgraph isomorphic to H?	**Question:** Is there a set $\mathcal{P}$ of (at least) k vertex-disjoint copies of H in G?

Of particular interest to us will be $H = T_r$, where a T_r is any tree with r edges. Notice that, for $r = 2$, T_2 is simply a path of length two, also known as P_2. The MAXIMUM T_1-PACKING problem is nothing else than the polynomial-time solvable MAXIMUM MATCHING problem, and COVERING-T_1'S is just the VERTEX COVER problem, which has been always a kind of testbed problem for the area of parameterized algorithms. Since our covering resp. packing problems can be modeled as r-HITTING SET resp. r-SET PACKING, it is clear that they are in $\mathcal{FPT}$, see [1,2].

(3) A more general motivation for studying this problem comes from the area of (integer) linear programming ((I)LP). For example, the natural minimization version of COVERING-T_2'S can be expressed as an ILP as follows: Minimize $\sum x_i$ subject to $x_i + x_j + x_\ell \geq 1$ for any path $v_i v_j v_\ell$ in the graph. As usual, the variables x_i model the vertices v_i, and we are only interested in those solutions that satisfy the integrality condition $x_i \in \{0, 1\}$. Relaxation and dualization produce the following problem: Maximize $\sum y_j$ subject to $\sum_{p_j \ni v} y_j \leq 1$ for all vertices v; so the variables y_j correspond to paths p_j in the graph, and this interpretation is in particular sound if we re-introduce integrality conditions $y_j \in \{0, 1\}$. What kind of graph problem does this formulation express? It is not hard to see that a well-studied problem appears this way: T_2-PACKING. In the sense just described, we will call T_2-PACKING the *combinatorial dual* of COVERING-T_2'S. Clearly, the size of an optimum solution to an instance I of COVERING-T_2'S is a lower bound to an optimum solution to I, viewed as an instance of T_2-PACKING, and this is true in general due to the well-known weak duality theorem of linear programming. Can we obtain good kernelizations for both types of problems? Accordingly, the combinatorial dual of a maximization problem is understood. The combinatorial dual of the combinatorial dual is the original problem.

Main Contributions.: Having the idea of combinatorial dual in mind, we ex-hibit how to transfer kernelization results from the packing to the covering sce-nario. We also develop a general notion of crown reductions and apply these to derive kernels for the T_r-PACKING and COVERING-T_r's problems that have only linearly many vertices (for each fixed r). These kernels are unlikely to have sub-quadratically many edges due to [9]. We then show with the example of T_2-packings how the general kernel bounds can be further lowered for more special cases. Namely, we get a $6k$ vertex kernel for T_2-PACKING that improves on earlier findings [24,25].

A Primer to Parameterized Complexity: Parameterized complexity offers a two-dimensional view on the complexity of a computational problem whose input contains some explicit parameter, as in our case, for instance, an upper bound on the number of vertices in a solution to a covering problem. A parameterized (decision) problem (Π, κ) is said to be *fixed-parameter tractable* [10,14,22], or belongs to $\mathcal{FPT}$, if there is a deterministic algorithm that, given an instance I of the problem, where $k = \kappa(I)$ is the parameter of the instance, correctly answers YES or NO in time $\mathcal{O}(f(k) \cdot p(\sigma(I)))$, where f is some arbitrary (computable) function, σ measures the size of an instance, and p is some polynomial.

Given a classical, non-parameterized problem, it can be parameterized in various ways, as indicated by the *parameterization function* κ. For decision problems derived from optimization problems, the *standard parameterization* refers to a bound k on the entity to be optimized, i.e., to an upper bound in case of minimization problems and to a lower bound in case of maximization problems. In this paper, this is the case with COVERING-T_r's and with T_r-PACKING, respectively. Hence, we will in the following denote instances as pairs (I, k) to make the concrete parameter values explicit.

A polynomial-time computable reduction that transforms an instance (I, k) to an instance $(\hat{I}, \hat{k})$ of the same problem is called a *kernelization* if, for some (computable) function g, $\max\{\sigma(\hat{I}), \hat{k}\} \leq g(k)$. $\hat{I}$ is also referred to as a *(problem) kernel* of I. It is well-known that a parameterized problem is in $\mathcal{FPT}$ if and only if it admits a kernel(ization). In the case of graph problems, it has become customary to be more specific on the size measure σ. So, we speak about a $6k$ vertex kernel. Kernelizations are often given as a collection of simple reduction rules that are to be exhaustively applied.

2 Re-cycling Kernel Results

In the context of parameterized complexity, the main reason why we advocate the notion of combinatorial dual is that we claim that often kernel results obtained for maximization problems (for instance packings with standard parameterization, bounding the entity to be maximized) can be read as kernel results for the combinatorial dual, i.e., a minimization problem (i.e., covering problems under standard parameterization).

This path seems to be promising, since kernelization results seem to be easier to obtain for maximization problems, due to the fact that extremal combinatorial arguments are relatively easy at hand, cf. [23].

Let $\Pi_{\max}$ be a maximization problem where the task is to find, under certain restrictions, a set P of objects with maximum cardinality $|P|$. We assume that

Table 1. Standard algorithms for maximization problems and their duals

Input: an instance (I, k) of $(\Pi, \kappa_{\max})$	Input: an instance (I_c, k_c) of $(\Pi_c, \kappa_{\min})$
1. efficiently compute a maximal solution P of instance I;	1. efficiently compute a maximal solution P of the dual instance I;
2. **if** $\|P\| \geq k$ **then** return YES;	2. **if** $\|P\| > k_c$ **then** return NO;
3. apply reduction rules in R; possibly restart.	3. apply the dual reduction rules in R'; possibly restart.
PRIMAL-KERNELIZATION$(\Pi, \kappa_{\max})$	DUAL-KERNELIZATION$(\Pi_c, \kappa_{\min})$

$\Pi_{\max}$ can be formulated as an ILP whose relaxation can be denoted, in matrix notation, as (†): "maximize $\mathbf{c}^{\mathrm{T}}\mathbf{x}$ subject to $A\mathbf{x} \leq \mathbf{b}$ and $\mathbf{x} \geq \mathbf{0}$". The dual linear program becomes (‡): "Minimize $\mathbf{b}^{\mathrm{T}}\mathbf{y}$ subject to $A^{\mathrm{T}}\mathbf{y} \geq \mathbf{c}$ and $\mathbf{y} \geq \mathbf{0}$". $\Pi_{\min}$ denotes a minimization problem that can be formulated as an ILP whose relaxation can be written as (‡). Let $(\Pi, \kappa_{\max})$ be the standard parameterization of $\Pi_{\max}$ and $(\Pi_c, \kappa_{\min})$ be the standard parameterization of $\Pi_{\min}$. $(\Pi_c, \kappa_{\min})$ will be called a *combinatorial dual* of $(\Pi, \kappa_{\max})$. Having fixed $\Pi_{\min}$, there is a unique mapping from an instance (I, k) of $(\Pi, \kappa_{\max})$ to the combinatorial dual instance (I_c, k_c) of $(\Pi_c, \kappa_{\min})$ and vice versa.

Kernelization algorithms for $(\Pi, \kappa_{\max})$ often take some *standard form* as sketched in the left column in Table 1 and described in more detail below. As it is usual with optimization problems, a *feasible solution* refers to any valid packing (in our case). Maximality in its simplest greedy form means that the current packing cannot be extended by adding further objects to it. However, one can develop local optimization procedures that lead to less trivial notions of maximality, as we will see in this paper.

The effectiveness of the algorithm PRIMAL-KERNELIZATION$(\Pi, \kappa_{\max})$ depends on the structures of the problem $\Pi_{\max}$. In many cases, a kernel of size bounded by $g(k) \geq k$ is obtained for the instance (I, k) of $(\Pi, \kappa_{\max})$ based on a *Boundary Lemma* with the following *standard structure*.

> **Boundary Lemma.** Let (I, k) be an instance of $(\Pi, \kappa_{\max})$ and let P be a maximal solution. If $\|P\| < k$ and $\sigma(I) > g(k)$, then some reduction rules in R apply to (I, k).

In particular, if in step 3 of algorithm PRIMAL-KERNELIZATION$(\Pi, \kappa_{\max})$, no reduction rules in R apply, then the Boundary Lemma derives immediately that $\sigma(I) \leq g(k)$, giving a kernel of size bounded by $g(k)$.

Now we study how the kernelization algorithm for $(\Pi, \kappa_{\max})$ is used for kernelization of its combinatorial dual $(\Pi_c, \kappa_{\min})$. Let (I, k) be an instance of $(\Pi, \kappa_{\max})$ and let (I_c, k_c) be the combinatorial dual of (I, k), an instance of $(\Pi_c, \kappa_{\min})$. Note that the size functions $\sigma(I)$ and $\sigma_c(I_c)$ may be different. We call the combinatorial dual $(\Pi_c, \kappa_{\min})$ *size-h bounded*, where h is a non-decreasing function, if $\sigma_c(I_c) \leq h(\sigma(I))$ for all instances I of $\Pi_{\max}$. Moreover, we say that the set R of reduction rules for $(\Pi, \kappa_{\max})$ *transfers* if there is a corresponding set R' of *dual reduction rules* for $(\Pi_c, \kappa_{\min})$ such that if some reduction rules in R are applicable to an instance (I, k) of $(\Pi, \kappa_{\max})$, then some reduction rules in R' are applicable to (I_c, k_c).

Based on the kernelization algorithm PRIMAL-KERNELIZATION for the problem $(\Pi, \kappa_{\max})$, we can propose a kernelization algorithm for its combinatorial dual $(\Pi_c, \kappa_{\min})$, as given in the right column in Table 1.

Theorem 1. *Suppose that $(\Pi, \kappa_{\max})$ satisfies the Boundary Lemma, that the set R of reduction rules in the algorithm* PRIMAL-KERNELIZATION *transfers, and that the combinatorial dual $(\Pi_c, \kappa_{\min})$ is size-h bounded. Then* DUAL-KERNELIZATION *is a kernelization algorithm for $(\Pi_c, \kappa_{\min})$ that produces a kernel of size bounded by $h(g(k_c + 1))$.*

Proof. The correctness of step 2 of DUAL-KERNELIZATION is guaranteed by the LP dualization theorem. Now suppose that the algorithm DUAL-KERNELIZATION stops with an instance (I'_c, k'_c) of $(\Pi_c, \kappa_{\min})$ and a packing P of the dual (I', k') of (I'_c, k'_c); (I', k') is an instance in $(\Pi, \kappa_{\max})$) such that $|P| \leq k'_c < k'_c + 1$ and that no reduction rules in R' are applicable to (I'_c, k'_c). Since the set R of reduction rules transfers, no reduction rules in R are applicable to (I', k'). Because $(\Pi, \kappa_{\max})$ satisfies the Boundary Lemma, we get $\sigma(I') \leq g(k'_c + 1)$. This directly implies $\sigma_c(I'_c) \leq h(\sigma(I')) \leq h(g(k'_c + 1))$, as $(\Pi_c, \kappa_{\min})$ is size-h bounded. $\qquad\square$

We give some applications of Theorem 1.

An *r-set* is a set of exactly r elements. Consider the *r*-SET PACKING problem (given a collection $\mathcal{C}$ of *r*-sets and a parameter k, decide if there are k disjoint *r*-sets in $\mathcal{C}$), and the *r*-HITTING SET problem (given a collection $\mathcal{C}$ of *r*-sets and a parameter k_c, decide if there is a k_c-set that intersects all *r*-sets in $\mathcal{C}$). It is easy to verify that *r*-HITTING SET is a combinatorial dual of *r*-SET PACKING. Define for both problems the *size* of an instance to be the cardinality of the union of all *r*-sets in the input collection $\mathcal{C}$. Thus, *r*-HITTING SET is a size-h bounded dual of *r*-SET PACKING, with h being the identity. Abu-Khzam [1] presented a greedy kernelization algorithm for *r*-SET PACKING using a set R of reduction rules that transfers, and proved that *r*-SET PACKING with the reduction rules in R satisfies the Boundary Lemma with $g(k) = \mathcal{O}(k^{r-1})$. By Theorem 1, this result immediately implies a kernel of size $\mathcal{O}((k_c + 1)^{r-1}) = \mathcal{O}(k_c^{r-1})$ for the *r*-HITTING SET problem, which is the main result of [2].

Theorem 1 also allows us to develop new kernelization results. For example, Moser [21] presented a greedy kernelization algorithm for the problem H-PACKING that satisfies the conditions of Theorem 1 and produces a kernel with at most $\mathcal{O}(k^{|V(H)|-1})$ many vertices. For a special case where H is a star with r edges, Prieto and Sloper [24] presented a greedy kernelization algorithm for the problem that produces a kernel of size $\mathcal{O}(k^2)$ where the algorithm also satisfies the conditions of Theorem 1. Using Theorem 1 we can directly translate these results to kernelization results for their combinatorial duals as follows:

Theorem 2. *Fix an undirected graph H with h vertices.* COVERING-H'S *admits a kernel with $\mathcal{O}(k_c^{h-1})$ many vertices. If H is a star with r edges, then* COVERING-H *admits a quadratic vertex kernel.* $\qquad\square$

So far, we have neglected three questions that are crucial for the success of our approach: (1) How to obtain reduction rules? (2) How to prove boundary lemmas? (3) How to define useful notions of maximality of packings in order to improve on kernel bounds. These questions will be examplarily answered in the following sections.

3   Crown Rules for Packing and for Covering

Set covering/packing problems can be modeled by domination and its dual. The natural combinatorial dual of r-RED/BLUE DOMINATION is r-RED/BLUE DISTANCE-3 PACKING, as defined below. Slightly abusing notation, we will call a set $D \subseteq R$ a *covering* for $B' \subseteq B$ iff $N(D) \supseteq B'$, so for each $x \in B'$, there is a $y \in D$ such that $\{x, y\} \in E$. If $|D| \leq k_c$, D is called a k_c-covering. A set $P \subseteq B$ is a *packing* in G if each $v \in R$ is neighbor of at most one $x \in P$. If $|P| \geq k$, then P is called a k-packing.

r-RED/BLUE DOMINATION	r-RED/BLUE DISTANCE-3 PACKING
Given: a bipartite graph $G = (R \uplus B, E)$, $\forall x \in B : deg(x) = r$, $k_c \in \mathbb{N}$	**Given:** a bipartite graph $G = (R \uplus B, E)$, $\forall x \in B : deg(x) = r$, $k \in \mathbb{N}$
Parameter: k_c	**Parameter:** k
Question: Is there a set $D \subseteq R$, $|D| \leq k_c$, with $N(D) = B$?	**Question:** Is there a set $P \subseteq B$, $|P| \geq k$, s.t. $\forall v \in R : |N(v) \cap P| \leq 1$?

Definition 1. *A generic crown decomposition (H, C, X) of a bipartite graph $G = (R \uplus B, E)$, with $\forall x \in B : deg(x) = r$, is given by a head $H \subseteq R$ and a crown $C \subseteq R \uplus B$ satisfying*

1. *H is a cutset in G, one collection of components being C.*
2. *H is a covering for $C \cap B$.*
3. *There is an injective mapping (matching) $M : H \to C \cap B$ such that (a) $\forall x \in H : \{x, M(x)\} \in E$, and (b) $M(H)$ is a packing in G.*
4. *$X := (R \uplus B) \setminus (H \cup C)$ are the remaining vertices.*

Notice that items 1 and 2 imply that $N(C) \subseteq H \cup C$. In particular, $N(C \cap R) \subseteq C \cap B$. Moreover, item 3 implies $|M(H)| = |H|$.

Theorem 3. *A bipartite graph $G = (R \uplus B, E)$ with a generic crown decomposition (H, C, X) has a k-covering (or k-packing, resp.) if and only if $G \setminus (H \cup C)$ has a $(k - |H|)$-covering (or $(k - |H|)$-packing, resp.).*

Proof. Let us first consider the covering part. By the definition of a generic crown decomposition, the "only-if"-part is clear. Now assume K is a k-covering of B, i.e., $N(K) = B$. In particular, K covers $M(H)$. Since $M(H)$ forms a packing, at least $|H| = |M(H)|$ vertices from K are needed to cover $M(H)$. Let these K-vertices be collected in $K' \subseteq K$. Since H forms a separator, $(K \setminus K') \cup H$ is also a covering of G, with $|K| \geq |(K \setminus K') \cup H|$. Hence, $G \setminus (H \cup C)$ has a $(k - |H|)$-covering.

For reasons of space, we omit the packing part proof. □

So, the *generic crown rule* would delete a crown (whenever identified) and change the parameters (of the covering or packing variant) accordingly. A special case is when R contains isolates that can be put into a crown with an empty head. How can we use this generic crown rule in other concrete situations? Let us explain this first with the most classical notion of a crown and then continue with other situations. As can be seen, the generic crown rule serves as a modelling tool rather than a concrete reduction rule. However, concrete and problem-dependent reduction rules can be derived from this generic rule. Then, it should be also possible to derive methods how to find crowns in polynomial time.

Example 1. The classical crown rule is meant to work for VERTEX COVER. Here, a *VC-crown decomposition* of a graph $G = (V, E)$ consists of a partition (H, C, X) of the vertex set of G, such that (1) H is a cutset, (2) C is an independent set, (3) there is an injective mapping (matching) $M : H \rightarrow C$ such that, for all $x \in H$, $\{x, M(x)\} \in E$.

This decomposition (and the derived reduction rule) is a special case of the generic crown decomposition, if we associate to V the *bipartite model graph* $B_G = (R \uplus B, E')$ with $R = V$, $B = E$, and $(x, e) \in E'$ iff x is endpoint of e in G. By construction, for all $x \in B$, $deg(x) = 2$. Moreover, $H \subseteq V$ is a cutset in G if and only if H is a cutset in B_G.

4   Small Kernels for T_r-PACKING & COVERING T_r's

We propose the following definition of a T_r-crown (decomposition), generalizing earlier works of Prieto and Sloper [23,24].

Definition 2. *A T_r-crown decomposition (H, C, X) in a graph $G = (V, E)$ is a partitioning of the vertices of the graph into three sets H, C, and X that have the following properties: (1) H (the head) is a separator in G such that there are no edges in G between vertices belonging to C and vertices belonging to X. (2) C (the crown) induces a collection of trees with $\leq r - 1$ edges in G. (3) There are r injective mappings $\pi_1, \ldots, \pi_r$ from H to C, called witness functions, such that $\forall 1 \leq i < j \leq r : \pi_i(H) \cap \pi_j(H) = \emptyset$, and for each $v \in H$, the vertex set $p(v) = \{v\} \cup \{\pi_i(v) \mid 1 \leq i \leq r\}$ forms a T_r-tree in G.*

Crowns are used in a kernelization strategies that simply eliminates crowns. The correctness of this strategy is implied by the following result:

Theorem 4. *Consider a graph $G = (V, E)$ with a T_r-crown decomposition (H, C, X). G has a T_r-covering with k vertices (or a T_r-packing with k trees, resp.) if and only if $G \setminus (H \cup C)$ has a T_r-covering with $(k - |H|)$ vertices (or T_r-packing with $(k - |H|)$ trees, resp.).*

The correctness of this theorem immediately follows from observing the connection to the natural bipartite graph model $B_G = (B \uplus R, E')$ for an instance to this problem: $R = V$, and B consists of all trees with d edges in G, where E' encodes whether a vertex participates in a tree.

When applying a crown elimination strategy, we should be able to find crowns fast. The according result is inspired by results of Prieto and Sloper [23,24], based on the following new notion:

Definition 3. *We call a T_r-crown decomposition (H, C, X) asteroidal if for the witness functions π_i, $\{v, \pi_i(v)\}$ is an edge for each $v \in H$ and each $1 \leq i \leq r$.*

Let $G = (V, E)$ be a graph. Then, $\mathcal{I} \subseteq 2^V$ is called *independent* if there are no edges between any two sets $I, J \in \mathcal{I}$. The *out-neighborhood* of $\mathcal{I}$ is $N(\mathcal{I}) = \bigcup_{I \in \mathcal{I}} \bigcup_{v \in I} (N(v) \setminus I)$.

Lemma 1. *Let $G = (V, E)$ be a graph. Let $\mathcal{I} \subseteq 2^V$ be independent and any $I \in \mathcal{I}$ obeys $|I| \leq r$, where $|\mathcal{I}| \geq r \cdot |N(\mathcal{I})|$. Then, G possesses an asteroidal T_r-crown decomposition that can be found in linear time.*

Proof. Construct an auxiliary bipartite graph $B = (V_B, E_B)$ as follows: $V_B = \mathcal{I} \cup (\{1, \ldots, r\} \times N(\mathcal{I}))$; $E_B = \{\{U, (i, v)\} \mid U \in \mathcal{I}, v \in N(\mathcal{I}), 1 \leq i \leq r\} \cup \{\{(i, v), (j, w)\} \mid v, w \in N(\mathcal{I}), \{v, w\} \in E, 1 \leq i, j \leq r\}$. So, the vertex set $\mathcal{I}$ has at least as many elements as its out-neighborhood in G. Hence, B has a VC-crown decomposition (H_B, C_B, X_B) that can be found in linear time, see Ex. 1 and [7]. To this decomposition, we can associate an injective mapping (matching) $\pi : H_B \to C_B$. Moreover, $C_B \subseteq \mathcal{I}$ and no proper subset of H_B is a cutset for C_B (+) (by the algorithm described in [7]). If $(i, v) \in H_B$ for some i, then $\{(j, v) \mid 1 \leq j \leq r\} \subseteq H_B$; otherwise, H_B would not be a cutset for C_B by the definition of E_B and due to (+). Associate to (H_B, C_B, X_B) the structure (H, C, X) of subsets of V, with $H = \{v \mid \exists i[(i, v) \in H_B]\}$, $C = \bigcup_{I \in C_B} I$, $X = V \setminus (H \cup C)$. Then, (H, C, X) is an asteroidal T_r-crown decomposition of G. Namely, define $\pi_i(v) = u$ for $\pi((i, v)) = I$ and some (arbitrary) $u \in I$ that testifies that v is in the neighborhood of I, where $I \in \mathcal{I}$. □

Lemma 1 justifies the correctness and applicability of the *asteroidal T_r-crown rule* whose exact definition (that should differentiate between the packing and the covering scenarios) is left to the reader. This definition enables us to establish:

Theorem 5. *Let $r \geq 2$. T_r-PACKING (COVERING T_r's, resp.) admits a kernel with at most $f_r \cdot k$ ($f_r \cdot (k+1)$, resp.) many vertices, with $f_r = ((r^2 + 1) \cdot (r + 1))$, that can be found in linear time.*

Proof. Run the algorithm described in Sec. 2, the only reduction rule being the asteroidal T_r-crown rule. In a reduced instance $((G, E), k)$, a maximal T_r-packing $\mathcal{P}$ has at most k elements, each containing at most $(r + 1)$ many vertices (let $U \subseteq V$ collect these), and there cannot be more than $r \cdot |U|$ components in $G - U$ (due to the reducedness property), where each component has at most r vertices (by the maximality of $\mathcal{P}$). □

Notice that for $r = 2$, we face the problems COVERING-P_2'S and P_2-PACKING (trees on two edges are just paths on two edges); for the latter problem, our kernel bound matches the $15k$-bound obtained in [23,24] by actually using two reduction rules. This bound was further improved in [25] to a bound of $7k$ vertices. We will further improve on this bound in the following section.

5   Improving on Kernel Size: A Case Study

The $7k$-kernel result optimizes packings that are found in a greedy fashion (step 1. of PRIMAL KERNELIZATION), by either increasing the packing or reducing the number of K_1-components (isolated vertices). These rules are visualized in Fig. 1 and 2. We have used solid circles and thick lines for vertices and edges, respectively, in the P_2-packing $\mathcal{P}$, and hollow circles and thin lines for vertices and edges not in $\mathcal{P}$. In particular, two hollow circles linked by a thin line represent a K_2-component (an isolated edge). Let us call a packing where none of these rules applies $(1, 2)$-*optimal*, and call a vertex from the packing that has neighbors outside the packing a *border point*. The following assertion was shown in [25].

Proposition 1. ([25]) *In a $(1, 2)$-optimal P_2-packing $\mathcal{P}$, each path from $\mathcal{P}$ contains at most two border points. If $p \in \mathcal{P}$ contains two border points, one of them is the middle point of the path p.*

A direct plug-in of this proposition into the proof of Theorem 5 yields a $10k$ kernel, but a more fine-grained analysis gives the mentioned result. A further challenge is to improve the kernel size $7k$. Due to its special shape, we will call an asteroidal T_2-crown decomposition *V-crown decomposition* in the following.

On top of Rules 1 and 2 described in Fig. 1 and 2, resp., we propose three further rules to optimize a maximal P_2-packing $\mathcal{P}$:

Rule 3. If a P_2-path in $\mathcal{P}$ has one of its end adjacent to a K_2 and another of its vertices adjacent to a K_1, then increase the size of the P_2-packing by 1, as illustrated in Figure 3.

Rule 4. If a P_2-path in $\mathcal{P}$ has its middle vertex adjacent to a K_2 and one of its ends adjacent to at least two K_1's, then increase the size of the P_2-packing by 1, as illustrated in the left figure in Figure 4.

Rule 5. If a P_2-path p in $\mathcal{P}$ has its middle vertex w adjacent to more than one K_2, and one of its ends adjacent to a K_1, then replace p by a new P_2-path that consists of w and a K_2 adjacent to w, as illustrated in the right figure in Figure 4.

Fig. 1. Rule 1: Two K_1's are adjacent to two vertices in a P_2-path

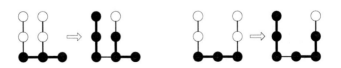

Fig. 2. Rule 2: Two K_2's are adjacent to two vertices in a P_2-path

Fig. 3. Rule 3: A K_2 is adjacent to an end of a P_2-path

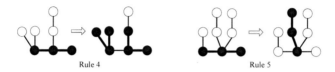

Rule 4   Rule 5

Fig. 4. Rule 4-5: K_2's are adjacent to the middle vertex of a P_2-path

Our proposed kernelization algorithm **P2PK** first computes a non-extendible P_2-packing and then optimizes this packing according to the given five rules. One important technical detail is that Rule 5 is applied on $\mathcal{P}$ only when all other rules are not applicable on $\mathcal{P}$. If no packing with more than k P_2's was found in this way, the rules provide enough structural information to claim a sharper kernel bound (boundary lemma).

Lemma 2. *Let $\mathcal{P}$ be an optimized P_2-packing of size k_0 obtained from our algorithm **P2PK**(G, k). If G contains at least $6k_0$ vertices, then there is a P_2-crown decomposition of G that can be constructed in linear time.*

Proof. Since $\mathcal{P}$ is a maximal P_2-packing, the complement $G - \mathcal{P}$ induces a collection I_1 of K_1's and a collection I_2 of K_2's. We partition the P_2-paths in $\mathcal{P}$ into two sets $\mathcal{P}_1$ and $\mathcal{P}_2$, the K_1's in I_1 into two sets I_1' and I_1'', and the K_2's in I_2 into two sets I_2' and I_2'', by the following procedure. 1. $\mathcal{P}_1 = \emptyset$; $I_1' = \emptyset$; $I_2' = \emptyset$;
2. $\mathcal{P}_2 = \mathcal{P}$; $I_1'' = I_1$; $I_2'' = I_2$;
3. **repeat**
   If a P_2-path p in $\mathcal{P}_2$ has its vertices adjacent to at most one K_1
       in I_1'' and at most one K_2 in I_2'',
   then move p from $\mathcal{P}_2$ to $\mathcal{P}_1$,
       move the K_1 adjacent to p from I_1'' to I_1',
       and move the K_2 adjacent to p from I_2'' to I_2'.
   until no change
Because none of the Rules 1-5 is applicable on the P_2-packing $\mathcal{P}$, the partitions $\mathcal{P} = \mathcal{P}_1 \cup \mathcal{P}_2$, $I_1 = I_1' \cup I_1''$, and $I_2 = I_2' \cup I_2''$ have a number of structural properties. Suppose that $\mathcal{P}_1 = \{p_1, \ldots, p_d\}$ and $\mathcal{P}_2 = \{p_{d+1}, \ldots, p_{k_0}\}$.

Property A. The total number of vertices in $I_1' \cup I_2'$ is bounded by $3d$.
Property B. No vertex in $I_1'' \cup I_2''$ is adjacent to any vertex in $\mathcal{P}_1$, or to any vertex in $I_1' \cup I_2'$.
Property C. No two K_1's in I_1'' can be adjacent to two different vertices in a P_2-path in $\mathcal{P}_2$.

Property D. No two K_2's in I_2'' can be adjacent to two different vertices in a P_2-path in $\mathcal{P}_2$.

Property E. No K_2 in I_2'' can be adjacent to more than one vertex in a P_2-path in $\mathcal{P}_2$.

Property F. No K_1 in I_1'' can be adjacent to more than one vertex in a P_2-path in $\mathcal{P}_2$.

Because of Property A, the collection $\mathcal{P}_2$ is not empty (assuming that the graph G is connected), $d < h_0$, and $I_1'' \cup I_2''$ is not empty. We claim:

Property G. Each P_2-path in $\mathcal{P}_2$ has at most one vertex adjacent to vertices in $I_1'' \cup I_2''$.

We omit the proof of the claims due to space restrictions.

Let $V(I_2'')$ be the vertex set that induces the K_2's in the collection I_2''.

If $|I_2''| \geq k_0 - d$, then $|V(I_2'')| \geq 2(k_0 - d)$. By Property B, the vertices in I_2'' can be adjacent to only vertices in $\mathcal{P}_2$, and by Property G, there are at most $k_0 - d$ vertices in $\mathcal{P}_2$ that are adjacent to vertices in I_2''. This gives $|N(V(I_2''))| \leq k_0 - d$, thus, $|V(I_2'')| \geq 2|N(V(I_2''))|$. By Property E and Lemma 1, a P_2-crown decomposition in G can be constructed in linear time given $V(I_2'')$.

Now suppose $|I_2''| = h_2 < k_0 - d$. Since the graph G has at least $6k_0$ vertices, and by Property A, the total number of vertices in $\mathcal{P}_2 \cup I_1'' \cup I_2''$ is more than $6k_0 - 6d$. Suppose that $|I_1''| = h_1$, and note that $|V(I_2'')| = 2h_2$ and the total number of vertices in $\mathcal{P}_2$ is $3(k_0 - d)$. Thus we have $h_1 + 2h_2 + 3(k_0 - d) > 6k_0 - 6d$. This gives immediately $h_1 + h_2 > 2(k_0 - d)$. Let I'' be the set of vertices in I_1'' and I_2'', then I'' induces the collection I_1'' of K_1's and the collection I_2'' of K_2's. By Properties B and G, the vertices in I'' are only adjacent to at most $k_0 - d$ vertices in $\mathcal{P}_2$, that is, $|I_1''| + |I_2''| > 2|N(I'') \setminus I''|$. By Lemma 1, a V-crown decomposition in the graph G can be constructed in linear time given I''.   □

We can now formulate the main result of this section.

Theorem 6. P_2-PACKING *has a kernel with* $\leq 6k - 6$ *many vertices.*

This yields an improved parameterized algorithm, using [13].

Theorem 7. P_2-PACKING *can be solved in time* $O^*(2.371^{3k})$.

Making use of the concept of combinatorial dual, Theorem 1 implies:

Corollary 1. COVERING-P_2'S *has a kernel with* $\leq 6k$ *many vertices.*

6   Prospects

We have initiated the study of combinatorial duals in parameterized complexity. We showed that kernel results (in particular, for maximization problems) can be transferred to their combinatorial duals, an observation that is quite rare in the field. We have been recently able to transfer these idea in a different field, namely that of moderately exponential-time algorithms.[2]

Our results also add to the list of problems where the parameterized dual has a linear vertex kernel, so that lower bound arguments apply [6].

---

[2] Slides of H. Fernau's talk at APEX 2012 are available at the workshop website.

References

1. Abu-Khzam, F.N.: An improved kernelization algorithm for r-Set Packing. IPL 110, 621–624 (2010)
2. Abu-Khzam, F.N.: A kernelization algorithm for d-Hitting Set. JCSS 76(7), 524–531 (2010)
3. Bodlaender, H.L., Downey, R.G., Fellows, M.R., Hermelin, D.: On problems without polynomial kernels. JCSS 75, 423–434 (2009)
4. Bodlaender, H.L., Fomin, F.V., Lokshtanov, D., Penninkx, E., Saurabh, S., Thilikos, D.M.: (Meta) kernelization. In: FOCS, pp. 629–638. IEEE Computer Society (2009)
5. Bodlaender, H.L., Thomassé, S., Yeo, A.: Kernel Bounds for Disjoint Cycles and Disjoint Paths. In: Fiat, A., Sanders, P. (eds.) ESA 2009. LNCS, vol. 5757, pp. 635–646. Springer, Heidelberg (2009)
6. Chen, J., Fernau, H., Kanj, Y.A., Xia, G.: Parametric duality and kernelization: lower bounds and upper bounds on kernel size. SIAM J. Comput. 37, 1077–1108 (2007)
7. Chor, B., Fellows, M., Juedes, D.W.: Linear Kernels in Linear Time, or How to Save k Colors in $O(n^2)$ Steps. In: Hromkovič, J., Nagl, M., Westfechtel, B. (eds.) WG 2004. LNCS, vol. 3353, pp. 257–269. Springer, Heidelberg (2004)
8. Dehne, F., Fellows, M., Rosamond, F.A., Shaw, P.: Greedy Localization, Iterative Compression, and Modeled Crown Reductions: New FPT Techniques, an Improved Algorithm for Set Splitting, and a Novel $2k$ Kernelization for Vertex Cover. In: Downey, R.G., Fellows, M.R., Dehne, F. (eds.) IWPEC 2004. LNCS, vol. 3162, pp. 271–280. Springer, Heidelberg (2004)
9. Dell, H., van Melkebeek, D.: Satisfiability allows no nontrivial sparsification unless the polynomial-time hierarchy collapses. In: STOC, pp. 251–260. ACM (2010)
10. Downey, R.G., Fellows, M.R.: Parameterized Complexity. Springer (1999)
11. Fellows, M.R.: Blow-Ups, Win/Win's, and Crown Rules: Some New Directions in FPT. In: Bodlaender, H.L. (ed.) WG 2003. LNCS, vol. 2880, pp. 1–12. Springer, Heidelberg (2003)
12. Fernau, H., Fomin, F.V., Philip, G., Saurabh, S.: The Curse of Connectivity: t-Total Vertex (Edge) Cover. In: Thai, M.T., Sahni, S. (eds.) COCOON 2010. LNCS, vol. 6196, pp. 34–43. Springer, Heidelberg (2010)
13. Fernau, H., Raible, D.: A parameterized perspective on packing paths of length two. J. Comb. Optim. 18, 319–341 (2009)
14. Flum, J., Grohe, M.: Parameterized Complexity Theory. Text in Theoretical Computer Science. Springer (2006)
15. Fomin, F.V., Lokshtanov, D., Saurabh, S., Thilikos, D.M.: Bidimensionality and kernels. In: SODA, pp. 503–510. SIAM (2010)
16. Fortnow, L., Santhanam, R.: Infeasibility of instance compression and succinct PCPs for NP. In: STOC, pp. 133–142. ACM (2008)
17. Guo, J., Niedermeier, R.: Linear Problem Kernels for NP-Hard Problems on Planar Graphs. In: Arge, L., Cachin, C., Jurdziński, T., Tarlecki, A. (eds.) ICALP 2007. LNCS, vol. 4596, pp. 375–386. Springer, Heidelberg (2007)
18. Kratsch, S.: Polynomial kernelizations for MIN $F^+\Pi_1$ and MAX NP. In: STACS. Leibniz International Proceedings in Informatics (LIPIcs), vol. 3, pp. 601–612. Schloss Dagstuhl–Leibniz-Zentrum für Informatik (2009)
19. Kratsch, S., Marx, D., Wahlström, M.: Parameterized Complexity and Kernelizability of Max Ones and Exact Ones Problems. In: Hliněný, P., Kučera, A. (eds.) MFCS 2010. LNCS, vol. 6281, pp. 489–500. Springer, Heidelberg (2010)

20. Kratsch, S., Wahlström, M.: Two Edge Modification Problems without Polynomial Kernels. In: Chen, J., Fomin, F.V. (eds.) IWPEC 2009. LNCS, vol. 5917, pp. 264–275. Springer, Heidelberg (2009)
21. Moser, H.: A Problem Kernelization for Graph Packing. In: Nielsen, M., Kučera, A., Miltersen, P.B., Palamidessi, C., Tůma, P., Valencia, F. (eds.) SOFSEM 2009. LNCS, vol. 5404, pp. 401–412. Springer, Heidelberg (2009)
22. Niedermeier, R.: Invitation to Fixed-Parameter Algorithms. Oxford University Press (2006)
23. Prieto, E.: Systematic Kernelization in FPT Algorithm Design. PhD thesis, The University of Newcastle, Australia (2005)
24. Prieto, E., Sloper, C.: Looking at the stars. TCS 351, 437–445 (2006)
25. Wang, J., Ning, D., Feng, Q., Chen, J.: An improved kernelization for P_2-packing. IPL 110, 188–192 (2010)

The Worst-Case Upper Bound for Exact 3-Satisfiability with the Number of Clauses as the Parameter

Junping Zhou and Minghao Yin

College of Computer, Northeast Normal University,
130117, Changchun, P.R. China
{zhoujp877,ymh}@nenu.edu.cn

Abstract. The rigorous theoretical analyses of algorithms for exact 3-satisfiability (X3SAT) have been proposed in the literature. As we know, previous algorithms for solving X3SAT have been analyzed only regarding the number of variables as the parameter. However, the time complexity for solving X3SAT instances depends not only on the number of variables, but also on the number of clauses. Therefore, it is significant to exploit the time complexity from the other point of view, i.e. the number of clauses. In this paper, we present algorithms for solving X3SAT with rigorous complexity analyses using the number of clauses as the parameter. By analyzing the algorithms, we obtain the new worst-case upper bounds $O(1.15855^m)$, where m is the number of clauses.

Keywords: X3SAT; upper bound; the worst case; connected-clauses principle.

1 Introduction

Exact satisfiability problem, abbreviated XSAT, is a problem of deciding whether there is a truth assignment satisfying exactly one literal in each clause. The exact 3-satisfiability (X3SAT) is the version in which each clause contains at most three literals. The X3SAT problem is an important variant of the well-known NP-complete problem of propositional satisfiability (SAT), which has played a key role in complexity theory as well as in automated planning. In fact, X3SAT is also a NP-complete problem even when restricted to all variables occurring only unnegated [1]. If P ≠ NP, it means that we can't solve the problem in polynomial time. Therefore, Improvements in the exponential time bounds are crucial in determining the size of NP-complete problem instances that can be solved. Even a slight improvement from $O(c^k)$ to $O((c-\varepsilon)^k)$ may significantly increase the size of the problem being tractable.

Recently, tremendous efforts have been made on analyzing of algorithms for X3SAT problems. Based on a recursive partitioning of the problem domain and a careful elimination of some branches, Drori and Peleg presented an algorithm running in $O(1.1545^n)$ for X3SAT, where n is the number of the variables [2]. By adapting and improving branching techniques, Porschen et al. proposed an algorithm for solving X3SAT running in $O(1.1487^n)$ [3]. According to exploit a perfect matching reduction and present a more involved deterministic case analysis, Porschen et al. prove a new upper bound for X3SAT ($O(1.1926^n)$) [4]. By providing a new transformation rule, Kulikov [5] simplified the proof of the bound for X3SAT ($O(1.1926^n)$) presented by Porschen et al. [4].

J. Snoeyink, P. Lu, K. Su, and L. Wang (Eds.): FAW-AAIM 2012, LNCS 7285, pp. 212–223, 2012.

Based on combining various techniques including matching and reduction, Dahllof et al. addressed an algorithm running in $O(1.1120^n)$ for X3SAT [6]. Further improved algorithms in [7] presented a new upper time bound for the X3SAT ($O(1.1004^n)$), which is the best upper bound so far.

Different from complexity analyses regarding the number of variables as the parameter, Skjernaa presented an algorithm for XSAT with a time bound $O(2^m)$ but using exponential space, where m is the number of clauses of a formula [8]. Bolette addressed an algorithm for XSAT with polynomial space usage and a time bound $O(m!)$ [9]. Similar to the XSAT problem, the time complexity of X3SAT problem is calculated based on the size of the X3SAT instances, which depends not only on the number of variables, but also on the number of clauses. Therefore, it is significant to exploit the time complexity from the other point of view, i.e. the number of clauses. However, so far all algorithms for solving X3SAT have been analyzed based on the number of variables. And to our best knowledge, it is still an open problem that analyzes the X3SAT algorithm with the number of clauses as the parameter.

The aim of this paper is to exploit new upper bounds for X3SAT using the number of clauses as the parameter. We provide an algorithm for solving X3SAT. This algorithm employs a new principle, i.e. the connected-clauses principle, to simplify formulae. This allows us to remove one sub-formula and therefore reduce as many clauses as possible in both branches. In addition, by improving the case analyses, we obtain the worst-case upper bound for solving X3SAT is $O(1.15855^m)$, where m is the number of clauses of a formula.

2   Problem Definitions

We describe some definitions used in this paper. A variable can take the values *true* or *false*. A literal of a variable is either the unnegated literal x, having the same truth value as the variable, or the negated literal $\neg x$, having the opposite truth value as the variable. A clause is a disjunction of literals, referred to as a k-clause if the clause is a disjunction on k literals. A k-SAT formula F in Conjunction Normal Form (CNF) is a conjunction of clauses, each of which contains at most k literals. A truth assignment for F is a map that assigns each variable a value. When a truth assignment makes the F *true*, we say the truth assignment is a satisfying assignment. The exact satisfiability problem (XSAT) is to find a truth assignment such that exactly one literal is true in each clause. The exact 3-satisfiability problem (X3SAT) is a version of the XSAT in which each clause contains at most three literals. We define m as the number of clauses in F, and n as the number of variables F contains. When a variable occurs once in F, it is referred to as singleton. The degree of a variable v, represented by $\varphi(v)$, is the number of times it occurs in a formula. The degree of a formula F, denoted by $\varphi(F)$, is the maximum degree of variables in F. A literal x is an (i, j)-literal if F contains exactly i occurrences of x and exactly j occurrences of $\neg x$. And a literal x is monotone if its complementary literal does not appear in F. Given a literal x, we say $var(x)$ is the variable that forms the literal and $\sim x$ indicates x or $\neg x$. We also use $F(\mu \leftarrow \eta)$ to denote the substitution of μ by η in the formula F, where μ is either a literal or a clause, and η is a literal, clause, or *false*. To avoid a tedious enumeration of trivialities, if more than one literal is substituted by *false*, μ is usually expressed as a set of literals. We use F/π to denote the formula obtained by removing from F, where is either a clause or a sub-formula. Given a formula F and a literal x, $NumC(F, N(x))$ is defined as follow.

$$NumC(F, N(x)) = |\{C|C \in F \wedge var(C) \cap N(x) \neq \emptyset\}|. \tag{1}$$

In Equation (1), $N(x)$ is the set of variables that appear in a clause with the literal x, and $var(C)$ is the set of variables occur in the clause C.

After specifying the definitions, we present some basic rules for solving X3SAT problem. Given a formula F, the basic strategy of Davis-Putnam-Logemann-Loveland (DPLL) is to arbitrarily choose a variable v that appears in F. Then,

$$F = (F \wedge v) \vee (F \wedge \neg v). \tag{2}$$

Given a formula F, if F can be partitioned into disjoint sub-formulae where any two sub-formulae have no common variables, then

$$F = F_1 \wedge F_2 \wedge ... \wedge F_n. \tag{3}$$

Thus, F can be evaluated by deciding the satisfiability of disjoint sub-formulae of F respectively.

2.1   Estimating the Running Time

In this subsection, we explain how to compute an upper bound on the running time of a DPLL-style algorithm. At first, we present a notion called branching tree. The branching tree is a hierarchical tree structure with a set of nodes, each of which is labeled with a formula [10]. Suppose there is a node labeled with a formula F, then its sons labeled with $F_1, F_2, ..., F_k$ are obtained by branching on one or more variables in the formula F, i.e., assigning values to the variable(s) such that the formula F is reduced to two or more sub-formulae $F_1, F_2, ..., F_k$ with fewer variables. Indeed, the construction of a branching tree can be viewed as an execution of a DPLL-style algorithm. Therefore, we use the branching tree to estimate the running time of our algorithm.

In the branching tree, every node has a branching vector. Let us consider a node labeled with F_0 and its sons labeled with $F_1, F_2, ... , F_k$. The branching vector of the node labeled with F_0 is $(r_1, r_2, ..., r_k)$, where $r_i = f(F_0) - f(F_i)$ ($f(F_0)$ is the number of clauses of F_0). The characteristic polynomial of the branching vector is defined as follows:

$$h(x) = 1 - \sum_{i=1}^{k} x^{-r_i}. \tag{4}$$

The positive root of this polynomial is called the branching number, denoted by $\lambda(r_1, r_2, , r_k)$. And we assume that the branching number of the leaves is 1. We define the maximum branching number of nodes in the branching tree as the branching number of the branching tree, expressed by $max \ \lambda(r_1, r_2, , r_k)$. The branching number of a branching tree has an important relationship with the running time ($T(m)$) of

DPLL-style algorithms. At first, we assume that the running time of DPLL-style algorithms performing on each node is in polynomial time. Then we obtain the following inequality.

$$T(m) \leq (max\lambda(r_1, r_2, , r_k))^m \times poly(F) = (max\sum_{i=1}^{k} T(m - r_i))^m \times poly(F). \quad (5)$$

In Equation (5), m is the number of clauses in the formula F, $ploy(F)$ is the polynomial time executing on the node F, and

$$\lambda(r_1, r_2, , r_k) = \sum_{i=1}^{k} T(m - r_i). \quad (6)$$

In addition, if a X3SAT problem recursively solved by the DPLL-style algorithms, the time required doesn't increase, for

$$\sum_{i=1}^{k} T(m_i) \leq T(m) \quad where \quad m = \sum_{i=1}^{k} m_i. \quad (7)$$

In the Equation (7), m is the number of clauses, m_i is the number of clauses in the sub-formula F_i ($1 \leq i \leq k$) of the formula F. Note that when analyzing the running time of our algorithms, we ignore the polynomial factor so that we assume that all polynomial time computations take $O(1)$ time in this paper.

3 Algorithm for Solving X3SAT

In this section, we present the algorithm X3SAT and prove an upper bound $O(1.15855^m)$, where m is the number of the clauses. Firstly we address some transformation rules used in the algorithm.

3.1 Transformation Rules

The transformation rules are applied before branching on one or more variables of the formula F. According to the complexity analysis described above, we just need to take into the difference value between the number of clauses of the input formula and the number of clauses of the formulae obtained from it by branching. The larger of the difference value, the smaller the upper bound obtained. In the following, we present the transformation rules (TR1) - (TR14) which are also used by [7].

(TR1). If F contains a variable x such that the number of negated occurrences is larger than the number of unnegated occurrences, then let $F = F(\neg x \leftarrow x)$.

(TR2). If F contains a 1-clause $C = x$, then $F = F(x \leftarrow true)$.

(TR3). If F contains a 2-clause $C = x \vee y$, then $F = F(x \leftarrow \neg y)$.

(TR4). If F contains a clause $C = x \vee x \vee y$, then $F = F(x \leftarrow false)$.

(TR5). If F contains a clause $C = x \vee \neg x \vee y$, then $F = F(y \leftarrow false)$.

(TR6). If F contains a clause $C = x \lor y \lor z$ where x and y are singletons, then $F = F(x \leftarrow false)$.

(TR7). If F contains clauses $C_1 = x \lor y \lor z$ and $C_2 = x \lor \neg y \lor z'$, then $F = F(x \leftarrow false)$.

(TR8). If F contains clauses $C_1 = x \lor y \lor z$ and $C_2 = \neg x \lor \neg y \lor z'$, then $F = F(y \leftarrow \neg x)$.

(TR9). If F contains clauses $C_1 = x_1 \lor y_1 \lor y_2, C_2 = x_2 \lor y_2 \lor y_3$, and $C_3 = x_3 \lor \neg y_3 \lor y_1$, then $F = F(C_3 \leftarrow (\neg x_1 \lor x_2 \lor x_3))$.

(TR10). If F contains clauses $C_1 = x_1 \lor \neg y_1 \lor y_2, C_2 = x_2 \lor \neg y_2 \lor y_3, ..., C_k = x_k \lor \neg y_k \lor y_1$, then $F = F(\{x_1, x_2, ..., x_k\} \leftarrow false)$.

(TR11). If F contains clauses $C_1 = x_1 \lor y_1 \lor y_2, C_2 = x_2 \lor y_2 \lor y_3$, and $C_3 = x_3 \lor \neg y_3 \lor y_1$ where x_1 is a singleton, then $F = F/C_1$.

(TR12). If F contains clauses $C_1 = x_1 \lor y_1 \lor y_2, C_2 = x_2 \lor y_2 \lor y_3$, and $C_3 = x_3 \lor y_3 \lor y_1$ where $val(x_3)$ is a singleton, then $F = F(C_3 \leftarrow (\neg x_1 \lor y_3 \lor x_3))$.

(TR13). If F contains clauses $C_1 = x_1 \lor y_1 \lor z_1, C_2 = x_1 \lor y_1 \lor z_2$, then $F = F(C_2 \leftarrow (\neg z_1 \lor z_2))$.

(TR14). If F contains a clause $C = x \lor y \lor z$, where x and y only occur unnegated and in clauses with a singleton in all other clauses, then $F = F(y \leftarrow false)$.

Actually, the above transformation rules are used in the *Reduce* function repeatedly until no transformation rule applies, which can be guaranteed to terminate in polynomial time. The function takes a CNF F as the input and returns a simplified X3SAT formula. In the following, we will show the character of the simplified X3SAT formula. From now on, unless otherwise stated, given a literal x, $Y_1 = \{y_1, y_2, ...\}$ is the set of literals that occur in a clause with x; $Y_2 = \{y_1', y_2', ...\}$ is the set of literals that occur in a clause with $\neg x$ and $Y = Y_1 \cup Y_2$; $Z = \{z_1, z_2, ...\}$ is the set of literals that don't occur in a clause with x. We use y's literals indicating the literals occur in Y. For example, if x is a $(2, 1)$ - literal, the clauses the literal x in are showed in Fig. 1.

Theorem 1.[7] A simplified X3SAT formula contains no 1-clauses or 2-clauses, and no two clauses have more than one variable in common; no clause has more than one singleton; all $(a, 0)$-literals and $(a, 1)$-literals that are not singletons are in a clause with no singletons.

Theorem 2. If a X3SAT formula contains clauses $C_1 = x \lor y_1 \lor y_2, C_2 = x \lor y_3 \lor y_4$, and $C_3 = \neg y_1 \lor y_3 \lor z_1$ where y_1 is a $(1, 1)$ - literal, then $F = F((C_1 \land C_3) \leftarrow (\neg y_4 \lor y_2 \lor z_1))$ and $\varphi(x) = \varphi(x) - 1$.

$$C_1 = x \lor y_1 \lor y_2 \quad C_2 = x \lor y_3 \lor y_4 \quad C_3 = \neg x \lor y_1' \lor y_2'$$

Fig. 1. The clauses that the literal x appears in when x is a $(2, 1)$ - literal

Proof. If a X3SAT formula contains clauses $C_1 = x \vee y_1 \vee y_2$, $C_2 = x \vee y_3 \vee y_4$, and $C_3 = \neg y_1 \vee y_3 \vee z_1$, then $F = F(C_3 \leftarrow (\neg y_4 \vee y_2 \vee z_1))$ by (TR9). Since y_1 is a (1, 1) - literal, y_1 is removed by (TR9). Thus, y_1 is a singleton in F. If a X3SAT formula contains clauses $C_1 = x \vee y_1 \vee y_2$, $C_2 = x \vee y_3 \vee y_4$, and $C_3 = \neg y_4 \vee y_2 \vee z_1$, then we can apply (TR11) and obtain $F = F/C_1$. Therefore, $F = F((C_1 \wedge C_3) \leftarrow (\neg y_4 \vee y_2 \vee z_1))$ and $\varphi(x) = \varphi(x) - 1$.

Theorem 3. When X3SAT formula F contains a clause $y_1' \vee y_3 \vee z_1$ and a (2, 1) - literal x, the formula F can be simplified and the literal x becomes a (2, 0) - literal.

Proof. Since x is a (2, 1) - literal, F contains clauses $C_2 = x \vee y_3 \vee y_4$ and $C_3 = \neg x \vee y_1' \vee y_2'$. Then we can apply (TR9) to $y_1' \vee y_3 \vee z_1$, $x \vee y_3 \vee y_4$, and $\neg x \vee y_1' \vee y_2'$, which can transform F to contain $\neg z_1 \vee y_4 \vee y_2'$ instead of $\neg x \vee y_1' \vee y_2'$. Therefore, F can be simplified and the literal x becomes a (2, 0) - literal.

Theorem 4. When X3SAT formula F contains a (3, 0) - literal x, a singleton y_4, and a clause $y_1 \vee y_3 \vee z_1$, the formula F can be simplified and the literal x becomes a (2, 0) - literal.

Proof. If x is a (3, 0) - literal, the formula F contains clauses $C_1 = x \vee y_1 \vee y_2$ and $C_2 = x \vee y_3 \vee y_4$. Using the (TR12) on $x \vee y_1 \vee y_2$, $x \vee y_3 \vee y_4$, and $y_1 \vee y_3 \vee z_1$ where y_4 is a singleton, we can replace $x \vee y_3 \vee y_4$ by $y_2 \vee y_3 \vee y_4$. Therefore, the formula F can be simplified and the literal x becomes a (2, 0) - literal.

Theorem 5. The transformation rules (TR1)~(TR14) run in polynomial time for a given X3SAT formula F.

Proof. Suppose that the formula F contains n variables and m clauses. In essence, (TR1), (TR9), (TR12), and (TR13) are aimed at making the formula have some good properties. (TR1) acts on the variables, while (TR9), (TR12), and (TR13) act on the clauses. For a given formula F, (TR1) runs at most n times, and (TR9), just the same to (TR12), (TR13), also runs at most m times. Therefore, the four transformation rules execute in $O(n + 3m)$. In order to obtain a better upper bound, (TR2), (TR3), (TR5), (TR6), (TR8), and (TR11) reduce as many clauses as possible. Since the formula F has m clause, these transformation rules run in $O(m)$. In addition, owing to the (TR4), (TR7), (TR10), and (TR14), the variables can be reduced from the given formula. So these transformation rules run in $O(n)$. In total, the transformation rules (TR1) (TR14) run in $O(2n + 4m)$, which are done in polynomial time.

3.2 Helper Principle

In this subsection, we concentrate on introducing the connected-clauses principle. Before presenting the details, we specify some notions used in this part. Given a simplified X3SAT formula F in CNF, F can be expressed as an undirected graph called connection graph. In the connection graph, the vertexes are the clauses of F and the edges between two vertexes if the corresponding clauses contain the same literal. We say that the clause

C is connected with C' if there is an edge connecting the corresponding vertexes in the connection graph. We call such two clauses the connected clauses. The character of connected clauses is showed in the following theorem.

Theorem 6. For any two connected clauses C_1 and C_2, there is only one edge connecting the corresponding vertexes in the connection graph.

Proof. In order to prove there is only one edge connecting the corresponding vertexes in the connection graph, we need to prove that C_1 and C_2 have only one common literal. By (TR2) and (TR3) we know that each clause has exactly three literal in a simplified X3SAT formula. If two clauses have common variables, the common variables must form the same literals based on (TR7) and (TR8). According to (TR13), there is at most only one common literal in any two clauses. Therefore, for any two connected clauses, there is only one edge connecting the corresponding vertexes in the connection graph.

Let us start to propose the connected-clauses principle. Suppose a connection graph G can be partitioned into two components G_1 and G_2 where there is only one edge l connecting a vertex in G_1 to a vertex in G_2, i.e. the formula F corresponding to G is partitioned into two sub-formulae F_1 and F_2 corresponding to the two components with only one common literal l. Then, we can determine the satisfiability of the X3SAT formula F as follows.

$$F = ((F_1 \wedge l) \wedge (F_2 \wedge l)) \vee ((F_1 \wedge \neg l)(F_2 \wedge \neg l)). \tag{8}$$

The aim of this principle is to partition the formula F into two sub-formulae. When F_1 contains a small number of clauses, it can be solved by exhaustive search in polynomial time. This allows us to remove F_1 from F and therefore reduce as many clauses as possible in both branches. The following theorem states that the principle in sound.

Theorem 7. The connected-clauses principle is sound.

Proof. To prove that the connected-clauses principle is sound, we just to prove after applying the connected-clauses principle do not change the satisfiability of the original formula. Suppose a connection graph G can be partitioned into two components G_1 and G_2 where there is only one edge l connecting a vertex in G_1 to a vertex in G_2, i.e. the formula F corresponding to G is partitioned into two sub-formulae F_1 and F_2 corresponding to the two components with only one common literal l. Then after applying the connected-clauses principle to the formula F, the formula F can be partitioned into two formulae F_1 and F_2.

Suppose F is satisfiable. Consider a satisfying assignment I for F. It is obvious that in the satisfying assignment the literal l either *true* or *false*. We assume that the literal l is fixed *true*. Then the satisfying assignment for F consists of a satisfying assignment for $F_1 \wedge l$ and a satisfying assignment for $F_2 \wedge l$. The similar situation is encountered when l is fixed *false*.

On the contrary, every satisfying assignment for $F_1 \wedge l$ (resp. $F_1 \wedge \neg l$) can combine with every satisfying assignment for $F_2 \wedge l$ (resp. $F_2 \wedge \neg l$), both of which have an assignment *true* (resp. *false*) for l. The combining satisfying assignments are indeed the satisfying assignments for F which has an assignment *true (false)* for l.

Therefore, the connected-clauses principle is sound.

3.3   Algorithm X3SAT for Solving Exact 3SAT

The algorithm X3SAT for exact 3SAT is based on the DPLL algorithm. The basic idea of the algorithm is to choose a variable and recursively determine whether the formula is satisfiable or not when variable is *true* or *false*. Before presenting the algorithm X3SAT, we address a function $\Omega(F, x)$ in Fig. 2, which recursively executes the propagation. The function takes a formula F and a literal x being assigned true as input. The detailed process of the function is presented as follows. (1) Remove all clauses containing literal x from F; (2) delete all literals occurring with x from the other clauses; (3) delete all occurrences of the negation of literal x from F; (4) perform the process as far as possible.

Function Ω (F, x)
1. If there exists a clause $x \vee y_1 \vee y_2$ in F,
   then remove the clause $x \vee y_1 \vee y_2$ and the literals y_1, y_2 from F.
2. If there exists a clause $\neg x \vee y'_1 \vee y'_2$ in F, remove $\neg x$ from $\neg x \vee y'_1 \vee y'_2$.
3. For $1 \leq i \leq 2$ do Ω $(F, \neg y_i)$.
4. Return $F = Reduce(F)$.

Fig. 2. The function Ω

Now let us start to describe the framework of our algorithm X3SAT in Fig. 3. The algorithm employs a new principle, i.e. the connected-clauses principle, to simplify formulae. It takes a simplified X3SAT formula F as the input. Note that in the algorithm ESX3SAT(F) is a function that solves the X3SAT by exhaustive search. As we all know, if a X3SAT instance is solved by exhaustive search, it will spend a lot of time. However, when the number of clauses that the formula F contains is so few, it may run in polynomial time. Therefore, we use the function ESX3SAT(F) only when the number of clauses isn't above 5, which can guarantee the exhaustive search runs in polynomial time. Prefect_Matching(F) is also a function that reduces the X3SAT instance to a perfect matching problem when $\varphi(F) \leq 2$, and this can be solved in polynomial time [11]. In Theorem 8, we analyze the algorithm X3SAT using the measure described above.

Theorem 8.   Algorithm X3SAT runs in $O(1.15855^m)$ time, where m is the number of the clauses.

Proof. Let us analyze the algorithm case by case. Note that when analyzing the running time of the algorithm, we ignore the polynomial factor so that we assume that all polynomial time computations take $O(1)$ time.

Case 1, 2 and 3 can solve the instances completely and run in $O(1)$.

Case 4 doesn't increase the time needed.

Case 5: When $x = true$, every clause containing x is removed and $\neg x$ is removed from clauses. More over, every clause containing $\neg x$ shrinks to 2-clause which can be removed by (TR3). Therefore, the current formula contains at least four clauses less than F and the same situation is encountered when $x = false$. In addition, when x is

---

Algorithm X3SAT(F)

Case 1: F has an empty clause. return *unsatisfiable*.

Case 2: F is empty. return *satisfiable*.

Case 3: $m < 6$. return ESX3SAT (F).

Case 4: F consists of disjoint sub-formulae $F_1, F_2, ..., F_k$.

      return X3SAT (F_1) $\wedge$ X3SAT (F_2) $\wedge$ ... $\wedge$ X3SAT (F_k).

Case 5: $\varphi(F) \geq 4$. Pick a maximum degree variable x.

return X3SAT ($\Omega(F, x)$) $\vee$ X3SAT ($\Omega(F, \neg x)$).

Case 6: $\varphi(F) = 3$ and there is a (2, 1)-literal x such that $C_1 = x \vee y_1 \vee y_2$, $C_2 = x \vee y_3 \vee y_4$, and $C_3 = \neg x \vee y'_1 \vee y'_2$.

  1. If two clauses C_4 and C_5 connect with $C_1 \sim C_3$.

    (1) C_4 connects with C_1 and C_2, C_5 connects with C_3, i.e., $C_4 = {\sim}y_1 \vee {\sim}y_3 \vee z_1$, and $C_5 = {\sim}y'_1 \vee z_2 \vee z_3$, where z_1 is a singleton.

      return X3SAT($\Omega(F_1, y'_1) \wedge \Omega(F_2, y'_1)$) $\vee$ X3SAT($\Omega(F_1, \neg y'_1) \wedge \Omega(F_2, \neg y'_1)$),

      where $F_1 = C_1 \wedge C_2 \wedge C_3 \wedge C_4$, $F_2 = F/F_1$.

    (2) C_4 connects with C_1 and C_2, C_5 connects with C_3, i.e., $C_4 = {\sim}y_1 \vee {\sim}y_3 \vee z_1$, and $C_5 = {\sim}y'_1 \vee z_2 \vee z_3$, where z_1 is not a singleton.

      return X3SAT ($\Omega(F, x)$) $\vee$ X3SAT ($\Omega(F, \neg x)$).

    (3) C_4 connects with C_1; C_5 connects with C_2 and C_3, i.e., $C_4 = {\sim} y_1 \vee z_1 \vee z_2$ and $C_5 = {\sim} y'_1 \vee {\sim}y_3 \vee z_3$, where z_3 is a singleton.

      return X3SAT($\Omega(F_1, y_1) \wedge \Omega(F_2, y_1)$) $\vee$ X3SAT($\Omega(F_1, \neg y_1) \wedge \Omega(F_2, \neg y_1)$),

      where $F_1 = C_1 \wedge C_2 \wedge C_3 \wedge C_5$, $F_2 = F/F_1$.

    (4) C_4 connects with C_1; C_5 connects with C_2 and C_3, i.e., $C_4 = {\sim} y_1 \vee z_1 \vee z_2$ and $C_5 = {\sim} y'_1 \vee {\sim}y_3 \vee z_3$, where z_3 is not a singleton.

      return X3SAT ($\Omega(F, x)$) $\vee$ X3SAT ($\Omega(F, \neg x)$).

    (5) otherwise, return X3SAT ($\Omega(F, x)$) $\vee$ X3SAT ($\Omega(F, \neg x)$).

  2. If three or more clauses connect with $C_1 \sim C_3$.

    return X3SAT ($\Omega(F, x)$) $\vee$ X3SAT ($\Omega(F, \neg x)$).

Case 7: $\varphi(F) = 3$ and there is a (3, 0)-literal x.

return X3SAT ($\Omega(F, x)$) $\vee$ X3SAT ($\Omega(F, \neg x)$).

Case 8: $\varphi(F) \leq 2$, return Prefect_Matching(F).

Fig. 3. The algorithm for solving X3SAT

fixed a value, the clauses containing the literals in Y can be also removed. Now we let $R = NumC(F, N(x))$ and $R' = NumC(F, N(\neg x))$. Then we have $T(m) = T(m - 4 - R) + T(m - 4 - R')$. By Theorem 1, we know that at least four literals in Y occur in other clauses. So we obtain $R + R' \geq 2$. Therefore, the worst case is when $T(m) = T(m - 6) + T(m - 4)$ with solution $O(1.15096_m)$.

Case 6.1.1: When z_1 is a singleton, the formula F can be partitioned into two formulae $F_1 = C_1 \wedge C_2 \wedge C_3 \wedge C_4$ and $F_2 = F/F_1$ with only one common literal y'_1. By the connected-clauses principle, we branch on the common literal y'_1. We know that when the number of clauses that a formula contains is less than 6, the formula can be solved by exhaustive search. This means that the formula F_1 can be solved in polynomial time. And when y'_1 is fixed a value, at least one clause containing y'_1 is removed from the formula F_2. So the current formulae contain at least five clauses less than F in both of the branches. Therefore, we have $T(m) = T(m - 5) + T(m - 5)$ with solution $O(1.14870^m)$.

Case 6.1.2: In this case, the $y's$ literals in C_4 must be unnegated based on Theorem 2. Thus, when $x = true$, every clause containing x or z_1 is removed and every clause containing x or z1 is also removed by (TR3). Since $var(z_1)$ occurs at least twice and $var(x)$ occurs three times in F, the current formula contains at least five clauses less than F. When $x = false$, every clause containing $var(x)$ or $var(y'_1)$ can be removed, which make y_1 and y_3 become singletons. So clause C_4 can be removed by (TR6). Therefore, the worst case is when $T(m) = T(m - 5) + T(m - 5)$ with solution $O(1.14870^m)$.

Case 6.1.3: This case is similar to the case 6.1.1. So the current formula contains at least five clauses less than F in both of the branches. Therefore, we have $T(m) = T(m - 5) + T(m - 5)$ with solution $O(1.14870^m)$.

Case 6.1.4: In this case, at least one of the $y's$ literals in C_5 must be negated based on Theorem 3. If we give true to x, at least four clauses containing $var(x)$ or $var(y_1)$, are removed. And simultaneously other clauses containing $var(z_3)$ are removed. As we know, z_3 is not a singleton and this means that $var(z_1)$ occurs at least twice. So the current formula contains at least six clauses less than F. When $x = false$, at least four clauses containing $var(x)$ or $var(y'_1)$ are removed. Therefore, the worst case is when $T(m) = T(m - 6) + T(m - 4)$ with solution $O(1.15096^m)$.

Case 6.1.5: Due to previous cases, we know that C_4 and C_5 both contain at least two $y's$ literals. When $x = true$, every clause containing x or y_i $(1 \le i \le 4)$ is removed. When $x = false$, every clause containing x or y'_j $(1 \le j \le 2)$ is removed. In addition, by Theorem 3, at least one of the $y's$ literals in the clause C_4 or C_5 with y'_j $(1 \le j \le 2)$ must be negated and therefore at least two clauses containing literals in Z can be also removed. Thus, it follows that $T(m) = T(m - 6) + T(m - 4)$ with solution $O(1.15096^m)$.

Case 6.2: Let us assume that $R = NumC(F, N(x))$ and $R' = NumC(F, N(\neg x))$. Since $\varphi(x) = 3$, the current formula contains at least three clauses less than F when x is fixed a value. Furthermore, when $x = true$, $y_i = false$ $(1 \le i \le 4)$ and the clauses containing y_i $(1 \le i \le 4)$ are removed by (TR3). The time needed in this case is thus bounded by $T(m) = T(m - 3 - R) + T(m - 3 - R')$ since exactly the similar situation arises when x is given the value $false$. It is easy to see that $R \ge 1$ and $R' \ge 1$ for there are three or more clauses connect with $C_1 \sim C_3$. Moreover, at least four literals in Y occur in the three or more clauses by Theorem 1. Consequently, $R + R' \ge 4$ and the worst case occurs when $R = 3, R' = 1$. Therefore, The time needed in this case is bounded by $T(m) = T(m - 6) + T(m - 4)$ and $T(m) \in O(1.15096^m)$.

Case 7: If x is a $(3, 0)$-literal, at least four variables in Y_1 must occur in other clauses by Theorem 1. And if F contains an unnegated and negated variable, it must be $(1, 1)$-literal, otherwise, the $(2, 1)$-literal case is met. Therefore, there are at least two clauses connected with $C_1 \sim C_3$. In the following, we analyze the complexity from three cases. (1) Two clauses C_4 and C_5 connect with $C_1 \sim C_3$. If F contains a clause with three variables in Y_1, then x must be given the value false. Otherwise, F can be simplified by Theorem 2 and 4. Therefore, the case (1) can be solved in $O(1)$. (2) Three clauses C_4, C_5, and C_6 connect with $C_1 \sim C_3$. Similarly, when F contains a clause with three variables in Y_1, the formula F can be solved in $O(1)$. When C_i $(4 \le i \le 6)$ contains two $y's$ literals, the literals must be unnegated according to Theorem 2 and 4. So when each clause C_i $(4 \le i \le 6)$ contains two $y's$ literals, we branch on x. If $x = true$, three clauses containing x are removed and three clauses containing $y's$ variables are also removed. If $x = false$, we substitute $\neg y_2$ for y_1; substitute $\neg y_4$ for y_3; and substitute $\neg y_6$ for y_5. Consequently, we obtain a formula F contains $\neg y_2 \vee \neg y_4 \vee z_1$, $y_2 \vee \neg y_6 \vee z_2$, and $y_6 \vee y_4 \vee z_3$. It is easy to see that the three clauses can be removed by (TR10). And when there is a clause containing only one $y's$ literal, the clause can be removed by (TR6 and TR4) when $x = false$. Therefore, The time needed in this case is bounded by $T(m) = T(m - 6) + T(m - 4)$ and $T(m) \in O(1.15096^m)$. (3) Four or more clauses connect with $C_1 \sim C_3$. In this case, we branch on x. When x is fix a value, the clauses containing x are removed. And the clauses containing $y's$ variables are removed when $x = true$. Therefore, this case is bounded by $T(m) = T(m - 7) + T(m - 3)$ and takes $O(1.15855^m)$ time.

Case 8: This case can solve the problems completely and run in $O(1)$.

In total, algorithm X3SAT runs in $O(1.15855^m)$ time, where m is the number of the clauses.

4   Conclusion

This paper addresses the worst-case upper bound for the X3SAT problem with the number of clauses as the parameter. The algorithm presented is a DPLL-style algorithm. In order to improve the algorithms, we put forward a new connected-clauses principle to simplify the formulae. After a skillful analysis of these algorithms, we obtain the worst-case upper bound $O(1.15855^m)$ for X3SAT.

References

1. Schaefer, T.J.: The Complexity of Satisfiability Problems. In: 10th Annual ACM Symposium on Theory of Computing, pp. 216–226 (1978)
2. Drori, L., Peleg, D.: Faster Exact Solutions for Some NP-hard Problems. Theoretical Computer Science 287(2), 473–499 (2002)
3. Porschen, S., Randerath, B., Speckenmeyer, E.: X3SAT is Decidable in time $O(2^{n/5})$. In: Fifth International Symposium on the Theory and Applications of Satisfiability Testing, pp. 231–235. Springer, Heidelberg (2002)
4. Porschen, S., Randerath, B., Speckenmeyer, E.: Exact 3-Satisfiability is Decidable in time $O(2^{0.16254n})$ (June 2002) (manuscript); Annals of Mathematics and Artificial Intelligence 43(1) 173-193 (2005)
5. Kulikov, A.S.: An Upper Bound $O(2^{0.16254n})$ for Exact 3-Satisfiability: a Simpler Proof. Zapiski Nauchnyh Seminarov POMI 293, 118–128 (2002)
6. Dahllof, V., Jonsson, P., Beigel, R.: Algorithms for Four Variants of the Exact Satisfiability Problem. Theoretical Computer Science 320(2-3), 373–394 (2004)
7. Byskov, J.M., Madsen, B.A., Skjernaa, B.: New Algorithms for Exact Satisfiability. Theoretical Computer Science 332(1-3), 515–541 (2005)
8. Skjernaa, B.: Exact Algorithms for Variants of Satisfiability and Colouring Problems. PhD thesis, Department of Computer Science, Aarhus University (2004)

9. Bolette, A.M.: An Algorithm for Exact Satisfiability Analysed with the Number of Clauses as Parameter. Information Processing Letters 97(1), 28–30 (2006)
10. Hirsch, E.A.: New Worst-Case Upper Bounds for SAT. J. Auto. Reasoning 24(4), 397–420 (2000)
11. Monien, B., Speckenmeyer, E., Vornberger, O.: Upper Bounds for Covering Problems. Methods Oper. Res. 43, 419–431 (1981)

Fixed-Parameter Tractability of almost CSP Problem with Decisive Relations

Chihao Zhang and Hongyang Zhang

BASICS, Department of Computer Science, Shanghai Jiao Tong University,
Shanghai, 200240, China
{chihao.zhang,hongyang90}@gmail.com

Abstract. Let I be an instance of binary boolean CSP. Consider the problem of deciding whether one can remove at most k constraints of I such that the remaining constraints are satisfiable. We call it the Almost CSP problem. This problem is **NP**-complete and we study it from the point of view of parameterized complexity where k is the parameter. Two special cases have been studied: when the constraints are inequality relations (Guo et al., WADS 2005) and when the constraints are OR type relations (Razgon and O'Sullivan, ICALP 2008). Both cases are shown to be fixed-parameter tractable (FPT). In this paper, we define a class of *decisive* relations and show that when all the relations are in this class, the problem is also fixed-parameter tractable. Note that the inequality relation is decisive, thus our result generalizes the result of the parameterized edge-bipartization problem (Guo et al., WADS 2005). Moreover as a simple corollary, if the set of relations contains no OR type relations, then the problem remains fixed-parameter tractable. However, it is still open whether OR type relations and other relations can be combined together while the fixed-parameter tractability still holds.

1 Introduction

Consider the following parameterized problem:

p-ALMOST-CSP
>   *Input:* An instance of binary boolean CSP, and a nonnegative integer k.
> *Parameter:* k.
> *Problem:* Decide whether one can delete at most k constraints such that the remaining constraints are satisfiable.

Many natural problems can be expressed in this setting. For example, p-Almost 2SAT problem[13], which asks whether a CNF formula φ can be satisfied if we are allowed to remove at most k clauses, is a special case of p-Almost CSP. It was noticed in [15] that the fixed-parameter tractability of p-Almost 2SAT is equivalent to the vertex cover problem parameterized above the perfect matching. Another special case which has received extensive attention in the literature is

J. Snoeyink, P. Lu, K. Su, and L. Wang (Eds.): FAW-AAIM 2012, LNCS 7285, pp. 224–234, 2012.
© Springer-Verlag Berlin Heidelberg 2012

the parameterized edge-bipartization problem[6,15], which asks whether one can remove at most k edges in an undirected graph such that the remaining graph is bipartite. Both problems have been shown to be fixed-parameter tractable.

The above special cases impose restriction on the type of relations. These results motivate us to explore the parameterized complexity of p-Almost-CSP under various other sets of relations. Let $\mathcal{R}$ be a set of relations. Let p-$\mathcal{R}$-Almost-CSP be the problem of p-Almost-CSP such that all the input CSP instance can only use relations in $\mathcal{R}$. Almost 2SAT is equivalent to the case that the constraints are restricted to OR type relations, which include $R_1(x, y) := "x \vee y"$, $R_2(x, y) := "x \vee \bar{y}"$, $R_3(x, y) := "\bar{x} \vee y"$, $R_4(x, y) := "\bar{x} \vee \bar{y}"$ (we denote the set of these four OR type relations by $\mathcal{R}_{or}$). Edge-bipartization problem corresponds to the case that $\mathcal{R}$ contains only inequality relation.

Our Results. We define a class of *decisive* relations. A binary relation R is decisive if for $x \in \{0, 1\}$, at most one of $(x, 0)$ and $(x, 1)$ is in R and at most one of $(0, x)$ and $(1, x)$ is in R. Intuitively, if we fixed one component of a pair (x, y) where $x, y \in \{0, 1\}$, there is at most one choice for the other component to make the pair in R. We denote the set of decisive relations by $\mathcal{R}_{decisive}$. Decisive relations are quite expressive, including AND type relations, equality relation and inequality relation.

We present an $O^*(4^{k^2})$ ($O^*(\cdot)$ suppresses the polynomial term) algorithm for p-$\mathcal{R}_{decisive}$-Almost-CSP, hence it is fixed-parameter tractable.

Interestingly, based on the algorithms for decisive relations, it easily follows that if $\mathcal{R}$ contains no OR type relations, then p-$\mathcal{R}$-Almost-CSP is fixed-parameter tractable.

Our approach is based on the technique of iterative compression, which was first introduced in [14] to deal with the odd cycle transversal problem. Following the standard routine of this technique, we reduce p-$\mathcal{R}_{decisive}$-Almost-CSP to a variant edge-separation problem on graphs, we call this problem p-MinMixedCut. We then show that p-MinMixedCut is fixed-parameter tractable. The most important ingredient of our algorithm is the edge version of *important separator* introduced in [10].

Related Work. The question of whether the Almost 2SAT problem is fixed-parameter tractable, as mentioned above, was regarded as a long standing open problem [9,12,3], and finally solved by Razgon and O'Sullivan[13]. For the parameterized edge-bipartization problem, a reduction to odd cycle transversal was first noticed in [15]. Guo et al. presented a better FPT algorithm in [6]. It is also shown in [8] that there is a parameterized reduction from the edge-bipartization problem to the Almost 2-SAT problem. All these algorithms rely on the framework of iterative compression, which was introduced in [14]. See [7] for a survey of this technique.

Important separator was first introduced in [10] but implicitly used in [2,1,13]. It has been widely used in designing algorithms for graph separation problems. See [11] for a gentle introduction to this concept.

This paper is organized as follows: In Section 2, we present the statement of the problem, give some necessary definitions and introduce the notations. In Section 3, we use iterative compression technique to reduce the problem to p-MinMixedCut and then present a $O^*(4^{k^2})$ algorithm to solve it in Section 4. In Section 5, we give an algorithm based on previous sections to prove the main theorem and evaluate its running time. Finally, we conclude in Section 6 with some open problems.

2   Preliminaries

2.1   Parameterized Problems and Fixed-Parameter Tractability

A parameterized problem is a pair (Q, κ), where $Q \subseteq \Sigma^*$ is a classic decision problems and $\kappa : \Sigma^* \to \mathbb{N}$ is a polynomial-time computable function. An instance of (Q, κ) is denoted by (x, k) where $k = \kappa(x)$. A fixed-parameter tractable (FPT) algorithm decides whether $x \in Q$ in time $O(f(k) \cdot |x|^c)$, where c is a constant and f is an arbitrary computable function that only depends on k. We may use $O^*(f(k))$ to suppress the polynomial term. The notion of FPT relaxes the polynomial-time tractability in the classic setting. Readers may refer to [4,5,12] for more information on parameterized complexity and algorithms.

2.2   Constraint Satisfaction Problem

An instance of Constraint Satisfaction Problem (CSP) is defined as a triple $I := (X, D, \mathcal{C})$ where X is a set of variables, D is a domain of values, and $\mathcal{C}$ is a set of constraints. Every constraint is a pair $\langle t, R \rangle$, where t is a c-tuple of variables and R is a c-ary relation on D. An evaluation of the variables is a function from the set of variables to the domain of values $v : X \to D$. An evaluation v satisfies a constraint $\langle (x_1, \dots, x_c), R \rangle$ if $(v(x_1), \dots, v(x_c)) \in R$. A solution is an evaluation that satisfies all constraints. An instance I is satisfiable if it has a solution.

In this paper, we only consider *binary boolean* CSP, namely $D = \{0, 1\}$ and $c \leq 2$ for all relations R.

To explain why we focus on binary boolean case, note that the decision version of CSP remains NP-hard when $|D| \geq 3$ and $c = 2$, or when $|D| = 2$ and $c \geq 3$, therefore in both cases p-Almost-CSP is not fixed-parameter tractable unless **PTIME = NP**.

2.3   Binary Boolean Relations

There are 16 different binary boolean relations in total, listed in Table 1.

We divide the relations into three categories, namely $\mathcal{R}_{or}, \mathcal{R}_{decisive}, \mathcal{R}_{other}$, as shown in the table. Let $\mathcal{R}_{decisive} := \{R_5, \dots, R_{11}\}$, a binary boolean relation R is decisive if $R \in \mathcal{R}_{decisive}$. This set of relations can be defined as follows in a more intuitive way:

Table 1. 16 binary boolean relations

	$\mathcal{R}_{or}$				$\mathcal{R}_{decisive}$							$\mathcal{R}_{other}$				
	R_1	R_2	R_3	R_4	R_5	R_6	R_7	R_8	R_9	R_{10}	R_{11}	R_{12}	R_{13}	R_{14}	R_{15}	R_{16}
(0,0)	0	1	1	1	1	0	1	0	0	0	0	1	1	1	0	0
(0,1)	1	0	1	1	0	1	0	1	0	0	0	1	1	0	1	0
(1,0)	1	1	0	1	0	1	0	0	1	0	0	1	0	1	0	1
(1,1)	1	1	1	0	1	0	0	0	0	1	0	1	0	0	1	1

Definition 1 (Decisive Relation). *Let R be a binary boolean relation. We say R is decisive if for every $u \in \{0,1\}$, $\neg(R(u,0) \wedge R(u,1))$ and $\neg(R(0,u) \wedge R(1,u))$.*

Intuitively, if we fix one component of the relation, there is at most one choice for the other component such that the pair is in R.

Decisive relations have very simple interpretations: $R_5(x,y) := \text{``}x = y\text{''}$, $R_6(x,y) := \text{``}x \neq y\text{''}$, $R_7 := \text{``}\bar{x} \wedge \bar{y}\text{''}$, $R_8 := \text{``}\bar{x} \wedge y\text{''}$, $R_9 := \text{``}x \wedge \bar{y}\text{''}$, $R_{10}(x,y) := \text{``}x \wedge y\text{''}$, $R_{11} := \varnothing$. Let $R_{and} := \{R_7, R_8, R_9, R_{10}\}$.

2.4 Problem Statement and Main Result

Let $\mathcal{R}$ be a set of relations, consider the problem

---

$p\text{-}\mathcal{R}\text{-}\textsc{Almost-CSP}$

Input: An instance of binary boolean CSP, and a nonnegative integer k.

Parameter: k.

Problem: Find a set of at most k constraints such that the remaining constraints are satisfiable after removing them, or report no such set exists.

The main result of this paper is

Theorem 1. *Let $\mathcal{R} = \mathcal{R}_{decisive}$ be the set of binary boolean decisive relations. Then $p\text{-}\mathcal{R}\text{-}\textsf{Almost-CSP}$ is fixed-parameter tractable.*

The relations in $\mathcal{R}_{other}$ are very special and easy to handle in our model. Based on the algorithm for decisive case, we obtain the following corollary:

Corollary 1. *Let $\mathcal{R} = \mathcal{R}_{decisive} \cup \mathcal{R}_{other}$, then $p\text{-}\mathcal{R}\text{-}\textsf{Almost-CSP}$ is fixed-parameter tractable.*

2.5 Graph and Separator

Let $G := (V, E)$ be an undirected graph, $U \subseteq V$ be a set of vertices, $S \subseteq E$ be a set of edges. A path $P := \{e_1, \ldots, e_s\}$ of length s from u to v is a set of s edge such that $u \in e_1, v \in e_s, e_i \cap e_{i+1} \neq \varnothing$ for $1 \leq i < s$.

We denote the set of vertices reachable from U in $G' := (V, E\backslash S)$ by $R(U, S)$. Let $X, Y \subset V$ and $X \cap Y = \varnothing$, a set of edges T is called an (X, Y)-separator if $Y \cap R(X, T) = \varnothing$. An (X, Y)-separator is minimal if none of its proper subsets is an (X, Y)-separator. An (X, Y)-separator S' dominates an (X, Y)-separator S if $|S'| \leq |S|$ and $R(X, S) \subsetneq R(X, S')$. For singleton set $\{u\}$, we may write it as u for simplicity.

3   Reduction by Iterative Compression

In this section, we use the method of iterative compression to reduce p-$\mathcal{R}$-Almost-CSP to a variant edge-separation problem. Similar reductions can be found in [13]. Unless otherwise specified, all the relations in this section belong to $\mathcal{R}_{decisive}\backslash\{R_{11}\}$ because constraints of type R_{11} are unsatisfiable and can be removed in advance.

Given a CSP instance $I = (X, \mathcal{C})$, where $\mathcal{C} = \{\langle t_1, R_1 \rangle, \ldots, \langle t_n, R_n \rangle\}$ consists of n decisive constraints and an integer $k \geq 0$. Then consider $n + 1$ instances $I_0, \ldots, I_n$ where $I_i = (X, \mathcal{C}_i)$ and $\mathcal{C}_i$ consists of first i constraints of $\mathcal{C}$. Note that $I_n = I$. We solve (I_i, k) for $i = 1, \ldots, n$ one by one.

Since $k \geq 0$, (I_0, k) is obviously a true instance. If for some $i \leq n$, (I_i, k) is a false instance, then we know that (I, k) is also a false instance. Now assume for some $m < n$ all (I_i, k) with $i \leq m$ are true instance, we need to decide (I_{m+1}, k).

We know that (I_m, k) is a true instance, let $\mathcal{S}$ be one of its solution sets where $|\mathcal{S}| \leq k$, then $\mathcal{S}' := \mathcal{S} \cup \{\langle t_{m+1}, R_{m+1} \rangle\}$ is a solution set for I_{m+1}. If $|\mathcal{S}'| \leq k$ then (I_{m+1}, k) is a true instance and we are done. Otherwise, we give an algorithm that either construct a solution set $\mathcal{T}$ of size at most k or report no such set exists.

To this end, we enumerate $\mathcal{ST} \subseteq \mathcal{S}'$ and consider the CSP instance $I' = (X, \mathcal{C}')$, where $\mathcal{C}' := \mathcal{C}_m \backslash \mathcal{ST}$. It is easy to see that the following holds:

Claim 1. *Let $\mathcal{T} \subseteq \mathcal{C}_{m+1}$ be a set of constraints. Then $\mathcal{T}$ is a solution set of I_{m+1} if and only if for $\mathcal{ST} := \mathcal{S}' \cap \mathcal{T}$, $\mathcal{T}\backslash\mathcal{ST}$ is a solution set of $I' = (X, \mathcal{C}')$ where $\mathcal{C}' := \mathcal{C}_m \backslash \mathcal{ST}$.*

Since $(\mathcal{T}\backslash\mathcal{ST}) \cap (\mathcal{S}'\backslash\mathcal{ST}) = \varnothing$, we come to the following problem:

PROBLEM 1
> *Input:* A binary boolean CSP I, a set of constraints $\mathcal{S}$ with $|\mathcal{S}| \leq k_1$ such that I is satisfiable after removing $\mathcal{S}$ and an integer $k_2 \geq 0$.
> *Parameter:* $k_1 + k_2$.
> *Problem:* Find a set of restrictions $\mathcal{T}$ with $|\mathcal{T}| \leq k_2$ such that $\mathcal{S} \cap \mathcal{T} = \varnothing$ and I is satisfiable after removing $\mathcal{T}$.

Lemma 1. *Problem 1 is fixed-parameter tractable.*

We first extend our terminologies. Let $\mathcal{S}$ be a set of constraints, then $V(\mathcal{S})$ is the set of variables appearing in $\mathcal{S}$. Let $I := (X, \mathcal{C})$ be a *satisfiable* binary boolean CSP instance, then I has a satisfiable assignment $F : X \to \{0, 1\}$.

Now let $(I := (X, \mathcal{C}), \mathcal{S}, k_1, k_2)$ be an instance of **Problem 1**, and let $I' := (X, \mathcal{C} \backslash \mathcal{S})$. We enumerate all the assignments $F : V(\mathcal{S}) \to \{0, 1\}$ such that F satisfies $\mathcal{S}$. The following claim is straightforward:

Claim 2. *The instance $(I := (X, \mathcal{C}), \mathcal{S}, k_1, k_2)$ has a solution set $\mathcal{T}$ if and only if for some $F : V(\mathcal{S}) \to \{0, 1\}$ that satisfies $\mathcal{S}$, I' contains a set of constraints $\mathcal{T}'$ such that (1) $|\mathcal{T}'| \leq k_2$ and (2) after removing $\mathcal{T}'$ in I', there exists a satisfiable assignment of I' consistent with F.*

Proof. For the forward direction, let $\mathcal{T}$ be a solution set of $(I, \mathcal{S}, k_1, k_2)$ and F_0 be a satisfiable assignment of $(X, \mathcal{C} \backslash \mathcal{T})$. Then $\mathcal{T}' := \mathcal{T}$ and F_0 fulfill our requirement.

Conversely, given $\mathcal{T}'$ and a satisfiable assignment F' of I' after removing $\mathcal{T}'$ such that the restriction of F' on $V(\mathcal{S})$ satisfies $\mathcal{S}$. Then $\mathcal{T} := \mathcal{T}'$ is a solution set of $(I, \mathcal{S}, k_1, k_2)$ since F' is a satisfiable assignment of $(X, \mathcal{C} \backslash \mathcal{T})$. □

Thus it suffices to solve **Problem 2** in FPT time:

PROBLEM 2

      Input: A satisfiable binary boolean CSP I, a partial assignment F and an integer $k \geq 0$.

   Parameter: k.

    Problem: Find a set of constraints $\mathcal{T}$ with $|\mathcal{T}| \leq k$ such that after removing $\mathcal{T}$ in I, there exists a satisfiable assignment of I consistent with F.

Lemma 2. *Problem 2 is fixed-parameter tractable.*

Since $I = (X, \mathcal{C})$ is satisfiable, let $A : X \to \{0, 1\}$ be one of its satisfiable assignment. If A is consistent with F, then we are done. Otherwise, let $D(F)$ be the domain of F, then for some variable $x \in D(F)$, we have $F(x) \neq A(x)$. Let $D \subseteq D(F)$ be the set of all such variables. Let $v \notin X$, for every $x \in D$, if $F(x) = 0$ then replace x by $\bar{v}$ in I; if $F(x) = 1$ then replace x by v in I. Let I' be the new instance after replacement.

Claim 3. *(I, F, k) contains a solution set $\mathcal{T}$ if and only if there is a set of constraints $\mathcal{T}'$ with $|\mathcal{T}'| \leq k$ and after removing $\mathcal{T}'$ in I', there is an assignment A' satisfying $A'(v) = 1$ and A' agrees with F on $D(F) \backslash D$.*

Proof. First assume (I, F, k) contains a solution set $\mathcal{T}$. We construct $\mathcal{T}'$ as follows: for every $C = ((x_1, x_2), R_C) \in \mathcal{T}$, if $x_1, x_2 \notin D$, then add C to $\mathcal{T}'$; otherwise, let C' be the constraint obtained from C by replacing the variable in D by v, and add C' to $\mathcal{T}'$. Let A be a satisfiable assignment of (I, F, k) after removing $\mathcal{T}$ and A is consistent with F. Define an assignment A' on $X \backslash D \cup \{v\}$ where A' agrees with A on $X \backslash D$ and $A'(v) = 1$. By the definition of I', A' is a satisfiable assignment of I' after removing $\mathcal{T}'$ and A' agrees with F on $D(F) \backslash D$.

The converse can be proved analogously and thus we omit it. □

Therefore we reduce Problem 2 to the following:

PROBLEM 3

    Input: A satisfiable binary boolean CSP $I := (X, \mathcal{C})$, a partial assignment F, a variable v and an integer $k \geq 0$. It is known that there is a satisfiable assignment A of I consistent with F and $A(v) = 0$.

    Parameter: k.

    Problem: Find a set of constraints $\mathcal{T}$ with $|\mathcal{T}| \leq k$ such that after removing $\mathcal{T}$ in I, there exists a satisfiable assignment of I, say A, such that A is consistent with F and $A(v) = 1$.

Next, we interpret Problem 3 as a graph separation problem.

Here each variable corresponds to a vertex and each constraint corresponds to an edge. An edge has an annotated type indicating the constraint upon the edge. Then a satisfiable assignment corresponds to a way to color each vertex with 0 or 1 such that all the edge constraints are satisfied.

First assume the graph is connected, without loss of generality, since between disconnected components there are no constraints. Since all the relations are decisive, if one vertex is assigned with some value, then to satisfy the constraints, the value of all the reachable vertices is determined. Our goal is to flip the value of v in a satisfiable assignment F while keeping the value of some other set of vertices S. To do this vertices set S should be separated from v. We denote this set of vertices by S_1. Furthermore, let $e = \{w, u\}$ be an edge where at least one of w and u is not in S and the type of e is in $\mathcal{R}_{and}$, then we have to either separate $\{w, u\}$ with v or remove edge e. We denote this set of edges by S_2. Therefore, the problem is equivalent to the following:

---

p-MINMIXEDCUT

    Input: An undirected graph $G := (V, E)$, a vertex $t \in V$, a set of vertices $S_1 := \{u_1, \ldots, u_p\}$ and a set of pairs of vertices $S_2 := \{\{v_1, w_1\}, \ldots, \{v_q, w_q\}\}$ where each $\{v_i, w_i\}$ is an edge in G. An integer $k \geq 0$.

    Parameter: k.

    Problem: Find a set of at most k edges T, such that (1) T is a separator with respect to S_1 and t; (2) For every pair $\{v, w\}$ in S_2, either edge $\{v, w\} \in T$ or T is a separator with respect to $\{v, w\}$ and t.

---

4 p-MinMixedCut Is Fixed-Parameter Tractable

The algorithm employs the method of bounded search tree. For each pair $\{v, w\} \in S_2$, we branch into two cases: either add $\{v, w\}$ to the solution set or separate them from t. To bound the width of each branch, we use the similar idea of *important separator* in [10].

Definition 2. *Let $G := (V, E)$ be an undirected graph. Let $X, Y \subset V$ and $X \cap Y = \varnothing$, a set of edges S is an important (X, Y)-separator if it is minimal and there is no (X, Y)-separator S' that dominates S.*

We show that it is enough to enumerate all the important separators in the branches.

Lemma 3. *Given an instance (G, t, S_1, S_2, k) of p-MinMixedCut, if there is a solution set T of size at most k, then there exists a solution set T' of size at most k such that (1) for every vertex $u \in S_1$, some subset of T' is an important (u, t)-separator and (2) for every pair $\{v, w\} \in S_2$, if the edge $\{v, w\} \notin T$, then some subset of T' is an important $(\{v, w\}, t)$-separator.*

Proof. We only prove (1), the proof of (2) is analogous. Let u be a vertex in S_1 and $S \subseteq T$ be a minimal (u, t)-separator. If S is an important (u, t)-separator, then we are done, otherwise, there is an edge set S' that dominates S, we show that $T' := (T \backslash S) \cup S'$ is also a solution set of size at most k.

Assume on the contrary that T' is not a solution, we distinguish between two cases:

(a) These is a vertex u' such that u' is separated from t by T but not by T'. This is impossible because every path P from u' to t intersects either $T \backslash S$ or S, and S' dominates S, hence P intersects with T'.

(b) For some edge $e := \{v, w\} \in S_2$, $e \in S$ and $e \notin S'$. Since $e \in S$ and S is minimal, v, w belong to different connected components after removing S. Without loss of generality, assume $v \in R(u, S)$ and $w \in R(t, S)$. Since S' dominates S, $v \in R(u, S')$, hence $w \in R(u, S')$. Therefore both v and w are separated from t by S' and by T' as well.

To prove (2), we can contract $\{v, w\}$ to a single vertex in G and use the same argument above.   □

This lemma implies that to separate every vertex u from t, it suffices to enumerate important (u, t)-separator, thus settling the correctness of our algorithm.

Next, the number of important separators can be bounded by a function of k. Essentially, this enables us to bound the number of branches in the search tree.

Lemma 4 ([11,2]). *Let $G := (V, E)$ be an undirected graph. There are at most 4^k important (X, Y)-separator of size at most k for every $X, Y \subseteq V$. Furthermore, all the important separators can be enumerated in $O^*(4^k)$ time.*

Therefore the following algorithm solves p-MinMixedCut in FPT time.

MinMixedCut(G, t, S_1, S_2, k)

Input: An undirected graph $G := (V, E)$, a vertex $t \in V$, a set of vertices S_1, and a set of pairs of vertices S_2. An integer $k \geq 0$.

Output: A set of edges T that fulfills our requirement, or return 'NO' if no such set exists.

   1.   if S_2 is nonempty and $k > 0$, choose $p = \{u, v\} \in S_2$ such that $p \in E$ and t is reachable from $\{u, v\}$ in G
     1.1 $T \leftarrow$ **MinMixedCut**$(G' := (V, E\backslash\{p\}), t, S_1, S_2\backslash\{p\}, k - 1)$
     1.2   if T is not 'NO' then return $T \cup \{p\}$
     1.3   for all important $(\{u, v\}, t)$-separator S such that $|S| \leq k$
       1.3.1 $T \leftarrow$ **MinMixedCut**$(G' := (V, E\backslash S), t, S_1, S_2\backslash S, k - |S|)$
       1.3.2   if T is not 'NO' then return $T \cup S$
     1.4   return 'NO'
   2. $T \leftarrow$ minimum edge cut from S_1 to $\{t\}$ in G
   3.   if $|T| \leq k$ return T else return 'NO'

To evaluate the running time of the above algorithm, consider its search tree T. The depth of T is at most k since in every recursive call for **MinMixedCut**, k decreases by 1 at least.

Next we consider the number of nodes in T. Since there are two branches in step 1.1, 1.3, respectively, and by Lemma 4 there are at most 4^{k+1} branches in step 1.3, so the total number of branches is at most $1 + 4^{k+1}$. Thus the size of T is $O(4^{k^2})$ and the total running time of the algorithm is $O^*(4^{k^2})$.

5 Main Theorem

In this section, we prove Theorem 1 and Corollary 1.

Proof (of Theorem 1).

Given a p-Almost-CSP instance (I, k), the main algorithm first reduces it to an instance of p-MinMixedCut (G, t, S_1, S_2, k'), following the procedure described in Section 3. Then it solves the instance by using the algorithm described in Section 4.

Now we evaluate the running time of above algorithm step by step:

1 p-Almost-CSP to Problem 1
  Let (I, k) be an instance of p-Almost-CSP. There are at most $|I|$ iterations. For each iteration, we enumerate at most 2^k $\mathcal{ST}$. The resulting instance $(I_1, \mathcal{S}, k_1, k_2)$ of Problem 1 satisfies $|I_1| \leq |I|, |\mathcal{S}| \leq k, k_1 + k_2 \leq k$.
2 Problem 1 to Problem 2
  Let $(I_1, \mathcal{S}, k_1, k_2)$ be an instance of Problem 1. We need to enumerate at most $2^{|\mathcal{S}|} \leq 2^k$ assignments F, and for each F, we get a new instance (I_2, F, k_3) of Problem 2 where $|I_2| \leq |I_1|, k_3 = k_2 \leq k$.

3 Problem 2 to Problem 3

Let (I_2, F, k_3) be an instance of Problem 2, we reduce it to an instance (I_3, F, v, k_4) of Problem 3 where $|I_3| \leq |I_2|, k_4 = k_3$ in $O(|I_2|)$ time.

4 Problem 3 to p-MinMixedCut

Let (I_3, F, v, k_4) be an instance of Problem 3, we reduce it to an instance (G, t, S_1, S_2, k') of p-MinMixedCut where $|G| + |S_1| + |S_2| = O(|I_3|))$ and $k' = k_4$ in $O(|I_3|)$ time.

So the total runtime of this procedure is $O(|I| \cdot 2^k \cdot 2^k \cdot |I|) = O(4^k |I|^2)$.

For every instance $(G := (V, E), t, S_1, S_2, k')$, we can solve it in $O^*(4^{k^2}) = O^*(4^{k^2})$. Therefore our algorithm for p-$\mathcal{R}$-Almost-CSP where $\mathcal{R}$ is the set of decisive relations runs in $O^*(4^k \cdot 4^{k^2}) = O^*(4^{k^2})$. $\qquad\square$

Now we prove Corollary 1.

Proof (of Corollary 1). Let $I := (X, \mathcal{C})$ be an instance of binary boolean CSP. We have five more relations now, i.e. $\mathcal{R}_{other}$. First R_{12} can be ignored since it is always satisfied. For other four relations, note that $R_{13}(x, y) = $ "$x = 0$", $R_{14}(x, y) = $ "$y = 0$", $R_{15}(x, y) = $ "$y = 1$", $R_{16}(x, y) = $ "$x = 1$", thus they can be reduce to equality relation by adding two variables $\mathbf{1}$ and $\mathbf{0}$ into X. For all constraints in $\mathcal{C}$ that is of type $R_{13}(x, y), R_{14}(x, y), R_{15}(x, y), R_{16}(x, y)$, replace them by $R_5(x, 0), R_5(y, 0), R_5(y, 1), R_5(x, 1)$ respectively. Then this instance can be solved in the same way as Theorem 1, except in the reduction from Problem 1 to Problem 2, we enumerate all $F : V(\mathcal{S}) \cup \{\mathbf{1}, \mathbf{0}\} \rightarrow \{0, 1\}$ such that $F(\mathbf{0}) = 0, F(\mathbf{1}) = 1$ instead. $\qquad\square$

6   Conclusions and Open Problems

In this paper we discussed the p-$\mathcal{R}$-Almost-CSP problem. By utilizing the powerful techniques of iterative compression and important separators, we solved for the case of decisive relations. To deal with the general case, however, the biggest technical challenge is about how to deal with OR type relations and decisive relations together.

Acknowledgements. This research was partially supported by the National Nature Science Foundation of China (60970011 & 61033002).

We are grateful to anonymous referees for pointing out some mistakes and their suggestion for presentation.

References

1. Chen, J., Liu, Y., Lu, S.: An improved parameterized algorithm for the minimum node multiway cut problem. Algorithmica 55(1), 1–13 (2009)
2. Chen, J., Liu, Y., Lu, S., O'Sullivan, B., Razgon, I.: A fixed-parameter algorithm for the directed feedback vertex set problem. Journal of the ACM (JACM) 55(5), 21 (2008)

3. Demaine, E., Gutin, G., Marx, D., Stege, U.: Open problems from dagstuhl seminar 07281, available electronically, Technical report,
   `http://drops.dagstuhl.de/opus/volltexte/2007/1254/pdf/07281`

4. Downey, R.G., Fellows, M.R.: Parameterized complexity. Springer, New York (1999)

5. Flum, J., Grohe, M.: Parameterized complexity theory. Springer-Verlag New York Inc. (2006)

6. Guo, J., Gramm, J., Huffner, F., Niedermeier, R., Wernicke, S.: Compression-based fixed-parameter algorithms for feedback vertex set and edge bipartization. Journal of Computer and System Sciences 72(8), 1386–1396 (2006)

7. Guo, J., Moser, H., Niedermeier, R.: Iterative compression for exactly solving np-hard minimization problems. Algorithmics of Large and Complex Networks, 65–80 (2009)

8. Khot, S., Raman, V.: Parameterized complexity of finding subgraphs with hereditary properties. Theoretical Computer Science 289(2), 997–1008 (2002)

9. Mahajan, M., Raman, V.: Parametrizing above guaranteed values: Maxsat and maxcut. In: Electronic Colloquium on Computational Complexity (ECCC), vol. 4 (1997)

10. Marx, D.: Parameterized graph separation problems. Theoretical Computer Science 351(3), 394–406 (2006)

11. Marx, D.: Important separators and parameterized algorithms (February 2011),
    `http://www.cs.bme.hu/~dmarx/papers/marx-mds-separators-slides.pdf`

12. Niedermeier, R.: Invitation to fixed-parameter algorithms, vol. 31. Oxford University Press, USA (2006)

13. Razgon, I., O'Sullivan, B.: Almost 2-sat is fixed-parameter tractable. Journal of Computer and System Sciences 75(8), 435–450 (2009)

14. Reed, B., Smith, K., Vetta, A.: Finding odd cycle transversals. Operations Research Letters 32(4), 299–301 (2004)

15. Wernicke, S.: On the algorithmic tractability of single nucleotide polymorphism (SNP) analysis and related problems. PhD thesis (2003)

On Editing Graphs into 2-Club Clusters*

Hong Liu**, Peng Zhang, and Daming Zhu

School of Computer Science and Technology, Shandong University, Shandong
Provincial Key Laboratory of Software Engineering, Jinan, China
{hong-liu,algzhang,dmzhu}@sdu.edu.cn

Abstract. In this paper, we introduce and study three graph modification problems: 2-CLUB CLUSTER VERTEX DELETION, 2-CLUB CLUSTER
EDGE DELETION, and 2-CLUB CLUSTER EDITING. In 2-CLUB CLUSTER
VERTEX DELETION (2-CLUB CLUSTER EDGE DELETION, and 2-CLUB
CLUSTER EDITING), one is given an undirected graph G and an inte-
ger $k \geq 0$, and needs to decide whether it is possible to transform G
into a 2-club cluster graph by deleting at most k vertices (by deleting
at most k edges, and by deleting and adding totally at most k edges).
Here, a 2-club cluster graph is a graph in which every connected component is of diameter 2. We first prove that all these three problems are
NP-complete. Then, we present for 2-CLUB CLUSTER VERTEX DELE-
TION a fixed parameter algorithm with running time $O^*(3.31^k)$[1], and
for 2-CLUB CLUSTER EDGE DELETION a fixed parameter algorithm with
running time $O^*(2.74^k)$.

Keywords: Fixed Parameter Tractability, Graph-based Data Clustering, 2-Club.

1   Introduction

Data clustering [15] is the process of partitioning data set into clusters so that the
data records within a cluster are highly interrelated, while there are less interrelations between elements in different clusters. It is an important task in many
areas, e.g., machine learning, data mining, decision-making, and exploratory
pattern-analysis. Various approaches have been proposed for data clustering. In
graph-based data clustering [2,14], data records are represented as vertices, there
is an edge between two vertices if and only if the interrelations of the corresponding data records exceeds a certain threshold, and a cluster is therefore interpreted
as a dense subgraph. Traditionally, complete graphs, also called cliques, are used
to model dense subgraphs. However, in various application scenarios, the requirement for clusters to be cliques is too restrictive. As alternatives, various relaxed

---

* Research supported by the National Natural Science Foundation of China (61070019)
  and the National Natural Science Foundation of China (60603007).
** Corresponding author.
[1] In the O^* notation we omit the polynomial terms, so that $O^*(f(k))$ stands for
$O^*(f(k)P(n))$ for some polynomial P.

J. Snoeyink, P. Lu, K. Su, and L. Wang (Eds.): FAW-AAIM 2012, LNCS 7285, pp. 235–246, 2012.
© Springer-Verlag Berlin Heidelberg 2012

clique models for dense subgraph have been proposed, e.g. μ-clique, s-clique, s-club, and s-plex, etc. [10].

A graph-based data clustering problem can be formulated as a *cluster graph modification problem*: Given an undirected graph G, one asks for a minimum-cardinality set of editing operations that transform G into a graph (called cluster graph) in which every connected component is a dense subgraph (cluster). Traditionally, there are three standard editing operations, i.e., adding edges, deleting edges, and deleting vertices. The most prominent problem in this context maybe CLUSTER EDITING [2,14], which asks whether a graph can be transformed into a collection of disjoint cliques by altogether at most k edge adding and edge deleting operations. A closed related problem is CLUSTER VERTEX DELETION [9], in which the editing operation is instead vertex deleting. In this line, a series of problems considering relaxed clique models are also studied, for examples, s-PLEX CLUSTER EDITING [8], s-PLEX CLUSTER VERTEX DELETION [7], s-DEFECTIVE CLIQUE EDITING [7], μ-CLIQUE EDITING [7], etc..

Extending and complementing previous work, we introduce three novel cluster graph modification problems: s-CLUB CLUSTER VERTEX DELETION, s-CLUB CLUSTER EDGE DELETION, and s-CLUB CLUSTER EDITING. A vertex subset $S \subseteq V$ of a graph $G = (V, E)$ is called s-club if the diameter of the induced subgraph $G[S]$ is at most s. Clearly, a clique is nothing but a 1-club. A s-club cluster graph is a graph whose connected components are s-clubs. For small values of s, the s-clubs have been a practical and popular choice to model clusters in the context of social networks [1] and protein interaction networks [3]. Thus, our problems also has much practice implications. This work provides the first theoretical study of these three problems. The decision versions of them are formulated as follows, respectively:

s-CLUB CLUSTER VERTEX DELETION
Input: A graph $G = (V, E)$ and an integer $k \geq 0$
Question: Can G be modified by up to k vertex deletions into an s-club cluster graph?

s-CLUB CLUSTER EDGE DELETION
Input: A graph $G = (V, E)$ and an integer $k \geq 0$
Question: Can G be modified by up to k edges deletions into an s-club cluster graph?

s-CLUB CLUSTER EDITING
Input: A graph $G = (V, E)$ and an integer $k \geq 0$
Question: Can G be modified by up to k edge deletions and insertions into an s-club cluster graph?

Particularly, 2-clubs have an outstanding characteristic, i.e., every pair of vertices in it either have a direct edge or have a common neighbor. This intuitive two-hop interpretation has encouraged the choice of 2-clubs in many applications where a "two-hop transitivity" is expected [12]. In this paper, we spend our effort mainly for the cases where $s=2$.

The rest of the paper is organized as follows. In the 2nd section, we introduce some useful notions and concepts. In the 3rd section, we present NP-completeness results for our problems. Then, in the 4th section, we present for 2-CLUB CLUSTER VERTEX DELETION a fixed parameter algorithm with running time of $O^*(3.31^k)$, and for 2-CLUB CLUSTER EDGE DELETION a fixed parameter algorithm with running time $O^*(2.74^k)$. Conclusion and future work are given in the last section.

2   Preliminaries

We only consider simple (i.e., with no loops or multiple edges) and undirected graphs $G = (V, E)$, where V is the set of vertices and E is the set of edges. For a graph, we also use V_G and E_G to denote its vertex and edge sets, respectively. The (open) neighborhood of a vertex $v \in V$ in G is $N_G(v) := \{u|\{u,v\} \in E\}$. The closed neighborhood of a vertex $v \in V$ in G is $N_G[v] := N_G(v) \cup \{v\}$. Moreover, for a subset $V' \subseteq V$, let $N_G(V') := \bigcup_{v \in V'} N_G(v) \setminus V'$ and $N_G[V'] := \bigcup_{v \in V'} N_G[v]$. The degree of a vertex v in G is $deg_G(v) := |N_G(v)|$. We say two vertices are connected, if there exists at least one path between them. The distance between two connected vertices is the length of a shortest path between them. The maximum distance between any pair of connected vertices in G is called the *diameter* of G. For a set $V' \subseteq V$, we denote by $G[V']$ the subgraph of G induced by the vertices in V'. A chordless path with four distinct vertices in G is called a P_4. Given a P_4 $stuv$, we call s and v *end vertices* of $stuv$, and t and u *internal vertices*. Particularly, the two end vertices s and v in $stuv$ neither have a direct edge nor have a common neighbor, it's called a *restricted* P_4 and denoted by P_{stuv}.

Let G be any graph. If S is a subset of V, such that $G' = (V - S, E)$ is a 2-club cluster graph, then S is called a *2-Club Cluster Vertex-Deletion Set* for G. If F is a subset of $V \times V$, such that $G' = (V, E \triangle F)$ is a 2-club cluster graph, where $E \triangle F = (E \setminus F) \cup (F \setminus E))$, then F is called a *2-Club Cluster Edge-Edition Set* for G. If in addition $F \subseteq E$, then F is called a *2-Club Cluster Edge-Deletion Set* for G.

Our algorithm is based on the *depth-bounded search tree* technique that is frequently successfully applied in the development of fixed-parameter algorithms [5,13]. A depth-bounded search tree algorithm works in a recursive manner. The number of recursion calls is the size of search tree. If the algorithm solves a problem instance with parameter k and calls itself recursively for problem instances with parameters $k - d_1, k - d_2, \cdots, k - d_i$, then $(d_1, \cdots, d_i)$ is called the *branching vector* of this recursion, and the overall search tree size reads as $T(k) = T(k - d_1) + T(k - d_2) + \cdots + T(k - d_i)$. If α is a solution of the recurrence which has maximum absolute value and is positive, then α is called the *branching number* corresponding to the branching vector $(d_1, \cdots, d_i)$ (all branching numbers that occur in this paper are single roots). The size of the search tree is therefore $O(\alpha^k)$, where α is the largest branching number that will occur.

3 NP-Hardness Proofs

3.1 NP-Complteness of 2-Club Cluster Vertex Deletion

Note that 2-club cluster graphs are not hereditary, that is, not closed under taking induced subgraphs. For instance, a 5-cycle is a 2-club cluster graph but deleting a vertex results in a P_4, which is not a 2-club cluster graph. Thus the general NP-completeness result for vertex deletion problems for hereditary graph properties [11] does not apply here. In this section we prove that, for any $s \geq 2$, s-CLUB CLUSTER VERTEX DELETION is NP-complete.

Theorem 1. s-CLUB CLUSTER VERTEX DELETION *is NP-complete.*

Proof. Membership in NP is trivial. In fact, one can verify in polynomial time that a connected graph is of diameter s by computing the distance between each pair of its vertices.

We prove NP-hardness by reduction from the well-known NP-complete problem VERTEX COVER [6]: Given a graph $G = (V, E)$ and an nonnegative integer $k \leq |V|$, determine if there is a subset $V_c \subseteq V$ such that $|V_c| \leq k$ and, for each edge $\{u, v\} \in E$, at least one of u and v belongs to V_c.

Let $G = (V, E)$ be an graph. For each vertex $v \in V$, we attach a P_{s-1} to obtain a new graph $G' = (V', E')$, and the corresponding P_s induced by vertices in this P_{s-1} together with v is denoted by P_s^v (see Fig. 1 for an illustration). We shall prove that G has a vertex cover of size at most k if and only if G' has a 2-Club Cluster Vertex-Deletion Set of size at most k.

$\Rightarrow$: Suppose that there exists a vertex cover $V_c \subseteq V$ of size k in G. Delete the k corresponding vertices from G'. It is obvious that very connected component in $G' - V_c$ is of diameter s.

$\Leftarrow$: Now suppose G' has a size-k 2-Club Cluster Vertex-Deletion Set V_s. Let $V_c = \{v \in V | P_s^v \cap V_s \neq \emptyset\}$. Obviously, $|V_c| \leq k$. We claim that V_c is a vertex cover of G. Otherwise, suppose there is an edge $\{u, v\} \in E$ with $u \notin V_c$ and $v \notin V_c$. This means that none of the vertices in both P_s^u and P_s^v is included in V_s. Consequently, there exists at least one pair of vertices in $G' - V_s$ whose distance is greater than s. Thus, a contradiction. $\square$

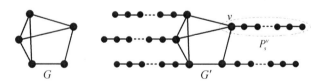

Fig. 1. An illustration of reduction from VERTEX COVER to s-CLUB CLUSTER VERTEX DELETION

3.2   NP-Complteness of 2-Club Cluster Editing

We show that 2-CLUB CLUSTER EDITING is NP-hard by a reduction from the NP-complete 3-EXACT 3-COVER (3X3C) problem [6]. A similar technique was used in [14] to prove NP-hardness of the CLUSTER EDITING problem.

Theorem 2. 2-CLUB CLUSTER EDITING *is NP-complete.*

Proof. Membership in NP is trivial. In the following, we present a reduction from 3X3C to 2-CLUB CLUSTER EDITING. A 3X3C instance includes a finite set of elements $U = \{u_1, u_2, \cdots u_{3n}\}$, and a set C of triplets of elements from U, i.e., $C = \{S_1, S_2, \cdots, S_r\}$, where, for every $1 \le i \le r$, $S_i \subseteq U$ and $|S_i| = 3$. Moreover, for every $1 \le j \le n$, u_j is included in at most three triplets. The question is whether there is a sub-collection $C' \subseteq C$ of size n that covers U, i.e., every element of U occurs in exactly one triplet in C'.

Let $\langle U, C \rangle$ be an instance of 3X3C. Assume $r > n > 1$. We build a graph $G = (V, E)$ according to $\langle U, C \rangle$ as follows (see Fig. 2 for an illustration): First, for each element u_i in U, we introduce a vertex, which is denoted also by u_i. For each triplet $S_i \in C$, we add necessary edges that link each pair of vertices in S_i. Denote by E_C the set of all such edges in G and define $t \equiv |E_C|$. It is not hard to see that, since every element of U is included in at most three triplets, we have $t \le 9n$. Let $k = 3(r - n)(6n + 1) + (t - 3n)$. Next, for each triplet $S_i \in C$, we introduce a gadget, denoted by G_i. In G_i, there are two cliques, denoted by Q_i^1 and Q_i^2 respectively. Q_i^1 contains $6n + 1$ vertices, while Q_i^2 contains $k + 2$ vertices. Then, we add to G necessary edges that join every vertex in Q_i^1 to every vertex in Q_i^2. Last, we add to G necessary edges that join every vertex in S_i to every vertex in Q_i^1, and the set of all such edges in G is denote by E_{UQ^1}. Clearly, $|E_{UQ^1}| = 3r(6n + 1)$.

In the following, we show that there is an exact cover of U if and only if there is a 2-Club Cluster Edge-Edition Set for G of size at most k.

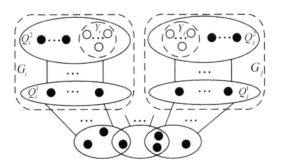

Fig. 2. The gadget for reduction from 3X3C to 2-CLUB CLUSTER EDITING

$\Rightarrow$: Suppose that $C' \subseteq C$ is an exact cover of U. Define edge set
$$A \equiv \{\{u, v\} | u \in S_i, S_i \notin C', v \in V_{Q_i^1}\}, \text{ and edge set}$$
$$B \equiv \{\{u, v\} | u \in S_i, v \in S_j, S_i \in C', S_j \in C', i \ne j\}.$$

Clearly, $|A| + |B| = k$. It is easy to verify that $A \cup B$ is a 2-Club Cluster Edge-Edition Set for G.

$\Leftarrow$: Suppose that G has one or more 2-Club Cluster Edge-Edition Set of size at most k and let M be one of minimum size. Clearly, $|M| \leq k$. We shall prove that one can derive from M an exact cover of U. Let $G' = (V, E \triangle F)$.

Claim 1. For every $1 \leq i \leq r$, $G'[V_{Q_i^2}]$ is connected.

Proof of Claim. For any fixed $1 \leq i \leq r$, since $G[V_{Q_i^2}]$ is a clique of size $k + 2$, if $G'[V_{Q_i^2}]$ is disconnected, then at least $k + 1 > |M|$ edges should be deleted from $G[V_{Q_i^2}]$. This is not allowed. Thus the correctness of the claim.   ◇

Claim 2. In G', for every $1 \leq i \leq r$, every vertex in $V_{Q_i^1}$ is adjacent to at least one vertex in $V_{Q_i^2}$.

Proof of Claim. Since any vertex v in $G[V_{Q_i^1}]$ is adjacent to $k + 2$ vertices in $G[V_{Q_i^2}]$, to make v disconnected to every vertices in $G[V_{Q_i^2}]$, $k + 2 > |M|$ edges incident with v should be deleted from G. This is not allowed. Thus the correctness of the claim.   ◇

Claim 3. For any pair $1 \leq i, j \leq r$, with $i \neq j$, there exist no edge between $G'[V_{Q_i^2}]$ and $G'[V_{Q_j^2}]$.

Proof of Claim. (By contradiction) Without loss of generality, suppose there are x edges between $G'[V_{Q_i^2}]$ and $G'[V_{Q_j^2}]$. (These x edges should be included in M.) It is easy to see that at most x vertices in $G'[V_{Q_i^2}]$ are adjacent to at most x vertices in $G'[V_{Q_j^2}]$. Since there are $k + 2$ vertices in both $V_{Q_i^2}$ and $V_{Q_j^2}$, then in G', at least $k + 2 - x$ vertices in $V_{Q_i^2}$ (this set of vertices is denoted by Y_i) are not adjacent to any vertex in $V_{Q_j^2}$, and symmetrically, at least $k + 2 - x$ vertices in $V_{Q_j^2}$ (this set of vertices is denoted by Y_j) are not adjacent to any vertex in $V_{Q_i^2}$. According to Claim 1, in G', vertices in Y_i (Y_j) are in the same connected component. Since G' is of diameter at most 2, for any pair of vertices, $y_i \in Y_i$ and $y_j \in Y_j$, they should share a common neighbor. This means that, besides the x edges between $G'[V_{Q_i^2}]$ and $G'[V_{Q_j^2}]$, at least $k + 2 - x$ edges should be added to G to achieve this. Clearly, all these edges should be included in M. Consequently, $|M|$ must be greater than $k + 2$. Thus a contradiction.   ◇

Claim 4. For any pair $1 \leq i, j \leq r$, with $i \neq j$, any vertex in $V_{Q_i^2}$ and any vertex in $V_{Q_j^2}$ can not be in the same connected component in G'.

Proof of Claim. (By contradiction) Suppose a vertex $v_i \in V_{Q_i^2}$ and a vertex $v_j \in V_{Q_j^2}$ are in the same connected component in G'. Since, according to Claim 1, $G'[V_{Q_j^2}]$ is connected, v_i and all vertices in $V_{Q_j^2}$ are in the same connected component in G'. Note that, according to Claim 3, v_i is not adjacent to any vertex in $V_{Q_j^2}$ in G'. However, since G' is of diameter at most 2, v_i and every vertex in $V_{Q_j^2}$ should share a common neighbor (outside $V_{Q_i^2} \cup V_{Q_j^2}$) in G'. To achieve this, since $|V_{Q_j^2}| = k+2$, at least $k+2$ edges should be added to G. Clearly, these edges should be included in M. Consequently, $|M|$ would be greater than $k + 2$. Thus a contradiction.   ◇

For any pair $1 \leq i, j \leq r$, with $i \neq j$, since (according to Claims 1 and 2) both $G'[V_{G_i}]$ and $G'[V_{G_i}]$ are connected subgraphs, but (according to Claim 4) any vertex in $V_{Q_i^2}$ can not be connected with any vertex in $V_{Q_j^2}$, we see that V_{G_i} and V_{G_j} can not be included in the same connected component in G'. Thus, for every vertex $u \in U$, it is adjacent in G' to vertices of at most one Q^1 (say Q_i^1 if any), and all edges (if any) that link u and vertices outside Q_i^1 must be deleted from G. Thus, at least $3r(6n+1) - 3n(6n+1) = 3(r-n)(6n+1)$ edges in E_{UQ^1} must be in M. However, since M is a minimum size 2-Club Cluster Edge-Edition Set for G, $t - 3n \leq 6n$, and $|M| \leq 3(r-n)(6n+1) + (t-3n)$, we see, for every vertex $u \in U$, u is adjacent in G' to *all* vertices of *exactly one* set in $\{Q_i^1 | u \in S_i, S_i \in C\}$. In other words, for each $1 \leq i \leq r$, all vertices of Q_i^1 are adjacent in G' to none or all three vertices in S_i, outside Q_i^1. In fact, there are exactly n vertex sets $S_1', S_2', \cdots, S_n'$, each of whose elements are adjacent in G' to vertices of the corresponding Q^1. Furthermore, for every set S' of such n sets, none of the edges in $E_{G[S']} \subset E_C$ is in M. However, all other $t - 3n$ edges in E_C must be in M. Thus, from M, a 2-Club Cluster Edge-Edition Set of size at most k, we successfully derive a exact cover $C' = \{S_1', S_2', \cdots, S_n'\}$ for U.   □

Note that in the reduction of Theorem 2, none edge adding operation is allowed. Hence the following corollary.

Corollary 1. 2-CLUB CLUSTER EDGE DELETION *is NP-complete.*

4   Parameterized Algorithms

The following Lemma is important for the algorithms presented in this section.

Lemma 3. *A graph G is an 2-club cluster graph if and only if there exists no restricted P_4 as induced subgraph in G.*

Proof. The proof is straightforward and is omitted here.   □

According to Lemma 3, 2-CLUB CLUSTER VERTEX DELETION can be solved by repeatedly finding (in polynomial time) a restricted P_4 and then branching into all possibilities of deleting one of its vertices. This yields a trivial search tree algorithm with running time $O^*(4^k)$. In analogy, there exists a trivial search tree algorithm to solve 2-CLUB CLUSTER VERTEX DELETION in $O^*(3^k)$ time. In this section, we present for each of the two problems an improved search tree algorithm, by adopting more complicated branching rules. Each of our branching rules is presented by a set of sets of deleted objects (vertices or edges). For example, given a graph G, a branching rule $\{\{o_1\}, \{o_2, o_3\}, \{o_4\}\}$ means there are three branches to destroy one or more certain restricted P_4 in G, i.e., deleting o_1, deleting o_2 and o_3, and deleting o_4.

4.1   An Improved Parameterized Algorithm for 2-Club Cluster Vertex Deletion

In this subsection, we present for 2-CLUB CLUSTER VERTEX DELETION an improved search tree algorithm with running time $O^*(3.31^k)$. The basic idea is as follows.

Let $\langle G = (V, E), k \rangle$ be a 2-CLUB CLUSTER VERTEX DELETION instance. We start with identifying a restricted P_4, P_{stuv}, and distinguish a number of cases according to adjacency relations among vertices in $N_G[\{s, t, u, v\}]$.

In detail, we first distinguish the following two cases:

(C1) $deg_G(s) = deg_G(v) = 1$.

(C2) At least one of $deg_G(s)$ and $deg_G(v)$ is greater than 1. W.l.o.g, we assume there exists a vertex $w \in V$ such that $\{w, v\} \in E$ and $w \neq u$.

Regarding case (C1), we distinguish two subcases:

(C1.1) $N_G(t) \setminus \{s, u\} = N_G(u) \setminus \{t, v\}$. To destroy P_{stuv}, we make a branching $\{\{t\}, \{u\}, \{v\}\}$. The branching vector is $(1, 1, 1)$ and the branching number is $\alpha < 3$.

(C1.2) $N_G(t) \setminus \{s, u\} \neq N_G(u) \setminus \{t, v\}$. W.l.o.g, we assume there exists a vertex $w \in V$, which is adjacent to t but not u. Here, we need to destroy two restricted P_4, i.e., P_{stuv} and P_{wtuv}. For this purpose, we make a branching $\{\{t\}, \{u\}, \{v\}, \{s, w\}\}$. The branching vector is $(1, 1, 1, 2)$ and the branching number $\alpha = 3.31$.

Regarding case (C2), we distinguish two subcases:

(C2.1) s, w have a common neighbor x with $x \neq t$. To destroy P_{stuv} and P_{sxwv}, we make a branching $\{\{s\}, \{v\}, \{t, x\}, \{t, w\}, \{u, x\}, \{u, w\}\}$. The branching vector is $(1, 1, 2, 2, 2, 2)$ and the branching number is $\alpha < 3.24$.

(C2.2) s, w do not have a common neighbor.

Regarding case (C2.2), we distinguish two subcases:

(C2.2.1) t is adjacent to w. To destroy P_{stuv} and P_{stwv}, we make a branching $\{\{s\}, \{t\}, \{v\}, \{u, w\}\}$. The branching vector is $(1, 1, 1, 2)$ and the branching number is $\alpha < 3.31$.

(C2.2.2) t is not adjacent to w.

Regarding subsubcase (C2.2.2), we distinguish two subcases:

(C2.2.2.1) u is adjacent to w. To destroy P_{stuv} and P_{stuw}, we make a branching $\{\{s\}, \{t\}, \{u\}, \{v, w\}\}$. The branching vector is $(1, 1, 1, 2)$ and the branching number is $\alpha < 3.31$.

(C2.2.2.2) u is not adjacent to w.

Regarding case (C2.2.2.2), we distinguish two subcases:

(C2.2.2.2.1) t, w have a common neighbor x with $x \neq u$. To destroy P_{stuv} and P_{stxw}, we make a branching $\{\{s\}, \{t\}, \{u, x\}, \{u, w\}, \{v, x\}, \{v, w\}\}$. The branching vector is $(1, 1, 2, 2, 2, 2)$ and the branching number is $\alpha < 3.24$.

(C2.2.2.2.2) t, w do not have a common neighbor. To destroy P_{stuv} and P_{tuvw}, we make a branching $\{\{t\}, \{u\}, \{v\}, \{s, w\}\}$. The branching vector is $(1, 1, 1, 2)$ and the branching number is $\alpha < 3.31$.

In summary, in the worst case, the branching vector is (1,1,1,2), yielding a branching number 3.31. This results in the following theorem:

Theorem 4. 2-CLUB CLUSTER VERTEX DELETION *can be solved in* $O^*(3.31^k)$ *time.*

4.2 An Improved Parameterized Algorithm for 2-Club Cluster Edge Deletion

In this subsection, we present an improved search tree algorithm with running time $O^*(2.74^k)$. The basic idea here is somewhat similar to that used in subsection 4.1.

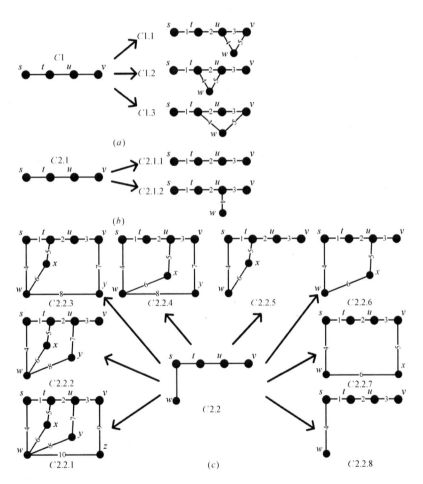

Fig. 3. Illustration of relationships of some cases considered in subsection 4.2. The edges are identified with numbers.

Let $\langle G = (V, E), k \rangle$ be a 2-CLUB CLUSTER EDGE DELETION instance. We start with identifying a restricted P_4, denoted by P_{stuv}, and distinguish the following two cases:

(C1) At least two vertices of P_{stuv} have a common neighbor in $V \setminus \{s, t, u, v\}$.
(C2) Any two vertices of P_{stuv} do not have a common neighbor in $V \setminus \{s, t, u, v\}$.

Regarding case (C1), we distinguish three subcases (See Fig. 3(a) for an illustration):

(C1.1) In P_{stuv}, an end vertex and its neighboring internal vertex have a common neighbor. W.l.o.g, assume u and v have a common neighbor w. To destroy P_{stuv}, we make a branching $\{\{1\}, \{2\}, \{3, 4\}, \{3, 5\}\}$.
(C1.2) In P_{stuv}, two internal vertices t and u have a common neighbor w. To destroy P_{stuv}, we make a branching $\{\{1\}, \{2, 4\}, \{2, 5\}, \{3, 4\}, \{3, 5\}\}$. the branching vector is $(1, 2, 2, 2, 2)$ and the branching number is $\alpha < 2.57$.

In both the above two cases, the branching vector is $(1, 1, 2, 2)$ and the branching number is $\alpha < 2.74$.

(C1.3) In P_{stuv}, an end vertex and an internal vertex, which are not adjacent to each other, have a common neighbor. W.l.o.g, assume u and v have a common neighbor w. To destroy P_{stuv}, we make a branching $\{\{1\}, \{2\}, \{3, 4\}, \{3, 5\}\}$.

Regarding case (C2), we distinguish two subcases:

(C2.1) Both s and v are leaves.
(C2.2) At least one of s and v are not leaves. W.l.o.g, assume there is a vertex $w \in V \setminus \{t, u, v\}$ which is adjacent to s.

Regarding case (C2.1), we distinguish two subcases (See Fig. 3(b) for an illustration):

(C2.1.1) Neither t nor u has a neighbor in $V \setminus \{s, t, u, v\}$. It is a trivial case, as an arbitrary edge in $\{1, 2, 3\}$ can be deleted to destroy P_{stuv}.
(C2.1.2) t and/or u have a neighbor in $V \setminus \{s, t, u, v\}$. W.l.o.g, assume $w \in V \setminus \{s, t, u, v\}$ is adjacent to u. To destroy P_{stuv} and P_{stuw}, we make a branching $\{\{1\}, \{2\}, \{3, 4\}\}$. The branching vector is $(1, 1, 2)$ and the branching number is $\alpha < 2.42$.

Regarding case (C2.2), we distinguish eight subcases according to wether or not w has a common neighbor (denoted by x, y and z, respectively, if any) with t, u and v, respectively. See Fig. 3(c) for the illustration of the eight subcases C2.2.1-C2.2.8. The corresponding branchings, branching vectors and branching numbers are listed in Tab. 1.

In summary, in the worst cases, the branching number is 2.74. In analogy to Theorem 4, we obtain the following theorem:

Theorem 5. 2-CLUB CLUSTER EDGE DELETION *can be solved in* $O^*(2.74^k)$ *time.*

Table 1. Branching rules for cases C2.2.1-C2.2.8, and their corresponding branching vectors and branching numbers

Cases	Branchings	Branching vectors	Branching numbers
C2.2.1	{{1,4},{1,6,8,10},{1,5,8,10},{1,6,7,10}, {1,5,7,10},{1,6,8,9},{1,5,8,9},{1,6,7,9}, {1,5,7,9},{2,4,5},{2,4,6},{2,7,9}, {2,7,10},{2,8,9},{2,8,10},{3,9},{3,10}, {3,4,5,7},{3,4,6,7},{3,4,6,8},{3,4,5,8}}	(2,4,4,4,4,4,4,4, 4,3,3,3,3,3,3, 2,2,4,4,4,4)	$\alpha < 2,65$
C2.2.2	{{3},{2,7},{2,8},{1,5,7}, {1,5,8},{1,6,7},{1,6,8}}	(1,2,2,4,4,4,4)	$\alpha < 2,47$
C2.2.3	{{1,5,7},{1,5,8},{1,6,7},{1,6,8}, {2,7},{2,8},{3,7},{3,8}}	(3,3,3,3,2,2,2,2)	$\alpha < 2,39$
C2.2.4	{{1,5,7},{1,5,8},{1,6,7},{1,6,8},{2,4}, {2,5,7},{2,5,8},{2,6,7},{2,6,8}, {3,7},{3,8},{3,4,5},{3,4,6}}	(3,3,3,3,2,3,3, 3,3,2,2,3,3)	$\alpha < 2,62$
C2.2.5	{{1,5},{1,6},{2},{3,4,5},{3,4,6}}	(2,2,1,3,3)	$\alpha < 2,27$
C2.2.6	{{1,5},{1,6},{2,5},{2,6},{3}}	(2,2,2,2,1)	$\alpha < 2,57$
C2.2.7	{{1,5},{1,6},{2,5},{2,6},{3,4,5},{3,4,6}}	(2,2,2,2,3,3)	$\alpha < 2,22$
C2.2.8	{{1},{2},{3,4}}	(1,1,2)	$\alpha < 2,42$

5 Conclusions

In this paper, we proved the NP-Completenesses of 2-CLUB CLUSTER VERTEX DELETION and 2-CLUB CLUSTER EDGE DELETION. We also presented fixed parameter algorithms for these two problems. There are still much work to be done. One is to find polynomial kernels for these two problems, or prove that no polynomial kernels exist (see, e.g., [4]). Another future work is to further improve the running times of the algorithms presented in this paper.

References

1. Alba, R.D.: A graph-theoretic definition of a sociometric clique. Journal of Mathematical Sociology 3, 113–126 (1973)
2. Bansal, N., Blum, A., Chawla, S.: Correlation clustering. Machine Learning 56(1-3), 89–113 (2004)
3. Balasundaram, B., Butenko, S., Trukhanov, S.: Novel approaches for analyzing biological networks. Journal of Combinatorial Optimization 10(1), 23–39 (2005)
4. Bodlaender, H.L., Downey, R.G., Fellows, M.R., Hermelin, D.: Onproblems without polynomial kernels. Journal of Computer and System Sciences 75(8), 423–434 (2009)
5. Downey, R.G., Fellows, M.R.: Parameterized complexity. Springer (1999)
6. Garey, M., Johnson, D.: Computers and intractability: A guide to the theory of NP-completeness. W. H. Freeman and Company (1979)

7. Guo, J., Kanj, I.A., Komusiewicz, C., Uhlmann, J.: Editing graphs into disjoint unions of dense clusters. Algorithmica 61, 949–970 (2011)
8. Guo, J., Komusiewicz, C., Niedermeier, R., Uhlmann, J.: A more relaxed model for graph-based data clustering: s-plex cluster editing. SIAM Journal on Discrete Mathematics 24(4), 1662–1683 (2010)
9. Hüffner, F., Komusiewicz, C., Moser, H., Niedermeier, R.: Fixed-parameter algorithms for cluster vertex deletion. Theory of Computing Systems 47(1), 196–217 (2010)
10. Lee, V.E., Ruan, N., Jin, R., Aggarwal, C.: A survey of algorithms for dense subgraph discovery. In: Aggarwal, C.C., Wang, H. (eds.) Managing and Mining Graph Data, pp. 303–336 (2010)
11. Lewis, J.M., Yannakakis, M.: The node-deletion problem for hereditary properties is NP-complete. Journal of Computer Systems and Science 20(2), 219–230 (1980)
12. Mahdavi, F., Balasundaram, B.: On inclusionwise maximal and max-imum cardinality k-clubs in graphs (2010),
   http://iem.okstate.edu/baski/files/DISCO-k-clubs-2010-02-11.pdf
13. Niedermeier, R.: Invitation to fixed-parameter algorithms. Oxford University Press (2006)
14. Shamir, R., Sharan, R., Tsur, D.: Cluster graph modification problems. Discrete Applied Mathematics 144(1-2), 173–182 (2004)
15. Xu, R., Wunsch II, D.: Survey of clustering algorithms. IEEE Transactions on Neural Networks 16(3), 645–678 (2005)

Solving Generalized Optimization Problems Subject to SMT Constraints

Feifei Ma[1], Jun Yan[1], and Jian Zhang[1,2]

[1] Institute of Software, Chinese Academy of Sciences
[2] State Key Laboratory of Computer Science
maff@ios.ac.cn, {junyan,jian_zhang}@acm.org

Abstract. In a classical constrained optimization problem, the logical relationship among the constraints is normally the logical conjunction. However, in many real applications, the relationship among the constraints might be more complex. This paper investigates a generalized class of optimization problems whose constraints are connected by various kinds of logical operators in addition to conjunction. Such optimization problems have been rarely studied in literature in contrast to the classical ones. A framework which integrates classical optimization procedures into the DPLL(T) architecture for solving Satisfiability Modulo Theories (SMT) problems is proposed. Two novel techniques for improving the solving efficiency w.r.t. linear arithmetic theory are also presented. Experiments show that the proposed techniques are quite effective.

1 Introduction

Many real-world and theoretical problems can be classified as optimization problems, i.e., minimizing or maximizing an objective function possibly subject to a set of constraints. A constraint is typically a mathematical equation or inequality. Optimization problems have long been the interest of mathematicians as well as engineers. Many efficient algorithms have been developed for various kinds of optimization problems over the past decades, for example, the simplex method for linear programming.

Although not explicitly stipulated, in a classical constrained optimization problem, the logical relationship among the constraints is the logical **AND** ($\wedge$), which means all of the equations or inequalities must be simultaneously satisfied. For such problems, solid theoretical foundations have been laid. However, in some applications, the relationship among the constraints might be more complex. The constraints may be connected by several kinds of logical operators so as to describe, for instance, a compound environment in a robot path planning problem or certain restrictions in task scheduling. In software testing and analysis, one strategy is to find out the conditions under which the software system uses resources heavily. This can also be formulated as an optimization problem with complex constraints. For some specific applications, tailored methods were sporadically proposed. But in the literature, such optimization problems have rarely been studied in the general form.

In this paper, we assume that the constraints are expressed as Satisfiability Modulo Theories (SMT) formulae. SMT, as an extension to the satisfiability (SAT) problem, has received more and more attention in recent years [1,4,12,10]. Instead of Boolean

J. Snoeyink, P. Lu, K. Su, and L. Wang (Eds.): FAW-AAIM 2012, LNCS 7285, pp. 247–258, 2012.
© Springer-Verlag Berlin Heidelberg 2012

formulae, SMT checks the satisfiability of logical formulae with respect to combinations of background theories. Examples of theories include real numbers, integers, bit vectors and even non-linear constraints. An SMT solver typically integrates a powerful SAT solver as the Boolean search engine and several theory solvers for deciding the consistency of theory fragments.

In the sequel, we show how to solve an optimization problem which is constrained by SMT formulae via modifying the SMT solving procedure. Two pruning techniques are proposed so as to accelerate the searching process. Although the basic ideas of our approach are applicable to multiple theories, we are especially interested in problems with a linear objective function and Boolean combination of linear arithmetic constraints, or, SMT(LAC) constraints.

2 Background

This section describes some basic concepts and notations. We also give a brief overview of some existing techniques that will be used later.

The main object of study in this paper can be regarded as an optimization task whose constraints involve variables of various types (including integers, reals and Booleans). There can be logical operators (like AND, OR) and arithmetic operators.

We use b_i ($i > 0$) to denote Boolean variables, x_j ($j > 0$), y, ..., to denote numeric variables. A *literal* is a Boolean variable or its negation. A *clause* is a disjunction of literals. A Boolean formula in conjunctive normal form is a conjunction of clauses. A linear arithmetic constraint (LAC) is a comparison between two linear arithmetic expressions. A simple example is $x_1 + 2x_2 < 3$.

In general, a constraint ϕ can be represented as a Boolean formula $PS_\phi(b_1, \ldots, b_n)$ together with definitions in the form: $b_i \equiv expr_{i_1} \ op \ expr_{i_2}$. That means, the Boolean variable b_i stands for the LAC: $expr_{i_1} \ op \ expr_{i_2}$. Here $expr_{i_1}$ and $expr_{i_2}$ are numeric expressions, while op is a relational operator like '$<$', '$=$', etc. The LAC is called the *theory predicate* corresponding to b_i. The Boolean formula PS_ϕ is the *propositional skeleton* of the constraint ϕ.

For a Boolean formula, a *model* is an assignment[1] of truth values to all the Boolean variables such that the formula is evaluated to TRUE. Usually the Davis-Putnam (DPLL) procedure can be used to decide whether a Boolean formula has a model.

2.1 DPLL(T) Framework

Most of the state-of-the-art SMT solvers are built upon the DPLL(T) architecture [6]. DPLL(T) is a generalization of DPLL for solving the fragment of a decidable theory T. It combines a DPLL-based SAT solver and theory-specific solving [10] through a well-defined interface.

To solve an SMT formula, a Boolean abstraction of the formula is derived by encoding each theory predicate with a new propositional variable. The SAT solver then explores the Boolean search space of the formula, and passes the conjunctions of theory

---

[1] In this paper we represent an assignment as a set of literals.

predicates on to the theory solver for feasibility checking. For this integration to work well, however, the theory solver must be able to participate in propagation and conflict analysis, i.e., it must be able to infer new facts from already established facts, as well as to supply succinct explanations of infeasibility when theory conflicts arise.

2.2   Optimization Problem with Complex Constraints

In classical constrained optimization problems, the constraints are conjunctively connected. For instance, in linear programming, all linear inequalities must hold simultaneously. However, in many real applications, the relationships among constraints are very complex, and such problems cannot be handled with traditional optimization methods.

Let us see a practical example. In real-time operating systems, an important research issue is rate monotonic scheduling. Given a set of n periodic tasks $\{\tau_1 \ldots \tau_i \ldots \tau_n\}$, where each task τ_i is associated with a computation time C_i and a release period T_i, the goal is to assign the computation time to each task such that all the tasks can be scheduled and the performance of the system is optimized as well. It is proved that the optimization problem can be formulated as follows:

$$\text{maximize} \qquad \Sigma_{i=1}^n \frac{C_i}{T_i}$$

$$\text{subject to} \qquad \forall i \bigvee_{t \in S_i} (\Sigma_{j=1}^i \lceil \frac{t}{T_j} \rceil C_j - t \leq 0)$$

$$C_i^{min} \leq C_i \leq C_i^{max}$$

where $S_i = \{rT_j | j \leq i, r = 1, \ldots, \lfloor \frac{T_i}{T_j} \rfloor\}$. The objective function represents the CPU utilization. The constraints are integer linear inequalities, connected by disjunctions and conjunctions, forming an SMT(LAC) formula. In this paper, we study the optimization of a given function subject to SMT constraints[2].

3   Solving Optimization Problems with DPLL(T)

3.1   A Straightforward Method

We know that an SMT instance ϕ is satisfiable if there is an assignment α to the Boolean variables in PS_ϕ such that:

1. α propositionally satisfies ϕ, or formally $\alpha \models PS_\phi$;
2. The conjunction of theory predicates under the assignment α, which is denoted by $\hat{T}h(\alpha)$, is consistent w.r.t. the addressed theory.

We call an assignment satisfying the above conditions a *feasible* assignment. For example, suppose PS_ϕ is $b_1 \vee b_2 \vee b_3$, where $b_1 \equiv (x > 10)$, $b_2 \equiv (y > 5)$, $b_3 \equiv (x + 2y < 18)$. Then $\{b_1, b_2, \neg b_3\}$ is a feasible assignment; but $\{b_1, b_2, b_3\}$ is not.

---

[2] Without loss of generality, in the sequel we assume all objective functions are to be minimized.

Suppose an SMT optimization problem is composed of an objective function $f(x)$ and an SMT formula ϕ. The solution space of ϕ is the union of feasible regions of all feasible assignments of ϕ. Denote the set of all feasible assignments of ϕ by $Mod(\phi)$. The minimum value of $f(x)$ over ϕ, denoted by $Min(f(x), \phi)$, is defined as follows.

$$Min(f(x), \phi) = min\{Min(f(x), \alpha) | \alpha \in Mod(\phi)\}$$

Since α is a conjunction of theory predicates, $Min(f(x), \alpha)$ is a normal procedure for computing the optimal solution of a classical optimization problem.

The DPLL(T) architecture can be directly adapted to compute the optimal solution of a given formula by replacing the theory solver with a theory optimizer. We ask the SMT solver to enumerate all feasible assignments to formula ϕ, compute the optimal solution of each assignment and pick the smallest one.

However, the problem with the straightforward approach is that the classical optimization procedure is called as many times as the number of feasible assignments are. It is vital to the efficiency of the algorithm to reduce the number of calls to classical optimization procedures. In the rest of the section, we shall present two such techniques. The ideas behind the two techniques have one thing in common: they are both concerned with how to compute the optimal solution of an area as large as possible through a single call to the classical optimization procedure.

3.2   Optimization in Bunches

In [11], when studying the volume computation problem of SMT formulae, Ma et al. proposed a strategy called "volume computation in bunches". We find that the same idea is applicable to the optimization problem.

Given an assignment α to the Boolean variables in formula ϕ, concerning the two conditions for feasible assignments, we can distinguish four cases:

(i) $\alpha \models PS_\phi$, and $\hat{T}h(\alpha)$ is consistent in the specific theory. (Here α is a feasible assignment.)
(ii) $\alpha \models PS_\phi$, while $\hat{T}h(\alpha)$ is inconsistent in the specific theory.
(iii) α falsifies ϕ propositionally, while $\hat{T}h(\alpha)$ is consistent in the specific theory.
(iv) α falsifies ϕ propositionally, and $\hat{T}h(\alpha)$ is inconsistent in the specific theory.

The situation that α is a feasible assignment is just one of the cases when both the conditions are `true`. The optimal solution of formula ϕ is the smallest one amongst those of all assignments in case (i). When $\hat{T}h(\alpha)$ is inconsistent, as in case (ii) or case (iv), its feasible region can be viewed as empty. So when searching through the combined feasible regions of feasible assignments, it will be safe to count in some theory-inconsistent assignments since they would not affect the result. The theory-inconsistent assignments, when properly selected, can be combined with the feasible assignments to form fewer assignments, reducing the number of classical optimizations.

Definition 1. *A set of full assignments S is called a* bunch *if there exists a partial assignment α_c such that for any full assignment α, $\alpha \in S \leftrightarrow \alpha_c \subseteq \alpha$. α_c is called the* cube *of S.*

In other words, the assignments in a bunch S share a partial assignment α_c, and for the Boolean variables which are not assigned by the cube α_c, these assignments cover all possibilities of value combinations. For the assignments in a bunch, optimization can be greatly simplified, as the following theorem reveals:

Proposition 1. *Given an SMT optimization problem which is to minimize $f(x)$ over ϕ, for a bunch S with the cube α_c, the following equation holds:*

$$min\{Min(f(x), \alpha) | \alpha \in S\} = Min(f(x), \alpha_c).$$

Example 1. Consider the optimization problem that minimizes $x - y$ subjected to ϕ, where $\phi = (((y + 3x < 3) \rightarrow (30 < y)) \vee (x \leq 60)) \wedge ((30 < y) \rightarrow \neg(x > 3) \wedge (x \leq 60))$.

We first introduce a Boolean variable for each linear inequality of ϕ and obtain its propositional skeleton $PS_\phi = ((b_1 \rightarrow b_2) \vee b_4) \wedge (b_2 \rightarrow \neg b_3 \wedge b_4)$, where $b_1 \equiv (y + 3x < 3)$, $b_2 \equiv (30 < y)$, $b_3 \equiv (x > 3)$ and $b_4 \equiv (x \leq 60)$.

There are seven feasible assignments, three of which are: $\alpha_1 = \{\neg b_1, \neg b_2, b_3, \neg b_4\}$, $\alpha_2 = \{\neg b_1, \neg b_2, \neg b_3, b_4\}$, and $\alpha_3 = \{\neg b_1, \neg b_2, b_3, b_4\}$. Their respective optimal solutions are[3]:

$$Min(x - y, \alpha_1) = 30 + \delta, \text{ where } x = 60 + \delta, y = 30$$
$$Min(x - y, \alpha_2) = -39, \text{ where } x = -9, y = 30$$
$$Min(x - y, \alpha_3) = -27 - \delta, \text{ where } x = 3 + \delta, y = 30$$

Now let's consider another assignment: $\alpha_4 = \{\neg b_1, \neg b_2, \neg b_3, \neg b_4\}$. It is easy to check that α_4 satisfies PS_ϕ, but $\hat{Th}(\alpha_4)$ is inconsistent in linear arithmetic. Also, these four assignments form a bunch whose cube is $\{\neg b_1, \neg b_2\}$. Noticing this, we have

$$min\{Min(x - y, \alpha_1), Min(x - y, \alpha_2), Min(x - y, \alpha_3)\}$$
$$= min\{Min(x - y, \alpha_1), Min(x - y, \alpha_2), Min(x - y, \alpha_3), Min(x - y, \alpha_4)\}$$
$$= Min(x - y, \{\neg b_1, \neg b_2\})$$
$$= -39$$

where $x = -9, y = 30$. As a result, we need to call a linear programming routine only once rather than four times.

It would be ideal to incorporate the assignments in both case (ii) and case (iv) to form larger bunches. However, we currently neglect case (iv) because a typical DPLL(T)-style solver doesn't provide decision procedures for assignments that falsifies the propositional skeleton. For convenience, we just handle the assignments in case (ii). A key point is that when the SMT solver finds a feasible assignment, we try to obtain a smaller one which still propositionally satisfies the formula. It is formally defined as follows.

Definition 2. *Suppose α is a feasible assignment for formula ϕ. An assignment α_{mc} is called a* minimum cube *of α if i) $\alpha_{mc} \subseteq \alpha$ and $\alpha_{mc} \models PS_\phi$ and ii) $\forall \alpha'(\alpha' \models PS_\phi \rightarrow \alpha' \not\subseteq \alpha_{mc})$.*

---

[3] δ represents an arbitrarily small value since there exist strict inequlities.

In fact, the minimum cube α_{mc} of an assignment α is the cube of a bunch $\mathcal{S}$ such that for any bunch $\mathcal{S}'$, $\alpha \in \mathcal{S}' \rightarrow \mathcal{S} \not\subset \mathcal{S}'$. Any assignment in $\mathcal{S}$ also satisfies PS_ϕ because only part of it has evaluated PS_ϕ to be true. As we have explained before, it is pretty safe to count in such an assignment while solving the SMT optimization problem, regardless of its consistency in the specific theory.

Note that an assignment might have several minimum cubes. Currently we use a simple method to find only one minimum cube, as described in [11].

3.3   Feasible Region Expansion

When solving an SMT constrained optimization problem, the standard optimization procedure is called each time the SAT engine reaches a feasible assignment. If it is possible to obtain more information other than the optimal solution in the optimization subroutine, the whole searching process might benefit a lot.

For a standard optimization problem, there might be certain constraints which do not influence the optimal solution, in other words, the optimal solution is not bounded or restricted by these constraints. They are called **redundant constraints**, defined formally as follows.

Definition 3. *Suppose an optimization problem is to minimize $f(x)$ over a set of constraints $\{c_1, c_2, \ldots, c_m\}$. c_k $(1 \leq k \leq m)$ is a redundant constraint if the following equation holds:*

$$Min(f(x), \bigwedge_{1 \leq i \leq m} c_i) = Min(f(x), \bigwedge_{1 \leq i \leq m, i \neq k} c_i) \tag{1}$$

By definition, a standard optimization problem can be reduced to a simplified version by removing redundant constraints, while preserving the optimal solution. The feasible region of the problem expands as the constraints are removed. As a result, when we get an optimal solution for a feasible region, possibly we can conclude that it is also the optimal solution for a larger one. Inspired by this observation, we propose the second pruning strategy, namely "feasible region expansion". It can be used simultaneously with the "optimization in bunches" strategy. The basic idea is as follows: each time a feasible assignment α is obtained, find the optimal solution of its minimum cube α_{mc}, then remove all redundant constraints from α_{mc}, and get a smaller partial assignment α'. α' has the same optimal solution as α_{mc}, while covers a larger region than α_{mc} does. So the negation of α' instead of α_{mc} is added to PS_ϕ. In this way, through one single call to the standard optimization subroutine, a piece of broader search space can be examined and excluded.

A key problem naturally arises: How to identify redundant constraints without extra calls to the optimization procedure? The problem is very general while the answer is closely related to the specific type of the constraints. We find that linear programming has a very fine property which makes the identification of some redundant constraints quite a simple task.

Proposition 2. *Given a linear programming problem with the objective $c^\top x$ and subject to $Ax \otimes b$, if it has an optimal solution x_{opt}, then the linear constraint $A_i x \otimes_i b_i$ is redundant if the point x_{opt} is not in the hyperplane $A_i x = b_i$, or formally $A_i x_{opt} \neq b_i$.*

Such a constraint $A_i x \otimes_i b_i$ is called a **non-binding constraint**. It does not affect the optimal solution.

Example 2. Given an SMT(LAC) optimization problem which is defined as:

$$\begin{aligned}
\text{minimize} \quad & x + y \\
\text{subject to} \quad & \phi : x \geq 0 \wedge y \geq 0 \\
& \wedge (x \geq 1 \vee \neg(y \geq 1)) \quad (2) \\
& \wedge (\neg(x \geq 1) \vee y \geq 1)
\end{aligned}$$

We have $PS_\phi = b_1 \wedge b_2 \wedge (b_3 \vee \neg b_4) \wedge (\neg b_3 \vee b_4)$, where

$$\begin{cases}
b_1 \equiv (x \geq 0); \\
b_2 \equiv (y \geq 0); \\
b_3 \equiv (x \geq 1); \\
b_4 \equiv (y \geq 1);
\end{cases}$$

The optimization problem is illustrated in the right figure. The solution space of ϕ is the combined area of feasible regions R1 and R2. Suppose we get a feasible assignment $\alpha = \{b_1, b_2, \neg b_3, \neg b_4\}$, whose feasible region is R1. Obviously the minimum cube of α is α itself. The optimal solution of $Min(x + y, \alpha)$ is $\{x = 0, y = 0\}$, with the objective function evaluating to 0. Since the point $\{x = 0, y = 0\}$ is located on the lines $x = 0$ and $y = 0$, while away from the lines $x = 1$ and $y = 1$, the linear constraints corresponding to $\neg b_3$ and $\neg b_4$ are redundant constraints, thus can be omitted in the linear programming problem. The solution $\{x = 0, y = 0\}$ is also the optimal solution of the partial assignment $\{b_1, b_2\}$, which covers the whole solution space of ϕ. As a result, we know $\{x = 0, y = 0\}$ is the optimal solution of the original problem, without calculating the other feasible assignment $\{b_1, b_2, b_3, b_4\}$ with feasible region R2.

3.4   The Algorithms

In this subsection, we present the pseudo-codes of our approach and the pruning techniques.

Figure 1 describes the algorithm to compute the minimum cube of a given assignment α. The algorithm flips the literals in α one by one. More specifically, if α with literal l_i removed still evaluates PS to true, then l_i can be removed from α. Finally α becomes the minimum cube of the original assignment. Note that in an assignment, some variables are decision variables, while others get assigned by BCP. These implied literals need not be checked when finding the minimum cube of an assignment. (It has been proved in [11].)

Figure 2 illustrates the "feasible region expansion" strategy. For a given assignment α and its optimal solution, FeaRegExpan removes the redundant constraints from α.

The DPLL(T) framework for SMT solving is adapted to SMT optimization. The detailed algorithm is presented in Figure 3. *OptSol* stands for the optimal solution

```
1: MiniCube( Assignment α, Boolean Formula PS )
2: Assignment α′;
3: for all Literal l_i ∈ α do
4:    if l_i is a decision variable or its negation then
5:       α′ = α − {l_i};
6:       if α′ ⊨ PS then
7:          α = α′;
8:       end if
9:    end if
10: end for
11: return α;
```

Fig. 1. Function: MiniCube

```
1: FeaRegExpan( Assignment α, solution )
2: for all literal l_i ∈ α do
3:    if solution is not in the hyperplane T̂h(l_i) then
4:       α = α − {l_i};
5:    end if
6: end for
7: return α;
```

Fig. 2. Function: FeaRegExpan

```
1: Boolean Formula PS = PS_φ;
2: OptSol = Null;
3: while TRUE do
4:    if BCP() == CONFLICT then
5:       backtrack-level = AnalyzeConflict();
6:       if backtrack-level < 0 then
7:          return OptSol;
8:       end if
9:       backtrack to backtrack-level;
10:   else
11:      α = current assignment;
12:      if α ⊨ PS then
13:         α = MiniCube(α, PS);
14:         CurSol = LinearProgramming(T̂h(α));
15:         if CurSol is unbounded then
16:            return UNBOUNDED;
17:         end if
18:         if CurSol is smaller than OptSol then
19:            OptSol = CurSol;
20:         end if
21:         α = FeaRegExpan(α, CurSol);
22:         Add ¬α to PS;
23:      else
24:         choose a Boolean variable and extend α;
25:      end if
26:   end if
27: end while
```

Fig. 3. DPLL(T) for Optimization

of the SMT optimization problem, and $CurSol$ is the optimal solution of the current assignment during the search.

At the beginning of the algorithm, the propositional skeleton of the SMT formula ϕ is extracted and denoted as PS. After Boolean constraint propagation (BCP()), if the current assignment α already satisfies the Boolean formula PS, the subroutine MiniCube is called to the minimum cube of α.

When the minimum cube is obtained, the algorithm calls a linear programming package to compute the optimal solution of the cube. If the optimal solution $CurSol$ exists, the algorithm replaces $OptSol$ with $CurSol$ if the latter is smaller, and call the subroutine FeaRegExpan to eliminate redundant constraints from α. Otherwise, α is inconsistent and PostCheck is called to reduce it. Then its negation $\neg\alpha$ is added to PS so that the feasible region associated with the reduced α would not be counted more than once. It is a blocking clause, ruling out all the assignments in the bunch related to the reduced α. The algorithm terminates when there is no model for PS and returns $OptSol$, or when an unbounded solution is discovered.

Although Figure 3 describes the optimization approach under the assumption that all linear constraints are defined over real numbers, it can also handle integer linear constraints with slight modification. Firstly, in line 14, the LP relaxation of the integer linear programming is solved. Then if $CurSol$ is smaller than $OptSol$, $T̂h(\alpha)$ is solved with integer linear programming, otherwise there is no need to compute the integer optimal solution, and "feasible region expansion" strategy is applied w.r.t. the real optimal solution $CurSol$.

4   Implementation and Experimental Results

The algorithm was implemented using the SAT solver MiniSat 2.0 [5], which serves as the search engine for the Boolean structure of the SMT(LAC) instance. The linear programming tool Cplex [8] is integrated for consistency checking and optimization w.r.t. a set of linear constraints. A laptop with Core 2 duo 2.10 GHz CPU running 32-bit linux was used for the experiments.

To study the effectiveness of the aforementioned techniques, we randomly generated a number of SMT(LAC) instances whose propositional skeletons are in CNF, and the length of clauses varies from 3 to 5. We compared both the running times and the numbers of calls to Cplex on these random instances in three settings: with no pruning strategy, with only "optimization in bunches", and with both the "bunch" and "feasible region expansion" employed. The results are listed in Table 1, where '-' represents a timeout of 30 minutes. Obviously each of the pruning techniques can reduce the running time and number of calls significantly, even by several orders of magnitude in some cases. The denotations #LC, #C, #V and #calls represent the number of linear constraints, clauses, numerical variables and calls to Cplex, respectively.

We also compared the performance of our program with Cplex on some random instances. The inputs for Cplex are mixed integer linear programming problems translated from SMT(LAC) instances by introducing "Big-M" constraints. Table 2 lists the running results (where OptSMT denotes our program), which indicate that our method outperforms MILP on instances with large number of boolean constraints.

Table 1. Comparison of Techniques

(#LC #C #V)	No Pruning		Bunch		Bunch&FRE	
	Time	#calls	Time	#calls	Time	#calls
(20 40 15)	-		124.34	15,228	0.05	52
(20 50 9)	26.66	17,595	0.92	1,175	0.15	262
(20 50 10)	-		227.73	19,453	0.06	79
(20 100 10)	89.42	35,463	8.66	5,236	0.32	383
(40 400 20)	-		219.61	23,192	18.03	4,908
(40 500 20)	-		1178.80	44,759	47.41	9,421
(100 800 60)	498.49	19,844	136.17	3,820	25.72	804
(100 700 80)	-		215.18	3,976	47.95	913
(100 1000 60)	209.57	5,160	24.75	693	5.53	139

Table 2. Comparison with Cplex

Parameters			Time	
#LC	#C	#V	OptSMT	Cplex
100	600	50	78.08	83.42
100	800	50	0.12	31.22
100	1000	50	0.05	14.52
200	1000	100	16.63	4.40
200	1000	120	0.19	4.91
200	1200	100	5.71	34.10
200	1400	100	242.35	-
200	1600	100	5.54	-
200	2000	100	0.96	-

The pruning techniques are effective for integer linear constraints as well. Here we just give an example instead of a thorough evaluation. Let us consider an instance of rate monotonic scheduling problem which has been introduced in the background section. It consists of 4 tasks, with characteristics defined as $\{T_1 = 100, T_2 = 150, T_3 = 210, T_4 = 400\}$, and $\{20 \leq C_1 \leq 60, 20 \leq C_2 \leq 75, 30 \leq C_3 \leq 100, 30 \leq C_4 \leq 150\}$. Our program finds an optimal solution $\{C_1 = 33, C_2 = 20, C_3 = 30, C_4 = 148\}$, with the optimal value 0.9762. Cplex is called 59 times, and the running time is 0.028s. If the "feasible region expansion" technique is disabled, there are 75 calls to Cplex and the running time is 0.040s. However, if both techniques are disabled, the number of calls grows to 2835, and it takes 4.123s to find the same solution.

We also did experiments on real optimization problems in software testing area. For instances with dozens of integer variables, our program can find the optimal solution within 0.1s, which is quite useful to software testing practitioners.

5   Related Works and Discussion

The problem studied in this paper is a generalization of SMT, and also a generalization of the traditional optimization (linear programming) problem.

As we mentioned at the beginning, there are quite some works on solving the SMT problem. But few of them care about optimization, except for two recent works in [3] and [13] which discuss a variant of weighted Max-SAT problem, namely weighted Max-SMT. Each clause of the SMT formula is associated with some weight or cost. The task is to find a feasible assignment such that the total weight of satisfied clauses are maximized, or a given cost function is minimized. It is possible to translate weighted Max-SMT into our optimization problem in general. However, weighted Max-SMT has its own features, and deserves independent study.

Many works on constraint solving and constraint programming (CP) can be extended to deal with optimization. In particular, Hooker and his collaborators have been advocating the tight integration of CP and classical optimization techniques (like mixed integer linear programming) [7]. One promising approach is mixed logic linear programming (MLLP) which extends MILP by introducing logic-based modeling and solutions [9]. However, it seems that the constraints they considered have simpler logical structures than those studied in this paper. We did not compare with their work for we can only find a demo version of their tool and our experimental instances cannot be translated into the demo tool's input automatically.

Cheng and Yap studied search space reduction methods for constraint optimization problems [2] where the constraints are of some special forms. The variables are all 0-1 integer variables and the constraints are basically linear equations. The methods perform quite well on the Still-Life problem. We will investigate whether they will be helpful on the more general problems.

6   Concluding Remarks

Optimization problems occur naturally in various important applications. Traditionally the constraint part in such a problem is a set of inequalities between arithmetic expressions. In this paper, we investigate a generalized class of optimization problems constrained by arbitrary Boolean combinations of linear inequalities. Based on the DPLL(T) framework for SMT, we present an exact optimization algorithm augmented with two efficient pruning techniques. The experimental results show that our approach is very promising. A significant future study is to incorporate more theories into the framework.

Acknowledgements. The work is supported by the National Natural Science Foundation of China (NSFC) under grant No. 61100064 and No. 60903049, and partially funded by the State Key Laboratory of Rail Traffic Control and Safety of Northern Jiaotong University. The authors are grateful to Tian Liu for his comments on an earlier version of the paper.

References

1. Barrett, C.W., Tinelli, C.: CVC3. In: Damm, W., Hermanns, H. (eds.) CAV 2007. LNCS, vol. 4590, pp. 298–302. Springer, Heidelberg (2007),
   http://www.cs.nyu.edu/acsys/cvc3
2. Cheng, K.C., Yap, R.H.C.: Search space reduction and Russian doll search. In: Proceedings of the 22nd AAAI Conference on Artificial Intelligence (AAAI 2007) (2007)
3. Cimatti, A., Franzén, A., Griggio, A., Sebastiani, R., Stenico, C.: Satisfiability Modulo the Theory of Costs: Foundations and Applications. In: Esparza, J., Majumdar, R. (eds.) TACAS 2010. LNCS, vol. 6015, pp. 99–113. Springer, Heidelberg (2010)
4. Dutertre, B., de Moura, L.: A Fast Linear-Arithmetic Solver for DPLL(T). In: Ball, T., Jones, R.B. (eds.) CAV 2006. LNCS, vol. 4144, pp. 81–94. Springer, Heidelberg (2006),
   http://yices.csl.sri.com/
5. Eén, N., Sorensson, N.: The MiniSat Page (2011), http://minisat.se/
6. Ganzinger, H., Hagen, G., Nieuwenhuis, R., Oliveras, A., Tinelli, C.: DPLL(T): Fast Decision Procedures. In: Alur, R., Peled, D.A. (eds.) CAV 2004. LNCS, vol. 3114, pp. 175–188. Springer, Heidelberg (2004)
7. Hooker, J.N.: Logic, optimization, and constraint programming. INFORMS Journal on Computing 14, 295–321 (2002)
8. IBM. Cplex, http://www-01.ibm.com/software/integration/optimization/cplex-optimization-studio/
9. Hooker, J.N., Osorio, M.A.: Mixed logical-linear programming. Discrete Appl. Math. 96-97, 395–442 (October 1999)
10. Kroening, D., Strichman, O.: Decision Procedures. Springer (2008)
11. Ma, F., Liu, S., Zhang, J.: Volume Computation for Boolean Combination of Linear Arithmetic Constraints. In: Schmidt, R.A. (ed.) CADE-22. LNCS, vol. 5663, pp. 453–468. Springer, Heidelberg (2009)
12. de Moura, L., Bjørner, N.S.: Z3: An Efficient SMT Solver. In: Ramakrishnan, C.R., Rehof, J. (eds.) TACAS 2008. LNCS, vol. 4963, pp. 337–340. Springer, Heidelberg (2008),
    http://research.microsoft.com/projects/z3/index.html
13. Nieuwenhuis, R., Oliveras, A.: On SAT Modulo Theories and Optimization Problems. In: Biere, A., Gomes, C.P. (eds.) SAT 2006. LNCS, vol. 4121, pp. 156–169. Springer, Heidelberg (2006)

A   Proofs

Proof of **Proposition** 1

Proof. Firstly, we show that if the left side of the equation is infeasible (doesn't have any solution), so is the right side. Assume that the left side is infeasible, which means that $\{Min(f(x), \alpha)|\alpha \in \mathcal{S}\}$ is an empty set, while the right side has an optimal solution, namely $x^\star$. For a Boolean variable b_i which is not assigned by α_c, its truth value can be determined w.r.t $x^\star$: if $x^\star$ satisfies $\hat{T}h(b_i)$, then b_i is true. Denote the set of such literals by $\alpha_c^\star$, we now get a full assignment $\alpha^\star = \alpha_c \cup \alpha_c^\star$. Obviously $x^\star$ satisfies $\hat{T}h(\alpha^\star)$, and $x^\star$ is the optimal solution of $Min(f(x), \alpha^\star)$. Since $\mathcal{S}$ is a bunch with a cube α_c, we have $\alpha^\star \in \mathcal{S}$. Therefore, the assumption that $\{Min(f(x), \alpha)|\alpha \in \mathcal{S}\}$ is empty is contradicted.

Now assume that $x^\star$ is the optimal solution to the left side of the equation, and its corresponding assignment is $\alpha^\star$, i.e., $min\{Min(f(x), \alpha)|\alpha \in \mathcal{S}\} = Min(f(x), \alpha^\star) = f(x^\star)$. Since $\alpha_c \subseteq \alpha^\star$, $x^\star$ is a feasible solution of $\hat{T}h(\alpha_c)$, $Min(f(x), \alpha_c)$ is not infeasible and we have $f(x^\star) \leq Min(f(x), \alpha_c)$. Suppose $Min(f(x), \alpha_c)$ has an optimal solution x', or formally $Min(f(x), \alpha_c) = f(x')$. A full assignment α' can be constructed w.r.t x' in the same way as we have mentioned above. Because x' satisfies $\hat{T}h(\alpha')$, we have $f(x') \leq Min(f(x), \alpha')$. Also since $\alpha_c \subseteq \alpha'$, we have $\alpha' \in \mathcal{S}$ and thus $Min(f(x), \alpha') \leq f(x^\star)$. Therefore, $f(x') = f(x^\star)$.

The situation that the optimal value is unbounded can be viewed as a special case of the second situation.   □

Proof of **Proposition** 2

Proof. A well-known fact about linear programming is that if the optimal solution exists, it must be one of vertices of the polytope defined by the linear constraints. This property is essential to our proof. We denote the polytope defined by $Ax \otimes b$ as $\mathcal{P}$, and the polytope without the constraint $A_i\mathbf{x} \otimes_i b_i$ as $\mathcal{P}'$.

We firstly explain that with $A_i\mathbf{x} \otimes_i b_i$ removed from the constraints, the linear programming problem still has an optimal solution. Assume without the constraint $A_i x \otimes_i b_i$, the problem is unbounded. There must exist a point x_1 such that $c^\top x_1 < c^\top x_{opt}$, and satisfies all constraints except for $A_i x \otimes_i b_i$. Since the two points x_{opt} and x_1 are on different sides of the hyperplane $A_i x = b_i$, the line connecting the two points will intersect with the hyperplane at some point, say x_2. Obviously x_2 satisfies all constraints including $A_i x \otimes_i b_i$. Moreover, because of the monotonicity of the objective function, we have $c^\top x_1 < c^\top x_2 < c^\top x_{opt}$. So x_2 is a solution and is smaller than x_{opt}, contradicting the precondition of the theorem. As a result, the linear programming problem is still bounded without the constraint $A_i x \otimes_i b_i$. Also, the possibility for the problem to be infeasible can be directly ruled out. There must be an optimal solution of the problem.

Suppose the new optimal solution is x'_{opt}. According to the property of linear programming, it is a vertex of $\mathcal{P}'$. Similarly, x_{opt} is a vertex of $\mathcal{P}$. Since $\mathcal{P}'$ is obtained via removing the hyperplane $A_i\mathbf{x} \otimes_i b_i$ from $\mathcal{P}$, each vertex of $\mathcal{P}'$ is also a vertex of $\mathcal{P}$. So we have $c^\top x_{opt} \leq c^\top x'_{opt}$. Because the point x_{opt} is out of the hyperplane $A_i\mathbf{x} = b_i$, it is also a vertex of $\mathcal{P}'$ and we have $c^\top x'_{opt} \leq c^\top x_{opt}$. As a result, x_{opt} and x'_{opt} are the same optimal solution, and the constraint $A_i x \otimes_i b_i$ is redundant.   □

Solving Difficult SAT Problems by Using OBDDs and Greedy Clique Decomposition[*]

Yanyan Xu[1], Wei Chen[2], Kaile Su[3,4], and Wenhui Zhang[5]

[1] School of Information Science and Technology, Beijing Forestry University, China
xuyyxu@gmail.com
[2] Naveen Jindal School of Management, The University of Texas at Dallas, USA
wei.chen@utdallas.edu
[3] College of Mathematics Physics and Information Engineering,
Zhejiang Normal University, Jinhua, China
[4] School of Electronics Engineering and Computer Science, Peking University, China
kailepku@gmail.com
[5] State Key Laboratory of Computer Science, Institute of Software,
Chinese Academy of Sciences
zwh@ios.ac.cn

Abstract. In this paper, we propose an OBDD-based algorithm called greedy clique decomposition, which is a new variable grouping heuristic method, to solve difficult SAT problems. We implement our algorithm and compare it with several state-of-art SAT solvers including Minisat, Ebddres and TTS. We show that with this new heuristic method, our implementation of an OBDD-based satisfiability solver can perform better for selected difficult SAT problems, whose conflict graphs possess a clique-like structure.

1 Introduction

Boolean formula satisfaction problems and SAT solving techniques play an extremely important role in theoretical computer science as well as practice. The question of whether there exists a complete polynomial time SAT algorithm is a key problem for theoretical computer science and is open for many years [1]. On the other hand, the practical use of the SAT solvers is also very important. Applications of SAT solving techniques range from microprocessor verification [2] and field-programmable gate array design [3] to solving AI planning problems by translating them into Boolean formulas [4].

Symbolic SAT solving is an approach where the clauses of a CNF formula are represented by OBDDs. These OBDDs are then conjoined, and checking satisfiability is reduced to the question of whether the resulting OBDD is identical to false or true. Using OBDDs for SAT is an active research area [5,6,7]. It turns

[*] Supported by the Beijing Forestry University Young Scientist Fund No. BLX2009013, the Chinese National 973 Plan (No.2010CB328103), the ARC grants FT0991785 and DP120102489, the National Natural Science Foundation of China under Grant No. 60833001, and the CAS Innovation Program.

J. Snoeyink, P. Lu, K. Su, and L. Wang (Eds.): FAW-AAIM 2012, LNCS 7285, pp. 259–268, 2012.
© Springer-Verlag Berlin Heidelberg 2012

out that OBDDs and search based techniques are complementary [8,9]. There are problems for which one works better than the other.

Excellent performance breakthroughs have been made in solving SAT problems for these years. New algorithms and implementation techniques focusing on real life SAT problems are proposed and many benchmark problems are solved by the state-of-the-art solvers [10,3] in time proportional to the size of the input. It seems that the difficulty of many SAT benchmark problems consists in their size only. A lot of smaller benchmark problems are solved in real-time by those state-of-the-art solvers. It is very hard to compete with these best SAT solvers on these benchmark problems. That is why we are concentrating on difficult SAT problems only, where the word *difficult* means these problems are difficult for these state-of-the-art SAT solvers.

In [11], we show that at least for pigeonhole principles PHP_n^{n+1}, which are difficult SAT problems, there exists a direct OBDD proof of polynomial size. A natural next step is to try to generalize the way we find the grouping of variables in the pigeonhole case to an automatic mechanism, and apply it to other structural similar problems with the hope that we can get a polynomial time algorithm for these instances. We get the inspiration from Pavel Surynek [12] and the technique of greedy clique decomposition is exactly what we need in this situation.

In this paper, we explore the possibility of applying greedy clique decomposition as the variables grouping heuristics to an OBDD-based SAT solver. Experiments show that this method is quite efficient for some typical problems which do possess a clique-like structural property in their conflict graph compared with some well-known satisfiability solvers. Moreover, our implementation can deal with both satisfiable and unsatisfiable instances. Original method proposed by [12] can only works well when the input is an unsatisfiable formula and it is only a preprocessing procedure. Moreover, it is not based on OBDDs.

The rest of this paper is organized as follows: First, we give the details of OBDD-based satisfiability solving in Section 2. Then, we introduce how to apply the idea of clique decomposition as the grouping of variables heuristics in Section 3. Thereafter, in Section 4 we present the experimental results. Finally, we conclude this paper in Section 5.

2   OBDD-Based Satisfiability Solving

In this section, we first introduce Ordered Binary Decision Diagrams(OBDDs), and then we give the OBDD-based satisfiability solving procedure.

2.1   Ordered Binary Decision Diagrams(OBDDs)

Ordered Binary Decision Diagrams (OBDDs) introduced by Bryant [13] can represent Boolean functions as DAGs (directed acyclic graphs). OBDD is a canonical and compact representation according to two reduction rules. During the last decade, powerful search techniques using OBDDs have been developed in the area of symbolic model checking [14] and replanning [15,16,17]. In the most common form as reduced OBDDs, each Boolean function is uniquely represented by an OBDD, and thus all semantically equivalent formulas share the same OBDD. OBDDs are based on the Shannon's Expansion

$$f = ITE(x, f_1, f_0) = (x \to f_1) \land (\neg x \to f_0),$$

decomposing f into its *co-factors* f_0 and f_1, where f_0 (resp. f_1) is obtained by setting variable x into false (resp. true) in f. By repeatedly applying Shannon's Expansion to a formula until no more variables are left, its OBDD representation is obtained. Merging equivalent nodes and deleting nodes with coinciding *co-factors* result in reduced OBDDs. Fig. 1 shows the OBDD for the formula $f = x \lor (y \land \neg z)$. In Fig. 1, the dashed line means the variable is set into 0 and the solid line means 1.

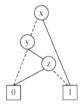

Fig. 1. OBDD for $f = x \lor (y \land \neg z)$ with the variable order x>y>z

---

Algorithm 1. *BDD-and(f,g)*

---

1 $BDD\text{-}and(BDD_f, BDD_g)$
   Input: BDD_f, BDD_g
   Output: $BDD_{f \land g}$
2 **begin**
3   if $BDD_f = 0$ or $BDD_g = 0$ then return 0;
4   if $BDD_f = 1$ then return BDD_g else if $BDD_g = 1$ then return BDD_f;
5      $(x, BDD_{f_0}, BDD_{f_1}) = decompose(BDD_f)$;
       $(y, BDD_{g_0}, BDD_{g_1}) = decompose(BDD_g)$;
6   if $x < y$ then return $newNode(y, BDD\text{-}and(f, g_0), BDD\text{-}and(f, g_1))$;
7   if $x = y$ then return $newNode(x, BDD\text{-}and(f_0, g_0), BDD\text{-}and(f_1, g_1))$;
8   if $x > y$ then return $newNode(x, BDD\text{-}and(f_0, g), BDD\text{-}and(f_1, g))$;
9 **end**

To generate OBDDs for formulas, we should build them *bottom-up* starting with basic OBDDs for variables or literals, and then constructing more complex OBDDs by using OBDD operations (e.g., *BDD-and*, *BDD-or*) for logical connectives. We give the *BDD-and* algorithm (Algorithm 1) explicitly as we need it in this paper. Here, *decompose* breaks down a non-terminal OBDD node into its constituent components, i.e., its variable and co-factors. The function *newNode* constructs a new OBDD node if it doesn't exist, and otherwise returns the already existent node. The algorithm is based on a given OBDD variable order.

2.2   OBDD-Based Satisfiability Solving

Given an unsatisfiable CNF formula, we first convert each of its clauses into corresponding OBDD, and then partition its variable set X into several disjoint buckets (sets): $X_1, X_2, ..., X_k$, each containing one or several variables. After that we join all the clauses in which the variables in X_i appear, and denote the OBDD we obtained by $F(X_i)$ (note that we follow the order $F(X_1)$, $F(X_2)$, ..., $F(X_k)$ when constructing these $F(X_i)$, and each clause is joined exactly once, i.e., if a clause is first joined in $F(X_k)$ with $k < i$, it won't be joined in $F(X_i)$ even if variables in X_i do appear in this clause), do the following:

$$\exists X_k.(...\exists X_2.(\exists X_1.(F(X_1)) \wedge F(X_2))... \wedge F(X_k)).$$

That is, we treat each bucket in turn, and eliminate all variables in that bucket at a time using existential quantification.

In this case, how we divide X into these buckets and the relative order these buckets are quantified out play an important role in affecting the sizes of the intermediate OBDDs. The existing implementation and tools known to us [18,19,20] treat each variable as a single bucket and no involved efforts are made to discover the more natural variable grouping of the input formulas. In this paper we provide a method to discover effective variable groupings automatically and utilize this information to improve the performance of OBDD-based satisfiability solving, at least for some particular subclasses of the input formulas. The greedy clique decomposition may suit this purpose well (note that in this case, we treat each clique as a bucket).

3   Greedy Clique Decomposition

A clique is a complete subgraph. The greedy clique decomposition decomposes the *conflict graph* of the input formula into different cliques using a simple greedy method.

Definition 1 (Conflict graph). *A conflict graph $G = (V, E)$ for a CNF formula is defined as follows: Let V be the set of all its variables [1], for different $x, y \in V$, $edge(x, y) \in E$ iff there is a length 2 clause containing exactly these two variables.*

From the definition, it is easy to see that constructing a conflict graph from a given CNF instance only takes time linear to the instance size.

The pseudo code for the greedy clique decomposition algorithm is given in Algorithm 2. Note that this is a quite straight-forward approximation algorithm. First

---

[1] The original description of the *consistent* graph [12] is more complicated and is based on the concept of arc-consistency and singleton arc-consistency. Our definition is a simplified version. However, the essential idea remains the same. Note also that in [12], V is defined to contain all literals. In our own experimental experience, this is not necessary.

we find the vertex with the largest degree in the conflict graph and then put this vertex and all its neighbors in the conflict graph into the first clique, then all vertices in the first clique and edges connecting from/to these vertices are removed from the conflict graph, and we find the vertex with the largest degree in the remaining graph, ..., keep doing this until we are left with only non-connected vertices in the remaining graph, then all such vertices are treated as a single clique.

Algorithm 2. Greedy Clique Decomposition

```
1  GCD(G)
   Input: the conflict graph G = (V, E)
   Output: clique[clique_num]
2  begin
3  |   clique_num=0;
4  |   find the vertex v with the largest degree n in graph G;
5  |   while (n !=0) do
6  |   |   add v and all its neighbors to clique[clique_num];
7  |   |   clique_num++;
8  |   |   delete v and its neighbors and all edges connecting from/to these
   |   |   vertices from G;
9  |   |   find the new vertex v with the largest degree in the remaining graph G;
10 |   |_  set n = this degree number;
11 |   add all remaining vertices in G to clique[clique_num];
12 end
```

We illustrate the concept of the conflict graph and this algorithm by an example: the pigeonhole problem PHP_3^4 with 4 pigeon and 3 holes expressing that we can put 4 pigeons into 3 holes without putting 2 pigeons into one and the same hole, which is obviously unsatisfiable. Let variable p_{ij} denotes that the pigeon i is in the hole j, we encode this problem using the following set of clauses (non-onto version):

$$\{p_{11} \lor p_{12} \lor p_{13}, p_{21} \lor p_{22} \lor p_{23}, p_{31} \lor p_{32} \lor p_{33}, p_{41} \lor p_{42} \lor p_{43},$$
$$\neg p_{11} \lor \neg p_{21}, \neg p_{11} \lor \neg p_{31}, \neg p_{21} \lor \neg p_{31}, \neg p_{11} \lor \neg p_{41}, \neg p_{21} \lor \neg p_{41}, \neg p_{31} \lor \neg p_{41},$$
$$\neg p_{12} \lor \neg p_{22}, \neg p_{12} \lor \neg p_{32}, \neg p_{22} \lor \neg p_{32}, \neg p_{12} \lor \neg p_{42}, \neg p_{22} \lor \neg p_{42}, \neg p_{32} \lor \neg p_{42},$$
$$\neg p_{13} \lor \neg p_{23}, \neg p_{13} \lor \neg p_{33}, \neg p_{23} \lor \neg p_{33}, \neg p_{13} \lor \neg p_{43}, \neg p_{23} \lor \neg p_{43}, \neg p_{33} \lor \neg p_{43}\}.$$

The corresponding conflict graph is shown in Fig. 2. It consists of three cliques:

$$\{p_{11}, p_{21}, p_{31}, p_{41}\}, \{p_{12}, p_{22}, p_{32}, p_{42}\}, \{p_{13}, p_{23}, p_{33}, p_{43}\}.$$

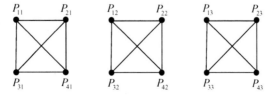

Fig. 2. The Conflict Graph for Instance PHP_3^4

It can be seen that the algorithm correctly decomposes the conflict graph in Fig. 2 to these three clique components. One can easily see that for PHP_n^{n+1}, there will be n cliques in the corresponding conflict graph:

$$\{p_{11}, p_{21}, ..., p_{(n+1)1}\}, \{p_{12}, p_{22}, ..., p_{(n+1)2}\}, ..., \{p_{1n}, p_{2n}, ..., p_{(n+1)n}\},$$

and this algorithm still correctly decomposes the corresponding conflict graph.

For pigeonhole problems, viewing each clique as a bucket, we know theoretically that we will get a polynomial size proof (in this case, the relative order of how these buckets are eliminated does not matter) [11]. Since the worst-case time complexity of this intuitive approximation algorithm is only quadratic to the conflict graph size, we are sure that we can implement the OBDD-based satisfiability solving algorithm for the pigeonhole problems in a practical sense (which is rather inspiring, since the existing OBDD-based solvers [18,19,20] can't handle pigeonhole cases efficiently).

It is natural to try this algorithm on other problems of which the conflict graphs are similar to the pigeonhole case, so we can improve the performance of the problems. With our experimental results we show that this is indeed the case. Readers are referred to Section 4 for more details. For the intuitive effectiveness of putting each clique as a bucket when doing variable elimination, we argue that when representing a Boolean function using an OBDD, the more two variables interact with each other, the less size OBDD we can get by putting these variables nearby each other in the variable order. In other words, the dependencies between these variables are the main cause for the OBDD size explosion. It follows that if we can put all these variables into a clique and eliminate them together, it will reduce the sizes of intermediate OBDDs.

4   Experimental Results

Our experiments[2] are carried out on a server machine with 32 GB main memory and four 3 GHz Xeon CPUs running Linux 2.6.9. No parallel mechanism are used. Our implementation makes use of the CUDD package [21]. We use the benchmark suite [22]. Timeout is set to 10 minutes (600 seconds). For some solvers (Ebddres and Minisat), we list both the runtime (seconds) and the peak memory usage (MB), for others (TTS, SSAT and our implementation), we list the runtime only because the peak memory usage is negligible.

4.1   The SAT Solvers Used in Our Experiments

Minisat is the winner for the SAT 2005 Competition for three industrial categories and one crafted category. It also won the SAT 2007 and 2009 Competition. It is considered one of the best SAT solvers.

---

[2] Our implementation along with the set of benchmarks we used can be downloaded from http://www.utdallas.edu/~wxc103020/bddsat.tar.bz2.

Table 1. Test instances and the corresponding clique decomposition

Instance	#vars/ #clauses	satisfiable?	clique decomposable?	cliques
hole 7	56/ 204	No	Yes	8*7
hole 8	72/ 297	No	Yes	9*8
hole 9	90/ 415	No	Yes	10*9
hole 10	110/ 561	No	Yes	11*10
hole 11	132/ 738	No	Yes	12*11
hole 12	156/ 949	No	Yes	13*12
hole 13	182/ 1197	No	Yes	14*13
hole 14	210/ 1485	No	Yes	15*14
hole 20	420/ 4221	No	Yes	21*20
hole 30	930/ 13981	No	Yes	31*30
chnl10_11	220/ 1122	No	Yes	11*20
chnl10_12	240/ 1344	No	Yes	12*20
chnl10_13	260/ 1586	No	Yes	13*20
chnl11_12	264/ 1476	No	Yes	12*22
chnl11_13	286/ 1742	No	Yes	13*22
chnl11_20	440/ 4220	No	Yes	20*22
fpga10_8_sat	120/ 448	Yes	Yes	8*10+5*8
fpga10_9_sat	135/ 549	Yes	Yes	9*10+5*9
fpga12_8_sat	144/ 560	Yes	Yes	8*12+6*8
fpga12_9_sat	162/ 684	Yes	Yes	9*12+6*9
fpga12_11_sat	198/ 968	Yes	Yes	11*12+6*11
fpga12_12_sat	216/ 1128	Yes	Yes	12*12+6*12
fpga13_9_sat	176/ 759	Yes	Yes	9*13+7*5+6*4
fpga13_10_sat	195/ 905	Yes	Yes	10*13+7*5+6*5
fpga13_12_sat	234/ 1242	Yes	Yes	12*13+7*6+6*5
Urq3_5	46/ 470	No	No	n/a
Urq4_5	74/ 694	No	No	n/a
Urq5_5	121/ 1210	No	No	n/a
Urq6_5	180/ 1756	No	No	n/a
Urq7_5	240/ 2194	No	No	n/a
Urq8_5	327/ 3252	No	No	n/a

Ternary Tree Solver (TTS) is the silver medal winner for the SAT 2007 Competition in the crafted category, UNSAT specialization, known for dealing with small but hard instances.

Ebddres is an OBDD-based satisfiability solver [19]. Note that the primary purpose of this solver is to generate the proof of unsatisfiability, which in turn is to be utilized in the hardware verification.

SSAT is a preprocessing tool based on the idea of clique consistency [12]. It can decide very fast if the input instance is unsatisfiable and clique decomposable. However, for satisfiable problems, it serves only as a preprocessor.

4.2   Test Instances and Corresponding Clique Decomposition

Detailed information about the instances we used and the corresponding clique decomposition of their conflict graphs are presented in Table 1. In the first column we list the instances. There are 4 sets of instances used in our evaluation: the pigeonhole cases ("hole"), the randomized urquhart instances ("Urq"), and the field programmable gate array routing instances ("chnl" and "fpga"). We list the number of variables and the number of clauses of each instance in the second column. We list whether an instance is satisfiable in the third column and whether an instance is clique decomposable using our algorithm in the fourth column. The detailed clique information of corresponding conflict graph of each instance is given in the last column. For example, "8*10" indicates that there are 10 cliques of size 8 in the corresponding conflict graph, and "8*10+5*8" indicates that there are 10 cliques of size 8 and 8 cliques of size 5 in the corresponding conflict graph. Careful readers might notice that the number of vertices in this column equals the number of variables of the instance. We would like to point out that for "Urq" cases, our algorithm fails to find useful clique information and hence as we can see from Table 2, the performance of our implementation is quite poor.

4.3   Results and Analysis

Analysis of the results For each $F(X_i)$ mentioned in Section 2, there are two ways of obtaining the corresponding OBDD: linear computation and tree computation. "–ve" denotes variable elimination with linear computation. "–tree" denotes tree computation without variable elimination. "–vetree" corresponds to variable elimination with tree computation. As we can see from Table 2, using Ebddres, going from –ve to –vetree we can get a constant factor speedup. The results are similar in our implementation. Going from linear to tree computation we can get an

Table 2. Experimental results

Instance	Ebddres –ve	Ebddres –tree	Ebddres –vetree	Minisat	TTS	SSAT	our	+tree
hole 7	0.03/3.5	0.00/0.6	0.03/3.5	0.05/0.11	0.09	0.01	0.02	0.01
hole 8	0.17/14.0	0.01/2.3	0.14/14.0	0.27/9.23	0.28	0.01	0.02	0.01
hole 9	0.76/53.1	0.01/2.0	0.65/56.1	1.56/9.49	0.08	0.03	0.02	0.02
hole 10	3.14/212.1	0.14/15.6	2.22/132.1	29.98/10.53	0.22	0.05	0.03	0.02
hole 11	12.40/816.1	0.08/7.8	7.86/448.1	598.36/14.06	0.43	0.09	0.04	0.02
hole 12	40.13/1696.1	0.29/27.3	27.91/1344.1	t/o	1.07	0.14	0.04	0.03
hole 13	150.97/6528.1	1.96/124.1	88.41/3584.1	t/o	2.59	0.25	0.06	0.03
hole 14	552.01/18944.1	1.74/114.1	320.70/10752.1	t/o	7.09	0.44	0.09	0.03
hole 20	t/o	t/o	t/o	t/o	t/o	5.46	0.58	0.05
hole 30	t/o	t/o	t/o	t/o	t/o	54.09	5.58	0.20
chnl10_11	3.14/212.1	0.24/18.1	2.25/132.1	29.49/18.61	0.20	0.23	0.03	0.03
chnl10_12	3.66/212.1	0.07/9.1	3.10/224.1	49.55/12.84	0.23	0.33	0.04	0.03
chnl10_13	5.37/408.1	0.09/9.2	3.45/224.1	46.90/13.52	0.26	0.48	0.05	0.03
chnl11_12	12.44/816.1	0.13/15.7	7.79/448.1	367.69/16.05	0.43	0.46	0.05	0.03
chnl11_13	14.07/816.1	0.22/15.7	9.03/448.1	t/o	0.51	0.62	0.05	0.03
chnl11_20	53.41/3104.3	0.81/62.4	23.73/1184.3	t/o	1.02	5.32	0.17	0.05
fpga10_8_sat	1.19/102.1	0.82/57.1	0.84/56.1	0.00/9.11	0.07	n/a	0.02	0.02
fpga10_9_sat	1.51/112.1	0.52/54.6	1.29/84.1	4.38/9.88	0.15	n/a	0.03	0.02
fpga12_8_sat	5.91/376.1	2.85/218.1	4.09/224.1	0.00/9.11	0.16	n/a	0.03	0.03
fpga12_9_sat	9.34/424.1	11.05/852.1	7.85/448.1	0.00/9.23	0.34	n/a	0.03	0.03
fpga12_11_sat	23.26/1184.1	10.91/872.1	16.80/896.1	t/o	1.27	n/a	0.04	0.03
fpga12_12_sat	33.07/1696.1	3.10/218.1	20.76/1056.1	49.56/12.28	0.61	n/a	0.05	0.03
fpga13_9_sat	20.41/848.1	31.55/1704.1	16.24/896.1	0.00/9.24	0.37	n/a	0.04	0.03
fpga13_10_sat	39.28/1696.1	37.58/1744.1	31.97/1792.1	0.00/9.23	0.73	n/a	0.04	0.03
fpga13_12_sat	78.38/3392.1	46.47/3368.1	51.55/2112.1	0.00/9.24	0.59	n/a	0.06	0.03
Urq3_5	0.01/0.4	0.01/1.2	0.01/0.3	36.40/10.15	0.30	t/o	0.03	0.02
Urq4_5	0.01/0.8	0.83/54.6	0.01/0.6	t/o	5.30	t/o	0.42	0.12
Urq5_5	0.03/2.2	t/o	0.02/1.2	t/o	t/o	t/o	5.11	0.93
Urq6_5	0.04/3.2	t/o	0.04/2.3	t/o	t/o	t/o	t/o	t/o
Urq7_5	0.06/6.2	t/o	0.05/3.3	t/o	t/o	t/o	t/o	t/o
Urq8_5	0.12/12.2	t/o	0.08/6.4	t/o	t/o	t/o	t/o	t/o

obvious speedup. Our implementation (*our*) can be seen as a slightly advanced version than Ebddres with –vetree, in the sense that we use advanced heuristic for dividing all variables into several subgroups (buckets), and we quantify out all variables in each bucket altogether each time. It can be seen in the experimental results that identifying subgroups in variables can be sometimes useful comparing to the performance of the DPLL-based solvers. Moreover, our implementation using the greedy clique decomposition plus tree computation(*+tree*) outperforms most other solvers in Table 2.

SSAT Solver As mentioned before, SSAT solver is merely a preprocessing tool. It can be seen from the results that putting greedy clique decomposition in the framework of OBDD-based satisfiability solving improves its performance, since we can decide satisfiable instance fast and unsatisfiable instances faster than SSAT. Note that our "fpga" instances are all satisfiable and therefore SSAT can not be used to run these instances.

5 Conclusions

We introduce an algorithm to solve difficult SAT problems by using OBDDs and greedy clique decomposition in this paper. Compared with some state-of-the-art SAT solvers, our implementation shows that for problems of whose conflict graphs possessing a clique-like structure, this algorithm is quite effective.

References

1. Stephen, A.: Cook. The Complexity of Theorem-Proving Procedures. In: STOC, pp. 151–158. ACM (1971)
2. Velev, M.N., Bryant, R.E.: Effective use of Boolean satisfiability procedures in the formal verification of superscalar and VLIW microprocessors. J. Symb. Comput. 35(2), 73–106 (2003)
3. Nam, G.-J., Sakallah, K.A., Rutenbar, R.A.: A new FPGA detailed routing approach via search-based Boolean satisfiability. IEEE Trans. on CAD of Integrated Circuits and Systems 21(6), 674–684 (2002)
4. Kautz, H.A., Selman, B.: Planning as Satisfiability. In: ECAI, pp. 359–363 (1992)
5. Franco, J., Kouril, M., Schlipf, J., Ward, J., Weaver, S., Dransfield, M., Vanfleet, W.M.: SBSAT: a State-Based, BDD-Based Satisfiability Solver. In: Giunchiglia, E., Tacchella, A. (eds.) SAT 2003. LNCS, vol. 2919, pp. 398–410. Springer, Heidelberg (2004)
6. Damiano, R.F., Kukula, J.H.: Checking satisfiability of a conjunction of BDDs. In: DAC, pp. 818–823. ACM (2003)
7. Pan, G., Vardi, M.Y.: Search vs. Symbolic Techniques in Satisfiability Solving. In: H. Hoos, H., Mitchell, D.G. (eds.) SAT 2004. LNCS, vol. 3542, pp. 235–250. Springer, Heidelberg (2005)
8. Rish, I., Dechter, R.: Resolution versus Search: Two Strategies for SAT. J. Autom. Reasoning 24(1/2), 225–275 (2000)
9. Groote, J.F., Zantema, H.: Resolution and binary decision diagrams cannot simulate each other polynomially. Discrete Applied Mathematics 130(2), 157–171 (2003)
10. Moskewicz, M.W., Madigan, C.F., Zhao, Y., Zhang, L., Malik, S.: Chaff: Engineering an Efficient SAT Solver. In: DAC, pp. 530–535. ACM (2001)

11. Chen, W., Zhang, W.: A direct construction of polynomial-size OBDD proof of pigeon hole problem. Inf. Process. Lett. 109(10), 472–477 (2009)
12. Surynek, P.: Solving Difficult SAT Instances Using Greedy Clique Decomposition. In: Miguel, I., Ruml, W. (eds.) SARA 2007. LNCS (LNAI), vol. 4612, pp. 359–374. Springer, Heidelberg (2007)
13. Bryant, R.E.: Graph-Based Algorithms for Boolean Function Manipulation. IEEE Trans. Computers 35(8), 677–691 (1986)
14. McMillan, K.L.: Symbolic Model Checking. Kluwer Academic Publishers ACM (1993)
15. Yue, W., Xu, Y., Su, K.: BDDRPA*: An Efficient BDD-Based Incremental Heuristic Search Algorithm for Replanning. In: Sattar, A., Kang, B.H. (eds.) AI 2006. LNCS (LNAI), vol. 4304, pp. 627–636. Springer, Heidelberg (2006)
16. Xu, Y., Yue, W.: A Generalized Framework for BDD-based Replanning A* Search. In: Kim, H.-K., Lee, R.Y. (eds.) SNPD, pp. 133–139. IEEE Computer Society (2009)
17. Xu, Y., Yue, W., Su, K.: The BDD-Based Dynamic A* Algorithm for Real-Time Replanning. In: Deng, X., Hopcroft, J.E., Xue, J. (eds.) FAW 2009. LNCS, vol. 5598, pp. 271–282. Springer, Heidelberg (2009)
18. Pan, G., Vardi, M.Y.: Symbolic Techniques in Satisfiability Solving. J. Autom. Reasoning 35(1-3), 25–50 (2005)
19. Jussila, T., Sinz, C., Biere, A.: Extended Resolution Proofs for Symbolic SAT Solving with Quantification. In: Biere, A., Gomes, C.P. (eds.) SAT 2006. LNCS, vol. 4121, pp. 54–60. Springer, Heidelberg (2006)
20. Sinz, C., Biere, A.: Extended Resolution Proofs for Conjoining BDDs. In: Grigoriev, D., Harrison, J., Hirsch, E.A. (eds.) CSR 2006. LNCS, vol. 3967, pp. 600–611. Springer, Heidelberg (2006)
21. Somenzi, F.: CUDD: CU Decision Diagram Package, Release 2.4.1. Technical report, University of Colorado at Boulder (2005)
22. Aloul, F.A., Ramani, A., Markov, I.L., Sakallah, K.A.: Solving difficult SAT instances in the presence of symmetry. In: DAC 2002: Proceedings of the 39th Conference on Design Automation, pp. 731–736. ACM, New York (2002)

Zero-Sum Flow Numbers of Regular Graphs

Tao-Ming Wang and Shih-Wei Hu

Department of Applied Mathematics,
Tunghai University,
Taichung, Taiwan 40704, R.O.C.
wang@thu.edu.tw

Abstract. As an analogous concept of a nowhere-zero flow for directed graphs, we consider zero-sum flows for undirected graphs in this article. For an undirected graph G, a **zero-sum flow** is an assignment of non-zero integers to the edges such that the sum of the values of all edges incident with each vertex is zero, and we call it a **zero-sum k-flow** if the values of edges are less than k. We define the **zero-sum flow number** of G as the least integer k for which G admitting a zero-sum k-flow. In this paper, among others we calculate the zero-sum flow numbers for regular graphs and also the zero-sum flow numbers for Cartesian products of regular graphs with paths.

1 Introduction to Zero-Sum Flows

Throughout this paper, all terminologies and notations on graph theory can be referred to the textbook by D. West[13]. Let G be a directed graph. A **nowhere-zero flow** on G is an assignment of non-zero integers to each edge such that for every vertex the Kirchhoff current law holds, that is, the sum of the values of incoming edges is equal to the sum of the values of outgoing edges. A **nowhere-zero k-flow** is a nowhere-zero flow using edge labels with maximum absolute value $k - 1$. Note that for a directed graph, admitting nowhere-zero flows is independent of the choice of the orientation, therefore one may consider such concept over the underlying undirected graph. A celebrated conjecture of Tutte in 1954 says that every bridgeless graph has a nowhere-zero 5-flow. F. Jaeger showed in 1979 that every bridgeless graph has a nowhere-zero 8-flow[6], and P. Seymour proved that every bridgeless graph has a nowhere-zero-6-flow[9] in 1981. However the original Tutte's conjecture remains open. There is another analogous and more general concept of a nowhere-zero flow that uses bidirected edges instead of directed ones, first systematically developed by Bouchet[4] in 1983. Bouchet raised the conjecture that every bidirected graph with a nowhere-zero integer flow has a nowhere-zero 6-flow, which is still unsettled. Recently another related nowhere-zero flow concept has been studied, as a special case of bi-directed one, over the undirected graphs by S. Akbari et al.[2] in 2009.

Definition 1. *For an undirected graph G, a **zero-sum flow** is an assignment of non-zero integers to the edges such that the sum of the values of all edges*

J. Snoeyink, P. Lu, K. Su, and L. Wang (Eds.): FAW-AAIM 2012, LNCS 7285, pp. 269–278, 2012.

incident with each vertex is zero. A **zero-sum** k**-flow** *is a zero-sum flow whose values are integers with absolute value less than* k.

S. Akbari et al. raised a conjecture for zero-sum flows similar to the Tutte's 5-flow Conjecture for nowhere-zero flows as follows:

Conjecture. (Zero-Sum 6-Flow Conjecture) If G is a graph with a zero-sum flow, then G admits a zero-sum 6-flow.

It was proved in 2010 by Akbari et al. [1] that the above Zero-Sum 6-Flow Conjecture is equivalent to the Bouchet's 6-Flow Conjecture for bidirected graphs, and the existence of zero-sum 7-flows for regular graphs were also obtained. Based upon the results, they raised another weaker conjecture for regular graphs:

Conjecture. (Zero-Sum 5-Flow Conjecture for Regular Graphs) If G is a r-regular graph with $r \geq 3$, then G admits a zero-sum 5-flow.

In literature a more general concept **flow number**, which is defined as the least integer k for which a graph may admit a k-flow, has been studied for both directed graphs and bidirected graphs. We extend the concept in 2011 to the undirected graphs and call it **zero-sum flow numbers**[12], and also considered general **constant-sum flows** for regular graphs[11]. In this paper, we study zero-sum flows over undirected graphs and calculate zero-sum flow numbers for regular graphs and zero-sum flow numbers for Cartesian products of regular graphs with paths.

2   Zero-Sum Flow Numbers

In the study of both nowhere-zero flows of directed graphs and bidirected graphs, one considers a more general concept **flow number**, namely, the least number of k for which a graph may admit a k-flow. Some authors used the term **flow index** instead in literature. In 2011 [12] we consider similar concepts for zero-sum k-flows:

Definition 2. *Let* G *be a undirected graph. The* **zero-sum flow number** $F(G)$ *is defined as the least number of* k *for which* G *may admit a zero-sum* k-*flow.* $F(G) = \infty$ *if no such* k *exists.*

Obviously the zero-sum flow numbers can provide with more detailed information regarding zero-sum flows. For example, we may restate the previously mentioned Zero-Sum Conjecture as follow: Suppose a undirected graph G has a zero-sum flow, then $F(G) \leq 6$.

In 2011, We showed some general properties of small flow numbers, so that the calculation of zero-sum flow numbers gets easier. It is well known that, for nowhere-zero flows over a graph with orientation, a graph G admits a nowhere-zero 2-flow if and only if it is Eulerian (every vertex has even degree). We obtain the following for zero-sum flows:

Lemma 3. (T. Wang and S. Hu [12]) *A graph G has zero-sum flow number $F(G) = 2$ if and only if G is Eulerian with even size (even number of edges) in each component.*

Proof. We put the proof here for completion:

"$\Rightarrow$"

Without loss of generality, we may assume G is connected. Since a graph G has flow index $F(G) = 2$ meaning it admits a zero-sum 2-flow, thus the edge function $f(e) \in \{1, -1\}$. For each vertex $v \in V(G)$, the number of incident edges labeled 1 must equal to the number of incident edges labeled -1. Note that both numbers are equal to $\frac{1}{2}deg(v)$, therefore $deg(v)$ must be even, and G is Eulerian. On the other hand, the number of all 1-edges (or (-1)-edges) in G is

$$\frac{1}{2} \sum_{v \in V(G)} (\frac{1}{2}deg(v)) = \frac{1}{2}|E(G)|$$

which is an integer, so $|E(G)|$ are even.

"$\Leftarrow$"

We label the edges in an Euler tour of G by 1 and -1 alternatively, then every vertex is incident with the same number of 1-edges and (-1)-edges, including the starting(ending) vertex, since the number of edges is even. Therefore it will be a zero-sum 2-flow in G. $\square$

Tutte obtained in 1949 that a cubic graph has a nowhere-zero 3-flow if and only if it is bipartite. Similarly for zero-sum flows we have the following:

Theorem 4. (T. Wang and S. Hu [12]) *A cubic graph G has zero-sum flow number $F(G) = 3$ if and only if G admits a perfect matching.*

Also we obtain the zero-sum flow numbers for wheel graphs and fan graphs, among others:

Theorem 5. (T. Wang and S. Hu [12]) *The flow numbers of wheel graphs W_n with $n \geq 3$ are as follows:*

$$F(W_n) = \begin{cases} 5, & n = 5. \\ 3, & n = 3k, \ k \geq 1. \\ 4, & otherwise. \end{cases}$$

Theorem 6. (T. Wang and S. Hu [12]) *The flow numbers of fan graphs F_n are as follows:*

$$F(F_n) = \begin{cases} \infty, & n = 1, 2, 3. \\ 3, & n = 3k + 1, \ k \geq 1. \\ 4, & otherwise. \end{cases}$$

3 Flow Numbers for Regular Graphs

In this section we calculate the zero-sum flow numbers of regular graphs, which is closely related the **Zero-Sum 5-Flow Conjecture for Regular Graphs**. We start with the following well known factorization theorem for even regular graphs:

Theorem 7. (Petersen, 1891[8]) *Every regular graph of even degree is 2-factorable.*

First we consider regular graphs of even degree. Note that by Lemma 3 we see that a 2-regular graph G has flow number $F(G) = 2$ if G consists of even cycles only, and $F(G) = \infty$ if G contains at least one odd cycle. For other even regular graphs, we obtain that the zero-sum flow numbers are either 2 or 3 in the following lemmas.

Lemma 8. *Let G be a connected even r-regular graph, $r \geq 4$. If either (1) $r \equiv 2 \pmod 4$ and G is of even size ($|E(G)|$ is even), or (2) $r \equiv 0 \pmod 4$, then $F(G) = 2$.*

Proof
In the first case, G is Eulerian with even size, by Lemma 3, $F(G) = 2$. In the second case, $r \equiv 0 \pmod 4$ implies the size of G must be even. By Lemma 3, $F(G) = 2$. □

Lemma 9. *If G is a connected even r-regular, $r \geq 4$, with odd size and $r \equiv 2 \pmod 4$, then $F(G) = 3$.*

Proof
Since $r \equiv 2 \pmod 4$, let $r = 4s + 6$, $s = 0, 1, 2, \cdots$. By Theorem 7, G can be decomposed into union of 2-factors as follows: $G = R_1 \oplus R_2 \oplus R_3 \cdots \oplus R_{2s+3}$, where R_i are 2-factors. We set f as follow:

$$f(e) = \begin{cases} 1 & , \quad e \in R_1 \cup \cdots \cup R_s, \text{ if } s > 0 \\ -1 & , \quad e \in R_{s+1} \cup \cdots \cup R_{2s}, \text{ if } s > 0 \\ 2 & , \quad e \in R_{2s+1} \\ -1 & , \quad e \in R_{2s+2} \cup R_{2s+3} \end{cases}$$

Therefore f is a zero-sum 3-flow. On the other hand, the size of G is odd, by Lemma 3, $F(G) \neq 2$, thus $F(G) = 3$. □

Now we consider regular graphs of odd degree. Note that $F(G) = \infty$ if G is 1-regular. Note that in Theorem 4 we characterize 3-regular graphs with flow number 3, which are those with perfect matchings. Naturally one will ask whether this is true for other odd r-regular graphs, $r \geq 5$. To answer the above question, first we have the following sufficient condition of odd regular graphs having flow numbers 3:

Lemma 10. *If G is an odd r-regular $(r \geq 3)$ graph with a perfect matching, then $F(G) = 3$.*

Proof

Let G be $(2k+3)$-regular, $k \geq 0$. Since G has a 1-factor, set $G = E \oplus R$, where E is 1-factor and R is $(2k+2)$-factor. By Theorem 7, R can be decomposed into union of 2-factors as follows: $G = E \oplus R_1 \oplus R_2 \cdots \oplus R_{k+1}$. We set f as follows if k is even:

$$f(e) = \begin{cases} 1 & , \quad e \in R_1 \cup \cdots \cup R_{\frac{k}{2}}, \text{ if } k > 0 \\ -1 & , \quad e \in R_{\frac{k}{2}+1} \cup \cdots \cup R_k, \text{ if } k > 0 \\ 1 & , \quad e \in R_{k+1} \\ -2 & , \quad e \in E \end{cases}$$

And if k is odd:

$$f(e) = \begin{cases} 1 & , \quad e \in R_1 \cup \cdots \cup R_{\frac{k-1}{2}}, \text{ if } k > 0 \\ -1 & , \quad e \in R_{\frac{k-1}{2}+1} \cup \cdots \cup R_{k-1}, \text{ if } k > 0 \\ 2 & , \quad e \in R_k \\ -1 & , \quad e \in R_{k+1} \\ -2 & , \quad e \in E \end{cases}$$

Now, f forms a zero-sum 3-flow. On the other hand, the degrees of G are odd, by Lemma 3 $F(G) \neq 2$, thus $F(G) = 3$. $\qquad\square$

However, on the other hand, we have an infinite family of examples of odd r-regular $(r \geq 5)$ graphs without perfect matching, and whose zero-sum flow numbers are 3.

Please see the following Figure 1 to the left, for the example of a 5-regular graph on 60 vertices, without perfect matching, having zero-sum flow number 3. Note that in the Figure 1, the component C represents the piece of graph to the right with indicated edge labels -1, 2, -2, and C' means the graph obtained from C by reversing the sign of each edge. The fact that this example is a 5-regular graph without perfect matching can be confirmed easily by the well known Tutte's characterization for graphs having a perfect matching.

With similar constructions, one may have examples of odd r-regular $(r \geq 7)$ graphs G, without perfect matching, whose flow numbers are 3. Therefore we can not expect the converse statement of Lemma 10 to be true.

In below is another result of zero-sum flow numbers for regular graphs, which helps to improve the results one obtain so far for the zero-sum 5-flow conjecture for regular graphs. First we need a result regarding regular factors in regular graphs by Gallai dated back to 1950:

Theorem 11. (Gallai, 1950 [5]) *Let r and k be integers such that $1 \leq k < r$, and G be a λ-edge connected r-regular general graph, where $\lambda \geq 1$. If r and k are both odd and $\frac{r}{\lambda} \leq k$, then G has a k-regular factor.*

By the above Theorem 11, we have the following improved result for the flow numbers of odd regular graphs:

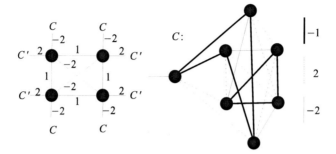

Fig. 1. An example of a 5-regular graph G without perfect matching and $F(G) = 3$

Theorem 12. *Let G be a 2-edge-connected $(2m + 1)$-regular graph with $m \geq 2$, then $F(G) = 3$ when $m \geq 3$, $F(G) = 3$ or 4 when G is 5-regular $(m = 2)$.*

Proof

By Theorem 11, we can see that G has a 3-factor for $m \geq 2$. Let $G = E \oplus R$ where E is a 3-factor, R is a $(2m - 2)$-factor. By Petersen's Theorem 7, R can be factored as $R_1 \oplus R_2 \oplus R_3 \cdots \oplus R_{m-1}$ where R_i are 2-factors for all $1 \leq i \leq m-1$.

Now we set up labeling function f as follows:

If $m \geq 4$ is even:

$$f(e) = \begin{cases} 1 & , \quad e \in R_1 \cup \cdots \cup R_{\frac{m-4}{2}}, \text{ for } m \geq 6 \\ -1 & , \quad e \in R_{\frac{m-4}{2}+1} \cup \cdots \cup R_{m-4}, \text{ for } m \geq 6 \\ 1 & , \quad e \in R_{m-3} \cup R_{m-2} \cup R_{m-1} \\ -2 & , \quad e \in E \end{cases}$$

If $m \geq 3$ is odd:

$$f(e) = \begin{cases} 1 & , \quad e \in R_1 \cup \cdots \cup R_{\frac{m-3}{2}}, \text{ for } m \geq 5 \\ -1 & , \quad e \in R_{\frac{m-3}{2}+1} \cup \cdots \cup R_{m-3}, \text{ for } m \geq 5 \\ 1 & , \quad e \in R_{m-2} \\ 2 & , \quad e \in R_{m-1} \\ -2 & , \quad e \in E \end{cases}$$

If $m = 2$:

$$f(e) = \begin{cases} 3 & , \quad e \in R_1 \\ -2 & , \quad e \in E \end{cases}$$

This f forms a zero-sum flow for G and since G is odd-regular, $F(G) \neq 2$. So $F(G) = 3$ when $m \geq 3$, $F(G) \leq 4$ when $m = 2$. $\qquad \square$

Note that recently S. Akbari et al. raised the **Zero-Sum 5-Flow Conjecture for Regular Graphs**, and proved the following partial results regarding the above conjecture:

Theorem 13. (S. Akbari et al., 2010 [1]) *Let G be an r-regular graph, $r \geq 3$. Then G has a zero-sum 7-flow. If r is a multiple of 3, then G has a zero-sum 5-flow.*

More recently, S. Akbari et al. announced [3] to confirm the above conjecture of existence of 5-flows for all r-regular graphs except $r = 5$, using the concept of constant-sum flows (will be introduced in next section) and a result of M. Kano regarding regular factors in regular graphs:

Theorem 14. (M. Kano, 1986 [7]) *Let $r \geq 3$ be an odd integer and let k be an integer such that $1 \leq k \leq \frac{2r}{3}$. Then every r-regular graph has a $[k - 1, k]$-factor each component of which is regular.*

We summarize results of zero-sum flow numbers for regular graphs and list two open questions in below from the above Lemmas and Theorems:

Theorem 15. *Suppose G is a r-regular graph, $r \geq 3$, then*

1. *$F(G) = 2$ if $r \equiv 0 \pmod 4$, or $r \equiv 2 \pmod 4$ with even size.*
2. *$F(G) = 3$ if $r \equiv 2 \pmod 4$ with odd size, or G r-regular with perfect matching and odd $r \geq 3$, or G a 2-edge-connected r-regular graph with odd $r \geq 7$.*
3. *$F(G) = 3$ or 4 if G 2-edge-connected 5-regular.*
4. *$F(G) \leq 5$ if $r \neq 5$, and $F(G) \leq 7$ for G 5-regular.*

Note that recently the Zero-Sum Flow Conjecture for Regular graphs is settled, except one open case that $F(G) \leq 5$ for G being 5-regular [3]. Therefore inspired from the above results, we post the following closely related open problems for further studies:

Problem. Characterize the (r-regular, r odd and $r \geq 5$) graphs with zero-sum flow number 3.

Problem. Calculate the flow number $F(G)$ when G is 5-regular and not 2-edge-connected (with bridges).

Problem. Calculate the flow number $F(G)$ when G is a general odd regular graph.

4   Flow Numbers for Cartesian Products

For calculating the flow numbers of Cartesian product of regular graphs with paths, we define a general constant-sum flows in below:

Definition 16. *For an undirected graph G, a **constant c-sum k-flow** of G is an assignment of non-zero integers to the edges such that the sum of the values of all edges incident with each vertex is constant c, and whose values are integers with absolute value less than k.*

The following is the formula for flow numbers of Cartesian product of regular graphs with paths:

Theorem 17. *Let R_r be a r-regular graph and P_s be a path with $|V(P_s)| = s$, $r \geq 2, s \geq 2$. Then*

$$F(R_r \times P_s) = \begin{cases} 2, & r \text{ is odd and } s = 2. \\ 3, & \text{otherwise} \end{cases}$$

Proof. We proceed with the following cases:

Case 1. r is odd, $r \geq 3$.

Note that the order of R_r is even, so $R_r \times P_2$ is an Eulerian graph with even size. By the Lemma 3, $F(R_r \times P_2) = 2$. Then we express $R_r \times P_s$, $s \geq 3$, as an obvious union of $s - 1$ copies of $R_r \times P_2$, so that the pairwise intersection is one copy of R_r. Within each copy of $R_r \times P_2$ in the Cartesian product, we name one copy of R_r as A and the other copy of R_r as B. Now fix a zero-sum 2-flow of $R_r \times P_2$ as mentioned, so we have fixed ± 1 edge labeling over A and B. Now we set up a zero-sum 3-flow of $R_r \times P_s$ by adding up edge labels over the union, in an order A, B, B, A, A, B, B, A, etc. such that the edge labels will be twice as much as ± 1, which is ± 2, over each A or B part, except two end parts. Again by Lemma 3, $F(R_r \times P_s) = 3$ where $s \geq 3$, since it has vertices of odd degree.

Case 2. r is even, $r \geq 4$.

We use the 2-sum 3-flow and 0-sum 3-flow to build the labels of a zero-sum 3-flow of $R_r \times P_s$. As the Figure 2 shows, we may get a 0-sum 3-flow by labeling a 2-sum 3-flow in the first copy of R_r, and labeling by 0-sum 3-flows in the following copies of R_r, until labeling either a 2-sum 3-flow or a (-2)-sum 3-flow in the last copy of R_r, depending on s is even or odd, respectively. Over the edges of the path P_s part within the Cartesian product we label -2, 2 alternatively as in the Figure.

In the following, we show the existence of 2-sum 3-flow f_2 (hence (-2)-sum 3-flow by reversing the sign of each edge label) and 0-sum 3-flow f_0 over R_r:

Sub-case 2.1: $r = 4l + 4$, $R_r = E_1 \oplus E_2 \cdots \oplus E_{2l+2}$, $l = 0, 1, 2 \cdots$.
Let

$$f_2(e) = \begin{cases} 1 & , \quad e \in E_1 \cup \cdots \cup E_l \\ -1 & , \quad e \in E_{l+1} \cup \cdots \cup E_{2l} \\ 2 & , \quad e \in E_{2l+1} \\ -1 & , \quad e \in E_{2l+2} \end{cases}$$

$$f_0(e) = \begin{cases} 1 & , \quad e \in E_1 \cup \cdots \cup E_{l+1} \\ -1 & , \quad e \in E_{l+2} \cup \cdots \cup E_{2l+2} \end{cases}$$

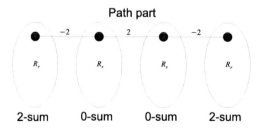

Fig. 2. A zero-sum 3-flow of $R_r \times P_s$

Sub-case 2.2: $r = 4l + 6$, $R_r = E_1 \oplus E_2 \cdots \oplus E_{2l+3}$, $l = 0, 1, 2 \cdots$.
Let

$$
f_2(e) = \begin{cases} 1 & , \quad e \in E_1 \cup \cdots \cup E_{l+1} \\ -1 & , \quad e \in E_{l+2} \cup \cdots \cup E_{2l+2} \\ 1 & , \quad e \in E_{2l+3} \end{cases}
$$

$$
f_0(e) = \begin{cases} 1 & , \quad e \in E_1 \cup \cdots \cup E_l \\ -1 & , \quad e \in E_{l+1} \cup \cdots \cup E_{2l} \\ 2 & , \quad e \in E_{2l+1} \\ -1 & , \quad e \in E_{2l+2} \cup E_{2l+3} \end{cases}
$$

Follows from the above two sub-cases and Lemma 3, $F(R_r \times P_s) = 3$.

Case 3. $r = 2$.
Note that clearly $R_2 \times P_2$ is a 3-regular graph with 1-factor, so we can label -1 in 2-regular part and 2 in 1-factor part. For $s \geq 3$, similar as Case 1, consider $R_2 \times P_s$ as the union of $s - 1$ copies of $R_2 \times P_2$'s so that they overlap with each other along R_2 part. We simply use the zero-sum 2-flow of $R_2 \times P_2$ given previously to build the label of $R_2 \times P_s$. By adding up the edge labels of ± 1 on the overlap part, we have the edge labels of ± 2, hence a zero-sum 3-flow. Therefore by Lemma 3 again, $F(R_2 \times P_s) = 3$ for $s \geq 2$, since it contains odd vertices. $\square$

Acknowledgment. The authors wish to express their sincere thanks for the referee's comments and corrections.

References

1. Akbari, S., Daemi, A., Hatami, O., Javanmard, A., Mehrabian, A.: Zero-Sum Flows in Regular Graphs. Graphs and Combinatorics 26, 603–615 (2010)
2. Akbari, S., Ghareghani, N., Khosrovshahi, G.B., Mahmoody, A.: On zero-sum 6-flows of graphs. Linear Algebra Appl. 430, 3047–3052 (2009)
3. Akbari, S., et al.: A note on zero-sum 5-flows in regular graphs. arXiv:1108.2950v1 [math.CO] (2011)
4. Bouchet, A.: Nowhere-zero integral flows on a bidirected graph. J. Combin. Theory Ser. B 34, 279–292 (1983)
5. Gallai, T.: On factorisation of grahs. Acta Math. Acad. Sci. Hung 1, 133–153 (1950)
6. Jaeger, F.: Flows and generalized coloring theorems in graphs. J. Combin. Theory Ser. B 26(2), 205–216 (1979)
7. Kano, M.: Factors of regular graph. J. Combin. Theory Ser. B 41, 27–36 (1986)
8. Petersen, J.: Die Theorie der regularen graphs. Acta Mathematica (15), 193–220 (1891)
9. Seymour, P.D.: Nowhere-zero 6-flows. J. Combin. Theory Ser. B 30(2), 130–135 (1981)
10. Tutte, W.T.: A contribution to the theory of chromatic polynomials. Can. J. Math. 6, 80–91 (1954)

11. Wang, T.-M., Hu, S.-W.: Constant Sum Flows in Regular Graphs. In: Atallah, M., Li, X.-Y., Zhu, B. (eds.) FAW-AAIM 2011. LNCS, vol. 6681, pp. 168–175. Springer, Heidelberg (2011)
12. Wang, T.-M., Hu, S.-W.: Nowhere-zero constant-sum flows of graphs. Presented in the 2nd India-Taiwan Conference on Discrete Mathematics, Coimbatore, Tamil Nadu, India (September 2011) (manuscript)
13. West, D.B.: Introduction to Graph Theory, 2nd edn. Prentice Hall, Englewood Cliffs (2001)

More Efficient Parallel Integer Sorting

Yijie Han[1] and Xin He[2]

[1] School of Computing and Engineering,
University of Missouri at Kansas City,
Kansas City, MO 64110, USA
hanyij@umkc.edu
[2] Department of Computer Science and Engineering,
University at Buffalo, The State University of New York,
201 Bell Hall,
Buffalo, NY 14260-2000, USA
xinhe@buffalo.edu

Abstract. We present a more efficient CREW PRAM algorithm for integer sorting. This algorithm sorts n integers in $\{0, 1, 2, ..., n^{1/2}\}$ in $O((\log n)^{3/2} / \log \log n)$ time and $O(n(\log n / \log \log n)^{1/2})$ operations. It also sorts n integers in $\{0, 1, 2, ..., n - 1\}$ in $O((\log n)^{3/2} / \log \log n)$ time and $O(n(\log n / \log \log n)^{1/2} \log \log \log n)$ operations. Previous best algorithm [13] on both cases has time complexity $O(\log n)$ but operation complexity $O(n(\log n)^{1/2})$.

Keywords: Algorithms, design of algorithms, bucket sorting, integer sorting, PRAM algorithms.

1 Introduction

Sorting is a classical problem which has been studied by many researchers [1][2][3][6][11][12][13][14][16][17][18][19]. For elements in an ordered set comparison sorting can be used to sort the elements. In the case when a set contains only integers both comparison sorting and integer sorting can be used to sort the elements. Since elements of a set are usually represented by binary numbers in a digital computer, integer sorting can, in many cases, replace comparison sorting. In this paper we study parallel integer sorting and present an algorithm which outperforms the operation complexity of the best previous result.

The parallel computation model we use is the PRAM model[15] which is used widely by parallel algorithm designers. Usually three variants of PRAM models are used in the design of parallel algorithms, namely the EREW (Exclusive Read Exclusive Write) PRAM, the CREW (Concurrent Read Exclusive Write) PRAM and the CRCW (Concurrent Read Concurrent Write) PRAM[15]. In a PRAM model a processor can access any memory cell. On the EREW PRAM simultaneous access to a memory cell by more than one processor is prohibited. On the CREW PRAM processors can read a memory cell simultaneously, but simultaneous write to the same memory cell by several processors is prohibited.

J. Snoeyink, P. Lu, K. Su, and L. Wang (Eds.): FAW-AAIM 2012, LNCS 7285, pp. 279–290, 2012.
© Springer-Verlag Berlin Heidelberg 2012

On the CRCW PRAM processors can simultaneously read or write to a memory cell. The CREW PRAM is a more powerful model than the EREW PRAM. The CRCW PRAM is the most powerful model among the three variants.

Parallel algorithms can be measured either by their time complexity and processor complexity or by their time complexity and operation complexity which is the time processor product. A parallel algorithm with small time complexity is regarded as fast while a parallel algorithm with small operation complexity is regarded as efficient. The operation complexity of a parallel algorithm can also be compared with the time complexity of the best sequential algorithm for the same problem. Let T_1 be the time complexity of the best sequential algorithm for a problem, T_p be the time complexity of a parallel algorithm using p processors for the same problem. Then $T_p \cdot p \geq T_1$. That is, T_1 is a lower bound for the operation complexity of any parallel algorithm for the problem. A parallel algorithm is said to be optimal if its operation complexity matches the time complexity of the best sequential algorithm, i.e. $T_p \cdot p = O(T_1)$.

On the CREW PRAM the best previous integer sorting algorithm [13] sorts n integers in $O(\log n)$ time and $O(n(\log n)^{1/2})$ operations. In this paper we study the problem of sorting n integers in $\{0, 1, ..., n^{1/2}\}$ and in $\{0, 1, ..., n-1\}$. The best previous result for these two cases due to Han and Shen [13] also sorts in $O(\log n)$ time and $O(n(\log n)^{1/2})$ operations. In this paper we present a CREW PRAM algorithm which sorts n integers in $\{0, 1, ..., n^{1/2}\}$ in $O((\log n)^{3/2}/\log \log n)$ time and $O(n(\log n/\log \log n)^{1/2})$ operations. It also sorts n integers in $\{0, 1, ..., n-1\}$ in $O((\log n)^{3/2} \log \log n)$ time and $O(n(\log n/\log \log n)^{1/2} \log \log \log n)$ operations.

When randomization is used usually better or even optimal algorithms can be achieved. Rajasekaran and Reif first achieved an optimal randomized parallel sorting algorithm [18]. Reif and Valiant first achieved an optimal randomized parallel network sorting algorithm [19].

Parallel integer sorting is such a fundamental problem in parallel algorithm design and many renowned researchers worked on this problem relentlessly. The milestones on parallel integer sorting on exclusive write PRAMs include 1997 Albers and Hagerup's paper [2] published on Information and Computation and 2002 Han and Shen's improvement [13] published on SIAM Journal on Computing. There are many results of many researchers published before Albers and Hagerup's work without much progress passing over the $O(n \log m)$ operations for sorting n integers in $\{0, 1, .., m - 1\}$. After Han and Shen's work there is virtually no progress ever since. We worked very hard and only achieved the not so big improvements presented in this paper. To our experience significant improvement over Han and Shen's work [13] on the operation complexity for parallel integer sorting is very difficult. So to speak that the results we have achieved and presented here is significant.

2   Nonconservative Sorting

First we will show the EREW PRAM algorithm in [13] to sort $n_1 = 2^{4(\log n)^{1/2}}$ integers in $\{0, 1, ..., 2^{(\log n)^{1/2}}\}$ with word length (the number of bits in a word)

$\log n$. This algorithm is based on the AKS sorting network[1], Leighton's column sort[16], Albers and Hagerup's test bit technique[2] and the Benes permutation network[4][5].

Because the word length is $O(\log n)$ we can store $c(\log n)^{1/2}$ integers in a word for a small constant c. Using the test bit technique[2][3] we can do pairwise comparison of the $c(\log n)^{1/2}$ integers in a word with the $c(\log n)^{1/2}$ integers in another word in constant time using one processor. Moreover, using the result of the comparison the $c(\log n)^{1/2}$ larger integers in all pairs in the two words under comparison can be extracted into one word and the $c(\log n)^{1/2}$ smaller integers in all pairs in these two words can be extracted into another word and this can also be done in constant time using one processor[2][3]. Without loss of generality we may also assume that $c(\log n)^{1/2}$ is a power of 2. We first pack n_1 input integers into $n_2 = n_1/(c(\log n)^{1/2})$ words with each word containing $c(\log n)^{1/2}$ integers. We then imagine an AKS sorting network [1] being built on these n_2 words. On the AKS sorting network we compare two words at each internal node of the network. Thus each node of the AKS sorting network can be used to compare the $c(\log n)^{1/2}$ integers in one word with the $c(\log n)^{1/2}$ integers in another word in parallel. At the output of the AKS sorting network we have sorted $c(\log n)^{1/2}$ sets with the i-th set containing i-th integers in all n_2 words. In terms of Leighton's column sort[16] we can view that we place n_1 integers in $c(\log n)^{1/2}$ columns with each column containing n_2 integers. The i-th column, $0 \le i < c(\log n)^{1/2}$, contains the i-th integer of every word. At the output of the AKS sorting network, every column is sorted. The principle of Leighton's column sort says that to sort n_1 integers we need only to sort all $c(\log n)^{1/2}$ columns independently and concurrently for a constant number of times (passes) and perform a fixed permutation among the n_1 integers after each pass. Besides, these fixed permutations are simple permutations such as shuffle, unshuffle and shift. Applying the column sort principle, we perform a fixed permutation among the n_1 integers when they are output from the AKS sorting network after each pass. The permutation can be done by disassembling the integers from the words, applying the permutation and then reassembling the integers into words. Thus each pass consisting of sorting on columns and then permutation can be done in $O((\log n)^{1/2})$ time and $O(n_1)$ operations. According to Leighton's column sort we need only a constant number of passes in order to have all the n_1 integers sorted. Thus the sorting of n_1 integers can be done in $O((\log n)^{1/2})$ time and $O(n_1)$ operations.

For our purpose (see later section that we have integers not in an array but in a linked list) we also need the following scheme to accomplish the permutation mentioned above. The permutation should be done by routing the integers through a network N which is the butterfly network in conjunction with a reverse butterfly network(see Fig. 1.). Network N can be used to emulate the Benes permutation network[4][5] to perform permutations.

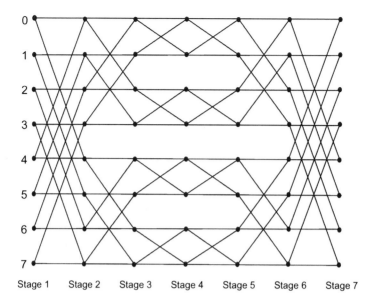

Fig. 1. A permutation network

Each stage of the butterfly network emulates the processor connection along a dimension on the hypercube and switches integers between words or within words (within words means each integer is switched with another integer in the same word. This is where we need $c(\log n)^{1/2}$ to be a power of 2). Because $c(\log n)^{1/2}$ is a power of 2 each stage of the butterfly network can be done in constant time even when integers are switched within words. Because butterfly network has $O((\log n)^{1/2})$ stages, the permutation can be done in $O((\log n)^{1/2})$ time. Because there are only n_2 words the operation complexity is time×processors= $O((\log n)^{1/2}) \times n_2 = O(n_1)$. Note that since the permutations we performed here are fixed permutations according to Leighton's column sort, the setting of the switches in the butterfly network can be precomputed (according to the way Benes permutation network is used to perform permutations).

The following Lemmas 1 and 2 are the cornerstone of the paper [13] on SIAM Journal on Computing.

Sorting integers into linked lists means, after sorting, integers of the same value are in the same linked list and integers of different values are in different linked lists. It does not imply integers of the same value are packed into consecutive locations.

Lemma 1 [13]: n integers in the range $\{0, 1, ..., 2^{(\log n)^{1/2}}\}$ can be sorted into linked lists on the EREW PRAM with word length $O(\log n)$ in $O((\log n)^{1/2})$ time using $O(n)$ operations and $O(n)$ space.   □

Lemma 2 [13]: n' integers in $\{0, 1, ..., 2^{t(\log n)^{1/2}}\}$ can be sorted into linked lists on the EREW PRAM with word length $\log n$ in $O(t(\log n)^{1/2})$ time and $O(tn')$ operations.   □

Here in Lemma 2 n' is not related to n. Lemma 2 is essentially the t iterations of execution of Lemma 1.

Note that the result of Cook et al. [7] says that if we sort these integers in an array it will need $\Omega(\log n)$ time. The property of sorting into linked lists and the small range of values for integers enabled Lemmas 1 and 2 to be proved in [13].

3 Sorting n Integers in $\{0, 1, ..., 2^{c(\log n \log \log n)^{1/2}}\}$

We consider the problem of sorting n integers in the range $\{0, 1, ..., 2^{c(\log n \log \log n)^{1/2}}\}$ on the CREW PRAM with word length $O(\log n)$, where c is a small constant. For our purpose we assume that $(\log n/\log \log n)^{1/2}$ is a power of 2.

In the first stage we pack every $(\log n/\log \log n)^{1/2}$ integer into a word (called original word later). This results in a set S_1 of $n_3 = n/(\log n/\log \log n)^{1/2}$ words. We now show how to sort these n_3 words in S_1.

The first step of this stage is to sort the integers (each having $c(\log n \log \log n)^{1/2}$ bits) within each word. This is done by a table lookup because we can precompute such a table of size n^c. This takes constant time (here we used concurrent read).

Then we take the most significant $(\log \log n)/4$ bits from each integer in each word and pack them together to obtain a word containing $(\log n/\log \log n)^{1/2}$ $(\log \log n)/4$ bits. We first use a mask to extract these bits as shown in the first step in Fig. 2 (Applying mask). We cannot pack these extracted bits in a word together independently for each word because of complexity considerations. Therefore we shift the bits in a word and then do bitwise OR with another word to combined two words into one word, and we repeatedly do this (repeat $\log \log \log n$ times) to combine $\log \log n$ words into one word. This is step 2 in Fig. 2 (Shift and bitwise OR) and takes $O(\log \log \log n)$ time and $O(n_3)$ operations. Now all the extracted bits are stores in $n_3/\log \log n$ words. Within each words there are null bits between two blocks of extracted bits and therefore we pack extracted bits to let them occupy consecutive bits in a word. We do this independently for each word and because there are $n_3/\log \log n$ words we can afford this. This is the step 3 in Fig. 2 (Compack). This step takes $O(\log \log n)$ time (Because there are $(\log n/\log \log n)^{1/2}$ blocks of extracted bits in one word. Using constant operations we can reduce the number of blocks in a word w by half by taking half of the blocks in w out and put them in another word w_1 then shift bits in w_1 and then do $w\,OR\,w_1$.) and $O(n_3/\log \log n \times \log \log n) = O(n_3)$ operations. Now although extracted bits are packed, the order they appear in a word is *extracted bits from original word1; extracted bits from original word2; ...; extracted bits from original word(*$\log \log n$*); extracted bits from original word1; extracted bits from original word2; ...; extracted bits from original word(*$\log \log n$*);....* Extracted bits come from different original words because of step 2 in Fig. 2. Our objective is to pack extracted bits in each original word and store them in one word. Therefore we now do step 4 in Fig 2. (Applying mask) and in $\log \log \log n$ steps and $O(n_3)$ operations we separate (disassemble) one word into $\log \log n$ words and extracted bits from each original word is now in an independent word. Because of step 3 in Fig. 2 the extracted bits are somewhat compacked in a word and therefore we can again combine words together. This times we can let extracted bits from one original word being consecutive but not compacked. This is step 5 in Fig. 2 (Shift and then bitwise OR). This step takes $O(\log \log \log n)$ time and $O(n_3)$ operations. Now again we have put all extracted bits in $n_3/\log \log n$ words. And now we do step 6 in Fig. 2 (compack) independently for each word. The complexity of this step is similar to that of step 3 (but now we have $(\log n/\log \log n)^{1/2} \log \log n$ blocks) and takes $O(\log \log n)$ time and $O(n_3)$ operations. Now we have extracted bits from each original word packed in consecutive bits of a word. Now we do step 7 in Fig. 2, i.e. separate extracted bits from each original word into an independent word. This step is similar to step 4 and takes $O(\log \log n)$ time and $O(n_3)$ operations.

Thus it takes $O(\log \log n)$ time and $O(n_3)$ operations for all the steps in Fig. 2. We call the set of these words obtained at the end of Fig. 2 S_2. Note that because many extracted bits in an original word have the same value (there are more integers in a word ($(\log n/\log \log n)^{1/2}$ of them) than the number of different values of extracted bits ($2^{(\log \log n)/4}$ of them) and integers within an original word has been sorted, therefore a word in S_2 of $(\log n/\log \log n)^{1/2}(\log \log n)/4$ bit can have only $\sum_{i=1}^{2^{(\log \log n)/4}-1} \binom{(\log n/\log \log n)^{1/2}}{i} < (\log n)^{1/2}$ values (different sorted situation corresponds to different ways of setting the position of first integer (extracted bits) among the integers (extracted bits') of the same value (except the first integer which assumes position 0)). Thus a word in S_2 can be uniquely represented by an integer i within 0 and $2^{(\log n)^{1/2}}-1$. Therefore i can be represented using no more than $(\log n)^{1/2}$ bits. Again we can use table lookup to convert a $(\log n/\log \log n)^{1/2}(\log \log n)/4$ bit integer in S_2 to an integer of $(\log n)^{1/2}$ bits. We let set S_3 to be the set of $(\log n)^{1/2}$-bit integers converted from integers in S_2. Each word in S_3 corresponds to a word in S_1.

We now partition the n_3 words of S_3 into $n_3/2^{4(\log n)^{1/2}}$ groups with each group containing $2^{4(\log n)^{1/2}}$ words. We then sort every group concurrently using the algorithm in Section 2. We spend $O((\log n)^{1/2})$ time and $O(n_3)$ operations.

We may assume that every integer value in $\{0, 1, ..., 2^{(\log n)^{1/2}}-1\}$ (for a word) exists in each group. If such an integer value does not exist within a group we add a dummy word to the group to represent this integer value. We thus added no more than $2^{(\log n)^{1/2}}$ dummy words to each group which account for a very small fraction of the total number of words in the group. Now because words in each group has been sorted we can make $2^{(\log n)^{1/2}}$ linked lists for each group with each linked list linking all integers with the same integer value in the group together. Then we join a linked list for integer value i in a group g with lined lists for integer value i of g's left and right neighboring groups. With the help of dummies we thus obtained $2^{(\log n)^{1/2}}$ linked lists for all groups.

Now we can link words in S_1 the same way as we link words in S_3 because each word in S_1 corresponds to a word in S_3. The time complexity is $O((\log n)^{1/2})$ and the operation complexity is $O(n_3)$.

This accomplishes the first stage.

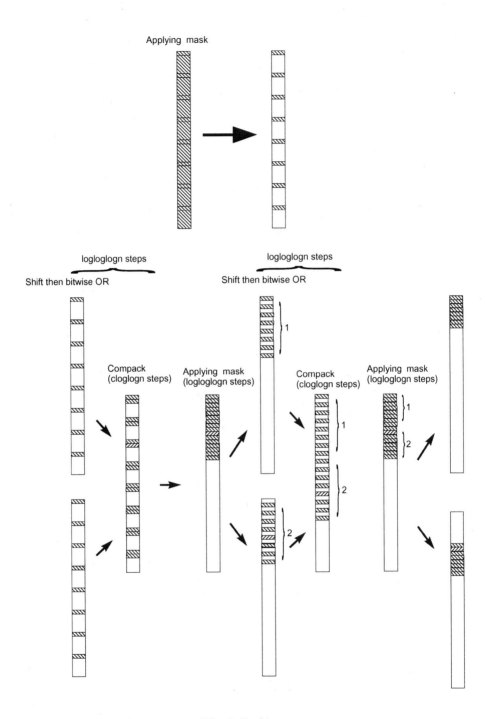

Fig. 2. Packing

In each subsequent stage we take the next $(\log \log n)/4$ bits from each integer in a word in S_1 to form a word in set S_2 (now there is a set S_2 for each linked list). Then from S_2 we obtain S_3 (again one for each linked list) and then we sort each group of S_3 of each linked list.

Now we discuss how each linked list is split in each stage. Elements in each linked list are sorted (using the sorting algorithm in Section 2 and here we need to do permutation in the sorting algorithm using the butterfly network, see also [13]) in each stage and this linked list is going to be split into multiple linked lists such that elements of the same value will be in the same linked list and elements of different value are sorted into different linked lists.

A linked list is short if it contains less than $2^{4(\log n)^{1/2}}$ elements (words), is long if it contains at least $2^{4(\log n)^{1/2}}$ elements. We first group every consecutive S elements in a linked list into one group. For a short linked list S is the number of total elements in the linked list. For a long linked list S varies from group to group but is at least $2^{4(\log n)^{1/2}}$ and no more than $2^{5(\log n)^{1/2}}$.

If the linked list is short there is only one group in the linked list. The sorting will then enable us to split the linked list into $t \le 2^{(\log n)^{1/2}}$ linked lists such that each linked list split contains all words whose integer values are the same, where t is the number of different integer values. Here we note that for short linked list t could be less than $2^{(\log n)^{1/2}}$ (for example if all integer values are the same t will be equal to 1).

If the linked list is long we will always split the linked list into exactly $2^{(\log n)^{1/2}}$ linked lists no matter how many different integer values are there. After sorting in each group, words in each group are split into $2^{(\log n)^{1/2}}$ linked lists. If an integer value among the $2^{(\log n)^{1/2}}$ values does not exist we create a linked list containing only one dummy element representing this integer value. Again as we stated above, no more than $2^{(\log n)^{1/2}}$ dummy elements will be created for each group. For consecutive (neighboring) groups on a long linked list we then join the split linked lists in the groups such that linked lists with the same integer values are joined together. With the help of those dummy elements we now have split a long linked list into exactly $2^{(\log n)^{1/2}}$ linked lists.

With the existence of dummy elements in the linked list, the splitting process should be modified a little bit. For a short linked list, after the grouping all dummy elements will be eliminated. For a long linked list, the dummy elements will also be eliminated after grouping, but new dummy elements could be created.

Since each group on a long linked list has at least $2^{4(\log n)^{1/2}}$ elements and since each such a group creates at most $2^{(\log n)^{1/2}}$ dummy elements, the total number of dummy elements created in a stage is at most $n_3/2^{3(\log n)^{1/2}}$. Dummy elements generated in a stage are eliminated in the next stage and new dummy elements are generated for the next stage. For a total of $O((\log n)^{1/2})$ stages the total number of dummy elements generated is no more than $O(n_3(\log n)^{1/2}/2^{3(\log n)^{1/2}})$.

Because integers are now on linked lists, linked list contraction is needed to form groups. This paragraph describes linked list contraction and is somewhat involved. Readers who are not very familiar with symmetry breaking and linked list contraction can skip this paragraph. We apply symmetry breaking schemes by Han[9][10] and Beame[8] to break a linked list into sublists of length no more than $\log^{(c)} n$ in $O(\log c)$ time for a constant c. Pointer jumping[20] is then executed for each sublist. When pointer jumping finishes the sublist is contracted into one node. Since the length of these sublists are different some sublists finish pointer jumping faster and some sublists finish pointer jumping slower. If a sublist is contracted into a single node v, the processor associated with v checks to see if the neighboring sublists also have been contracted into single nodes. If one of its neighboring sublist is contracted into a single node then nodes representing the sublists form a new list and symmetry breaking and pointer jumping can be applied to this new list. And therefore the contraction process continues. If v finds out that both of its neighboring sublist have not finished pointer jumping then v becomes inactive. In this case v will be picked up (activated and contracted together with) by the contracted node representing the neighboring sublist which first finishes pointer jumping. We define one step for a node as first picking up its inactive neighbors and then if it is still active performing symmetry breaking and a pointer jump. This whole contraction process can be viewed as contracting a linked list of length l to a linked list of length $2l/3$ in a step because if a node is inactive then both of its neighbors are active in the contraction process. Thus to contract S elements into a node takes only $O(\log S)$ time. For a long linked list each group can be kept between $2^{6(\log n)^{1/2}}$ and $2^{7(\log n)^{1/2}}$. Thus for each stage the contraction can thus be done in $O((\log n)^{1/2})$ time with $O(S \log^{(c+1)} n)$ operations for each group ($O(n_3 \log^{(c+1)} n)$ operations for all linked lists). This factor of $\log^{(c+1)} n$ is introduced because of pointer jumping. We can remove this $\log^{(c+1)} n$ factor because we can pack $c(\log n)^{1/2}$ words in S_3 into one word and therefore the pointer jumping needs not to be done by every word in S_3. Thus the linked list contraction takes $O((\log n)^{1/2})$ time and $O(n_3)$ operations.

More complications of this process such as where to store dummy elements, how to move words to sorted position, etc., are explained in [13].

Let us estimate the complexity. Because each stage removes $c(\log n \log \log n)^{1/2}$ bits. there are $O((\log n/ \log \log n)^{1/2})$ stages. Because each stage takes $O((\log n)^{1/2})$ time the time for our algorithm in this section is $O(\log n/ (\log \log n)^{1/2})$. Each stage takes $O(n_3)$ operations and therefore for all stages it has $O(n_3 \log n/(\log \log n)^{1/2}) = O(n)$ operations.

Now for each linked list L, the words of S_3 on L are all having the same value (i.e. the j-th integer in all these words are the same. However, the i-th integer and the j-th integer may be different.). We can group every $(\log n/ \log \log n)^{1/2}$ words on L together and do a transposition (put the j-th integer in all these words in one word). This takes $O(\log \log n)$ time and $O(n_3 \log \log n)$ operations (this should be simple and readers can work it out or see [13]). After that we sort the transposed words into linked lists in $O((\log n \log \log n)^{1/2})$ time and $O(n_3(\log \log n)^{1/2})$ operations using Lemma 2 (note that now each word contains $(\log n/ \log \log n)^{1/2}$ integers of the same value which is in $\{0, 1, ..., 2^{c(\log n \log \log n)^{1/2}} - 1\}$).

Thus we have:

Theorem 1: n integers in $\{0, 1, ..., 2^{c(\log n \log \log n)^{1/2}}\}$ can be sorted into linked lists on the CREW PRAM with word length $\log n$ in $O(\log n/(\log \log n)^{1/2})$ time and $O(n)$ operations.   □

4   Sorting Integers in $\{0, 1, ..., n^{1/2}\}$ and in $\{0, 1, ..., n-1\}$

To sort n integers in $\{0, 1, ..., n^{1/2}\}$ We apply Theorem 1 $(1/(2c))(\log n/\log \log n)^{1/2}$ times and reach

Theorem 2: n integers in $\{0, 1, ..., n^{1/2}\}$ can be sorted on the CREW PRAM with word length $\log n$ in $O((\log n)^{3/2}/\log \log n)$ time and $O(n(\log n/\log \log n)^{1/2})$ operations.   □

The situation for sorting n integers in $\{0, 1, ..., n-1\}$ is different. For sorting the most significant $\log n/2$ bits we can apply Theorem 2. After that n integers are partitioned into $n^{1/2}$ sets and we have to sort every set concurrently and independently. Here on the average each set has $n^{1/2}$ integers. When we are sorting $n^{1/2}$ integers we cannot pack every $(\log n/\log \log n)^{1/2}$ integers to form words of $\log n$ bits in paragraph 2 of Section 3 (if the algorithm in Section 3 is well understood then one can see that n integers corresponds to $\log n$ bits for sorting). To sort $n^{1/2}$ integers we can use only $\log n/2$ bits and therefore we can pack only $(1/2)(\log n/\log \log n)^{1/2}$ integers in $\{0, 1, ..., 2^{c(\log n \log \log n)^{1/2}}\}$ into one word. However, because the number of bits is reduced by half the number of stages in the algorithm of Theorem 1 is also reduced by half. Thus sorting the next $\log n/4$ bits has half the time complexity but the same operation complexity as sorting the most significant $\log n/2$ bits. Again sorting the next $\log n/8$ bits takes the $1/4$ time complexity and the same operation complexity as sorting the most significant $\log n/2$ bits.

Thus if we iterate t times we spend $O((\log n)^{3/2}/\log \log n)$ time and $O(tn(\log n/\log \log n)^{1/2})$ operations and we have $\log n/2^t$ bits left to be sorted. By Lemma 2 the remaining $\log n/2^t$ bits can be sorted in $O((\log n)^{1/2}/2^t)$ time and $O(n(\log n)^{1/2}/2^t)$ operations. Now to pick the optimal t let

$$tn(\log n/\log \log n)^{1/2} = n(\log n)^{1/2}/2^t$$

and we obtain that $t = (\log \log \log n)/2$. Thus we have that

Theorem 3: n integers in $\{0, 1, ..., n-1\}$ can be sorted on the CREW PRAM with word length $\log n$ in $O((\log n)^{3/2}/\log \log n)$ time and $O(n(\log n/\log \log n)^{1/2} \log \log \log n)$ operations.   □

5   Conclusions

We presented a CREW integer sorting algorithm which outperforms the operation complexity of previous best result. Many problems remains open such as: can we remove concurrent read from our algorithm? can time complexity be lowered to $O(\log n)$? can we sort integers with value larger than n? etc.. Note that

Han proved before [11] that n integers in $\{0, 1, ..., m - 1\}$ can be sorted on the EREW PRAM in $O((\log n)^2)$ time and $O(n(\log \log n)^2 \log \log \log n)$ operations provided that $\log m \geq (\log n)^2$. This provides a partial solution to one of the open problems mentioned here. Our hunch is that removing the restriction of integers being bounded by n probably should be the next target to achieve. We hope our future research will resolve some of the open problems mentioned here.

References

1. Ajtia, M., Komlós, J., Szemerédi, E.: Sorting in $c \log n$ parallel steps. Combinatorica 3, 1–19 (1983)
2. Albers, S., Hagerup, T.: Improved parallel integer sorting without concurrent writing. Information and Computation 136, 25–51 (1997)
3. Andersson, A., Hagerup, T., Nilsson, S., Raman, R.: Sorting in linear time? In: Proc. 1995 Symposium on Theory of Computing, pp. 427–436 (1995)
4. Benes, V.E.: On rearrangeable three-stage connecting networks. Bell Syst. Tech. J. 41, 1481–1492 (1962)
5. Benes, V.E.: Mathematical Theory of Connecting Networks and Telephone Traffic. Academic, New York (1965)
6. Chen, S., Reif, J.H.: Using difficulty of prediction to decrease computation: fast sort, priority queue and convex hull on entropy bounded inputs. In: 34th Annual IEEE Conference on Foundations of Computer Science (FOCS 1993) Proceedings, Palo Alto, CA, pp. 104–112 (November 1993)
7. Cook, S., Dwork, C., Reischuk, R.: Upper and Lower Time Bounds for Parallel Random Access Machines without Simultaneous Writes. SIAM J. Comput. 15(1), 87–97 (1986)
8. Goldberg, A.V., Plotkin, S.A., Shannon, G.E.: Parallel symmetry-breaking in sparse graphs. SIAM J. on Discrete Math. 1(4), 447–471 (1988)
9. Han, Y.: Matching partition a linked list and its optimization. In: Proc. 1989 ACM Symposium on Parallel Algorithms and Architectures (SPAA 1989), Santa Fe, Mexico, pp. 246–253 (June 1989)
10. Han, Y.: An optimal linked list prefix algorithm on a local memory computer. In: Proc. 1989 Computer Science Conference (CSC 1989), pp. 278–286 (February 1989)
11. Han, Y.: Improved fast integer sorting in linear space. Information and Computation 170(1), 81–94 (2001)
12. Han, Y.: Deterministic sorting in $O(n \log \log n)$ time and linear space. Journal of Algorithms 50, 96–105 (2004)
13. Han, Y., Shen, X.: Parallel integer sorting is more efficient than parallel comparison sorting on exclusive write PRAMs. SIAM J. Comput. 31(6), 1852–1878 (2002)
14. Hightower, W.L., Prins, J., Reif, J.H.: Implementations of randomized sorting on large parallel machines. In: 4th Annual ACM Symposium on Parallel Algorithms and Architectures (SPAA 1992), San Diego, CA, pp. 158–167 (July 1992)
15. JáJá, J.: An Introduction to Parallel Algorithms. Addison-Wesley (1992)
16. Leighton, T.: Tight bounds on the complexity of parallel sorting. IEEE Trans. Comput. C-34, 344–354 (1985)
17. Reif, J.H.: An $n^{1+\epsilon}$ processor, $O(\log n)$ time probabilistic sorting algorithm. In: SIAM 2nd Conference on the Applications of Discrete Mathematics, Cambridge, MA, pp. 27–29 (June 1983)

18. Rajasekaran, S., Reif, J.H.: An optimal parallel algorithm for integer sorting. In: 26th Annual IEEE Symposium on Foundations of Computer Science, Portland, OR, pp. 496–503 (October 1985); Published as Optimal and sublogarithmic time randomized parallel sorting algorithms. SIAM Journal on Computing 18(3), 594–607 (1989)
19. Valiant, L.G., Reif, J.H.: A Logarithmic Time Sort for Linear Size Networks. In: 15th Annual ACM Symposium on Theory of Computing, Boston, MA, pp. 10–16 (April 1983); Published in Journal of the ACM(JACM) 34(1), 60–76 (1987)
20. Wyllie, J.C.: The complexity of parallel computation, TR 79-387, Department of Computer Science, Cornell University, Ithaca, NY (1979)

Fast Relative Lempel-Ziv Self-index
for Similar Sequences

Huy Hoang Do[1], Jesper Jansson[2], Kunihiko Sadakane[3], and Wing-Kin Sung[1]

[1] National University of Singapore, COM 1, 13 Computing Drive, Singapore 117417
{hoang,ksung}@comp.nus.edu.sg
[2] Ochanomizu University, 2-1-1 Otsuka, Bunkyo-ku, Tokyo 112-8610, Japan
Jesper.Jansson@ocha.ac.jp
[3] National Institute of Informatics, 2-1-2 Hitotsubashi, Tokyo 101-8430, Japan
sada@nii.ac.jp

Abstract. Recent advances in biotechnology and web technology are generating huge collections of similar strings. People now face the problem of storing them compactly while supporting fast pattern searching. One compression scheme called *relative Lempel-Ziv compression* uses textual substitutions from a reference text as follows: Given a (large) set S of strings, represent each string in S as a concatenation of substrings from a reference string R. This basic scheme gives a good compression ratio when every string in S is similar to R, but does not provide any pattern searching functionality. Here, we describe a new data structure that supports fast pattern searching.

1   Introduction

There is an increasing need for indexing methods that can store collections of *similar* strings (or repetitive text) compactly while supporting fast pattern searching queries. For example, in genomic applications, the sequencing of individual genomes is becoming a feasible task. The "1000 Genomes Project" [1], aimed at characterizing common human genetic variations, has already sequenced the partial genomes of a large number of persons from various populations. In the near future, researchers will face the problem of storing those individual (and highly similar) genomic sequences compactly and indexing them efficiently. As another example, Wikipedia documents are continually modified and snapshots are taken every day to remember older versions of the data. Typically, changes between versions are small. Hence, fast indexing methods for similar texts may allow people to search archived versions of Wikipedia documents quickly.

To compress a single string S of length n, methods that are guaranteed to achieve the empirical k-order entropy $nH_k(S)$ are often used. However, this entropy measurement may not be a good bound for repetitive texts whose repeats are longer than k. For example, the entropy for storing the text SS (where $|S| > k$) is greater than $2nH_k(S)$. On the other hand, one can easily encode the text in $nH_k(S)+O(1)$ space. Thus, there are methods which achieve the empirical k-order entropy, yet perform poorly for repetitive texts [24]. As a consequence,

J. Snoeyink, P. Lu, K. Su, and L. Wang (Eds.): FAW-AAIM 2012, LNCS 7285, pp. 291–302, 2012.

compression methods have been designed for specific types of repetitive texts in biology. Christley *et al.* [5] compressed DNA sequences with respect to a reference sequence, and BioCompress [12] and XM [3] are other repetitive compressors designed specifically for DNA. Alternative approaches include methods based on grammar compression and LZ77 compression [26] for general repetitive texts. These methods can store repetitive texts compactly, but do not allow random access to the compressed text directly. Some previous work has addressed this issue. Kreft and Navarro [14] provided the first efficient random access operations for the LZ77 method. Bille *et al.* [2] built additional data structures on top of an existing grammar-based compression scheme to allow random access of any region with only logarithmic extra time per query.

However, one important operation on large text databases is *indexing*, in which the occurrences of an arbitrary pattern inside the stored text need to be located quickly. Some specialized data structures for indexing repetitive texts have appeared recently. In a pioneering paper of Mäkinen *et al.* [19], a repetitive text is defined as a collection of strings of total length N, where the strings are assumed to be highly similar, each string length is approximately n, and the strings share an alphabet of size σ. They employed run-length encoding to reduce the redundancy of a suffix array structure. Their approach shrinks the total index size greatly, but the space of the index is still proportional to the number of strings. In another paper, Huang *et al.* [13] assumed that every string contains at most m' point mutations with respect to a reference string. They designed a space-efficient data structure of size $O(n \log \sigma + m' \log m')$ bits to encode all such strings. Although the resulting data structure is small, their approach cannot index certain other types of similar strings such as *genome rearrangements*, formed by swapping substrings in genomic sequences, efficiently. (When only a few such rearrangements have occurred, long substrings of the genomic sequences will be preserved; they just occur in a different order.) Kreft and Navarro [15] built a self-index based on LZ77 compression. If the text of length N can be compressed using m LZ77 phrases, their data structure is of size $2m \log N + m \log m + 5m \log \sigma + O(m) + o(N)$ bits, but the query time is $O(\ell^2 h + (\ell + occ) \log N)$, i.e., quadratic in the pattern length ℓ and also dependent on the maximal depth of the phrases $h \le m$. In another line of research, Claude and Navarro [6] proposed a self-index for grammar-based compression methods. It uses $O(r \log r) + r \log N$ bits, where r is the number of rules in their grammar, and the resulting query time is quadratic $(O((\ell^2 + h(\ell + occ)) \log r))$. Some results for LZ78 compression and FM-index were given in [8]; on the negative side, these methods require $O(NH_k)$ bits in the worst case, and they may not be good enough to index a repetitive text in practice [24] or in theory [23]. In summary, existing indexes for a set of similar strings either require: (1) a lot of space, (2) that the indexed text has some special structure, or (3) quadratic query time.

New Results: Our main contribution is a data structure that stores a set $\mathcal{S}$ of strings and a reference string R in asymptotically almost optimal space, while providing almost linear-time pattern searching queries, as follows:

Theorem 1. *Given a reference string R of length n over an alphabet Σ of size σ and a set of strings $\mathcal{S} = \{S_1, \ldots, S_t\}$ over Σ, let m be the smallest possible number of factors to represent $\mathcal{S}$ with respect to R. All exact occurrences of any query pattern P of length ℓ can be reported within either of the following space and time complexities:*

(a) $2nH_k(R) + 5.55n + O(m \log n)$ bits and $O(\ell(\log \sigma + \frac{\log n}{\log \log n}) + occ \cdot (\log n + \frac{\log m}{\log n}))$ query time; or, alternatively,

(b) $2nH_k(R) + 5.55n + O(m \log n \log \log n)$ bits and $O(\ell(\log \sigma + \log \log n) + occ \cdot (\log n + \frac{\log m}{\log n}))$ query time,

where occ is the number of occurrences of P.

In this paper, we assume that the reference R is given. In case no such R is available, we can apply the method of Kuruppu *et al.* [17] to find a suitable one.

We compress each sequence in $\mathcal{S}$ using a new variant of the relative Lempel-Ziv (RLZ) compression scheme from [16]. RLZ represents each $S_i \in \mathcal{S}$ as a concatenation of substrings of R (referred to as *factors*) obtained from the LZ77-like factorization of R. See Fig. 1 for an example. Experiments on large scale genomic data in [16] have shown that this method yields good compression ratios for repetitive texts even when parts of the sequence are rearranged.

Our pattern searching algorithm follows a "standard" strategy for strings decomposed into factors. It considers two cases: case 1, where the pattern P is a substring of a single factor; and case 2, where P crosses at least one boundary between two factors. (See Fig. 2.) For case 2, the pattern is usually divided into two parts: left and right. The left part ends at one end of a factor, and the right part begins at the start of another factor. Each part is searched independently and then joined together by an appropriate 2D range query data structure.

Although using the same basic strategy, the currently existing methods require quadratic time for pattern searching due to the fact that they need to re-search the left part and right part for each possible boundary between two factors. In our approach, we deploy multiple tricks to search all the possible left parts and right parts in only one run, and combine the results effectively. Notably, for the left part search (Section 4), we observe a mapping between the suffix array of the factors and the reference sequence, and then simulate the search in factors using the data structure for the reference. To implement the right part search (Sections 5-6), we use dynamic programming and backward search to utilize the results of previous searches. Note that these techniques are only valid because of the properties provided by the RLZ compression scheme. According to Theorem 1 above, the total space used by our data structure which supports fast pattern searching queries is only $O(nH_k(R) + n + m \log n)$ bits, where m is the minimal number of encoding factors. This is very close to the minimal space required by any RLZ variant, which is $\Omega(nH_k(R) + m \log n)$ bits.

We remark that recently, Gagie *et al.* [10] independently proposed a similar method to index a set of sequences. Their space complexity is $O(nH_k(R) + n + m(\log n + \log m \log \log m))$ bits, and the query time is $O((\ell + occ) \log^\epsilon n)$, where $\epsilon > 0$. (Thus, their method always uses more space than the method in our Theorem 1 (a) above but is faster, and is incomparable to the method in Theorem 1 (b).) Also note that in their method, the reference must be equal to one of the sequences in $\mathcal{S}$ since otherwise false occurrences may be reported.

The paper is organized as follows. Section 2 defines the notation used throughout the paper and outlines the framework of our new data structure. Section 3 describes some auxiliary data structures from the literature used in our construction. Section 3 also presents a new data structure for answering a restricted type of 2D range queries. Sections 4 – 6 describe further technical details of our main data structure. Due to space limitations, the focus will be on describing the construction of our new data structure; correctness proofs and additional intermediate explanations will be available in the full version of the paper.

2   Data Structure Framework

2.1   The Relative Lempel-Ziv (RLZ) Compression Scheme

Let R be a reference sequence of length n over an alphabet Σ and let $\mathcal{S} = \{S_1, \ldots, S_t\}$ be a given set of strings over Σ. Each sequence $S_i \in \mathcal{S}$ is compressed based on R by relative Lempel-Ziv (RLZ) compression [16], defined next: Given two strings S and R, where R contains all the symbols in S, the *Lempel-Ziv factorization* (or *parsing*) of S relative to R, denoted by $LZ(S|R)$, is a way to express S as a concatenation of substrings of the form $S = w_0 w_1 w_2 \ldots w_z$ such that: (1) w_0 is an empty string; and (2) w_i for $i > 0$ is a non-empty substring of S and w_i is the longest prefix of $S[(|w_0..w_{i-1}| + 1)..|S|]$ that occurs in R. Each substring w_i is called a *factor* (or *phrase*), and can be represented by a pair of numbers (p_i, l_i), where p_i is a starting position of w_i in R and l_i denotes the length of w_i. $LZ(S|R)$ can be computed in linear time [16]. By definition, the decomposition guarantees that no factor can be expanded any further to the right. Furthermore, the RLZ compression scheme has the following property:

Lemma 1. $LZ(S|R)$ *represents* S *using the smallest possible number of factors.*

$R = \text{ACGTGATAG}$

$S_1 = \text{TGATAGACG} = \text{TGATAG, ACG}\quad = 8\ 2$
$S_2 = \text{GAGTACTA}\quad = \text{GA, GT, AC, TA} = 5\ 6\ 1\ 7$
$S_3 = \text{GTACGT}\quad\quad = \text{GT, ACGT}\quad = 6\ 3$
$S_4 = \text{AGGA}\quad\quad\quad = \text{AG, GA}\quad\quad = 4\ 5$

(a)

$T[...]$	Factor	Pos. in R
1	AC	1..2
2	ACG	1..3
3	ACGT	1..4
4	AG	8..9
5	GA	5..6
6	GT	3..4
7	TA	7..8
8	TGATAG	4..9

(b)

Fig. 1. (a) A reference string R and a set of strings $\mathcal{S} = \{S_1, S_2, S_3, S_4\}$ decomposed into the smallest possible number of factors from R. (b) The array $T[1..8]$ (to be defined in Section 2) consists of the distinct factors sorted in lexicographical order.

We will need some more definitions. Let m be the minimum number of factors required to represent all of $\mathcal{S}$ with respect to R. Denote the Lempel-Ziv factorization of each S_i relative to R by $S_i = S_{i1}S_{i2}\ldots S_{ic_i}$ for $i = 1, 2, \ldots, t$. Next, take all the s *distinct factors* that appear in the factorizations for $\mathcal{S}$ and let $T[1..s]$ be an array containing these factors sorted in lexicographical order (see Fig. 1 (b)). Define $m = \sum_{i=1}^{t} c_i$. Note that $s \leq \min\{n^2, m\}$. Our data structure stores $T[1..s]$ in $O(s \log n)$ bits by encoding each $T[j]$ by its starting and ending positions in the reference string R, and the set $\mathcal{S}$ in $O(m \log s) = O(m \log n)$ bits by representing each $S_i \in \mathcal{S}$ as a list of indices from $T[1..s]$ (see Fig. 1 (a)).

Let $F[1..m]$ be the lexicographically sorted array of all non-empty suffixes in $\mathcal{S}$ that start with a factor; i.e., each element $F[y]$ is of the form $S_{ip}S_{i(p+1)}\ldots S_{ic_i}$, and is called a *factor suffix* from here on. For any string x, $\bar{x}$ denotes its reverse. Let $\overline{T}[1..s]$ be an array of all reversed distinct factors $\overline{S}_{ij}$ sorted lexicographically.

2.2   Pattern Searching

To find the occurrences of a query pattern P in $\mathcal{S}$, we follow the basic strategy briefly mentioned in Section 1. Suppose P is a query pattern. Each occurrence of P in $S_1, \ldots, S_t$ belongs to one of the following two main cases; see Fig. 2:

- **Case 1:** P lies completely inside one factor, denoted by S_{ip}.
- **Case 2:** P is not a substring of a single factor, i.e., $P = XS_{ip}\ldots S_{iq}Y$, where X is a suffix of $S_{i(p-1)}$ and Y is a prefix of $S_{i(q+1)}$.

(Observe that the case $P = XY$ is an instance of case 2.) To locate all occurrences of P, our data structure uses a number of auxiliary data structures, as explained next, to report all occurrences of P in $\mathcal{S}$ according to case 1 and case 2 separately. Summing all their complexities together yields Theorem 1 above. Let occ_1 and occ_2 be the number of occurrences of P as in case 1 and case 2, respectively.

Case 1: [P occurs inside a factor] We use the data structure $\mathcal{I}(T)$ defined in Section 4 to find all occurrences of P in $O(|P| + occ_1 \log n)$ time (Theorem 2). The data structure is of size $2n + o(n) + O(s \log n)$ bits. See Section 4 for details.

Case 2: [P is not a substring of a single factor] As illustrated in Fig. 2, in this case, every occurrence of P can be divided into two parts: the *left part* (the suffix of a factor), and the *right part* (starting with a factor). We use three additional

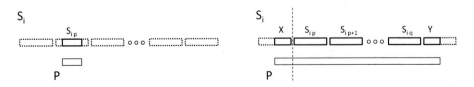

Fig. 2. When P occurs in string S_i, there are two possibilities, referred to as case 1 and case 2. In case 1 (shown on the left), P is contained inside a single factor S_{ip}. In case 2 (shown on the right), P stretches across two or more factors $S_{i(p-1)}, S_{ip}, \ldots, S_{i(q+1)}$.

data structures: (i) $\mathcal{X}(\overline{T})$ to find the left parts; (ii) $\mathcal{Y}(F,T)$ to find the right parts by dynamic programming; and (iii) $\mathcal{M}$ to report the correct combinations of the left parts and right parts. The technical details of $\mathcal{X}(\overline{T})$, $\mathcal{Y}(F,T)$, and $\mathcal{M}$ are given in Sections 5, 6, and 3, respectively. Their usage is summarized as follows:

(i) $\mathcal{X}(\overline{T})$ in Section 5 uses $O(s \log n) + o(n)$ bits space. It finds all occurrences of prefixes of P that are equal to a suffix of a factor $S_{i(p-1)}$ in $O(|P| \log \log n)$ time, More precisely, $\mathcal{X}(\overline{T})$ returns, for every j, the maximal range $st_j..ed_j$ in $\overline{T}$ such that $\overline{P[1..j]}$ is a prefix of every element in $\overline{T}[st_j], \ldots, \overline{T}[ed_j]$.

(ii) $\mathcal{Y}(F,T)$ in Section 6 uses $2.55n + 2nH_k(R) + O(m \log n)$ bits space. It finds all occurrences of suffixes of P that are equal to a prefix of a factor suffix in F, i.e., $S_{ip} \ldots S_{iq}Y$, where Y is a prefix of $S_{i(q+1)}$, in $O(|P| \log \sigma \log \log n)$ time. More precisely, $\mathcal{Y}(F,T)$ returns, for every j, the maximal range $st'_j..ed'_j$ such that $P[(j+1)..|P|]$ is a prefix of every element in $F[st'_j], \ldots, F[ed'_j]$.

(iii) Encode all combinations of X and $S_{ip} \ldots S_{iq}Y$ that are adjacent in some $S_i \in \mathcal{S}$ as follows: Define M to be a binary $(s \times m)$-matrix where $M[x,y] = 1$ iff $\overline{T}[x]$ is the preceding factor of the suffix $F[y]$, i.e., $F[y] = S_{ip}S_{i(p+1)} \ldots S_{ic_i}$ and $\overline{S}_{i(p-1)} = \overline{T}[x]$ is the x-th lexicographically smallest in $\overline{T}$. Note that each column of the matrix M contains exactly one 1. All case 2 occurrences of P can be found by listing the entries equal to 1 in the rectangles $[st_j, ed_j] \times [st'_j, ed'_j]$ in M, for all j. Section 3 gives two alternative 2D range query data structures $\mathcal{M}$ that support the operation query_2d$(M, [st, ed], [st', ed'])$ on M for finding these entries: If $\mathcal{M}$ is of size $O(m \log s \log \log s)$ bits, all entries equal to 1 can be found in $O((1 + occ) \log \log s)$ time, and if $\mathcal{M}$ is of size $O(m \log s)$ bits, the query takes $O(\log s / \log \log s + occ \cdot \log^\epsilon s)$ time.

As a final step, we decode all occurrences of case 1 and 2 to find their actual locations in $\mathcal{S}$. A simple array of $m \log n$ bits is used to store sampled occurrences and an extra $O(\log m / \log n)$ time for reporting each occurrence is required. (Due to space constraints, the details are deferred to the full version of this paper.)

3   Some Useful Auxiliary Data Structures

***Rank* and *Select* and Integer Data Structures:** Let $B[1..n]$ be a bit vector of length n with k ones and $n - k$ zeros. The *rank* and *select* data structure supports two operations: $\text{rank}_B(i)$ returns the number of ones in $B[1..i]$; and $\text{select}_B(i)$ returns the position in B of the ith one. Given an array $A[1..n]$ of non-negative integers, where each element is at most m, we are interested in the following operations: $\text{max\_index}_A(i,j)$ returns $\arg\max_{k \in i..j} A[k]$, and $\text{range\_query}_A(i,j,v)$ returns the set $\{k \in i..j : A[k] \geq v\}$. We also need one more operation for the case when $A[1..n]$ is sorted in non-decreasing order, called $\text{successor\_index}_A(v)$, which returns the smallest index i such that $A[i] \geq v$. The data structure for this operation is called the *y-fast trie* [25]. The complexities of some existing data structures supporting the above operations are listed in the next table.

Operation	Extra space	Time	Reference	Remark
$\mathrm{rank}_B(i)$, $\mathrm{select}_B(i)$	$\log\binom{n}{k} + o(n)$	$O(1)$	[22]	
$\mathrm{max\_index}_A(i,j)$	$2n + o(n)$	$O(1)$	[9]	
$\mathrm{range\_query}_A(i,j,v)$	$O(n\log m)$	$O(1 + occ)$	[20], p. 660	
$\mathrm{successor\_index}_A(v)$	$O(n\log m)$	$O(\log\log m)$	[25]	A is sorted

The Suffix Array and BWT Index: Consider any ·string R with a special terminating character \$ which is lexicographically smaller than all the other characters. The *suffix array* SA_R is the array of all suffixes of R sorted lexicographically. Any substring x of R can be represented by a pair of indices (st, ed), called a *suffix range* or SA_R-*range*. For any given string P specified by its suffix range (st, ed) in SA_R, a *BWT (Burrows-Wheeler transform) index of* R supports the following operations: $\mathrm{lookup}_R(i)$ returns the value of $SA_R[i]$; $\Psi_R(i)$ returns the index j such that $SA_R[j] = SA_R[i] + 1$; and $\mathrm{backward\_search}_R(c, (st, ed))$, where c is any character, returns the suffix range in SA_R of the string cP.

Given any string R of length n over an alphabet of size σ, [7,18] showed how to construct a BWT index of R that uses $nH_k(R) + o(n)$ bits and supports $\mathrm{backward\_search}_R$ in $O(\log\sigma)$ time and Ψ_R in $O(1)$ time. Using an additional $n + o(n)$ bits, lookup_R can be supported in $O(\log n)$ time.

A *general BWT index* is a BWT index extended to alphabets of unbounded size. The next lemma is our simple extension of the normal BWT to the general BWT case, obtained by applying the result from [11] and some additional arrays:

Lemma 2. *Given any string S of length m over an alphabet of size s, there exists a general BWT index of S that uses $m\log s + o(m\log s)$ bits and supports $\mathrm{backward\_search}_S$ in $O(\log\log s)$ time and Ψ_S in $O(1)$ time. Using an additional $m\log s + o(m\log s)$ bits, lookup_S can be supported in $O(\log m/\log s)$ time.*

A New Data Structure for a Special Case of 2D Range Queries: We now describe the *2D range query* data structure mentioned in Section 2 for case 2. This data structure, called $\mathcal{M}$, helps to combine the results of $\mathcal{X}(\overline{T})$ and $\mathcal{Y}(F, T)$ to form the final answers for case 2. Let M be a binary $(s \times m)$-matrix. We define $M[x, y] = 1$ if $\overline{T}[x]$ is the preceding factor of the factor suffix $F[y]$. The operation $\mathrm{query\_2d}(M, [a_1, a_2], [b_1, b_2])$ reports all points in the rectangle $[a_1, a_2] \times [b_1, b_2]$ in M whose values are 1. Here, $[a_1, a_2]$ and $[b_1, b_2]$ specify consecutive rows and consecutive columns of M, respectively. Using existing results by Chan *et al.* [4] and Nekrich [21], we can improve the time complexity for 2D range queries for the special case when each column of M contains exactly one 1. We obtain:

Lemma 3. *Let M be a given binary matrix of size $s \times m$, where $s \leq m$ and every column contains exactly one entry equal to 1. We can store M while supporting $\mathrm{query\_2d}(M, [a_1, a_2], [b_1, b_2])$ within the following space and time complexities:*

1. *$O(m\log s\log\log s)$ bits and $O((1+occ)\log\log s)$ query time; or, alternatively,*
2. *$O(m\log s)$ bits and $O(\log s/\log\log s + occ \cdot \log^\epsilon s)$ query time,*

where $\epsilon > 0$ is a constant and occ is the number of 1s in the specified rectangle.

4   The Data Structure $\mathcal{I}(T)$ for Case 1

Recall from Section 2 that the array $T[1..s]$ stores the s distinct factors of R that occur in the factorizations of $\mathcal{S}$ in lexicographical order. Here, we define a data structure named $\mathcal{I}(T)$ and apply it to locate all occurrences of a query pattern P that lie entirely inside single factors in $T[1..s]$ (case 1 in Section 2). The main result of this section is summarized in the following theorem:

Theorem 2. *The data structure $\mathcal{I}(T)$ uses $2n + o(n) + O(s \log n)$ bits. Given the suffix range $st..ed$ of a query pattern P in SA_R, it reports all occurrences of P inside factors stored in $T[1..s]$ using $O(occ_1 \log n)$ time, where occ_1 is the number of answers.*

A naive solution is to concatenate all the factors in $T[1..s]$ and then build a suffix tree or an FM-index, but the space used by such an approach would be proportional to the total size of $\mathcal{S}$. Instead, we formulate the problem as an interval cover problem. For each $i \in \{1, 2, \ldots, s\}$, define sp_i and ep_i as the starting and ending positions of the factor $T[i]$ inside the reference string R, i.e., $T[i] = R[sp_i..ep_i]$. We say that any factor $T[i]$ *covers a position* p if $sp_i \leq p \leq ep_i$. Also, factor $T[i]$ is to the *left of* factor $T[j]$ if either: (1) $sp_i < sp_j$; or (2) $sp_i = sp_j$ and $ep_i < ep_j$. Let $G[1..s]$ be an array of indices such that $G[i] = j$ if $T[j]$ is the i-th leftmost factor. To be able to convert between indices, we define $I_s[j] = sp_{G[i]}$ and $I_e[j] = ep_{G[i]}$. Note that $I_s[1]$ is the starting position of the leftmost factor and that the values of $I_s[1..s]$ are non-decreasing.

Next, for every $p \in \{1, 2, \ldots, n\}$, define $D[p] = \max_{j=1..s}\{I_e[j]-p+1 : I_s[j] \leq p\}$. Intuitively, $D[p]$ measures the distance from position p to the rightmost ending position of all factors that cover p. Let $D'[1..n]$ be an array such that $D'[p] = D[SA_R[p]]$. (For an example, see Fig. 3 (a).) $D'[p]$ tells us the length of the longest interval whose starting position equals $SA_R[p]$. Hence, we can check if a substring of R is covered by at least one factor according to the next lemma:

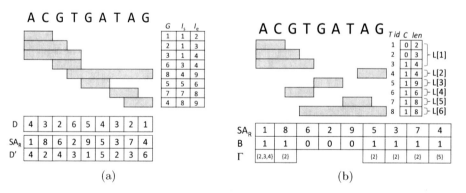

Fig. 3. (a) The factors (displayed as grey bars) from the example in Fig. 1 listed in left-to-right order, and the arrays G, I_s, I_e, D, and D' that define the data structure $\mathcal{I}(T)$ in Section 4. (b) The same factors ordered lexicographically from top to bottom, and the arrays B, C, and Γ that define the data structure $\mathcal{X}(\overline{T})$ in Section 5.

Algorithm Search_Pattern(st, ed)

Input: The data structure $\mathcal{I}(T)$ and the suffix range $st..ed$ of the pattern P in SA_R.
Output: Every factor $T[j]$ in which P occurs.

1: Compute $q = \text{max\_index}_{D'}(st, ed)$
2: **if** $D'[q] \geq |P|$ **then**
3:     Report all factors that cover $SA_R[q]..(SA_R[q] + |P| - 1)$ using Lemma 5
4:     Search_Pattern($st, q - 1$)
5:     Search_Pattern($q + 1, ed$)
6: **end if**

Fig. 4. Algorithm for computing all occurrences of P in $T[1..s]$

Lemma 4. *For any index p and length ℓ, there exists a factor $T[j]$ that covers positions $SA_R[p]..(SA_R[p] + \ell - 1)$ in R if and only if $D'[p] \geq \ell$.*

Now, we describe the new data structure $\mathcal{I}(T)$. It consists of: (i) The array $G[1..s]$, using $s \log n$ bits; (ii) A successor data structure (see Section 3) for I_s, using $s \log n + o(n)$ bits; (iii) A range maximum data structure (see Section 3) for I_e, using $2s + o(s)$ bits; and (iv) A range maximum data structure for D', using $2n + o(n)$ bits. Note that we do not explicitly store the arrays $D[1..n]$, $D'[1..n]$, $I_s[1..s]$, and $I_e[1..s]$. Lemma 5 shows how to recover the values of $D[p]$ and $D'[p]$ for any position $p \in \{1, 2, \ldots, n\}$ from the data structure $\mathcal{I}(T)$. Also, $I_s[i]$ and $I_e[i]$ can be computed in $O(1)$ time given $G[i]$ and information about the factors.

Lemma 5. *Given two positions p and q in R, we can: (i) Compute $D[p]$ in $O(1)$ time and $D'[p]$ in $O(\log n)$ time; and (ii) Report all factors that cover positions $p..q$ in $O(1 + occ)$ time.*

Based on $\mathcal{I}(T)$ and the suffix range for the query pattern P, Algorithm Search_Pattern in Fig. 4 computes all occurrences of P in factors from $T[1..s]$. Let $st..ed$ be the suffix range of P in SA_R. The algorithm recursively finds every index q such that $st \leq q \leq ed$ (lines 4 and 5) and $D'[q] \geq |P|$ (lines 1 and 2). By Lemma 4, this condition guarantees that $SA_R[q]$ and $SA_R[q] + |P| - 1$ are covered by at least one factor. Since $st \leq q \leq ed$, it holds that $R[SA_R[q]..(SA_R[q] + |P| - 1)]$ is an occurrence of P in R. Then, the algorithm reports every $T[j]$ that contains P by using Lemma 5 on line 3.

5   The Data Structure $\mathcal{X}(\overline{T})$ for Case 2

We now turn our attention to case 2 in Section 2 (see Fig. 2 (b)). This section gives the details of the data structure $\mathcal{X}(\overline{T})$ which supports the following query: for any given pattern P, locate every occurrence of a prefix of P that equals a suffix X of a factor of $\mathcal{S}$.

First, note that each of the $|P| - 1$ non-empty proper prefixes of P may be considered separately as a query pattern for $\mathcal{X}(\overline{T})$. Therefore, we only consider how to locate the occurrences of the entire P as suffixes of factors. Secondly, we assume that P is specified by the corresponding suffix range $st_P..ed_P$ in the suffix array SA_R for the reference string R, along with the length of P. Thirdly, recall that the array $T[1..s]$ stores the s distinct factors of the form $S_{ij} \in \mathcal{S}$ sorted lexicographically, and that $\overline{T}[1..s]$ stores all reversed distinct factors $\overline{S}_{ij}$ sorted lexicographically. Thus, $\mathcal{X}(\overline{T})$ will output the maximal range $p..q$ in $\overline{T}$ such that $\overline{P}$ is a prefix of every element in $\overline{T}[p], \ldots, \overline{T}[q]$. In Section 6, we will also need the symmetric data structure $\mathcal{X}(T)$ which, for any given query pattern P, outputs the maximal range $p..q$ in T such that P is a prefix of every element in $T[p], \ldots, T[q]$. To simplify the presentation, we only describe $\mathcal{X}(T)$ below.

Theorem 3. *The data structure $\mathcal{X}(T)$ uses $O(s \log n) + o(n)$ bits. For any suffix range $st..ed$ in SA_R of a query pattern P, it can report the maximal range $p..q$ such that P is a prefix of all $T[j]$, where $p \leq j \leq q$, in $O(\log \log n)$ time.*

Since the factor $T[j]$ is a substring of R, let $st_j..ed_j$ denote the corresponding suffix range of $T[j]$ in SA_R. For every $i = 1, \ldots, n$, define $\Gamma(i) = \{|T[j]| : st_j = i$ and $st_j..ed_j$ is the suffix range of $T[j]$ in $SA_R\}$. In other words, $\Gamma(i)$ is the set of lengths of factors whose suffix ranges start at i in SA_R. We use $\Gamma(i)$ to map a suffix range in SA_R to a range of factors in T according to:

Lemma 6. *Suppose $st_P..ed_P$ is the suffix range of P in SA_R. Then, $p..q$ is the range in $T[1..s]$ such that P is a prefix of all $T[j]$ where $p \leq j \leq q$, where $p = 1 + \sum_{i=1}^{st_P-1} |\Gamma(i)| + |\{x \in \Gamma(st_P) : x < |P|\}|$ and $q = \sum_{i=1}^{ed_P} |\Gamma(i)|$.*

Now, we define $\mathcal{X}(T)$ based on Lemma 6. First, let $B[1..n]$ be a bit vector such that $B[i] = 1$ if $\Gamma(i)$ is non-empty, and $B[i] = 0$ otherwise. Next, suppose $\Gamma(i)$ is the r-th non-empty set, and let $L[r]$ be a y-fast trie [25] for $\Gamma(i)$ (see Section 3). Let $C[1..s]$ be a bit vector such that $C[\sum_{i=1}^{r} |\Gamma(i)|] = 1$, and 0 otherwise. See Fig. 3 (b). We define $\mathcal{X}(T)$ to consist of three parts: (i) The *rank* data structure for the bit vector $B[1..n]$ ($s \log n + o(n)$ bits); (ii) The *select* data structure for the bit vector $C[1..s]$ ($s \log n + o(n)$ bits); and (iii) The y-fast trie data structure $L[r]$ for $\Gamma(i)$ if $\Gamma(i)$ is the r-th non-empty set ($O(s \log n)$ bits). In total, $\mathcal{X}(T)$ requires $O(s \log n) + o(n)$ bits.

Note that, for any ℓ, we have $\sum_{i=1}^{\ell} |\Gamma(i)| = \text{select}_C(\text{rank}_B(\ell))$ and $|\{x \in \Gamma(\ell) : x < c\}| = \text{successor\_index}(L[\text{rank}_B(\ell)], c)$. Using $\mathcal{X}(T)$, they can be computed in $O(\log \log n)$ time. Hence, the values of p and q in Lemma 6 can be computed in $O(\log \log n)$ time. Theorem 3 follows.

6   The Data Structure $\mathcal{Y}(F, T)$ for Case 2

Our next task is: Given any pattern P, compute the range of $P[i..|P|]$ in F for $1 \leq i \leq |P|$, i.e., the range $st..ed$ in F such that $P[i..|P|]$ is a prefix of $F[st], \ldots, F[ed]$. Let $Q[i]$ denote the range for each i. This section introduces a data structure $\mathcal{Y}(F, T)$ which allows us to compute these ranges efficiently:

Theorem 4. *The data structure $\mathcal{Y}(F, T)$ uses $2.55n + 2nH_k(R) + O(m \log n)$ bits. It can find all suffix ranges of F that match some suffix of a query pattern P in $O(|P|(\log \sigma + \log \log n))$ time.*

For any $F[i]$, define the *head of $F[i]$* to be the first factor of $F[i]$. Let $\mathbb{S}$ be the concatenation of the factor representations of all strings in $\mathcal{S}$, and let $\mathcal{B}$ be a general BWT index of $\mathbb{S}$ (see Section 3) supporting $backward\_search_{\mathbb{S}}(T[i], (st, ed))$.

The array $Q[i]$ can be computed as follows. Define $A[i] = P[i..j]$, where j is the largest index such that $P[i..j]$ is a factor of $\mathcal{S}$, if one exists, and nil otherwise. Let $Y[i]$ be the range $st..ed$ in F such that $P[i..|P|]$ is the prefix of all the heads of factor suffixes $F[st]..F[ed]$, if one exists, and nil otherwise. Then:

$$
Q[i] = \begin{cases}
Y[i] & \text{if } Y[i] \neq nil \\
\text{backward\_search}_s(A[i], Q[i + |A[i]|]) & \text{if } Y[i] = nil \ \& \ A[i] \neq nil \\
nil & \text{otherwise}
\end{cases} \quad (1)
$$

By Equation (1), $Q[1..|P|]$ can be computed in three steps: (a) Compute $A[i]$ for $i = 1$ to $|P|$; (b) Compute $Y[i]$ for $i = |P|$ to 1; and (c) Compute $Q[i]$ for $i = |P|$ to 1. Next, we present the data structure $\mathcal{Y}(F, T)$ and discuss steps (a)–(c). The data structure $\mathcal{Y}(F, T)$ consists of:

- The BWT of R and the BWT of $\overline{R}$. Used to compute $A[1..|P|]$.
- The data structure $\mathcal{X}(T)$ (see Section 5). Used to compute $A[1..|P|], Y[1..|P|]$.
- The $select$ data structure for a bit-vector $V[1..m]$, defined by $V[i] = 1$ if the head of $F[i]$ differs from the head of $F[i + 1]$, and $V[i] = 0$ otherwise. Used to compute $Y[1..|P|]$.
- The general BWT index $\mathcal{B}$ of $\mathbb{S}$. Used to compute $Q[1..|P|]$.

In step (a), we compute $A[1..|P|]$ in $O(|P|(\log \sigma + \log \log n))$ time by using $\mathcal{X}(T)$ along with a bi-directional BWT index. In step (b), we compute $Y[1..|P|]$ in two phases. The first phase computes another array $Y'[1..|P|]$, defined as follows: $Y'[i]$ is the range $st'..ed'$ in T such that $P[i..|P|]$ is the prefix of $T[st'], \ldots, T[ed']$. By using the $\mathcal{X}(T)$ data structure from Section 5, we can obtain $Y'[1..|P|]$. Then, given $Y'[1..|P|]$, the second phase computes $Y[1..|P|]$ with the $select$ data structure for V as follows: $Y[i] = (\text{select}_V(st - 1) + 1, \text{select}_V(ed))$, where $(st, ed) = Y'[i]$. Finally, we apply Equation (1) to compute $Q[1..|P|]$.

Acknowledgments. JJ, KS, and WKS were supported in part by the Special Coordination Funds for Promoting Science and Technology (Japan), KAKENHI 23240002, and the MOE's AcRF Tier 2 funding R-252-000-444-112, respectively.

References

1. The 1000 Genomes Project Consortium: A map of human genome variation from population-scale sequencing. Nature 467(7319), 1061–1073 (2010)
2. Bille, P., Landau, G.M., Raman, R., Sadakane, K., Satti, S.R., Weimann, O.: Random access to grammar-compressed strings. In: SODA, pp. 373–389 (2011)
3. Cao, M.D., Dix, T.I., Allison, L., Mears, C.: A simple statistical algorithm for biological sequence compression. In: DCC, pp. 43–52 (2007)
4. Chan, T.M., Larsen, K.G., Pătraşcu, M.: Orthogonal range searching on the RAM, revisited. In: SoCG, pp. 1–10 (2011)
5. Christley, S., Lu, Y., Li, C., Xie, X.: Human genomes as email attachments. Bioinformatics 25(2), 274–275 (2009)

6. Claude, F., Navarro, G.: Self-indexed Text Compression Using Straight-Line Programs. In: Královič, R., Niwiński, D. (eds.) MFCS 2009. LNCS, vol. 5734, pp. 235–246. Springer, Heidelberg (2009)

7. Ferragina, P., Manzini, G.: Compression boosting in optimal linear time using the Burrows-Wheeler Transform. In: SODA, pp. 655–663 (2004)

8. Ferragina, P., Manzini, G.: Indexing compressed text. Journal of the ACM 52(4), 552–581 (2005)

9. Fischer, J., Heun, V.: A New Succinct Representation of RMQ-Information and Improvements in the Enhanced Suffix Array. In: Chen, B., Paterson, M., Zhang, G. (eds.) ESCAPE 2007. LNCS, vol. 4614, pp. 459–470. Springer, Heidelberg (2007)

10. Gagie, T., Gawrychowski, P., Kärkkäinen, J., Nekrich, Y., Puglisi, S.J.: A Faster Grammar-Based Self-index. In: Dediu, A.-H., Martín-Vide, C. (eds.) LATA 2012. LNCS, vol. 7183, pp. 240–251. Springer, Heidelberg (2012)

11. Golynski, A., Munro, J.I., Rao, S.S.: Rank/select operations on large alphabets: a tool for text indexing. In: SODA, pp. 368–373 (2006)

12. Grumbach, S., Tahi, F.: Compression of DNA sequences. In: DCC, pp. 340–350 (1993)

13. Huang, S., Lam, T.W., Sung, W.K., Tam, S.L., Yiu, S.M.: Indexing Similar DNA Sequences. In: Chen, B. (ed.) AAIM 2010. LNCS, vol. 6124, pp. 180–190. Springer, Heidelberg (2010)

14. Kreft, S., Navarro, G.: LZ77-like compression with fast random access. In: DCC, pp. 239–248 (2010)

15. Kreft, S., Navarro, G.: Self-indexing Based on LZ77. In: Giancarlo, R., Manzini, G. (eds.) CPM 2011. LNCS, vol. 6661, pp. 41–54. Springer, Heidelberg (2011)

16. Kuruppu, S., Puglisi, S.J., Zobel, J.: Relative Lempel-Ziv Compression of Genomes for Large-Scale Storage and Retrieval. In: Chavez, E., Lonardi, S. (eds.) SPIRE 2010. LNCS, vol. 6393, pp. 201–206. Springer, Heidelberg (2010)

17. Kuruppu, S., Puglisi, S.J., Zobel, J.: Reference Sequence Construction for Relative Compression of Genomes. In: Grossi, R., Sebastiani, F., Silvestri, F. (eds.) SPIRE 2011. LNCS, vol. 7024, pp. 420–425. Springer, Heidelberg (2011)

18. Mäkinen, V., Navarro, G.: Implicit Compression Boosting with Applications to Self-indexing. In: Ziviani, N., Baeza-Yates, R. (eds.) SPIRE 2007. LNCS, vol. 4726, pp. 229–241. Springer, Heidelberg (2007)

19. Mäkinen, V., Navarro, G., Sirén, J., Välimäki, N.: Storage and retrieval of highly repetitive sequence collections. J. of Computational Biology 17(3), 281–308 (2010)

20. Muthukrishnan, S.: Efficient algorithms for document retrieval problems. In: SODA, pp. 657–666 (2002)

21. Nekrich, Y.: Orthogonal range searching in linear and almost-linear space. Computational Geometry 42(4), 342–351 (2009)

22. Pătraşcu, M.: Succincter. In: FOCS, pp. 305–313 (2008)

23. Rytter, W.: Application of Lempel-Ziv factorization to the approximation of grammar-based compression. Theoretical Computer Science 302, 211–222 (2003)

24. Sirén, J., Välimäki, N., Mäkinen, V., Navarro, G.: Run-Length Compressed Indexes Are Superior for Highly Repetitive Sequence Collections. In: Amir, A., Turpin, A., Moffat, A. (eds.) SPIRE 2008. LNCS, vol. 5280, pp. 164–175. Springer, Heidelberg (2008)

25. Willard, D.E.: Log-logarithmic worst-case range queries are possible in space $\Theta(N)$. Information Processing Letters 17(2), 81–84 (1983)

26. Ziv, J., Lempel, A.: A universal algorithm for sequential data compression. IEEE Transactions on Information Theory 23(3), 337–343 (1977)

A Comparison of Performance Measures via Online Search[*]

Joan Boyar, Kim S. Larsen, and Abyayananda Maiti

Department of Mathematics and Computer Science, University of Southern Denmark,
Campusvej 55, DK-5230 Odense M, Denmark
{joan,kslarsen,abyaym}@imada.sdu.dk

Abstract. Since the introduction of competitive analysis, a number of alternative measures for the quality of online algorithms have been proposed, but, with a few exceptions, these have generally been applied only to the online problem for which they were developed. Recently, a systematic study of performance measures for online algorithms was initiated [Boyar, Irani, Larsen: WADS 2009], first focusing on a simple server problem. We continue this work by studying a fundamentally different online problem, online search, and the Reservation Price Policies in particular. The purpose of this line of work is to learn more about the applicability of various performance measures in different situations and the properties that the different measures emphasize. We investigate the following analysis techniques: Competitive, Relative Worst Order, Bijective, Average, Relative Interval, and Random Order. In addition, we have established the first optimality proof for Relative Interval Analysis.

1 Introduction

An optimization problem is *online* if input is revealed to an algorithm one piece at a time and the algorithm has to commit to the part of the solution involving the current piece before seeing the rest of the input [3]. The first and most well-known analysis technique for determining the quality of online algorithms is *competitive analysis* [16]. The competitive ratio expresses the asymptotic ratio of the performance of an online algorithm compared to an optimal offline algorithm with unlimited computational power. Though this works well in many contexts, researchers realized from the beginning [16] that this "unfair" comparison would sometimes make it impossible to distinguish between online algorithms of quite different quality in practice.

In recent years, researchers have considered alternative methods for comparisons of online algorithms, some of which compare algorithms directly, as opposed to computing independent ratios in a comparison to an offline algorithm. See references below and [10] for a fairly recent survey. Most of the new methods have been designed with one particular online problem in mind, trying to fix problems with competitive analysis for that particular problem. Not that much is known

---

[*] Supported by the Danish Council for Independent Research.

J. Snoeyink, P. Lu, K. Su, and L. Wang (Eds.): FAW-AAIM 2012, LNCS 7285, pp. 303–314, 2012.

about the strengths and weaknesses of these alternatives in comparison with each other. In [6], a systematic study of performance measures was initiated by fixing a (simple) online server problem and applying a collection of performance measures. Partial conclusions were obtained in demonstrating which measures focus on greediness as an algorithmic quality. It was also observed that some measures could not distinguish between certain pairs of algorithms where the one performed at least as well as the other on every sequence.

We continue this systematic study here by investigating a fundamentally different problem which has not yet been studied as an online problem other than with competitive analysis, the *online search problem* [13,12]. Online search is a very simple online (profit) maximization problem; the online algorithm tries to sell a specific item for the highest possible price. Prices, between the minimum price of m and the maximum price of M, arrive online one at a time, and each time a price is revealed, the algorithm can decide to accept that price and terminate or decide to wait. The length of the input sequence is not known to the algorithm in advance, but is revealed only when the last price is given, and the algorithm must accept that price, if it has not accepted one earlier.

This simple model of a searching problem has enormous importance due to its simplicity and its application in the much more complex problems of lowest or highest price searching in various real-world applications in the fields of Economics and Finance [15]. The online search problem is very similar to that of the one-way trading problem [8,12,13]. In fact, one-way trading can be seen as randomized searching. Note that the assumption of a known minimum and maximum price is often used for these types of problems because of the difficulties of defining and analyzing algorithms without them. Reasonable bounds can often be chosen by observing high and low values (of stock prices, currency exchange rates, or whatever is being bought or sold) over an appropriate period of time.

The long-term goal of systematically comparing performance measures is to be able to determine, based on characteristics of an online problem, how online algorithms should be theoretically analyzed so as to accurately predict the relative quality of the algorithms in practice. Online search differs from the server problem studied earlier in many respects, particularly in its consisting of a "one-shot" choice, as opposed to incremental decisions, so the greediness studied in [6] is not relevant here. In addition, online search is a maximization problem, instead of minimization, and its last request has a different requirement than the others (if nothing was chosen before then, the last value must be chosen). Thus, the findings obtained here are complementary to the results obtained in [6]. The difference between online search and many other problems also forced us to extend earlier definitions for some of the measures so that they could be applicable here as well.

Our primary study is of the class of Reservation Price Policy (RPP) algorithms [12,13]. This is a parameterized class, where the behavior of $\mathcal{R}_p$ is to accept the first price greater than or equal to the so-called reservation price p.

Here we compare the six different quality measures on RPP algorithms with different parameters. We have considered having an integral interval of possible prices between m and M as well as a real-valued scenario; for the most part, the results are similar. The following discussion in this introduction is assuming a real-valued scenario, allowing us to state the results better typographically, without rounding.

For the performance measures below, note that since profit is a constant (between m and M), independent of the sequence length, for measures of an asymptotic nature, we use the strict version since asymptotic results (allowing an additive constant) would deem all algorithms optimal (ratio one compared with an optimal algorithm—up to the additive constant). For the same reason, the measures which were originally defined using limits on the profit (or cost) achieved are modified here. In each section of the paper, we give the precise definition of the measures used.

We find that Competitive Analysis as well as Random Order Analysis favor $\mathcal{R}_{\sqrt{mM}}$, the reason being that they focus on limiting the worst case ratio compared to an optimal algorithm, independent of input length. Relative Interval Analysis favors $\mathcal{R}_{\frac{m+M}{2}}$, similarly limiting the worst case difference, as opposed to ratio. Average Analysis favors $\mathcal{R}_M$. This is basically due to focusing on the limit, i.e., when input sequences become long enough, any event will occur eventually. In Bijective Analysis, basically all algorithms are incomparable. Finally Relative Worst Order Analysis deems the algorithms incomparable, but gives indication that $\mathcal{R}_{\sqrt{mM}}$ is the best algorithm.

As an additional check that the different performance measures work "correctly", we have also considered simple "bad" algorithms, such as $\mathcal{R}_p^2$, which accepts the *second* price greater than or equal to p. As expected, this algorithm turns out to be worse than $\mathcal{R}_p$ for most of the measures.

In addition to these findings, this paper contains the first optimality result for Relative Interval Analysis, where we prove that no $\mathcal{R}_p$ algorithm can be better than $\mathcal{R}_{\frac{m+M}{2}}$. For Relative Worst Order Analysis, we refine the discussion of which algorithm is best through the concept of "superiority", which seems to be interesting for classes of parameterized algorithms. A first use of this concept, without naming it, appeared when analyzing a parameterized variant of Lazy Double Coverage for the server problem in [6].

Finally, we have investigated the sensitivity of the different measures with regards to the choice of integral vs. real-valued domains, and most of the measures seem very stable in this regard. Not surprisingly, using real values, Bijective Analysis indicates that all RPP algorithms are equivalent. Average analysis is inapplicable for a real-valued interval, but a generalization, which we call Expected Analysis, can be applied, giving similar results to what Average Analysis gives for integral values. Expected Analysis may be useful for other problems as well.

The rest of this paper is organized as follows. Section 2 defines the notation used and each subsequent section treats one of the measures described above. Due to space constraints, some of the proofs and results have been omitted. The full version of this paper in `arXiv` [7] contains the omitted material.

2   Problem Preliminaries

Unless otherwise stated, we assume that the prices are integral and drawn from some integral interval $[m, M]$ with $0 < m \leq M$. In any time step, any value from this closed interval can be drawn as a price, and there will be $N = M - m + 1$ possible prices. This assumption is made for the sake of consistency; some methods of analysis are uninteresting for real-valued intervals; see Section 4, for example. Also, this assumption is compatible with the real-world problems of online search as the set of prices is generally finite (the market decides on an agreed-upon number of digits after the decimal point).

We denote the length of the price sequence by n. Denote by $\mathcal{I}_n$ the set of all input sequences of length n. Thus, the total number of possible input sequences of length n is N^n. For an online algorithm $\mathcal{A}$ and an input sequence I, let $\mathcal{A}(I)$ be the profit gained by $\mathcal{A}$ on I, i.e., the price chosen. In some analyses (for example in Relative Worst Order Analysis), we need to permute the input sequences. We always use σ as a permutation and denote the permuted sequence by $\sigma(I)$.

Some of the analysis methods compare the online algorithms with a hypothetical optimal offline algorithm which receives the input in its entirety in advance and has unlimited computational power in determining a solution. We denote this optimal algorithm by OPT and the profit gained by it from an input sequence I as $OPT(I)$, which is the maximum price in that sequence.

To denote the relative performance of two online algorithms $\mathcal{A}$ and $\mathcal{B}$ according to an analysis method, x, we use the following notation. If $\mathcal{B}$ is better than $\mathcal{A}$, then we write $\mathcal{A} \prec_x \mathcal{B}$, and if $\mathcal{B}$ is no worse than $\mathcal{A}$, this is denoted by $\mathcal{A} \preceq_x \mathcal{B}$. If the measure deems the algorithms equivalent, then this is denoted by $\mathcal{A} \equiv_x \mathcal{B}$. Usually, we merely define either $\prec_x$ or $\preceq_x$ and the other relations follow in the standard way from that.

If $n = 1$, any algorithm must take the only price, so all online search algorithms are equivalent. To streamline the presentation of results, we always assume that $n \geq 2$. The core of this paper is concerned with the comparison of $\mathcal{R}_p$ and $\mathcal{R}_q$ for $p \neq q$. To avoid stating this every time, we always assume that $m \leq p < q \leq M$.

3   Competitive Analysis

The online search problem was first studied from an online algorithms perspective using Competitive Analysis by El-Yaniv et al. [12]. Competitive Analysis evaluates an online algorithm in comparison to an optimal offline algorithm.

Definition 1. *An online search algorithm $\mathcal{A}$ is* strictly c-competitive *if for all finite input sequences I, $OPT(I) \leq c \cdot \mathcal{A}(I)$.*
The competitive ratio *of algorithm $\mathcal{A}$ is* $\inf\{c \mid \mathcal{A}$ *is c-competitive*$\}$.

Denote the competitive ratio of an online algorithm $\mathcal{A}$ by c_A. If $c_A > c_B$, $\mathcal{B}$ is better than $\mathcal{A}$ according to Competitive Analysis and we denote this by $\mathcal{A} \prec_c \mathcal{B}$.

In [12], El-Yaniv formulated the Reservation Price Policy algorithm and proved that for real-valued prices, the *reservation price* $p^* = \sqrt{Mm}$ is the optimal price according to Competitive Analysis, and using this price, the competitive ratio is $\sqrt{M/m}$. A very similar result and proof holds for integer-valued prices.

Theorem 1. *According to Competitive Analysis, $\mathcal{R}_p \prec_c \mathcal{R}_q$, $\mathcal{R}_p \equiv_c \mathcal{R}_q$ and $\mathcal{R}_q \prec_c \mathcal{R}_p$ if and only if $Mm > p(q-1)$, $Mm = p(q-1)$ and $Mm < p(q-1)$, respectively.*

Proof. In any price sequence for an RPP algorithm $\mathcal{R}_p$, we consider two cases: (i) all the prices are less than p, in which case the performance ratio, offline to online, will be at most $\frac{p-1}{m}$ with equality when there is a price $p-1$ and the last price is m; and (ii) at least one price is greater than or equal to p, in which case the offline to online performance ratio would be at most $\frac{M}{p}$ with equality when the first price greater than or equal to p is exactly p and there is another price M somewhere later. So, the competitive ratio of $\mathcal{R}_p$ will be $c_{\mathcal{R}_p} = max(\frac{p-1}{m}, \frac{M}{p})$. It is easy to observe that $c_{\mathcal{R}_p} > c_{\mathcal{R}_q}$ if and only if $\frac{M}{p} > \frac{q-1}{m}$ since $\frac{p-1}{m} < \frac{q-1}{m}$ and $\frac{M}{p} > \frac{M}{q}$. This argument proves that $\mathcal{R}_p \prec_c \mathcal{R}_q$ if and only if $Mm > p(q-1)$. Similarly, we can conclude the other two results. □

Corollary 1. *Let $s = \left\lceil \sqrt{Mm} \right\rceil$. According to Competitive Analysis, the best RPP algorithm is $\mathcal{R}_s$.*

4   Bijective Analysis

In the Bijective Analysis model [1], we construct a bijection on the set of possible input sequences. In this bijection, we aim to pair input sequences for online algorithms $\mathcal{A}$ and $\mathcal{B}$ in such a way that the cost of $\mathcal{A}$ on every sequence I is no more than the cost of $\mathcal{B}$ on the image of I, or vice versa, to show that the algorithms are comparable. We present a version of the definition from [1] which is suitable for profit maximization problems such as online search.

Definition 2. *We say that an online search algorithm $\mathcal{A}$ is no better than an online search algorithm $\mathcal{B}$ according to Bijective Analysis if there exists an integer $n_0 \geq 1$ such that for each $n \geq n_0$, there is a bijection $b : \mathcal{I}_n \leftrightarrow \mathcal{I}_n$ satisfying $\mathcal{A}(I) \leq \mathcal{B}(b(I))$ for each $I \in \mathcal{I}_n$. We denote this by $\mathcal{A} \preceq_b \mathcal{B}$.*

Theorem 2. *According to Bijective Analysis, $\mathcal{R}_p \prec_b \mathcal{R}_q$, if $p = m$ and $m < q \leq M$. Otherwise, $\mathcal{R}_p$ and $\mathcal{R}_q$ are incomparable.*

Proof. Consider the sequences with $m < p$. Note that with $\mathcal{R}_p$, m will be chosen as output if and only if it is the last price of the sequence and all the preceding prices are smaller than p. As there are $p - m$ such prices and, not counting the last price, there are $n-1$ prices in the sequence, the number of possible sequence with m as output is $(p-m)^{n-1}$. With the same reasoning, for algorithm $\mathcal{R}_p$, each price in the range from m to $p-1$ will be the output for $(p-m)^{n-1}$ sequences.

For any prices in the range from p to M, algorithm $\mathcal{R}_p$ chooses this price as output at its first occurrence in the price sequence if no price greater than or equal to p has occurred before it. So all the preceding prices before this first occurrence should be smaller than p (specifically in the range from m to $p-1$) and the following prices can have any value. For example, the number of sequences where price p comes in the 3rd place in the sequence as well as taken as output will be $(p-m)^2 N^{n-3}$. So the number of sequences which give output k if the reservation price is p is

$$
N_{p,k} = \begin{cases} (p-m)^{n-1}, & \text{for } m \le k < p \\ \sum_{i=1}^{n} (p-m)^{i-1} N^{n-i}, & \text{for } p \le k \le M \end{cases} \tag{1}
$$

Recall the assumption throughout the paper that $q > p$. We consider two cases depending on p:

Case $p > m$: From Eq. (1), we can derive the fact that when $p > m$, the number of sequences with lowest output for algorithm $\mathcal{R}_q$ ($N_{q,m}$) is greater than that for algorithm $\mathcal{R}_p$ ($N_{p,m}$) since $(q-m)^{n-1} > (p-m)^{n-1}$. Thus, we cannot have any bijective mapping $b : \mathcal{I}_n \leftrightarrow \mathcal{I}_n$ that shows $\mathcal{R}_p(I) \le \mathcal{R}_q(b(I))$ for every $I \in \mathcal{I}_n$. On the other hand, it is also the case that the number of sequences with highest output (M) for algorithm $\mathcal{R}_q$ is greater than that of algorithm $\mathcal{R}_p$ since $N_{q,M} > N_{p,M}$. So there is no bijection b such that $\mathcal{R}_p(I) \ge \mathcal{R}_q(b(I))$ for every $I \in \mathcal{I}_n$. Thus for this case, $\mathcal{R}_p$ and $\mathcal{R}_q$ are incomparable according to the Bijective Analysis.

Case $p = m$: For algorithm $\mathcal{R}_m$, since the first price will be accepted, each price will be the output for exactly N^{n-1} sequences. We can derive the number of sequences with specific output for algorithm $\mathcal{R}_q$ using Eq. (1). In this case, each price in the range from m to $q-1$ will emerge as output in $(q-m)^{n-1}$ sequences and the number of sequences with output in the range from q to M will be $N^{n-1} + (q-m)N^{n-2} + (q-m)^2 N^{n-3} + \cdots + (q-m)^{n-1}$. Clearly, here we can construct a bijective mapping $b : \mathcal{I}_n \leftrightarrow \mathcal{I}_n$ where each sequence with output $k < q$ of algorithm $\mathcal{R}_m$ is mapped to sequences with the same output for algorithm $\mathcal{R}_q$. Let E_m denote the number of excess sequences with output $k < q$ of $\mathcal{R}_m$ which cannot be mapped in the above way. We map each sequence with output $k \ge q$ of algorithm $\mathcal{R}_m$ to sequences with the same output in algorithm $\mathcal{R}_q$. Let E_q denote the number of excess sequences with output $k \ge q$ of $\mathcal{R}_q$ which can not be mapped in above way. Clearly, $E_m = E_q$. Note that, for all of these E_m sequences, we can construct a mapping such that $\mathcal{R}_m(I) < \mathcal{R}_q(b(I))$. This mapping shows that $\mathcal{R}_m(I) \le \mathcal{R}_q(b(I))$ for each $I \in \mathcal{I}_n$, but there is no bijection b' such that $\mathcal{R}_m(I) \ge \mathcal{R}_q(b'(I))$ for all $I \in \mathcal{I}_n$. This shows that if $p = m$, $\mathcal{R}_m$ and $\mathcal{R}_q$ are comparable according to Bijective Analysis and $\mathcal{R}_m \prec_b \mathcal{R}_q$. □

4.1 Real-Valued Price Interval

The result of comparing the two algorithms using Bijective Analysis changes significantly when the values of the prices are real numbers: Bijective Analysis cannot differentiate between algorithms when the number of sequences is uncountable.

Theorem 3. $\mathcal{R}_p$ *and* $\mathcal{R}_q$ *are equivalent according to Bijective Analysis if the prices are drawn from real space in* $[m, M]$.

The same problem clearly arises for other online problems with real-valued inputs.[1]

5   Average Analysis

In general, using Bijective Analysis, algorithms could be incomparable because it is impossible to find a bijection showing that one algorithm dominates the other. In some of these cases, if we take the average performance of the algorithms, then we can still get an indication of which algorithm is better. In [1], Average Analysis is defined with that aim and is formulated here in terms of online search.

Definition 3. *We say that an online search algorithm* $\mathcal{A}$ *is no better than an online search algorithm* $\mathcal{B}$ *according to* Average Analysis *if there exists an integer* $n_0 \geq 1$ *such that for each* $n \geq n_0$, $\sum_{I \in \mathcal{I}_n} \mathcal{A}(I) \leq \sum_{I \in \mathcal{I}_n} \mathcal{B}(I)$. *We denote this by* $\mathcal{A} \preceq_a \mathcal{B}$.

Theorem 4. *For all* $n \geq \left\lfloor \frac{\log(N/(q-p))}{\log(N/(N-1))} \right\rfloor + 1$, $\sum_{I \in \mathcal{I}_n} \mathcal{R}_p(I) < \sum_{I \in \mathcal{I}_n} \mathcal{R}_q(I)$. *Thus, according to Average Analysis,* $\mathcal{R}_p \prec_a \mathcal{R}_q$.

Proof. Let $S_{p,n}$ denote the summation $\sum_{I \in \mathcal{I}_n} \mathcal{R}_p(I)$. We can derive the value of $S_{p,n}$ using Eq. (1) and that $N = M - m + 1$. $S_{p,n}$ equals

$$
\sum_{i=m}^{p-1} i N_{p,i} + \sum_{i=p}^{M} i N_{p,i} = (p-m)^{n-1} \sum_{i=m}^{p-1} i + \left(\sum_{i=1}^{n} (p-m)^{i-1} N^{n-i} \right) \sum_{i=p}^{M} i
$$

$$
= \frac{(p-m)^n (p+m-1)}{2} + \left(\frac{N^n - (p-m)^n}{N - (p-m)} \right) \frac{(N+m+p-1)(N+m-p)}{2}
$$

$$
= \frac{N^{n+1} + pN^n + mN^n - N^n - N(p-m)^n}{2} \tag{2}
$$

To compare $\mathcal{R}_p$ and $\mathcal{R}_q$, we show that the difference between the two corresponding sums $(S_{q,n} - S_{p,n})$ is greater than zero for some $n_0 \geq 1$ and for each $n \geq n_0$. Using Derivation (2), we have

$$
S_{q,n} - S_{p,n} > 0 \iff N^{n-1} > \frac{(q-m)^n - (p-m)^n}{q-p} \tag{3}
$$

Since $q - m \leq M - 1$ and $q > p$, Eq. (3) holds for any n_0. Thus, it holds for any n_0 satisfying $N^{n_0-1} > \frac{(N-1)^{n_0}}{q-p}$. Solving for n_0 gives

$$
(n_0 - 1) \log N > n_0 \log(N-1) - \log(q-p) \iff n_0 > \frac{\log(N/(q-p))}{\log(N/(N-1))}
$$

Therefore, for all $n_0 \geq \left\lfloor \frac{\log(N/(q-p))}{\log(N/(N-1))} \right\rfloor + 1$, $\sum_{I \in \mathcal{I}_n} \mathcal{R}_p(I) < \sum_{I \in \mathcal{I}_n} \mathcal{R}_q(I)$.   □

---

[1] The authors are thankful to Leah Epstein, Asaf Levin and Alejandro López-Ortiz for earlier discussions concerning this subject.

Corollary 2. *According to Average Analysis, the best RPP algorithm is $\mathcal{R}_M$.*

In Average Analysis, algorithms are compared by comparing the sums of their outputs on all possible sequences. For integral valued problems, this is equivalent to comparing the the sum of the outputs and the expected outputs over a uniform distribution on all input sequences. In contrast, in the case of real-valued problems, calculating the sum of the outputs of the infinitely many sequences is impossible. However, if we know the distribution of the input prices in the sequences, then we can derive the expected output of a sequence. We generalize Average Analysis to *Expected Analysis*. For a detail analysis of this new measure see arXiv [7]. This generalization may prove useful for other online problems as well.

Definition 4. *We say that an online search algorithm $\mathcal{A}$ is no better than an online search algorithm $\mathcal{B}$ according to* Expected Analysis *if there exists an integer $n_0 \geq 1$ such that for each $n \geq n_0$, $E_{I \in \mathcal{I}_n}[\mathcal{A}(I)] \leq E_{I \in \mathcal{I}_n}[\mathcal{B}(I)]$. We denote this by $\mathcal{A} \preceq_e \mathcal{B}$.*

6 Random Order Analysis

Kenyon [14] proposed another method for comparing the average behaviors of online algorithms by considering the expected result of a random ordering of an input sequence and comparing that to *OPT*'s result on the same sequence. In [14], Kenyon defines the random order ratio in the context of the bin packing problem which is a cost minimization problem.

Definition 5. *The* random order ratio *$RC(A)$ of an online bin packing algorithm A is*

$$RC(A) = \limsup_{OPT(I) \to \infty} \frac{E_\sigma[A(\sigma(I))]}{OPT(I)}$$

where the expectation is taken over all permutations of I.

An online algorithm B is better than an online algorithm A according to Random Order Analysis if $RC(A) > RC(B)$. We denote this by $A \prec_r B$. Since the value of $OPT(I)$ is bounded above by the constant M, the following definition, a maximization version of the definition of random order ratio in [9], is used here in place of the original definition.

$$RC(\mathcal{R}_p) = \limsup_{n \to \infty} \frac{OPT(I)}{E_\sigma[\mathcal{R}_p(\sigma(I))]} \tag{4}$$

Theorem 5. *The random order ratio of the RPP algorithm $\mathcal{R}_p$ is $\max(\frac{M}{p}, \frac{p-1}{m})$ when $p > 1$ and $p > m$. Consequently, $\mathcal{R}_p \prec_r \mathcal{R}_q$ if and only if $Mm > p(q-1)$.*

Corollary 3. *Let $s = \left\lceil \sqrt{Mm} \right\rceil$. According to Random Order Analysis, the best RPP algorithm is $\mathcal{R}_s$.*

7   Relative Interval Analysis

Dorrigiv et. al. [11] proposed another analysis method, Relative Interval Analysis, in the context of paging. Relative Interval Analysis also compares two online algorithms directly, i.e., it does not use the optimal offline algorithm as the baseline of the comparison. It compares two algorithms on the basis of the rate of the outcomes over the length of the input sequence rather than their worst case behavior. Here we define this analysis for profit maximization problems for two algorithms $\mathcal{A}$ and $\mathcal{B}$, following [11].

Definition 6. *Let*

$$Min_{\mathcal{A},\mathcal{B}}(n) = \min_{|I=n|}\{\mathcal{A}(I) - \mathcal{B}(I)\} \quad and \quad Max_{\mathcal{A},\mathcal{B}}(n) = \max_{|I=n|}\{\mathcal{A}(I) - \mathcal{B}(I)\}.$$

These functions are used to define the following two measures:

$$Min(\mathcal{A},\mathcal{B}) = \liminf_{n\to\infty} \frac{Min_{\mathcal{A},\mathcal{B}}(n)}{n} \quad and \quad Max(\mathcal{A},\mathcal{B}) = \limsup_{n\to\infty} \frac{Max_{\mathcal{A},\mathcal{B}}(n)}{n}. \quad (5)$$

Note that $Min(\mathcal{A},\mathcal{B}) = -Max(\mathcal{B},\mathcal{A})$ and $Max(\mathcal{A},\mathcal{B}) = -Min(\mathcal{B},\mathcal{A})$. The relative interval of $\mathcal{A}$ and $\mathcal{B}$ is defined as $l(\mathcal{A},\mathcal{B}) = [Min(\mathcal{A},\mathcal{B}), Max(\mathcal{A},\mathcal{B})]$. If $Max(\mathcal{A},\mathcal{B}) > |Min(\mathcal{A},\mathcal{B})|$, then $\mathcal{A}$ is said to have better performance than $\mathcal{B}$ in this model. In particular, if $l(\mathcal{A},\mathcal{B}) = [0,\beta]$ for $\beta > 0$, then it is said that $\mathcal{A}$ dominates $\mathcal{B}$ since $Min(\mathcal{A},\mathcal{B}) = 0$ indicates that $\mathcal{A}$ is never worse than $\mathcal{B}$ and $Max(\mathcal{A},\mathcal{B}) > 0$ says that $\mathcal{A}$ is better at least for some case(s).

Given the finite nature of the online search problem, the above limits are always zero. We propose a modification of Relative Interval Analysis to make it suitable for finite problems.

Definition 7. *$Min_{\mathcal{A},\mathcal{B}}(n)$ and $Max_{\mathcal{A},\mathcal{B}}(n)$ are as in Definition 6.*
   These functions are used to define the following two measures:

$$Min(\mathcal{A},\mathcal{B}) = \inf\{Min_{\mathcal{A},\mathcal{B}}(n)\} \quad and \quad Max(\mathcal{A},\mathcal{B}) = \sup\{Max_{\mathcal{A},\mathcal{B}}(n)\}. \quad (6)$$

The pair, $fl(\mathcal{A},\mathcal{B}) = [Min(\mathcal{A},\mathcal{B}), Max(\mathcal{A},\mathcal{B})]$, is used to denote the Finite Relative Interval of $\mathcal{A}$ and $\mathcal{B}$. Relative performance and dominance with regards to $fl(\mathcal{A},\mathcal{B})$ are defined as for $l(\mathcal{A},\mathcal{B})$ from Definition 6.

Theorem 6. *According to Finite Relative Interval Analysis, $fl(\mathcal{R}_q, \mathcal{R}_p) = [m - q + 1, M - p]$.*

Proof. The minimum value of $\mathcal{R}_q(I) - \mathcal{R}_p(I)$ is obtained by any sequence of prices with all the prices smaller than q, where the first price is $q - 1$ and the last price is m. In this case, $Min_{\mathcal{R}_q, \mathcal{R}_p}(N) = m - q + 1$. The maximum value of $\mathcal{R}_q(I) - \mathcal{R}_p(I)$ is $M - p$, which is obtained when the first price is p and the second price is M. This proves that $fl(\mathcal{R}_q, \mathcal{R}_p) = [m - q + 1, M - p]$.   □

Corollary 4. *Let $s = \lceil \frac{M+m}{2} \rceil$. According to Finite Relative Interval Analysis, the best RPP algorithm is $\mathcal{R}_s$.*

8   Relative Worst Order Analysis

Relative Worst Order Analysis [4] compares two online algorithms directly. It compares two algorithms on their worst orderings of sequences which have the same content, but possibly in different orders. The definition of this measure is somewhat more involved; see [5] for more intuition on the various elements. Here we use the definitions for the strict Relative Worst Order Analysis for profit maximization problems.

Definition 8. *Let I be any input sequence, and let n be the length of I. Let A be any online search algorithm. Then $A_W(I) = \min_\sigma A(\sigma(I))$.*

Definition 9. *For any pair of algorithms A and B, we define*

$$c_l(A, B) = \sup\{c \mid \forall I : A_W(I) \geq cB_W(I)\} \quad and$$
$$c_u(A, B) = \inf\{c \mid \forall I : A_W(I) \leq cB_W(I)\}.$$

If $c_l(A, B) \geq 1$ or $c_u(A, B) \leq 1$, the algorithms are said to be comparable *and the* strict relative worst order ratio $WR_{A,B}$ *of algorithm A to algorithm B is defined. Otherwise, $WR_{A,B}$ is undefined.*

$$\text{If } c_l(A, B) \geq 1 \text{ then } WR_{A,B} = c_u(A, B), \quad and$$
$$\text{if } c_u(A, B) \leq 1 \text{ then } WR_{A,B} = c_l(A, B).$$

If $WR_{A,B} > 1$, algorithms A and B are said to be comparable in A's favor. *Similarly, if $WR_{A,B} < 1$, algorithms are said to be* comparable in B's favor.

When two algorithms happen to be incomparable, Relative Worst Order Analysis can still be used to express their relative performance.

Definition 10. *If at least one of the ratios $c_u(A, B)$ and $c_u(B, A)$ is finite, the algorithms A and B are $(c_u(A, B), c_u(B, A))$-related.*

Theorem 7. *According to Relative Worst Order Analysis, the algorithms $\mathcal{R}_q$ and $\mathcal{R}_p$ are $(\frac{M}{p}, \frac{q-1}{m})$-related. They are comparable in $\mathcal{R}_q$'s favor if $p = m$ and $q = m + 1$.*

Proof. For the maximum value of the ratio of $\mathcal{R}_{q_w}(I)$ and $\mathcal{R}_{p_w}(I)$, we can construct a sequence I with only one p and one M and all the other prices smaller than q. Among all the permutations of I, the worst output for $\mathcal{R}_q$ is M and that of $\mathcal{R}_p$ is p. This gives the value of the upper bound $c_u(\mathcal{R}_q, \mathcal{R}_p)$ as $\frac{M}{p}$. For the lower bound, assume I has only one $q - 1$ and one m and all the other prices are smaller than p. Then, $\mathcal{R}_p$ takes $q - 1$ as its output on every permutation of I, but the worst output of $\mathcal{R}_q$ gives m. On this sequence, $\mathcal{R}_q$ performs worse than $\mathcal{R}_p$, and the ratio of the outputs of the two algorithms can never be lower than that. So,

$$c_l(\mathcal{R}_q, \mathcal{R}_p) = \frac{m}{q-1} \begin{cases} = 1, \text{ for } q = m + 1 \text{ and } p = m \\ < 1, \qquad \text{otherwise} \end{cases}$$
$$c_u(\mathcal{R}_q, \mathcal{R}_p) = \frac{M}{p} > 1$$

From the above expressions and the definitions of strict Relative Worst Order Analysis, we can see that $\mathcal{R}_q$ and $\mathcal{R}_p$ are comparable when $p = m$ and $q = m+1$. For all the other cases, they are incomparable. For this single feasible condition $(c_l(\mathcal{R}_q, \mathcal{R}_p) = 1)$, we have $WR_{\mathcal{R}_q, \mathcal{R}_p} = \frac{M}{p} > 1$, and we can say that algorithms $\mathcal{R}_q$ and $\mathcal{R}_p$ are comparable in $\mathcal{R}_q$'s favor. Using Definition 10, since all the ratios are finite, $c_u(\mathcal{R}_p, \mathcal{R}_q)$ is $\frac{q-1}{m}$ and the algorithms $\mathcal{R}_q$ and $\mathcal{R}_p$ are $(\frac{M}{p}, \frac{q-1}{m})$-related.

$\square$

Note that this relatedness result gives the same conditions indicating which algorithm is better as Competitive and Random Order Analysis. Although the original definition of relatedness in Relative Worst Order Ratio does not tell explicitly which algorithm is better, we can get a strong indication regarding this issue from the next corollary, using the concept of *better performance* [11] from Relative Interval Analysis.

Corollary 5. *Let* $s = \left\lceil \sqrt{Mm} \right\rceil$. *Then* $\forall q > s$, *if* $\mathcal{R}_q$ *and* $\mathcal{R}_s$ *are (c,c′)-related, then* $c \leq c′$; *and* $\forall p < s$, *if* $\mathcal{R}_s$ *and* $\mathcal{R}_p$ *are (c,c′)-related, then* $c > c′$.

This corollary shows that whatever the value of x ($x \neq s$), $c_u(\mathcal{R}_s, \mathcal{R}_x) \geq c_u(\mathcal{R}_x, \mathcal{R}_s)$. A similar result on a parameterized family of algorithms can be found in [6]. This could be defined as a weak form of optimality within a class of algorithms, and we will say that $\mathcal{R}_s$ is *superior* to any other RPP algorithm.

9   Concluding Remarks

With regards to the concrete results, for Competitive and Random Order Analysis, $\mathcal{R}_{\sqrt{mM}}$ is the best online algorithm. Relative Worst Order and Relative Interval have more nuanced answers, but point to $\mathcal{R}_{\sqrt{mM}}$ and $\mathcal{R}_{\frac{m+M}{2}}$, respectively. Bijective and Average Analysis seem to provide the least interesting information in this context; Average Analysis indicates $\mathcal{R}_M$ as the best algorithm, and Bijective Analysis deems most algorithms incomparable.

This points to three choices for the online player with regards to the optimal reservation prices, namely $\sqrt{mM}$, $\frac{m+M}{2}$, and M, depending on the different analysis methods, i.e., the geometric mean, the arithmetic mean, and the maximum M of all possible values. This clearly shows that the objectives of the different performance measures vary greatly, trying to limit poor performance in a proportional or additive sense, or focusing equally on all scenarios, including the possibly non-occurring upper bound of M. Thus, the different measures are tailored towards different degrees of risk aversion—cautiousness vs. aggressiveness. The observations above complement the findings regarding greediness and laziness from [6].

Studying performance measures and disclosing their properties and differences from each other is work in progress. With this study, we have added Online Search to the collection of problems that have been investigated with a spectrum of measures. More online problem scenarios must be analyzed this broadly before strong conclusions concerning the different performance measures can be drawn.

Another interesting direction for future work would be to incorporate other aspects of financial problems into the analysis in the context of other the performance measures, as it has been done for competitive analysis of financial games in the risk-reward framework of al-Binali [2].

References

1. Angelopoulos, S., Dorrigiv, R., López-Ortiz, A.: On the separation and equivalence of paging strategies. In: 18th Symposium on Discrete Algorithms, pp. 229–237. Philadelphia, PA, USA (2007)
2. Al-Binali, S.: A risk-reward framework for the competitive analysis of financial games. Algorithmica 25(1), 99–115 (1999)
3. Borodin, A., El-Yaniv, R.: Online Computation and Competitive Analysis. Cambridge University Press, Cambridge (1998)
4. Boyar, J., Favrholdt, L.M.: The relative worst order ratio for online algorithms. ACM Transactions on Algorithms 3(2), article 22 (2007)
5. Boyar, J., Favrholdt, L.M., Larsen, K.S.: The relative worst order ratio applied to paging. Journal of Computer and System Sciences 73(5), 818–843 (2007)
6. Boyar, J., Irani, S., Larsen, K.S.: A Comparison of Performance Measures for Online Algorithms. In: Dehne, F., Gavrilova, M., Sack, J.-R., Tóth, C.D. (eds.) WADS 2009. LNCS, vol. 5664, pp. 119–130. Springer, Heidelberg (2009), arXiv: 0806.0983v1 [cs.DS]
7. Boyar, J., Larsen, K.S., Maiti, A.: A comparison of performance measures via online search. Tech. Rep. arXiv:1106.6136v1 [cs.DS], arXiv (2011)
8. Chen, G.-H., Kao, M.-Y., Lyuu, Y.-D., Wong, H.-K.: Optimal buy-and-hold strategies for financial markets with bounded daily returns. SIAM Journal on Computing 31(2) (2001)
9. Coffman Jr., E.G., Csirik, J., Rónyai, L., Zsbán, A.: Random-order bin packing. Discrete Applied Mathematics 156(14), 2810–2816 (2008)
10. Dorrigiv, R., López-Ortiz, A.: A survey of performance measures for on-line algorithms. SIGACT News 36(3), 67–81 (2005)
11. Dorrigiv, R., López-Ortiz, A., Munro, J.I.: On the relative dominance of paging algorithms. Theoretical Computer Science 410(38–40), 3694–3701 (2009)
12. El-Yaniv, R.: Competitive solutions for online financial problems. ACM Computing Surveys 30(1), 28–69 (1998)
13. El-Yaniv, R., Fiat, A., Karp, R.M., Turpin, G.: Optimal search and one-way trading online algorithms. Algorithmica 30(1), 101–139 (2001)
14. Kenyon, C.: Best-fit bin-packing with random order. In: 7th Symposium on Discrete Algorithms, pp. 359–364 (1996)
15. Rothschild, M.: Searching for the lowest price when the distribution of prices is unknown. Journal of Political Economy 82(4), 689–711 (1974)
16. Sleator, D.D., Tarjan, R.E.: Amortized efficiency of list update and paging rules. Communications of the ACM 28(2), 202–208 (1985)

Online Exploration of All Vertices in a Simple Polygon*

Yuya Higashikawa and Naoki Katoh

Department of Architecture and Architectural Engineering, Kyoto University,
Nishikyo-ku, Kyoto 615-8540 Japan
{as.higashikawa,naoki}@archi.kyoto-u.ac.jp

Abstract. This paper considers an online exploration problem in a simple polygon where starting from a point in the interior of a simple polygon, the searcher is required to explore a simple polygon to visit all its vertices and finally return to the initial position as quickly as possible. The information of the polygon is given online. As the exploration proceeds, the searcher gains more information of the polygon. We give a 1.219-competitive algorithm for this problem. We also study the case of a rectilinear simple polygon, and give a 1.167-competitive algorithm.

Keywords: online algorithm, exploration, competitive analysis.

1   Introduction

The Tohoku Earthquake attacked East Japan area on March 11, 2011. When such a big earthquake occurs in an urban area, it is predicted that many buildings and underground shopping areas will be heavily damaged, and it is seriously important to efficiently explore the inside of damaged areas in order to rescue human beings left there. With this motivation, this paper deals with an *online exploration problem* (**OEP** for short) in a simple polygon. Given a simple polygon P, suppose the searcher is initially in the interior of P. Starting from the origin o, the aim of the searcher is to visit all vertices of P at least once and to return to the starting point as quickly as possible. The information of the polygon is given online. Namely, at the beginning, the searcher has only the information of a visible part of the polygon. As the exploration proceeds, the visible area changes. However, the information of the region which has once become visible is assumed to be accumulated. So, as the exploration proceeds, the searcher gains more information of the polygon, and determines which vertex to visit next based on the information obtained so far.

In general, the performance of an online algorithm is measured by a *competitive ratio* which is defined as follows. Let $\mathcal{S}$ denote a class of objects to be explored. When an online exploration algorithm ALG is used to explore an object $S \in \mathcal{S}$, let $|ALG(S)|$ denote the tour length (cost) required to explore S by ALG.

---

* Supported by JSPS Grant-in-Aid for Scientific Research(B)(21300003).

J. Snoeyink, P. Lu, K. Su, and L. Wang (Eds.): FAW-AAIM 2012, LNCS 7285, pp. 315–326, 2012.

Let $|\mathsf{OPT}(S)|$ denote the tour length (cost) required to explore S by the offline optimal algorithm. Then the competitive ratio of ALG is defined as follows.

$$\sup_{S \in \mathcal{S}} \frac{|\mathsf{ALG}(S)|}{|\mathsf{OPT}(S)|}.$$

Previous Work: OEP has been extensively studied for the case of graphs. Kalyanasundaram et al. [10] presented a 16-competitive algorithm for planar undirected graphs. Megow et al. [8] recently extended this result to undirected graphs with genus g and gave a $16(1+2g)$-competitive algorithm. For the case of a cycle, Miyazaki et al. [9] gave an optimal 1.37-competitive algorithm. All these results are concerned with a single searcher. For the case of $p(> 1)$ searchers, there are some results. Fraigniaud et al. [3] gave an $O(p/\log p)$-competitive algorithm for the case of a tree. Higashikawa et al. [6] gave $(p/\log p + o(1))$-competitive algorithm for this problem. Dynia et al. [2] showed a lower bound $\Omega(\log p/\log\log p)$ for any deterministic algorithm for this problem.

There are some papers that deal with OEP in geometric regions (see survey paper [5]). Kalyanasundaram et al. [10] studied the case of a polygon with holes where all edges are required to traverse. They gave a 17-competitive algorithm for this case. Hoffmann et al. [7] studied the problem that asks to find a tour in a simple polygon such that every vertex is visible from some point on the tour, and gave a 26.5-competitive algorithm.

Our Results: We will show 1.219-competitive algorithm for OEP in a simple polygon. We also study the case of a rectilinear simple polygon, and give a 1.167-competitive algorithm. We will give a lower bound result that the competitive ratio is at least 1.040 within a certain framework of exploration algorithms.

2   Strategy of AOE

In this paper, we define a *simple polygon* as a closed polygonal chain with no self-intersction in the plane. In the followings, we use the term *polygon* to stand for a simple polygon. Also an *edge of a polygon* (or a *polygon edge*) is defined as a line segment forming a part of the polygonal chain, a *vertex of a polygon* (or a *polygon vertex*) as a point where two polygon edges meet and the *boundary of polygon* as a polygonal chain. Let P be a polygon and o be the origin. Sometimes we abuse the notation P to stand for the interior (including the boundary) of P. Let $V = \{v_1, v_2, ..., v_n\}$ be a polygon vertex set of P sorted in clockwise order along the boundary and $E = \{e_1, e_2, ..., e_n\}$ be a polygon edge set of P composed of $e_i = (v_i, v_{i+1}) = (v_{e_i}^1, v_{e_i}^2)$ with $1 \leq i \leq n$ ($v_{n+1} = v_1$ is assumed). let $|e|$ denote the length of edge $e \in E$ and $L = \sum_{e \in E} |e|$ be the boundary length of P. For any two points $x, y \in P$, let $sp(x, y)$ denote the shortest path from x to y that lies in the inside of P, $|sp(x, y)|$ be its length and $|xy|$ be the Euclidean distance from x to y. Note that $sp(x, y) = sp(y, x)$ and $|xy| \leq |sp(x, y)|$. Furthermore, for any two vertices $x, y \in V$, let $bp(x, y)$ denote the clockwise path along the boundary of P from x to y and $|bp(x, y)|$ be its length. The cost of a tour is defined to be its length.

For a point $x \in P$ and an edge $e \in E$, let

$$cost(x, e) = |sp(x, v_e^1)| + |sp(x, v_e^2)| - |e|.$$

In the offline version of this problem, we will prove below that an optimal strategy is that starting from the origin o, the searcher first goes to one endpoint of some edge e, namely v_e^2, then follows the boundary path $bp(v_e^2, v_e^1)$ and finally comes back to o.

Lemma 1. *For offline exploration problem in a polygon P, the cost of the offline optimal algorithm satisfies the following.*

$$|\mathsf{OPT}(P)| = L + \min_{e \in E} cost(o, e).$$

Let $e_{opt} \in E$ be an edge satisfying the following equation.

$$cost(o, e_{opt}) = \min_{e \in E} cost(o, e). \tag{1}$$

For two points $x, y \in P$, we say that y is *visible* from x if the line segment xy lies in the inside of P. Then the *visibility polygon* $VP(P, x)$ is

$$VP(P, x) := \{y \in P \mid y \text{ is visible from } x\}.$$

Note that an edge of the visibility polygon is not necessarily an edge of P (see Fig. 1). For a polygon vertex b and a point $x \in P$, we call b a *blocking vertex* with respect to x if b is visible from x and there is the unique edge incident to b such that any point on the edge except b is not visible from x. Let b^* be a point where the extension of the line segment xb towards b first intersects the boundary of P. Then we call b^* a *virtual vertex* and the line segment bb^* a *cut edge*. Without loss of generality we assume that b^* does not coincide with any vertex in V. Also let $\hat{e}$ be an edge of P containing b^* then we regard a visible part of $\hat{e}$ as a new edge, which we call a *virtual edge*. Note that a cut edge bb^* divides P in two areas, a polygon which contains $VP(P, x)$ and the other not. We call the latter area the *invisible polygon* $IP(P, x, b)$. Notice that $VP(P, x)$ and $IP(P, x, b)$ share a cut edge bb^*.

We assume that there is a blocking vertex b with respect to the origin o since otherwise an optimal solution can be found by Lemma 1. Then we have the following lemma.

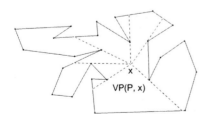

Fig. 1. Visibility polygon

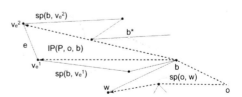

Fig. 2. Illustration of $sp(b, v_e^1)$, $sp(b, v_e^2)$ and $sp(o, w)$

Lemma 2. *For an invisible polygon $IP(P, o, b)$ defined by a blocking vertex b, let $e \in E$ be a polygon edge both endpoints of which are in $IP(P, o, b)$, and $w \in V$ be the polygon vertex adjacent to b which is not in $IP(P, o, b)$. Then*

$$cost(o, (b, w)) < cost(o, e).$$

Proof. First, we remark a simple fact. Let x, y, z be points in P such that both line segments xz and zy are lying the inside of P. Then the following inequality obviously holds.

$$|sp(x, y)| \leq |xz| + |zy|. \tag{2}$$

Notice the equality holds only when either (i) $sp(x, y)$ is a line segment xy and z is on xy, or (ii) $sp(x, y)$ is composed of two line segments xz and zy, i.e., y is not visible from x (z is a blocking vertex with respect to x). See Fig. 2. From the above observation and since b is visible from o, i.e., $|sp(o, b)| = |ob|$,

$$|sp(o, w)| < |ob| + |bw| = |sp(o, b)| + |bw|. \tag{3}$$

Besides, from the triangle inequality with respect to b, v_e^1 and v_e^2,

$$cost(b, e) = |sp(b, v_e^1)| + |sp(b, v_e^2)| - |e| \geq 0. \tag{4}$$

Furthermore both $sp(o, v_e^1)$ and $sp(o, v_e^2)$ pass through b. Hence, we have

$$|sp(o, b)| + |sp(b, v_e^1)| = |sp(o, v_e^1)| \text{ and } |sp(o, b)| + |sp(b, v_e^2)| = |sp(o, v_e^2)|. \tag{5}$$

Thus,

$$
\begin{aligned}
cost(o, (b, w)) &= |sp(o, b)| + |sp(o, w)| - |bw| \\
&< |sp(o, b)| + |sp(o, b)| + |bw| - |bw| && \text{(from (3))} \\
&\leq 2|sp(o, b)| + |sp(b, v_e^1)| + |sp(b, v_e^2)| - |e| && \text{(from (4))} \\
&= cost(o, e) && \text{(from (5))}
\end{aligned}
$$

holds. □

For e_{opt} defined by (1), the following corollary is immediate from Lemma 2.

Corollary 1. *For an invisible polygon $IP(P, o, b)$ defined by a blocking vertex b, let $e \in E$ be a polygon edge both endpoints of which are in $IP(P, o, b)$. Then e cannot be e_{opt} satisfying (1).*

Based on Corollary 1, candidates of e_{opt} are edges of $VP(P, o)$.

In what follows, we propose an online algorithm, AOE(Avoiding One Edge). By Lemma 1, the offline optimal algorithm chooses the edge e_{opt} which satisfies (1). But we cannot obtain the whole information about P. So, the seemingly best strategy based on the information of $VP(P, o)$ is to choose an edge in the same way as the offline optimal algorithm, assuming that there is no invisible polygon, namely $P = VP(P, o)$. Let E_1^* denote an edge set composed of all $e \in E$

such that both endpoints of e are visible from o, E_2^* denote a set of virtual edges on the boundary of $VP(P,o)$ and $E^* = E_1^* \cup E_2^*$. Also for a virtual edge $e \in E_2^*$, endpoints of e are labeled as v_e^1, v_e^2 in clockwise order around o and let $cost(o,e)$ denote the value of $|sp(o, v_e^1)| + |sp(o, v_e^2)| - |e|$. Let $e^* \in E^*$ be an edge satisfying the following equation.

$$cost(o, e^*) = \min_{e \in E^*} cost(o, e) \tag{6}$$

Then Algorithm AOE is described as follows.

Step 1: Choose $e^* \in E^*$ satisfying (6).
Step 2: If $e^* \in E_1^*$ then let $\hat{e} = e^*$, else let $\hat{e}$ be an edge of P containing e^*.
Step 3: Follow the tour $sp(o, v_{\hat{e}}^2) \to bp(v_{\hat{e}}^2, v_{\hat{e}}^1) \to sp(v_{\hat{e}}^1, o)$.

3   Competitive Analysis of AOE

First, we show the following lemma.

Lemma 3. *Let x be a point on the boundary of P and e^* be an edge satisfying (6). If x is visible from the origin o, then*

$$\frac{cost(o, e^*)}{2} \le |ox|.$$

Proof. Let $e' \in E^*$ be an edge of $VP(P, o)$ containing x. Then from (2), we have $|ox| \ge |sp(o, v_{e'}^1)| - |xv_{e'}^1|$ and $|ox| \ge |sp(o, v_{e'}^2)| - |xv_{e'}^2|$. Therefore, we obtain

$$2|ox| \ge |sp(o, v_{e'}^1)| + |sp(o, v_{e'}^2)| - |xv_{e'}^1| - |xv_{e'}^2|$$
$$= |sp(o, v_{e'}^1)| + |sp(o, v_{e'}^2)| - |e'| \ge cost(o, e^*),$$

namely $|ox| \ge cost(o, e^*)/2$.   □

Furthermore, we show a lemma which plays a crucial role in our analysis.

Lemma 4. *Let L be the length of the boundary of P and e^* be an edge satisfying (6). Then the following inequality holds.*

$$L \ge \pi \cdot cost(o, e^*). \tag{7}$$

Proof. Let C be a circle centered at the origin o with the radius of $cost(o, e^*)/2$. From Lemma 3, any edge of P does not intersect C. Thus L is greater than the length of the circumference of C, namely

$$L \ge 2\pi \cdot \frac{cost(o, e^*)}{2} = \pi \cdot cost(o, e^*)$$

holds.   □

Theorem 1. *The competitive ratio of Algorithm AOE is at most 1.319.*

Proof. The cost of Algorithm AOE obviously satisfies

$$|\text{AOE}(P)| = L + cost(o, e^*).$$

On the other hand, the cost of the offline optimal algorithm satisfies $|\mathsf{OPT}(P)| = L + cost(o, e_{opt})$ holds from Lemma 1. By the triangle inequality, $cost(o, e_{opt}) \geq 0$, namely $|\mathsf{OPT}(P)| \geq L$ holds. Thus we have

$$\frac{|\mathsf{AOE}(P)|}{|\mathsf{OPT}(P)|} \leq \frac{L + cost(o, e^*)}{L} = 1 + \frac{cost(o, e^*)}{L}.$$

From this and (7),

$$\frac{|\mathsf{AOE}(P)|}{|\mathsf{OPT}(P)|} \leq 1 + \frac{cost(o, e^*)}{\pi \cdot cost(o, e^*)} = 1 + \frac{1}{\pi} \leq 1.319$$

is obtained.   □

Theorem 1 gives an upper bound of the competitive ratio. In the followings, we will obtain a better bound by detailed analysis. First, we improve a lower bound of $|\mathsf{OPT}(P)|$. Note that for some points $x, y, z \in P$ such that both y and z are visible from x and the line segment yz is lying in P, we call $\angle yxz$ the *visual angle* at x formed by yz.

Lemma 5. *For an edge* $e^* \in E^*$ *satisfying (6), let* $d = cost(o, e^*)$ *and* θ *(0 $\leq$ $\theta \leq \pi$) be a visual angle at* o *formed by a visible part of* e_{opt}. *Then*

$$|\mathsf{OPT}(P)| \geq L + d - d \sin \frac{\theta}{2}. \tag{8}$$

Proof. We first show the following claim.

Claim 1. *Let* $b_1 \in V$ *(resp.* b_2*) be the vertex visible from* o *such that the path* $sp(o, v^1_{e_{opt}})$ *(resp.* $sp(o, v^2_{e_{opt}})$*) passes through* b_1 *(resp.* b_2*) (see Fig. 3). Then*

$$cost(o, e_{opt}) \geq |ob_1| + |ob_2| - |b_1 b_2|. \tag{9}$$

Proof. This follows from $|sp(o, v^1_{e_{opt}})| = |ob_1| + |sp(b_1, v^1_{e_{opt}})|$, $|sp(o, v^2_{e_{opt}})| = |ob_2| + |sp(b_2, v^2_{e_{opt}})|$ and $|e_{opt}| = |sp(v^1_{e_{opt}}, v^2_{e_{opt}})| \leq |sp(b_1, v^1_{e_{opt}})| + |b_1 b_2| + |sp(b_2, v^2_{e_{opt}})|$.   □

From (9), we have

$$|\mathsf{OPT}(P)| = L + cost(o, e_{opt}) \geq L + |ob_1| + |ob_2| - |b_1 b_2|. \tag{10}$$

Furthermore b_1 and b_2 satisfy $|ob_1| \geq d/2$ and $|ob_2| \geq d/2$ from Lemma 3. Hence there exist points u_1, u_2 on line segments ob_1, ob_2 such that $|ou_1| = |ou_2| = d/2$ (see Fig. 4). Then, from the triangle inequality with respect to u_1, u_2 and b_1,

$$|u_1 u_2| \geq |u_2 b_1| - |b_1 u_1| = |u_2 b_1| - (|ob_1| - \frac{d}{2})$$

holds. Similarly we have

$$|u_2 b_1| \geq |b_1 b_2| - |u_2 b_2| = |b_1 b_2| - (|ob_2| - \frac{d}{2}).$$

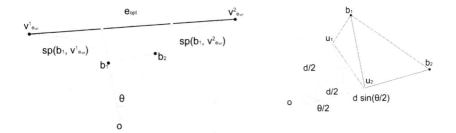

Fig. 3. A visible part of e_{opt} from o **Fig. 4.** u_1 and u_2

Thus we have

$$d - |u_1 u_2| \leq d - \{|u_2 b_1| - (|ob_1| - \frac{d}{2})\} = \frac{d}{2} + |ob_1| - |u_2 b_1|$$

$$\leq \frac{d}{2} + |ob_1| - \{|b_1 b_2| - (|ob_2| - \frac{d}{2})\} = |ob_1| + |ob_2| - |b_1 b_2|. \quad (11)$$

In addition, the length of $u_1 u_2$ satisfies the following equation.

$$|u_1 u_2| = \frac{d}{2} \cdot 2 \sin \frac{\theta}{2} = d \sin \frac{\theta}{2}. \quad (12)$$

By (10), (11) and (12),

$$|\mathsf{OPT}(P)| \geq L + d - |u_1 u_2| = L + d - d \sin \frac{\theta}{2}.$$

is shown. $\square$

Secondly, we show a better lower bound of L.

Lemma 6. *Let d and θ as defined in* Lemma 5. *Then*

$$L \geq d(\pi - \frac{\theta}{2} + \tan \frac{\theta}{2}). \quad (13)$$

Proof. Let C be a circle centered at o with radius $d/2$. From Lemma 3, any edge of P does not intersect C. Also let endpoints of a visible part of e_{opt} from o be w_1, w_2 in clockwise order around o. Then, we consider two cases; (Case 1) $\angle ow_1 w_2 \leq \pi/2$ and $\angle ow_2 w_1 \leq \pi/2$ and (Case 2) $\angle ow_1 w_2 > \pi/2$ and $\angle ow_2 w_1 \leq \pi/2$ (see Fig. 5, 6). Note that the case of $\angle ow_1 w_2 \leq \pi/2, \angle ow_2 w_1 > \pi/2$ can be treated in a manner similar to Case 2.

Case 1: Let w_1^* (resp. w_2^*) be a point on the line segment ow_1 (resp. ow_2) such that $w_1 w_2$ is parallel to $w_1^* w_2^*$ and the line segment $w_1^* w_2^*$ touches the circle C and let h be a tangent point of $w_1^* w_2^*$ and C. Also let $\angle w_1 oh = x\theta$ and $\angle w_2 oh = (1-x)\theta$ with some x $(0 \leq x \leq 1)$. Then the length of $w_1^* w_2^*$ satisfies

$$|w_1^* w_2^*| = \frac{d}{2} \tan x\theta + \frac{d}{2} \tan(1-x)\theta.$$

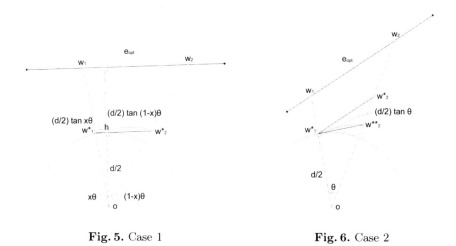

Fig. 5. Case 1   **Fig. 6.** Case 2

The right-hand side of this equation attains the minimum value when $x = 1/2$. Thus

$$|w_1^* w_2^*| \geq \frac{d}{2} \tan \frac{\theta}{2} + \frac{d}{2} \tan \frac{\theta}{2} = d \tan \frac{\theta}{2}. \tag{14}$$

Furthermore the sum of the visual angle at o formed by a visible part of the boundary other than $w_1 w_2$ is equal to $2\pi - \theta$. Hence we have

$$L \geq \frac{d}{2}(2\pi - \theta) + |w_1 w_2|. \tag{15}$$

Since $|w_1 w_2| \geq |w_1^* w_2^*|$ obviously holds, from (14) and (15), we obtain

$$L \geq \frac{d}{2}(2\pi - \theta) + d \tan \frac{\theta}{2} = d(\pi - \frac{\theta}{2} + \tan \frac{\theta}{2}).$$

Case 2: Let w_1^* (resp. w_2^*) be a point on the line segment ow_1 (resp. ow_2) such that $w_1 w_2$ is parallel to $w_1^* w_2^*$ and $|ow_1^*| = d/2$ (the circumference of C passes through w_1^*). Also let w_2^{**} an intersection point of the line segment ow_2 and the lineperpendicular to the line segment ow_1 through w_1^*. Then

$$|w_1^* w_2^*| > |w_1^* w_2^{**}| = \frac{d}{2} \tan \theta \geq d \tan \frac{\theta}{2}.$$

In the same way as Case 1, we obtain $L \geq d(\pi - \theta/2 + \tan(\theta/2))$.   □

By Lemma 5 and 6, we prove the following theorem.

Theorem 2. *The competitive ratio of Algorithm* AOE *is at most* 1.219.

Proof. Let d and θ as defined in Lemma 5. Since $|\text{AOE}(P)| = L + d$ holds, from (8), (13), we have

$$\frac{|\text{AOE}(P)|}{|\text{OPT}(P)|} \leq \frac{L+d}{L+d-d\sin\frac{\theta}{2}} \leq \frac{d(\pi - \frac{\theta}{2} + \tan\frac{\theta}{2}) + d}{d(\pi - \frac{\theta}{2} + \tan\frac{\theta}{2}) + d - d\sin\frac{\theta}{2}}$$

$$= \frac{\pi - \frac{\theta}{2} + \tan\frac{\theta}{2} + 1}{\pi - \frac{\theta}{2} + \tan\frac{\theta}{2} + 1 - \sin\frac{\theta}{2}} \qquad (0 \leq \theta \leq \pi). \qquad (16)$$

In the followings, we compute the maximum value of (16),

$$\max_{0 \leq \theta \leq \pi} \left\{ z(\theta) = \frac{\pi - \frac{\theta}{2} + \tan\frac{\theta}{2} + 1}{\pi - \frac{\theta}{2} + \tan\frac{\theta}{2} + 1 - \sin\frac{\theta}{2}} \right\}. \qquad (17)$$

Generally the following fact about the fractional program is known [1,11].

Fact 1. *Let $X \subseteq \mathbb{R}^n$, $f : \mathbb{R}^n \to \mathbb{R}$ and $g : \mathbb{R}^n \to \mathbb{R}$. Let us consider the following fractional program formulated as*

$$\text{maximize} \left\{ h(x) = \frac{f(x)}{g(x)} \,\middle|\, x \in X \right\}, \qquad (18)$$

where $g(x) > 0$ is assumed for any $x \in X$. Let $x^* \in \text{argmax}_{x \in X} h(x)$ denote an optimal solution of (18) and $\lambda^* = h(x^*)$ denote the optimal value. Furthermore, with a real parameter λ, let $h_\lambda(x) = f(x) - \lambda g(x)$ and $M(\lambda) = \max_{x \in X} h_\lambda(x)$. Then $M(\lambda)$ is monotone decreasing for λ and the followings hold.
(i) $M(\lambda) < 0 \Leftrightarrow \lambda > \lambda^*$, (ii) $M(\lambda) = 0 \Leftrightarrow \lambda = \lambda^*$, (iii) $M(\lambda) > 0 \Leftrightarrow \lambda < \lambda^*$.

In the same way as Theorem 2, with a real parameter λ, we define $z_\lambda(\theta)$ and $M(\lambda)$ for $z(\theta)$ as follows.

$$z_\lambda(\theta) = \pi - \frac{\theta}{2} + \tan\frac{\theta}{2} + 1 - \lambda(\pi - \frac{\theta}{2} + \tan\frac{\theta}{2} + 1 - \sin\frac{\theta}{2}) \quad (0 \leq \theta \leq \pi),$$

$$M(\lambda) = \max_{0 \leq \theta \leq \pi} z_\lambda(\theta).$$

From Fact 1 (ii), λ^* satisfying $M(\lambda^*) = 0$ is equal to (17), i.e., the maximum value of $z(\theta)$. Hence we only need to compute λ^*.

Finally, let $\theta_\lambda^* \in \text{argmax}_{0 \leq \theta \leq \pi} z_\lambda(\theta)$, then we show θ_λ^* is unique. A derivative of $z_\lambda(\theta)$ is calculated as

$$\frac{dz_\lambda}{d\theta} = -\frac{\lambda - 1}{2} \tan^2\frac{\theta}{2} + \frac{\lambda}{2}\cos\frac{\theta}{2}.$$

This derivative is monotone decreasing in the interval $0 \leq \theta \leq \pi$, therefore $z_\lambda(\theta)$ is concave in this interval, then θ_λ^* is unique. Indeed when $\lambda = 1.219$, $\theta_\lambda^* \simeq 2.0706$ then $M(1.219) \simeq -0.0010 < 0$. Also when $\lambda = 1.218$, $\theta_\lambda^* \simeq 2.0718$ then $M(1.218) \simeq 0.0029 > 0$. Thus we obtain $1.218 < \lambda^* < 1.219$. $\qquad \square$

3.1   Lower Bound

Theorem 3. *The competitive ratio of Algorithm* AOE *is at least 1.040 (see Fig. 7).*

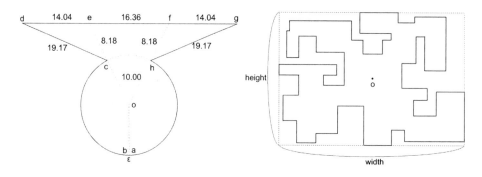

Fig. 7. Worst case polygon   **Fig. 8.** A rectilinear polygon

4   Competitive Analysis for Rectilinear Polygon

In this section, we analyze the competitive ratio of AOE for a rectilinear polygon (see Fig. 8). Generally a rectilinear polygon is defined as a simple polygon all of whose interior angles are $\pi/2$ or $3\pi/2$. Edges of the rectilinear polygon are classified as horizontal or vertical edges. Let R be a rectilinear polygon and R' be the minimum enclosing rectangle of R. Then we define the height of R' as the height of R and also the width of R' as the width of R. Note that the searcher follows the Euclidean shortest path even if he/she is in the rectilinear polygon.

Lemma 7. *For an edge $e^* \in E^*$ satisfying (6), let $d = cost(o, e^*)$ and θ $(0 \leq \theta \leq \pi)$ be a visual angle at o formed by a visible part of e_{opt}. Then*

$$L \geq \max\{4d, 2d + 2d\tan\frac{\theta}{2}\}. \tag{19}$$

Proof. First, we show $L \geq 4d$. Let C be a circle centered at o with the radius of $d/2$. From Lemma 3, any edge of R does not intersect C (see Fig. 9). Thus each of the height and width of R is greater than d (the diameter of C), namely $L \geq 4d$ holds.

Secondly, we show $L \geq 2d + 2d\tan(\theta/2)$. Note that we should just consider the case of $4d \leq 2d + 2d\tan(\theta/2)$, namely $\pi/2 \leq \theta \leq \pi$ because $L \geq 4d$ has been proved. Without loss of generality we assume that e_{opt} is a horizontal edge.

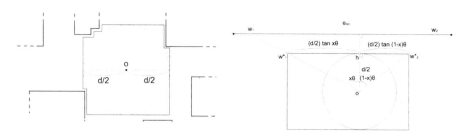

Fig. 9. $L \geq 4d$ **Fig. 10.** $L \geq 2d + 2d\tan(\theta/2)$

We label endpoints of a visible part of e_{opt} from o as w_1, w_2 in clockwise order around o. Let w_1^* (resp. w_2^*) be a point on the line segment ow_1 (resp. ow_2) such that $w_1 w_2$ is parallel to $w_1^* w_2^*$ and the line segment $w_1^* w_2^*$ touches the circle C and h be a tangent point of $w_1^* w_2^*$ and C (see Fig. 10). Also let $\angle w_1 o h = x\theta$ and $\angle w_2 o h = (1 - x)\theta$ with some x ($0 \le x \le 1$). Then the length of $w_1^* w_2^*$ satisfies

$$|w_1^* w_2^*| = \frac{d}{2} \tan x\theta + \frac{d}{2} \tan(1 - x)\theta \ge \frac{d}{2} \tan \frac{\theta}{2} + \frac{d}{2} \tan \frac{\theta}{2} = d \tan \frac{\theta}{2}.$$

Thus the width of R is greater than $d\tan(\theta/2)$ and the height of R is greater than d, then $L \ge 2d + 2d \tan(\theta/2)$ holds. $\square$

Theorem 4. *For a rectilinear polygon, the competitive ratio of Algorithm* AOE *is at most 1.167.*

Proof. Based on (19), we consider two cases; (Case 1) $0 \le \theta < \pi/2$ and (Case 2) $\pi/2 \le \theta \le \pi$. Note that $4d > 2d + 2d\tan(\theta/2)$ holds in Case 1 and $4d \le 2d + 2d\tan(\theta/2)$ holds in the other.

Case 1: From $L \ge 4d$ and (8), we obtain

$$\frac{|\mathsf{AOE}(P)|}{|\mathsf{OPT}(P)|} \le \frac{4d + d}{4d + d - d\sin\frac{\theta}{2}} = \frac{5}{5 - \sin\frac{\theta}{2}} < \frac{5}{5 - \sin\frac{\pi}{4}} \le 1.165.$$

Case 2: From $L \ge 2d + 2d\tan(\theta/2)$ and (8), we obtain

$$\frac{|\mathsf{AOE}(P)|}{|\mathsf{OPT}(P)|} \le \frac{2d + 2d\tan\frac{\theta}{2} + d}{2d + 2d\tan\frac{\theta}{2} + d - d\sin\frac{\theta}{2}} = \frac{3 + 2\tan\frac{\theta}{2}}{3 + 2\tan\frac{\theta}{2} - \sin\frac{\theta}{2}}. \tag{20}$$

We will compute the maximum value of (20) as in the proof of Theorem 2 by defining $z_\lambda(\theta)$ and $M(\lambda)$ for a real parameter λ as follows.

$$z_\lambda(\theta) = 3 + 2\tan\frac{\theta}{2} - \lambda\left(3 + 2\tan\frac{\theta}{2} - \sin\frac{\theta}{2}\right) \quad \left(\frac{\pi}{2} \le \theta \le \pi\right)$$

$$M(\lambda) = \max_{\frac{\pi}{2} \le \theta \le \pi} z_\lambda(\theta)$$

Let $\theta_\lambda^* \in \mathrm{argmax}_{0 \le \theta \le \pi} z_\lambda(\theta)$, then a derivative of $z_\lambda(\theta)$ is calculated as

$$\frac{dz_\lambda}{d\theta} = -(\lambda - 1)\frac{1}{\cos^2\frac{\theta}{2}} + \frac{\lambda}{2}\cos\frac{\theta}{2}.$$

This derivative is monotone decreasing in the interval $\pi/2 \le \theta \le \pi$, therefore $z_\lambda(\theta)$ is concave in this interval, then θ_λ^* is unique. Indeed when $\lambda = 1.167$, $\theta_\lambda^* \simeq 1.7026$ then $M(1.167) \simeq -0.0044 < 0$. Also when $\lambda = 1.166$, $\theta_\lambda^* \simeq 1.7056$ then $M(1.166) \simeq 7.6 \times 10^{-5} > 0$. Thus we obtain $1.166 < \lambda^* < 1.167$. $\square$

5   Discussion and Open Problems

We believe that the upper bound of the competitive ratio can be improved: the least upper bound could be close to the lower bound 1.04 given in Section 3.1.

As one of many variations of OEP, we could consider OEP with multiple searchers. In this problem, all searchers are initially at the same origin $o \in P$. The goal of the exploration is that each vertex is visited by at least one searcher and that all searchers return to the origin o. We regard the time when the last searcher comes back to the origin as the cost of the exploration. Note that our algorithm can be easily adapted to the case of OEP with 2-searchers. For an offline exploration problem with k-searchers, Frederickson et al. [4] proposed a $(e + 1 - 1/k)$-approximation algorithm, where e is the approximation ratio of some 1-searcher algorithm. Their idea is splitting a tour given by some 1-searcher algorithm into k parts such that the cost of each part is equal, where the cost of a part is the length of the shortest tour from o which passes along the part. When $k = 2$, we can apply this idea to our algorithm as follows. First, choose similarly $e^* \in E^*$ satisfying (6). Then let one searcher go to $v_{e^*}^1$ and walk counterclockwise along the boundary of P, and let symmetrically the other go to $v_{e^*}^2$ and walk clockwise. When two searchers meet at a point on the boundary, two searchers come back together to o along the shortest path in the inside of P. In this case, we obtain an upper bound 1.719. However, when $k \geq 3$, the above-mentioned idea cannot be directly applied. So, it remains open.

References

1. Dinkelbach, W.: On nonlinear fractional programming. Management Science 13(7), 492–498 (1967)
2. Dynia, M., Łopuszański, J., Schindelhauer, C.: Why Robots Need Maps. In: Prencipe, G., Zaks, S. (eds.) SIROCCO 2007. LNCS, vol. 4474, pp. 41–50. Springer, Heidelberg (2007)
3. Fraigniaud, P., Gsieniec, L., Kowalski, D.R., Pelc, A.: Collective tree exploration. Networks 48(3), 166–177 (2006)
4. Frederickson, G.N., Hecht, M.S., Kim, C.E.: Approximation algorithms for some routing problems. SIAM J. Comput. 7, 178–193 (1978)
5. Ghosh, S.K., Klein, R.: Online algorithms for searching and exploration in the plane. Computer Science Review 4(4), 189–201 (2010)
6. Higashikawa, Y., Katoh, N., Langerman, S., Tanigawa, S.: Online Graph Exploration Algorithms for Cycles and Trees by Multiple Searchers. In: Proc. 3rd AAAC Annual Meeting (2010)
7. Hoffmann, F., Icking, C., Klein, R., Kriegel, K.: The polygon exploration problem. SIAM J. Comput. 31(2), 577–600 (2002)
8. Megow, N., Mehlhorn, K., Schweitzer, P.: Online Graph Exploration: New Results on Old and New Algorithms. In: Aceto, L., Henzinger, M., Sgall, J. (eds.) ICALP 2011, Part II. LNCS, vol. 6756, pp. 478–489. Springer, Heidelberg (2011)
9. Miyazaki, S., Morimoto, N., Okabe, Y.: The online graph exploration problem on restricted graphs. IEICE Trans. Inf. & Syst. E92-D(9), 1620–1627 (2009)
10. Kalyanasundaram, B., Pruhs, K.R.: Constructing competitive tours from local information. Theoretical Computer Science 130, 125–138 (1994)
11. Schaible, S., Ibaraki, T.: Fractional programming. European Journal of Operational Research 12, 325–338 (1983)

In-Place Algorithms for Computing a Largest Clique in Geometric Intersection Graphs

Minati De[1], Subhas C. Nandy[1], and Sasanka Roy[2]

[1] Indian Statistical Institute, Kolkata - 700108, India
{minati_r,nandysc}@isical.ac.in
[2] Indian Institute of Science Education and Research, Kolkata, India
sasanka.ro@gmail.com

Abstract. In this paper, we study the problem of designing in-place algorithms for finding the maximum clique in the *intersection graphs* of axis-parallel rectangles and disks in 2D. We first propose $O(n^2 \log n)$ time in-place algorithms for finding the maximum clique of the intersection graphs of a set of axis-parallel rectangles of arbitrary sizes. For the rectangle intersection graph of fixed height rectangles, the time complexity can be slightly improved to $O(n \log n + nK)$, where K is the size of the maximum clique. For disk graphs, we consider two variations of the maximum clique problem, namely geometric clique and graphical clique. The time complexity of our algorithm for finding the largest geometric clique is $O(n^2 \log n)$, and it works for disks of arbitrary radii. For graphical clique, our proposed algorithm works for unit disks (i.e., of same radii) and the worst case time complexity is $O(n^2 + mK^4)$; m is the number of edges in the unit disk intersection graph, and K is the size of the largest clique in that graph. It uses $O(n^4)$ time in-place computation of maximum matching in a bipartite graph, which is of independent interest. All these algorithms need $O(1)$ work space in addition to the input array $\mathcal{R}$.

1 Introduction

Due to the wide applications of sensor networks, nowadays there is a high demand of space-efficient algorithms in the embedded software. So, designing space efficient algorithms for different practical problems have become an important area of research. Detailed survey on in-place algorithms for the geometric optimization and search problems is available in [2,4]. We consider in-place algorithms for the optimization problems on intersection graph of geometric objects.

In sophisticated database query and VLSI physical design, several optimization problems are formulated using the intersection graph of axis-parallel rectangles. Similarly, the disk graph plays an important role in formulating different problems in mobile ad hoc networks. We concentrate on finding the largest clique on several variations of the intersection graph of axis-parallel rectangles and disks. An array of size n containing n objects is given as input; each array element represents the corresponding object in minimum amount of space. During execution, the array elements can change their positions; but at the end of

J. Snoeyink, P. Lu, K. Su, and L. Wang (Eds.): FAW-AAIM 2012, LNCS 7285, pp. 327–338, 2012.
© Springer-Verlag Berlin Heidelberg 2012

the execution, each object will be available in the array. We are allowed to use at most $O(1)$ space for storing temporary results. From now onwards, by rectangle intersection graph we will mean the intersection graph of axis-parallel rectangles.

The first polynomial time algorithm for finding the maximum clique in a rectangle intersection graph was proposed in [8]. The best known algorithm for this problem runs in $O(n \log n)$ time and $O(n)$ space [7,10]. The best known algorithm for finding the maximum clique of a set of axis-parallel rectangles in d-dimensional space runs in $O(n^{d-1})$ time [9].

Let us consider the intersection graph of a set of disks. Let C be a subset of disks such that each pair of members in C intersect. In the aforesaid graph the nodes corresponding to C define a clique. However, since the disks do not satisfy Helly property, the members of C may not have a common intersection. Thus, a clique in a disk graph is usually referred to as *graphical clique*. In particular, if the members in a clique have common intersection region, then that clique is referred to as a *geometric clique*. For demonstration, see Fig. 1.

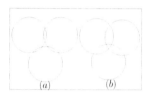

Fig. 1. (a) Graphical clique and (b) Geometric clique of disk graph

Given a set of points in 2D, the best known algorithm for computing the largest (graphical) clique of the corresponding unit disk graph takes $O(n^{3.5} \log n)$ time [1]. However, the position of an unit disk containing maximum number of points can be mapped to finding the maximum geometric clique of the unit disk graph. Using a plane sweep method, this can be solved in $O(n^2)$ time [3].

Our Results

We first propose an in-place $O(n \log n)$ time algorithm for computing the maximum clique of an intersection graph of a set of n intervals on the real line. We use this algorithm to design an in-place algorithm for finding the maximum clique of the intersection graph of a set of n axis-parallel rectangles of arbitrary size in $O(n^2 \log n)$ time. For fixed height rectangles, the time complexity can be slightly improved to $O(n \log n + nK)$, where K is the size of the largest clique.

Next, we consider the maximum clique problem for the disk graph. Our proposed in-place algorithm for computing the largest geometric clique of the intersection graph of a set of disks of arbitrary radii needs $O(n^2 \log n)$ time. For graphical clique, our in-place algorithm works for unit disks only, and it runs in $O(n^2 + mK^4)$ time, where n and m are the number of vertices and edges in the unit disk graph, and K is the size of the maximum clique in that graph. To solve this problem, we proposed an $O(n^4)$ time in-place algorithm for computing maximum matching in a bipartite graph $G = (V_1, V_2, E)$ where the two sets of nodes V_1 and V_2 are stored in two arrays, and the existence of an edge between a pair of nodes can be checked on demand by an oracle in $O(1)$ time. This is of independent interest, since to our knowledge there does not exist any in-place algorithm for computing the maximum matching in a bipartite graph in the literature.

The following table summarizes the results presented in the paper along with the comparative study with the best known algorithm available in the literature.

Notice that, the (*time* × *extra-work-space*) is less than or equal to the best-known results for all the problems we have considered excepting the last one.

Table 1. Comparative study of our algorithms with the existing algorithms in the literature for the largest clique problem of geometric intersection graphs

Problem	Existing algorithm			Our algorithm	
Largest Clique for	Time	Work-space	Reference	Time	Work-space
Interval graph	$O(n \log n)$	$O(n)$	[6]	$O(n \log n)$	$O(1)$
Intersection graph of arbitrary size rectangles	$O(n \log n)$	$O(n)$	[7,10]	$O(n^2 \log n)$	$O(1)$
Intersection graph of fixed height rectangles	$O(n \log n)$	$O(n)$	[7,10]	$O(n \log n + nK)$	$O(1)$
Disk graph of arbitrary radii (geometric clique)	$O(n^2)$	$O(n)$	[3]	$O(n^2 \log n)$	$O(1)$
Unit disk graph (graphical clique)	$O(n^{3.5} \log n)$	$O(n)$	[1]	$O(n^2 K^4)$	$O(1)$

2 Maximum Clique of an Interval Graph

Here a set of intervals $\mathcal{I} = \{I_1, I_2, \ldots, I_n\}$ is given in an array. Each element I_i of the array is a tuple (ℓ_i, r_i), that represents the coordinates of the left and right end points of I_i on the real line. The goal is to design an in-place algorithm for finding the maximum clique of the interval graph corresponding to the intervals in $\mathcal{I}$. In other words, we need to compute a point on the real line at which maximum number of intervals in $\mathcal{I}$ overlap. From now onwards, $\mathcal{I}$ will denote the array containing the given set of intervals.

Let L be the list of intervals in $\mathcal{I}$ sorted in increasing order of their ℓ coordinates. We maintain the list L and a heap H in the same array $\mathcal{I}$ during the execution. Initially, $\mathcal{I}$ contains the ordered list L and the heap H is empty. The elements in L are considered for processing in order. When an element is considered, it is deleted from L, and put into the heap with respect to its r coordinate. The root of the heap is the first element of the array $\mathcal{I}$.

The algorithm considers the intervals in L in increasing order of their left-end point (ℓ_i values) until the first right-end point of an interval (say $I_j = (\ell_j, r_j)$) is encountered. It is available at the root of H. Thus, a maximal clique is observed at r_j. Note that, I_j will never contribute to any other clique; thus I_j needs to be deleted from the data structure. Also note that, I_j is already deleted from L; we delete I_j from H. The array $\mathcal{I}$ has three parts: (i) the heap H, (ii) the array L, and (iii) the elements that are deleted from both L and H. We maintain two index variables λ and μ; $H = \mathcal{I}[1, 2, \ldots, \lambda]$, $L = \mathcal{I}[\mu + 1, \mu + 2, \ldots, n]$, and the elements $D = \mathcal{I}[\lambda + 1, \lambda + 2, \ldots, \mu]$ are deleted from both H and L. While deleting an element in L, we increment λ and μ, swap $\mathcal{I}[\mu]$ and $\mathcal{I}[\lambda]$, and then position $\mathcal{I}[\lambda]$ in its appropriate position in the heap H. We use two scalar variables χ and π to store the size and the point on the real line representing the maximum clique. Algorithm terminates after considering all the elements in L.

Theorem 1. *The time complexity of our algorithm for finding a clique of maximum size is $O(n \log n)$. It uses $O(1)$ extra space apart from the input array.*

3   Maximum Clique for Arbitrary Size Rectangles

A set $\mathcal{R}$ of n axis-parallel rectangles of arbitrary size is given in $\mathbb{R}^2$, and the objective is to find a point in $\mathbb{R}^2$ on which maximum number of rectangles overlap. Each rectangle R_i is specified by a tuple (α_i, β_i), where $\alpha_i = (x_{\alpha_i}, y_{\alpha_i})$ and $\beta_i = (x_{\beta_i}, y_{\beta_i})$ are the coordinates of the top-left and bottom-right corners of R_i. We use $\mathcal{R}$ as an array storing the set of rectangles $\{R_1, R_2, \ldots, R_n\}$.

The plane sweep algorithm for finding the maximum clique of a set of axis-parallel rectangles works as follows [10]. Sweep a horizontal line L from top to bottom. When the top boundary of a rectangle R_i is faced by the line L, it becomes active. When the bottom boundary of R_i is faced by the line L, we consider only those active rectangles which overlap on the bottom boundary of R_i. Each of these rectangles defines an interval corresponding to its portion of overlap with the bottom boundary of R_i. We compute the maximum clique among those intervals. Sweep continues until all the bottom boundaries are processed.

Our in-place algorithm follows the same technique. It executes n iterations to considers the rectangles in $\mathcal{R}$ in a top-to-bottom order of their bottom boundary. A variable γ is maintained to store the y coordinate of the bottom boundary of the last processed rectangle. In each iteration, we compare all the entries in array $\mathcal{R}$ with γ to identify the bottom boundary of a rectangle R_i having maximum y coordinate among the rectangles in $\mathcal{R}$ that are not yet processed. Next we swap R_i with the first element of the array $\mathcal{R}$ and execute the following steps.

- Identify all the rectangles in $\mathcal{R}$ that overlap on the bottom boundary of R_i, and bring them at the beginning of the array $\mathcal{R}$. If there are m_i such rectangles, they will occupy the positions $\mathcal{R}[2, \ldots, m_i + 1]$. This can be done in $O(n)$ time. We use two pointers π_1 and π_2, where π_1 moves from $\mathcal{R}[2]$ towards right until it finds a rectangle R_j that does not overlap on R_i, and π_2 moves from $\mathcal{R}[n]$ towards left until it finds a rectangle R_k that overlaps on R_i. R_j and R_k are swapped; the move proceeds until $\pi_1 = \pi_2$ is attained.
- Each of these m_i rectangles define an interval on the bottom boundary of R_i. We compute the clique of maximum size in this interval graph using the algorithm proposed in Section 2 in $O(m_i \log m_i)$ time.

The fact that $\sum_{i=1}^{n} m_i = O(n^2)$ lead to the following theorem.

Theorem 2. *The maximum clique of the intersection graph of a set of n axis-parallel rectangles can be computed in $O(n^2 \log n)$ time and $O(1)$ extra space.*

4   Maximum Clique for Fixed Height Rectangles

We now consider a constrained version of the maximum clique problem of rectangle intersection graph, where the height of all the rectangles in $\mathcal{R}$ are same, say δ.

Here each rectangle R_i is represented by a triple $(\alpha(R_i), \beta(R_i), \omega(R_i))$; $\alpha(R_i)$, $\beta(R_i)$ and $\omega(R_i)$ represent the x-coordinates of left, right vertical boundaries and the y-coordinate of the top boundary of R_i respectively.

Observation 1. *If we split the plane into horizontal strips of width δ, such that the horizontal boundary of no rectangle is aligned with any of the horizontal lines defining the strips, then each member in $\mathcal{R}$ spans exactly two consecutive strips.*

We split the region into horizontal strips satisfying Observation 1 and retain only those strips that contain at least one member of $\mathcal{R}$. We compute the maximum clique in each strip, and report the one having maximum size. We maintain two global counters χ and π to contain the size of the largest clique C and a point in the region representing the largest clique, during the entire execution. We first sort the members in $\mathcal{R}$ with respect to their ω-values. The elements stabbed by the top (resp. bottom) boundary of a strip are stored in consecutive locations of $\mathcal{R}$. We now describe the method of processing a strip.

4.1   Processing of a Strip S

Let $\mathcal{R}_t$ and $\mathcal{R}_b$ denote the two sub-arrays of $\mathcal{R}$ that contain all the rectangles stabbed by the top and bottom boundary of the strip S. We can designate the sub-arrays $\mathcal{R}_t$ and $\mathcal{R}_b$ in $\mathcal{R}$ using two integer tuples (m_1, m_2) and (n_1, n_2) respectively. We compute the largest clique on the right boundary of each rectangle in $\mathcal{R}_t \cup \mathcal{R}_b$, and finally note down the largest one among them as C_S.

Consider the processing of right boundary of a rectangle $\rho \in \mathcal{R}_t \cup \mathcal{R}_b$. Let $R_{1^*}, R_{2^*}, \ldots, R_{k^*} \in \mathcal{R}_t$ and $R_{1'}, R_{2'}, \ldots, R_{\ell'} \in \mathcal{R}_b$ be the set of rectangles whose left boundaries are to the *left* of the right boundary of ρ and whose right boundaries are to the *right* of right boundary of ρ. If the line containing the right boundary of ρ intersects the top and bottom boundaries of S at the points α and β respectively, then $R_{1^*}, R_{2^*}, \ldots, R_{k^*} \in \mathcal{R}_t$ and $R_{1'}, R_{2'}, \ldots, R_{\ell'} \in \mathcal{R}_b$ form clique at α and β, respectively. Thus, we have the following result.

Lemma 1. $|C_S| > \max\{k, \ell\}$.

We process the members in $\mathcal{R}_t \cup \mathcal{R}_b$ in increasing order of their α-values. This needs sorting of $\mathcal{R}_t$ and $\mathcal{R}_b$ in increasing order of their α-values (left boundaries). During the execution, we arrange the rectangles in $\mathcal{R}_t$ (resp. $\mathcal{R}_b$) into three portions as stated below (see Fig. 2 (a)).

A: All the rectangles, whose both left and right boundaries are processed. This portion stays at the left side of $\mathcal{R}_t$ (resp. $\mathcal{R}_b$),

B: The rectangles whose left boundary is processed but the right boundary is not yet processed. This portion stays at the middle of $\mathcal{R}_t$ (resp. $\mathcal{R}_b$) in decreasing order of their ω-values.

C: The rectangles whose both left and right boundaries are not processed. This portion stays at the end of $\mathcal{R}_t$ (resp. $\mathcal{R}_b$) in increasing order of their α-values.

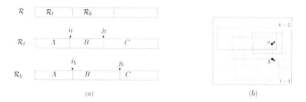

Fig. 2. (a) Arrangement of $\mathcal{R}_t \cup \mathcal{R}_b$, (b) Processing of Case 3

We maintain four index variables i_t, i_b, j_t and j_b, where i_t and j_t ($i_t \le j_t$) indicate the portion B of the array $\mathcal{R}_t$, and i_b and j_b ($i_b \le j_b$) indicate the portion B of the array $\mathcal{R}_b$. First, we sort both $\mathcal{R}_t$ and $\mathcal{R}_b$ in increasing order of their α-values. We also use two scalar variables ρ_t and ρ_b to maintain the rectangle having leftmost right boundary (β-value) among the members in the portion B in $\mathcal{R}_t$ and $\mathcal{R}_b$ respectively. Next, we start processing the elements of $\mathcal{R}_t$ and $\mathcal{R}_b$ in a merge like fashion. Initially, i_t, i_b, j_t, j_b are all set to 1; $\rho_t = \mathcal{R}_t[1]$ and $\rho_b = \mathcal{R}_b[1]$. At each step, we compare the $\alpha(\mathcal{R}_t[j_t + 1])$, $\alpha(\mathcal{R}_b[j_b + 1])$, $\beta(\rho_t)$ and $\beta(\rho_b)^1$. Here the following four situations may arise,

Case 1: $\alpha(\mathcal{R}_t[j_t + 1])$ **is minimum:** $\mathcal{R}_t[j_t + 1]$ is moved (from the C part) to the appropriate position in the B part of $\mathcal{R}_t$ with respect to its ω-value using a sequence of swap operations. If the β-value of the $\mathcal{R}_t[j_t + 1]$ is less than that of ρ_t, then ρ_t is updated by $\mathcal{R}_t[j_t + 1]$. Finally, j_t is incremented by 1.

Case 2: $\alpha(\mathcal{R}_b[j_b + 1])$ **is minimum:** This situation is similar to Case 1.

Case 3: $\beta(\rho_t)$ **is minimum:** Below, we explain the processing of this situation.

Case 4: $\beta(\rho_b)$ **is minimum:** This situation is handled as in Case 3.

Processing of Case 3

As mentioned in the proof of Lemma 1, all the rectangles $R_{i_t^*}, \ldots, R_{j_t^*} \in \mathcal{R}_t$ overlap at the point of intersection of ρ_t and the top-boundary of S. We initialize a variable $count$ with ($j_t - i_t$). Next, we process the members of the B portion of $\mathcal{R}_t$ (i.e., from the index position i_t to j_t), and the members of the B portion of $\mathcal{R}_b$ (i.e., from the index position i_b to j_b) together in decreasing order of their ω-values in a merge like fashion. We initialize two index variables θ_t and θ_b with i_t and i_b. At each step, if $\omega(\mathcal{R}_t[\theta_t]) - \delta > \omega(\mathcal{R}_b[\theta_b])$ then $count$ is decreased by one, and θ_t is incremented; otherwise (i) $count$ is increased by one, (ii) if $count > \chi$, then the global counters χ and π are set with $count$, the point of intersection of ρ_t and the rectangle $\mathcal{R}_b[\theta_b]$, and then (iii) θ_b is incremented. The process terminates when (i) θ_t reaches j_t or (ii) θ_b reaches j_b, or (iii) $\max(\omega(\mathcal{R}_t[\theta_t]) - \delta, \omega(\mathcal{R}_b[\theta_b])) < \omega(\rho_t)$. At the end of processing ρ_t we need to perform the following operations: (i) move ρ_t to the i_t-th position of $\mathcal{R}_t$ (i.e., the A part of $\mathcal{R}_t$) using a sequence of swap operations, (ii) set ρ_t by sequentially inspecting the members in $\mathcal{R}_t$ from index position ($i_t + 1$) to j_t, and (iii) increment i_t. See Fig. 2(b) for the demonstration.

---

[1] $\mathcal{R}_t[j_t + 1]$ and $\mathcal{R}_b[j_b + 1]$ are the first element of the C part of the respective arrays.

Lemma 2. *The time complexity of processing a strip is $O(n_S \log n_S + n_S|C_S|)$, where n_s is the number of rectangles in R that intersects the strip S.*

Proof. While processing a strip S, initial sorting of the members in $\mathcal{R}_t$ (resp. $\mathcal{R}_b$) with respect to their α-values need $O(|\mathcal{R}_t| \log |\mathcal{R}_t|)$ (resp. $O(|\mathcal{R}_b| \log |\mathcal{R}_b|)$) time. For each occurance of Case 1 (resp. Case 2) (i.e, a left boundary of a member in $R \in \mathcal{R}_t$ (resp. $\mathcal{R}_b$)) we need to position R in the B part of $\mathcal{R}_t$ (resp. $\mathcal{R}_b$) with respect to its ω-value. This needs at most $O(j_t - i_t)$ (resp. $O(j_b - i_b)$) swaps. By Lemma 1, the size of the B part of both $\mathcal{R}_t$ and $\mathcal{R}_b$ is at most $O(|C_S|)$ at any instant of time. We now analyze the time complexity of processing an instance of Case 3, i.e., the right boundary of a rectangle $\rho_t \in \mathcal{R}_t$.

While computing the largest clique along the right boundary of ρ_t, we inspect the members of $\mathcal{R}_t$ from index position i_t to j_t, and the members of $\mathcal{R}_b$ from index position i_b to j_b whose top boundaries are above the bottom boundary of ρ_t. Both these numbers are less than $|C_S|$ (by Lemma 1). Next, moving ρ_t to the end of the portion A of $\mathcal{R}_t$ needs at most $j_t - i_t$ swaps. The setting ρ_t with the existing members of B part of $\mathcal{R}_t$ for further processing in strip S needs another $j_t - i_t$ computations. Thus, the total time complexity for processing the right boundary of ρ_t is $O(|C_S|)$. The same arguments hold for processing the right boundary of a member in $\mathcal{R}_b$. Thus, processing the entire strip S needs $O(n_S \log n_S + n_S|C_S|)$ time.   □

Note that, each rectangle appears in exactly two strips, and if K be the size of largest clique then $K \geq |C_S| \; \forall$ strips S. Now, we have the following theorem.

Theorem 3. *The maximum size clique of an intersection graph of fixed height rectangles can be computed in $O(n \log n + nK)$ time and $O(1)$ extra space.*

5   Geometric Clique for Disks of Arbitrary Radii

We now follow the same method as in Section 2 to compute the maximum size geometric clique of a set of disks of arbitrary radii. Here the input is an array containing a set of disks $\mathscr{C} = \{C_1, C_2, \ldots, C_n\}$. Each element $C_i \in \mathscr{C}$ is a triple (α_i, β_i, r_i); (α_i, β_i) is the coordinate of the center of C_i and r_i is its radius.

Observation 2. *Any geometric clique of a disk graph corresponds to a closed convex region bounded by arc segments of some/all the disks participating in it.*

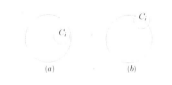

Let us consider a disk C_i; $\Delta(C_i)$ be the boundary of C_i. If a disk $C_j \in \mathscr{C} \setminus \{C_i\}$ properly contains C_i, then it contributes a *closed* arc along $\Delta(C_i)$ (see Fig. 3(a)). If $C_j \in \mathscr{C} \setminus \{C_i\}$ properly intersects C_i, it contributes a *non-closed* arc along $\Delta(C_i)$ (see Fig. 3(b)).

However, if a disk $C_j \in \mathscr{C} \setminus \{C_i\}$ is properly contained in C_i, it does not contribute any

Fig. 3. Closed and non-closed arc

arc along $\Delta(C_i)$. Thus, we have a circular arc graph (see [6]) G_i with the closed and non-closed arcs around $\Delta(C_i)$. By Observation 2, the maximum clique of the disk graph with the set of disks $\mathscr{C}$ corresponds to the maximum clique of G_i for some $i = 1, 2, \ldots, n$. We consider each disk $C_i \in \mathscr{C}$, and compute η_i, the size of the maximum clique of G_i. Finally we report $\eta = \max_{i=1}^n \eta_i$.

5.1   Finding Maximum Clique of the Circular-Arc Graph around C_i

While processing C_i, it is swapped with the first element of the array $\mathscr{C}$. Next, a scan among the elements of $\mathscr{C}$ is performed to accumulate all the disks that properly intersect C_i. Let us name this set of disks as $\mathcal{C}_i$, $n_i = |\mathcal{C}_i|$. These are all placed in the locations $\mathscr{C}[2 \ldots n_i]$. During this traversal, we count the number μ_i of disks that properly contain C_i. Next, we compute $\nu_i = $ the size of the maximum clique of the circular-arc graph of the non-closed arcs around $\Delta(C_i)$ as stated below. Finally, we compute $\eta_i = \mu_i + \nu_i$.

We fix a point θ on $\Delta(C_i)$. For each disk $C \in \mathcal{C}_i$, the *left* and *right* end-points (κ_{left} and κ_{right}) of the arc κ generated by C on $\Delta(C_i)$ is computed as follows:

If $\theta \in C$, then κ_{left} (resp. κ_{right}) is the point of intersection of C and C_i in anti-clockwise (resp. clockwise) direction from θ.

If $\theta \notin C$, then κ_{left} (resp. κ_{right}) is the closest (resp. farthest) point of intersection of C and C_i from θ in the clockwise direction.

We sort the members of $\mathcal{C}_i$ in clockwise order of their left end-points. Note that, we do not store the arcs along $\Delta(C_i)$. While comparing a pair of arcs, we compute the points of intersection of the corresponding disks with C_i. Next, we process the end-points of the arcs in an ordered manner as in Section 2, implementing both a heap H and a list L in the portion $\mathcal{C}_i$ of the array $\mathcal{C}$ along with a list of elements deleted from both H and L. However unlike Section 2, after processing all the left end-points, if H contains some non-deleted elements, the algorithm does not stop. It again sorts all the deleted elements in H in clockwise order of their left end-point, and continues the processing considering this list as L. The processing continues until all the elements of H are processed. Thus, $\eta_i = $ size of the maximum clique around C_i, is computed.

Theorem 4. *The geometric clique of maximum size among a set of n disks of arbitrary radii can be computed in $O(n^2 \log n)$ time, and it uses $O(1)$ extra space.*

6   Graphical Clique in the Unit Disk Graph

Let $\mathcal{C} = \{C_1, C_2, \ldots, C_n\}$ be a set of unit disks stored in an array of size n. Each element $C_i \in \mathcal{C}$ stores the coordinate of its center $c_i = (\alpha_i, \beta_i)$. We show that the algorithm proposed in [5] for computing the largest clique in the intersection graph $G = (\mathcal{C}, E)$ can be made in-place. Here the vertices in G correspond to the members in $\mathcal{C}$, and $(C_i, C_j) \in E$ indicates that the disks C_i and C_j intersect.

Let $\chi \in \mathcal{C}$ be a set of disks forming the largest clique, and c_i, c_j be the farthest pair of centers among the disks in χ. Now, the centers of all the members in χ lie in the region R_{ij} formed by the intersection of circles of radius $d(c_i, c_j)$ centered at c_i and c_j as shown in Fig. 4(a) [5]. We use R_{ij}^1 snd R_{ij}^2 to denote the parts of R_{ij} lying above and below the line segment $[c_i, c_j]$ respectively, and $\mathcal{C}_{ij}^1$ and $\mathcal{C}_{ij}^2$ to denote the centers of $\mathcal{C}$ that lie in R_{ij}^1 and R_{ij}^2 respectively. Note that, the Euclidean distance between each pair of centers in $\mathcal{C}_{ij}^k \cup \{C_i, C_j\}$ is less than or equal to 1 for both $k = 1, 2$. Thus, if we form a bipartite graph $G_B = (\mathcal{C}_{ij}^1 \cup \mathcal{C}_{ij}^2, E_B)$, where an edge between a pair of vertices implies that their distance is greater than 1, then the set χ corresponds to the maximum independent set in the graph G_B [5]. Note that, $\mathcal{C}_{ij}^k \cup \{C_i, C_j\}$ itself forms an independent set in G_B for each $k = 1, 2$.

Our algorithm starts with $\chi = 0$, and considers each pair of disks $C_i, C_j \in \mathcal{C}$. If $C_i, C_j \in \mathcal{C}$ intersect, then we compute $\mathcal{C}_{ij}^1$ and $\mathcal{C}_{ij}^2$. If χ denotes the size of the largest clique obtained so far, and $|\mathcal{C}_{ij}^1 \cup \mathcal{C}_{ij}^2 \cup \{C_i, C_j\}| \leq \chi$, then the disks centered at R_{ij} will not produce a clique of size greater than χ. Otherwise, we consider the graph G_B using $\mathcal{C}_{ij}^1$ and $\mathcal{C}_{ij}^2$ as the two sets of vertices. and compute the maximum matching in G_B. Finally, we compute $\chi' = |U| + |M|$, where U is the set of unmatched vertices in G_B, and M is the number of matched edges in G_B. If $|\chi'| > |\chi|$, then replace χ by χ', and remember i, j in a pair of integer locations i', j'. After considering all the pairs of vertices, we need to execute the same algorithm for the pair of disks $C_{i'}$ and $C_{j'}$ to report the largest clique.

6.1   Bipartite Matching in G_B

We first accumulate the centers of all the members $\mathcal{C}_{ij}^1 \cup \mathcal{C}_{ij}^2 = \{c_k | d(c_i, c_k) \leq 1 \ \& \ d(c_j, c_k) \leq 1\}$, and move them at the begining of the array $\mathcal{C}$. Next, we arrange the members in $\mathcal{C}_{ij}^1$ and $\mathcal{C}_{ij}^2$ in R_{ij} with respect to the position of their centers, i.e., above or below the line $[c_i, c_j]$. Let $\mathcal{C}_{ij}^1 = \{\mathcal{C}[k], k = 1, 2, \ldots \mu\}$, and $\mathcal{C}_{ij}^2 = \{\mathcal{C}[k], k = \mu + 1, \mu + 2, \ldots m\}$. At any instant of time, we use A_{ij}^1 (resp. A_{ij}^2) to denote the set of matched vertices, and B_{ij}^1 (resp. B_{ij}^2) to denote the set of unmatched (exposed) vertices in $\mathcal{C}_{ij}^1$ (resp. $\mathcal{C}_{ij}^2$), $|A_{ij}^1| = |A_{ij}^2| = \alpha$. Thus, $A_{ij}^1[k] = \mathcal{C}[k]$ and $A_{ij}^2[k] = \mathcal{C}[\mu + k]$ for $k = 1, 2, \ldots, \alpha$; The sets B_{ij}^1 and B_{ij}^2 start from the locations $\mathcal{C}[\alpha + 1]$ and $\mathcal{C}[\mu + \alpha + 1]$ respectively.

Let $|B_{ij}^1| \leq |B_{ij}^2|$. We consider each (exposed) vertex $w = \mathcal{C}[k] \in B_{ij}^1$ ($k \in \{\alpha + 1, \ldots, \mu\}$) and compute an augmenting path with a sequence of matched and unmatched edges that starts at w, and ends at an exposed vertex $w' = \mathcal{C}[\ell] \in B_{ij}^2$ (i.e., $\ell \in \{\mu + \alpha + 1, \ldots, m\}$) [11]. If an augmenting path is found, the matching is augmented. The cardinality of both A_{ij}^1 and A_{ij}^2 are increased by one, and the cardinality of both B_{ij}^1 and B_{ij}^2 are decreased by one. Otherwise, w is a *useless* vertex in the sense that it will never appear in any augmenting path in subsequent iterations (Corollary of Theorem 10.5 [12]). We move w at the end of the list B_{ij}^1. We use a variable γ such that the nodes of B_{ij}^1 stored in $\mathcal{C}[\gamma], \mathcal{C}[\gamma + 1], \ldots, \mathcal{C}[\mu]$ are *useless*. Initially γ is set with $\mu + 1$. The algorithm stops when there does not exist augmenting path starting from any exposed

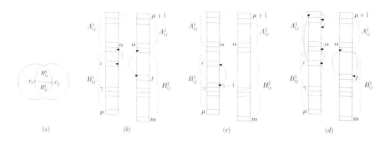

Fig. 4. (a) Lens formed by two circles, (b-d) Different steps for augmenting a matching

vertex in B_{ij}^1, i.e., $\gamma = \alpha + 1$ for the current values of α and γ. The procedures for finding an augmenting path and updating the matching are stated below.

Computing an Augmenting Path. Let $w \in B_{ij}^1$ be an exposed vertex stored at $\mathcal{C}[i]$, $\alpha < i \leq \gamma$. Our algorithm for finding the augmenting path consists of two stages. In Stage 1, we test whether there exists an exposed vertex $w' \in B_{ij}^2$ such that $(w, w') \in E_B$. If such a vertex exists (say at location $\mathcal{C}[j]$, $\mu + \alpha < j \leq m$), we increment α by 1, and move w and w' in $A_{ij}^1[\alpha]$ and $A_{ij}^2[\alpha]$ by executing swap($\mathcal{C}[i], \mathcal{C}[\alpha]$) and swap($\mathcal{C}[j], \mathcal{C}[\mu + \alpha]$) (see Fig. 4(b)). If no such vertex is found, we execute Stage 2 to get an augmenting path of length greater than 1.

In Stage 2, we first test whether w has an edge with any member of A_{ij}^2. If no such vertex is observed, then w is an *useless* vertex. We move w at the end of B_{ij}^1 by decrementing γ by 1, and executing swap($\mathcal{C}[i], \mathcal{C}[\gamma]$) (see Fig. 4(c)).

If a vertex $u = \mathcal{C}[\mu + \ell] \in A_{ij}^2$ is observed such that $(w, u) \in E_B$, then $w \to u \to v$ is an alternating path, where $v = \mathcal{C}[\ell] \in A_{ij}^1$, and (u, v) is a matched edge. We move u and v at the beginning of the array A_{ij}^2 and A_{ij}^1 respectively. We use a scalar variable β for this purpose; its initial value 0. To store a matched edge (u, v) on the alternating path, we increment β by 1, and store $u = \mathcal{C}[\mu + \ell]$ and $v = \mathcal{C}[\ell]$ in $\mathcal{C}[\mu + \beta]$ and $\mathcal{C}[\beta]$ respectively by executing swap($\mathcal{C}[\mu + \ell], \mathcal{C}[\mu + \beta]$) and swap($\mathcal{C}[\ell], \mathcal{C}[\beta]$). In the next step, we take $v = \mathcal{C}[\beta]$, and try to finish the augmenting path by searching a vertex $w' \in B_{ij}^2$ such that $(v, w') \in E_B$ as we did at the beginning of Step 2. If no such w' is found, then we search for a vertex $u' \in A_{ij}^2$ such that $(u', v) \in E_B$. If such a vertex u' is found, then we extend the alternating path with the matched edge (u', v') (where $v' \in A_{ij}^1$) by incrementing β by 1, and storing u' and v' in $\mathcal{C}[\mu + \beta]$ and $\mathcal{C}[\beta]$ respectively as earlier. However, if no such vertex u' is found, then from v the alternating path can not be extended. So, we decrement β by 1 to explore the other edges of $\mathcal{C}[\beta]$ for the current value of β. Here it needs to be mentioned that, (i) the first part of the array A_{ij}^2 behaves like a stack with β as its top pointer, and (ii) since the neighbors of a vertex in G_B are not available directly, we need to be careful (as explained below) in choosing an edge of $\mathcal{C}[\beta]$ next time which was not chosen earlier.

In order to get a new vertex $u' \in A_{ij}^2$ adjacent to $v \in A_{ij}^1$, we may need to inspect all the members in $u'' \in A_{ij}^2$ irrespective of whether $(u'', v) \in E_B$ or not. There may be several vertices in A_{ij}^2 that are adjacent to v. At an instant of time, we choose one having minimum distance from the line segment $[c_i, c_j]$, among those which are not yet considered. We use a variable *dist* for this purpose. This helps in avoiding the choice of same neighbor of v many times. The choice of new edge (v, u') is guided by the distance of u' from the line segment $[c_i, c_j]$ and the coordinate of u^*, where (v, u^*) is the current edge that failed to produce augmenting path. Thus, choosing a neighbor of v in the set A_{ij}^2 needs $O(|A_{ij}^2|)$ time in the worst case.

Finally, if $\beta = 0$ is observed, we explore other neighbors of w in B_{ij}^2. If we can complete an augmenting path, say $w = C[i] \to A_{ij}^2[1] \to A_{ij}^1[1] \to A_{ij}^2[2] \to A_{ij}^1[2] \to \ldots \to A_{ij}^2[\beta - 1] \to A_{ij}^1[\beta - 1] \to A_{ij}^2[\beta] \to A_{ij}^1[\beta] \to w' = C[j]$ $(w \in B_{ij}^1$ and $w' \in B_{ij}^2)$, then the matching is updated as follows:

- Execute $swap(C[i], C[k])$ for $k = 1, 2, \ldots, \beta$, and then execute $swap(C[i], C[\alpha + 1])$ and $swap(C[j], C[\mu + \alpha + 1])$. Finally increment α by 1. (see Fig. 4(d)).

However, if there exists no other neighbor of w in the set A_{ij}^2, no augmenting path is possible from w; so w is moved at the end of B_{ij}^1 using the index variable γ as stated earlier.

6.2   Complexity Analysis

Theorem 5. *Given a set P of n points in 2D, the time complexity of our proposed algorithm for computing the largest clique in the intersection graph G of unit disks centered at the points in P is $O(n^2 + mK^4)$; m is the number of edges in G and K is the size of the largest clique in G. The space complexity is $O(1)$.*

Proof. The first term in the time complexity is for testing the intersection of the unit disks corresponding to each pair of points.

Let S be the number of points in a lens. The time complexity of our proposed in-place algorithm for the bipartite matching in the graph G_B formed with the points in the aforesaid lens is $O(S^4)$ in the worst case. The reason is as follows: (i) we check every exposed vertex once for augmenting the matching, (ii) the process of checking the feasibility of augmenting the matching from an exposed vertex has to visit all the edges of the graph, and (iii) getting an edge needs needs $O(S)$ time in the worst case. If there exists an augmenting path, then the time required to augment the matching is proportional to its length, which is $O(S)$ in the worst case. Since S is the number of points in the lens, the size of the clique is at least $\frac{S}{2}$. The reason is that the points lying in one side of the line joining p_i and p_j inside the lens always form a clique. The result follows from the fact that the number of intersecting pairs of unit disks is m, and $S \leq 2K$ (K is the size of the largest clique) for each intersecting pairs of unit disks. The space complexity follows from the fact that apart from the points in the array C, we have used a constant number of index variables and a location *dist* in our algorithm. $\square$

References

1. Breu, H.: Algorithmic Aspects of Constrained Unit Disk Graphs, Ph.D. Thesis, University of British Columbia, Canada, Tech. Report No. TR-96-15 (1996)
2. Bronnimann, H., Chan, T.M., Chen, E.Y.: Towards in-place geometric algorithms and data structures. In: Symp. on Computational Geometry, pp. 239–246 (2004)
3. Chazelle, B.M., Lee, D.T.: On a circle placement problem. Computing 36, 1–16 (1986)
4. Chan, T.M., Chen, E.Y.: Optimal in-place and cache-oblivious algorithms for 3-d convex hulls and 2-d segment intersection. Comput. Geom. 43, 636–646 (2010)
5. Clark, B.N., Colbourn, C.J., Johnson, D.S.: Unit disk graph. Discrete Mathematics 86, 165–177 (1990)
6. Golumbic, M.C.: Algorithmic Graph Theory and Perfect Graphs. Academic Press (1980)
7. Imai, H., Asano, T.: Finding the connected components and maximum clique of an intersection graph of rectangles in the plane. Journal of Algorithms 4, 310–323 (1983)
8. Lee, D.T., Preparata, F.P.: An improved algorithm for the rectangle enclosure problem. Journal of Algorithms 3, 218–224 (1982)
9. Lee, D.T.: Maximum clique problem of rectangle graphs. In: Preparata, F.P. (ed.) Advances in Computing Research, pp. 91–107. JAI Press (1983)
10. Nandy, S.C., Bhattacharya, B.B.: A unified algorithm for finding maximum and minimum point enclosing rectangles and cuboids. Computers and Mathematics with Applications 29(8), 45–61 (1995)
11. Preparata, F.P., Shamos, M.I.: Computational Geometry - an Introduction. Springer (1990)
12. Papadimitriou, C.H., Steiglitz, K.: Combinatorial Optimization: Algorithms and Complexity. Prentice Hall of India Pvt. Ltd., New Delhi (1997)

The Black-and-White Coloring Problem
on Distance-Hereditary Graphs
and Strongly Chordal Graphs

Ton Kloks[1,*], Sheung-Hung Poon[1], Feng-Ren Tsai[2], and Yue-Li Wang[3]

[1] Department of Computer Science
[2] Institute of Information Systems and Applications,
National Tsing Hua University, No. 101, Sec. 2, Kuang Fu Rd., Hsinchu, Taiwan
spoon@cs.nthu.edu.tw, mevernom@gmail.com
[3] Department of Information Management,
National Taiwan University of Science and Technology,
No. 43, Sec. 4, Keelung Rd., Taipei, 106, Taiwan
ylwang@cs.ntust.edu.tw

Abstract. Given a graph G and integers b and w. The black-and-white coloring problem asks if there exist disjoint sets of vertices B and W with $|B| = b$ and $|W| = w$ such that no vertex in B is adjacent to any vertex in W. In this paper we show that the problem is polynomial when restricted to cographs, distance-hereditary graphs, interval graphs and strongly chordal graphs. We show that the problem is NP-complete on splitgraphs.

Keywords: Black-and-white coloring, Cographs, Distance-hereditary graphs, Strongly chordal graphs, Threshold graphs, Interval graphs.

1 Introduction

Definition 1. *Let $G = (V, E)$ be a graph and let b and w be two integers. A black-and-white coloring of G colors b vertices black and w vertices white such that no black vertex is adjacent to any white vertex.*

In other words, the black-and-white coloring problem asks for a complete bipartite subgraph M in the complement $\bar{G}$ of G with b and w vertices in the two color classes of M.

The black-and-white coloring problem is NP-complete for graphs in general [22]. That paper also shows that the problem can be solved for trees in $O(n^3)$ time. In a recent paper [6] the worst-case timebound for an algorithm on trees was improved to $O(n^2 \log^3 n)$ time [6]. The paper [6] mentions, among other things, a manuscript by Kobler, *et al.*, which shows that the problem can be solved in polynomial time

* National Science Council of Taiwan Support Grant NSC 99–2218–E–007–016.

J. Snoeyink, P. Lu, K. Su, and L. Wang (Eds.): FAW-AAIM 2012, LNCS 7285, pp. 339–350, 2012.

for graphs of bounded treewidth. All classes that we consider in this paper either contain arbitrarily large cliques or arbitrarily large complete bipartite graphs $K_{t,t}$. Hence none of these classes has bounded treewidth.

In this paper we investigate the complexity of the problem for some graph classes. We start our analysis for the class of cographs.

A P_4 is a path with four vertices.

Definition 2 ([13]). *A graph is a cograph if it has no induced P_4.*

There are various characterizations of cographs. For algorithmic purposes the following characterization is suitable.

Theorem 1. *A graph is a cographs if and only if every induced subgraph H is disconnected or the complement $\bar{H}$ is disconnected.*

It follows that a cograph has a tree decomposition which is called a cotree. A cotree is a pair (T, f) comprising a rooted binary tree T together with a bijection f from the vertices of the graph to the leaves of the tree. Each internal node of T, including the root, has a label $\otimes$ or $\oplus$. The $\otimes$ operation is called a join operation, and it makes every vertex that is mapped to a leaf in the left subtree adjacent to every vertex that is mapped to a leaf in the right subtree. The operator $\oplus$ is called a union operation. In that case the graph is the union of the graphs defined by the left - and right subtree. A cotree decomposition can be obtained in linear time [14].

2   Black-and-White Colorings of Cographs

In this section we show that the black-and-white coloring problem can be solved in polynomial time for cographs.

Theorem 2. *There exists an $O(n^3)$ algorithm which solves the black-and-white coloring problem on cographs.*

Proof. Let $f_G(b)$ be the maximum number of white vertices in a black-and-white coloring of G with b black vertices. We prove that the function f_G can be computed in $O(n^3)$ time for cographs.

Let G be a cograph with n vertices. We write f instead of f_G. By convention,

$$f(b) = 0 \quad \text{when } b < 0 \text{ or } b > n.$$

Assume that G has one vertex. Then

$$f(b) = \begin{cases} 1 & \text{if } b = 0 \\ 0 & \text{in all other cases.} \end{cases}$$

Assume that G is the join of two cographs G_1 and G_2. We write f_i instead of f_{G_i}, for $i \in \{1, 2\}$. We have that $f(0) = n$, where n is the number of vertices in G. When $b > 0$ we have

$$f(b) = \max\{\, f_1(b),\ f_2(b)\,\}.$$

Assume that G is the union of two cographs G_1 and G_2. Then

$$f(b) = \max_{0 \leqslant k \leqslant b}\ f_1(k) + f_2(b - k).$$

A cotree T has $O(n)$ nodes and it can be computed in linear time [13]. Consider a node i in T. Let G_i be the subgraph of G induced by the vertices that are mapped to leaves in the subtree rooted at i. By the previous observations, the function f_i for the graph G_i can be computed in $O(n^2)$ time.

Since T has $O(n)$ nodes this proves the theorem.   □

2.1 Threshold Graphs

A subclass of the class of cographs are the threshold graphs.

Definition 3 ([12]). *A graph* $G = (V, E)$ *is a* threshold graph *if there is a real number* T *and a real number* $w(x)$ *for every vertex* $x \in V$ *such that a subset* $S \subseteq V$ *is an independent set if and only if*

$$\sum_{x \in S} w(x) \geqslant T.$$

There are many ways to characterize threshold graphs [30]. For example, a graph is a threshold graph if it has no induced P_4, C_4 or $2K_2$.

Fig. 1. A graph is a threshold graph if it has no induced C_4, P_4 or $2K_2$

Another characterization is that a graph is a threshold graph if every induced subgraph has a universal vertex or an isolated vertex [12, Theorem 1].

In [12, Corollary 1B] appears also the following characterization. A graph $G = (V, E)$ is a threshold graph if and only if there is a partition of V into two sets A and B, of which one is possibly empty, such that the following holds true.

1. A induces a clique, and
2. B induces an independent set, and
3. there is an ordering $b_1, \dots, b_k$ of the vertices in B such that

$$N(b_1) \subseteq \dots \subseteq N(b_k).$$

We use the notation $N[x]$ to denote the closed neighborhood of a vertex x. Thus $N[x] = N(x) \cup \{x\}$.

Theorem 3. *There exists a linear-time algorithm which, given a threshold graph G and integers b and w, decides if there is a black-and-white coloring of G with b vertices colored black and w vertices colored white.*

Proof. Let $x_1, \ldots, x_n$ be an ordering of the vertices in G such that for all $i < n$

(a) $N(x_i) \subseteq N(x_{i+1})$ if x_i and x_{i+1} are not adjacent, and
(b) $N[x_i] \subseteq N[x_{i+1}]$ if x_i and x_{i+1} are adjacent.

Assume that there exists a black-and-white coloring which colors b vertices black and w vertices white. Assume that there is an index $k \leqslant b + w$ such that x_k is uncolored. Then there exists an index $\ell > b + w$ such that x_ℓ is colored black or white. Then we may color x_k with the color of x_ℓ and uncolor x_ℓ instead. Thus we may assume that there exists a coloring such that $x_1, \ldots, x_{b+w}$ are colored and all other vertices are uncolored.

Assume that $b \leqslant w$. We prove that there exists a coloring f such that

$$f(x_i) = \begin{cases} \text{black} & \text{if } 1 \leqslant i \leqslant b, \text{ and} \\ \text{white} & \text{if } b + 1 \leqslant i \leqslant b + w. \end{cases}$$

We may assume that $b \geqslant 1$ and that $w \geqslant 1$. Assume that x_i is adjacent to x_j for some $i \leqslant b < j$. Then

$$\{x_j, \ldots, x_{b+w}\} \subseteq N(x_i) \quad \text{and} \quad \{x_i, \ldots, x_j\} \subseteq N[x_j].$$

Thus all vertices in

$$\{x_i, \ldots, x_{b+w}\}$$

are the same color. If they are all black then there are at least $w + 1$ black vertices in the coloring, which contradicts $b \leqslant w$. If they are all white then we have at least $w + 1$ white vertices, which is a contradiction as well. Thus no two vertices x_i and x_j with $i \leqslant b < j$ are adjacent, which proves that the coloring above is valid.

This proves the theorem, since an algorithm only needs to check if x_b is adjacent to x_{b+w} or not. $\qquad\qquad\square$

2.2   Difference Graphs

Definition 4 ([21]). *A graph $G = (V, E)$ is a difference graph if there exists a positive real number T and a real number $w(x)$ for every vertex $x \in V$ such that $w(x) \leqslant T$ for every $x \in V$ and such that for any pair of vertices x and y*

$$\{x, y\} \in E \quad \text{if and only if} \quad |w(x) - w(y)| \geqslant T.$$

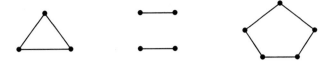

Fig. 2. A graph is a difference graph if it has no induced triangle, $2K_2$ or C_5

Difference graphs are sometimes called chain graphs [34].

Difference graphs can be characterized in many ways [21]. For example, a graph is a difference graph if and only if it has no induced K_3, $2K_2$ or C_5 [21, Proposition 2.6]. Difference graphs are bipartite. Let X and Y be a partition of V into two color classes. Then the graph obtained by making a clique of X is a threshold graph and this property characterizes difference graphs [21, Lemma 2.1].

Theorem 4. *There exists a linear-time algorithm which, given a difference graph* G *and integers* b *and* w, *decides if there is a black-and-white coloring of* G *with* b *black vertices and* w *white vertices.*

Proof. An argument, similar to the one given in Theorem 3, provides the proof.
$\square$

3 Distance-Hereditary Graphs

Definition 5 ([24]). *A graph* G *is distance hereditary if for every pair of nonadjacent vertices* x *and* y *and for every connected induced subgraph* H *of* G *which contains* x *and* y, *the distance between* x *and* y *in* H *is the same as the distance between* x *and* y *in* G.

In other words, a graph G is distance hereditary if for every nonadjacent pair x and y of vertices, all chordless paths between x and y in G have the same length.

There are various characterizations of distance-hereditary graphs. One of them states that a graph is distance hereditary if and only if it has no induced house, hole, domino or gem [4,24]. Distance-hereditary graphs are also characterized by the property that every induced subgraph has either an isolated vertex, or a pendant vertex, or a true or false twin [4].

Distance-hereditary graphs are the graphs of rankwidth one. This implies that they have a special decomposition tree which we describe next.

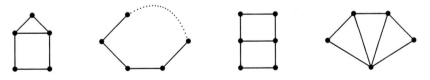

Fig. 3. A graph is distance hereditary if it has no induced house, hole, domino or gem

A decomposition tree for a graph $G = (V, E)$ is a pair (T, f) consisting of a rooted binary tree T and a bijection f from V to the leaves of T.

When G is distance hereditary it has a decomposition tree (T, f) with the following three properties [11].

Consider an edge $e = \{p, c\}$ in T where p is the parent of c. Let $W_e \subset V$ be the set of vertices of G that are mapped by f to the leaves in the subtree rooted at c. Let $Q_e \subseteq W_e$ be the set of vertices in W_e that have neighbors in $G - W_e$. The set Q_e is called the twinset of e. The first property is that the subgraph of G induced by Q_e is a cograph for every edge e in T.

Consider an internal vertex p in T. Let c_1 and c_2 be the two children of p. Let $e_1 = \{p, c_1\}$ and let $e_2 = \{p, c_2\}$. Let Q_1 and Q_2 be the twinsets of e_1 and e_2. The second property is that there is a join- or a union-operation between Q_1 and Q_2. Thus every vertex of Q_1 has the same neighbors in Q_2.

Let p be an internal vertex of T which is not the root. Let e be the line that connects p with its parent. Let Q_e be the twinset of e. Let c_1 and c_2 be the two children of p in T. Let $e_1 = \{p, c_1\}$ and let $e_2 = \{p, c_2\}$. Let Q_i be the twinset of e_i, for $i \in \{1, 2\}$. The third, and final, property is that

$$Q_e = Q_1 \quad \text{or} \quad Q_e = Q_2 \quad \text{or} \quad Q_e = Q_1 \cup Q_2.$$

When G is distance hereditary then a tree-decomposition for G with the three properties described above can be obtained in linear time [11].

Notice that the first property is a consequence of the other two. As an example, notice that cographs are distance hereditary. A cotree is a decomposition tree for a cograph with the three properties mentioned above.

Theorem 5. *There exists a polynomial-time algorithm that solves the black-and-white coloring problem on distance-hereditary graphs.*

The proof of this theorem can be found in [27].

4   Interval Graphs

In this section we show that there is an efficient algorithm to solve the black-and-white coloring problem on interval graphs.

Definition 6 ([29]). *A graph G is an interval graph if it is the intersection graph of a collection of intervals on the real line.*

There are various characterizations of interval graphs. For example, a graph is an interval graph if and only if it is chordal and it has no asteroidal triple. Also, a graph is an interval graph if and only if it has no C_4 and the complement $\bar{G}$ has a transitive orientation [19].

For our purposes the following characterization of interval graphs is suitable.

Theorem 6 ([19]). *A graph G is an interval graph if and only if there is a linear ordering L of its maximal cliques such that for every vertex, the maximal cliques that contain that vertex are consecutive in L.*

Interval graphs can be recognized in linear time. When G is an interval graph then G is chordal and so it has at most n maximal cliques. A linear ordering of the maximal cliques can be obtained in $O(n^2)$ time [7].

Theorem 7. *There exists an $O(n^6)$ algorithm that solves the black-and-white coloring problem on interval graphs.*

Proof. Let $[C_1, \ldots, C_t]$ be a linear ordering of the maximal cliques of an interval graph $G = (V, E)$ such that for every vertex x, the maximal cliques that contain x appear consecutively in this ordering.

Consider a black-and-white coloring of G. First assume that the first clique C_1 contains no black or white vertices. Then we may remove the vertices that appear in C_1 from the graph and consider a black-and-white coloring of the vertices in cliques of the linear ordering

$$[C_2^*, \ldots, C_t^*], \quad \text{where, for } i > 1, \quad C_i^* = C_i \setminus C_1.$$

Now assume that C_1 contains some black vertices. Then, obviously, C_1 contains no white vertices. Let i be the maximal index such that all the cliques C_ℓ with $1 \leqslant \ell \leqslant i$ contain no white vertices. Remove all the vertices that appear in $C_1, \ldots, C_i$ from the remaining cliques and consider the ordering

$$[C_{i+1}^*, \ldots, C_t^*] \quad \text{where, for } \ell > i, \quad C_\ell^* = C_\ell \setminus \bigcup_{k=1}^{i} C_k.$$

Then we may take an arbitrary black-and-white coloring of the graph induced by the vertices $\cup_{\ell=i+1}^{t} C_\ell^*$ and color an arbitrary number of vertices in $\cup_{\ell=1}^{i} C_\ell$ black. For this purpose define, for $p \leqslant q$,

$$X_{p,q} = \{ x \in V \mid x \in C_k \quad \text{if and only if} \quad p \leqslant k \leqslant q \}.$$

Thus $X_{p,q}$ consists of the vertices of which the indices of the first and the last clique that contain the vertex are both in the interval $[p, q]$.

For $i \geqslant 1$ let G_i be the graph with vertices in

$$\bigcup_{k=i}^{t} C_k^i, \quad \text{where, for } k \geqslant i, \quad C_k^i = C_k \setminus \bigcup_{\ell=1}^{i-1} C_\ell.$$

The algorithm keeps a table with entries $b, w \in \{1, \ldots, n\}$ and the boolean value $\gamma_i(b, w)$ which is true if and only if there exists a black-and-white coloring of G_i with b black vertices and w white vertices. Then we have, for $i = 1, \ldots, t$,

$$\gamma_i(b, w) = \text{true} \quad \text{if and only if} \quad \exists_{j \geqslant i} \, \exists_k \, 0 \leqslant k \leqslant |X_{i,j}| \quad \text{and}$$

$$\begin{cases} (b, w) \in \{(k, 0), (0, k)\} & \text{if } j = t, \text{ and} \\ \gamma_{j+1}(b - k, w) \text{ or } \gamma_{j+1}(b, w - k) & \text{if } j < t. \end{cases}$$

To implement this algorithm one needs to compute the cardinalities $|X_{p,q}|$. Initialize $|X_{p,q}| = 0$. We assume that we have, for each vertex x, the index $F(x)$ of the first clique that contains x and the index $L(x)$ of the last clique that contains x. Consider the vertices one by one. For a vertex x, add one to $|X_{p,q}|$ for all $p \leqslant F(x)$ and all $q \geqslant L(x)$. For each vertex x we need to update $O(n^2)$ cardinalities $|X_{p,q}|$. Thus computing all cardinalities $|X_{p,q}|$ can be done in $O(n^3)$ time.

For each $i = 1, \ldots, t$, the table for G_i contains $O(n^2)$ boolean values $\gamma_i(b, w)$. For the computation of each $\gamma_i(b, w)$ the algorithm searches the tables of G_j for all $j > i$. Thus the computation of $\gamma_i(b, w)$ takes $O(n^3)$ time. Thus the full table for G_i can be obtained in $O(n^5)$ time and it follows that the algorithm can be implemented to run in $O(n^6)$ time.

There exists a black-and-white coloring of G with b black vertices and w white vertices if and only if $\gamma_1(b, w) = \text{true}$. This proves the theorem.   □

5   Strongly Chordal Graphs

The class of interval graphs is contained in the class of strongly chordal graphs. In this section we generalize the results of Section 4 to the class of strongly chordal graphs.

Definition 7. *Let $C = [x_1, \ldots, x_{2k}]$ be a cycle of even length. A chord (x_i, x_j) in C is an odd chord if the distance in C between x_i and x_j is odd.*

Recall that a graph is chordal if it has no induced cycle of length more than three [15,20].

Definition 8 ([16]). *A graph G is strongly chordal if G is chordal and each cycle in G of even length at least six has an odd chord.*

Farber discovered the strongly chordal graphs as a subclass of chordal graph for which the weighted domination problem is polynomial. The class of graphs is closely related to the class of chordal bipartite graphs [9].

There are many ways to characterize strongly chordal graphs. For example, a graph is strongly chordal if and only if its closed neighborhood matrix, or also, its clique matrix, is totally balanced [2,3,9,16,23,28]. Strongly chordal graphs are also characterized by the property that they have no induced cycles of length more than three and no induced suns [9,16]. For $k \geqslant 3$, a k-sun consists of a clique $C = \{c_1, \ldots, c_k\}$ and an independent set $S = \{s_1, \ldots, s_k\}$. Each vertex s_i, with $1 \leqslant i < k$, is adjacent to c_i and to c_{i+1} and s_k is adjacent to c_k and c_1.

Another way to characterize strongly chordal graphs is by the property that every induced subgraph has a simple vertex.

Definition 9. *A vertex x in a graph G is simple if for all $y, z \in N[x]$*

$$N[y] \subseteq N[z] \quad or \quad N[z] \subseteq N[y].$$

Notice that a simple vertex is simplicial, that is, its neighborhood is a clique.

Fig. 4. A chordal graph is strongly chordal if it has no sun. The figure shows a 3-sun and a 4-sun.

Theorem 8 ([10,16]). *A graph is strongly chordal if and only if every induced subgraph has a simple vertex.*

The proof of the following theorem can be found in [27].

Theorem 9. *There exists a polynomial-time algorithm that solves the black-and-white coloring problem on strongly chordal graphs.*

6 Splitgraphs

In this section we show that the black-and-white coloring problem on splitgraphs is NP-complete.

Definition 10. *A graph* $G = (V, E)$ *is a splitgraph if there exists a partition of the vertices in two sets* C *and* S *such that* C *induces a clique in* G *and* S *induces an independent set in* G. *Here, one of the two sets* C *and* S *may be empty.*

A splitgraph can be characterized in various ways. Notice that, if G is a splitgraph then G is chordal and, furthermore, its complement $\bar{G}$ is also a splitgraph. Actually, this property characterizes splitgraphs [17]; a graph G is a splitgraphs if and only if G and its complement $\bar{G}$ are both chordal. Splitgraphs are exactly the graphs that have no induced C_4, C_5 or $2K_2$ [17].

Theorem 10. *The black-and-white coloring problem is NP-complete for the class of splitgraphs.*

Proof. Since splitgraphs are closed under complementation, we can formulate the problem as a black-and-white coloring problem with all black vertices adjacent to all white vertices. We call this the 'inverse black-and-white coloring problem.'

We adapt a proof of Johnson, which proves the NP-completeness of finding a balanced complete bipartite subgraph in a bipartite graph [25, Page 446].

Fig. 5. A graph is a splitgraph if it has no C_4, C_5 or $2K_2$

Let $G = (V, E)$ be a graph with $|V| = n$. Construct a splitgraph H as follows. The clique of the splitgraph consists of the set V. The independent set of the splitgraph consists of the set E. In the splitgraph, make a vertex $x \in V$ adjacent to an edge $\{y, z\} \in E$ if and only if x is NOT an endpoint of $\{y, z\}$.

This completes the description of H.

Assume that the clique number of G is ω. We may assume that n is even and $n > 6$, and that $\omega = \frac{n}{2}$ [25].

To get an inverse black-and-white coloring, we color the vertices of the clique white and the rest of V black. The edges of the clique are also colored white. Then we have an inverse black-and-white coloring of H with

$$b = \omega \quad \text{and} \quad w = \omega + \binom{\omega}{2} = \binom{\omega + 1}{2}. \tag{1}$$

For the converse, assume that H has an inverse black-and-white coloring with the numbers of black and white vertices as in Equation 1. Since E is an independent set in H the colored vertices in E must all have the same color. First assume that E contains no white vertices. Then V contains a set W of white vertices, and $V \setminus W$ is black. Since

$$w = \omega + \binom{\omega}{2} > n = 2\omega \quad \text{if } n > 6,$$

this is not possible. Thus the inverse black-and-white coloring has white vertices in E.

Assume that the inverse black-and-white coloring has a set E' of white vertices in E and a set of V' of ω black vertices in V. By the construction, no edge of E' has an endpoint in V'. Now $|V \setminus V'| = \omega$ and all the endpoints of E' are in $V \setminus V'$. The only possibility is that E' is the set of edges of a clique $V \setminus V'$ of cardinality ω in G.

This proves the theorem. □

References

1. Acharya, B., Las Vergnas, M.: Hypergraphs with cyclomatic number zero, triangulated graphs, and an inequality. Journal of Combinatorial Theory, Series B 33, 52–56 (1982)
2. Anstee, R.: Hypergraphs with no special cycles. Combinatorica 3, 141–146 (1983)
3. Anstee, R., Farber, M.: Characterizations of totally balanced matrices. Journal of Algorithms 5, 215–230 (1984)
4. Bandelt, H., Mulder, H.: Distance-hereditary graphs. Journal of Combinatorial Theory, Series B 41, 182–208 (1986)
5. Beineke, L., Pippert, R.: The number of labeled k-dimensional trees. Journal of Combinatorial Theory 6, 200–205 (1969)
6. Berend, D., Zucker, S.: The black-and-white coloring problem on trees. Journal of Graph Algorithms and Applications 13, 133–152 (2009)

7. Booth, K., Lueker, G.: Linear algorithms to recognize interval graphs and test for the consecutive ones property. In: Proceedings STOC 1975, pp. 255–265. ACM (1975)

8. Broersma, H., Kloks, T., Kratsch, D., Müller, H.: Independent sets in asteroidal triple-free graphs. SIAM Journal on Discrete Mathematics 12, 276–287 (1999)

9. Brouwer, A., Duchet, P., Schrijver, A.: Graphs whose neighborhoods have no special cycle. Discrete Mathematics 47, 177–182 (1983)

10. Brouwer, A., Kolen, A.: A super-balanced hypergraph has a nest point. Technical Report ZW 146, Mathematisch Centrum, Amsterdam (1980)

11. Chang, M., Hsieh, S., Chen, G.: Dynamic Programming on Distance-Hereditary Graphs. In: Leong, H.-V., Jain, S., Imai, H. (eds.) ISAAC 1997. LNCS, vol. 1350, pp. 344–353. Springer, Heidelberg (1997)

12. Chvátal, V., Hammer, P.: Aggregation of inequalities in integer programming. Technical Report STAN-CS-75-518, Stanford University, California (1975)

13. Corneil, D., Lerchs, H., Stewart-Burlingham, L.: Complement reducible graphs. Discrete Applied Mathematics 3, 163–174 (1981)

14. Corneil, D., Perl, Y., Stewart, L.: A linear recognition algorithm for cographs. SIAM Journal on Computing 14, 926–934 (1985)

15. Dirac, G.: On rigid circuit graphs. Abhandlungen aus dem Mathematischen Seminar der Universität Hamburg 25, 71–76 (1961)

16. Farber, M.: Characterizations of strongly chordal graphs. Discrete Mathematics 43, 173–189 (1983)

17. Földes, S., Hammer, P.: Split graphs. Congressus Numerantium 19, 311–315 (1977)

18. Gavril, F.: The intersection graphs of subtrees in trees are exactly the chordal graphs. Journal of Combinatorial Theory, Series B 16, 47–56 (1974)

19. Gilmore, P., Hoffman, A.: A characterization of comparability graphs and of interval graphs. The Canadian Journal of Mathematics 16, 539–548 (1964)

20. Hajnal, A., Surányi, J.: Über die Auflösung von Graphen in vollständige Teilgraphen. Annales Universitatis Scientiarum Budapestinensis de Rolando Eötvös Nominatae – Sectio Mathematicae 1, 113–121 (1958)

21. Hammer, P., Peled, U., Sun, X.: Difference graphs. Discrete Applied Mathematics 28, 35–44 (1990)

22. Hansen, P., Hertz, A., Quinodos, N.: Splitting trees. Discrete Mathematics 165, 403–419 (1997)

23. Hoffman, A., Kolen, A., Sakarovitch, M.: Totally-balanced and greedy matrices. Technical Report BW 165/82, Mathematisch Centrum, Amsterdam (1982)

24. Howorka, E.: A characterization of distance-hereditary graphs. The Quarterly Journal of Mathematics 28, 417–420 (1977)

25. Johnson, D.: The NP-completeness column: An ongoing guide. Journal of Algorithms 8, 438–448 (1987)

26. Kloks, T.: Treewidth – Computations and Approximations. LNCS, vol. 842. Springer, Heidelberg (1994)

27. Kloks, T., Poon, S., Tsai, F., Wang, Y.: The black-and-white coloring problem on distance-hereditary graphs and strongly chordal graphs. Manuscript on ArXiv: 1111.0867v1 (2011)

28. Lehel, J.: A characterization of totally balanced hypergraphs. Discrete Mathematics 57, 59–65 (1985)

29. Lekkerkerker, C., Boland, D.: Representation of finite graphs by a set of intervals on the real line. Fundamenta Mathematicae 51, 45–64 (1962)

30. Mahadev, N., Peled, U.: Threshold graphs and related topics. Elsevier Series Annals of Discrete Mathematics 56 (1995)
31. Moon, J.: The number of labeled k-trees. Journal of Combinatorial Theory 6, 196–199 (1969)
32. Rose, D.: Triangulated graphs and the elimination process. Journal of Mathematical Analysis and Applications 32, 597–609 (1970)
33. Rose, D.: On simple characterizations of k-trees. Discrete Mathematics 7, 317–322 (1974)
34. Yannakakis, M.: The complexity of the partial order dimension problem. SIAM Journal on Algebraic and Discrete Methods 3, 351–358 (1982)

An Improved Approximation Algorithm
for the Bandpass Problem

Weitian Tong[1], Randy Goebel[1], Wei Ding[2], and Guohui Lin[1,*]

[1] Department of Computing Science, University of Alberta,
Edmonton, Alberta T6G 2E8, Canada
[2] Zhejiang Water Conservancy and Hydropower College,
Hangzhou, Zhejiang, China
{weitian,rgoebel,guohui}@ualberta.ca
dingweicumt@163.com

Abstract. The general Bandpass-B problem is NP-hard and can be approximated by a reduction into the B-set packing problem, with a worst case performance ratio of $O(B^2)$. When $B = 2$, a maximum weight matching gives a 2-approximation to the problem. The Bandpass-2 problem, or simply the Bandpass problem, can be viewed as a variation of the maximum traveling salesman problem, in which the edge weights are dynamic rather than given at the front. We present in this paper a $\frac{36}{19}$-approximation algorithm for the Bandpass problem, which is the first improvement over the simple maximum weight matching based 2-approximation algorithm.

Keywords: Bandpass problem, approximation algorithm, edge coloring, maximum weight matching, worst case performance ratio.

1 Introduction

In optical communication networks, a sending point uses a binary matrix $A_{m \times n}$ to send m information packages to n different destination points, in which the entry $a_{ij} = 1$ if information package i is *not* destined for point j, or $a_{ij} = 0$ otherwise. To achieve the highest cost reduction via wavelength division multiplexing technology, an optimal packing of information flows on different wavelengths into groups is necessary [2]. Under this binary matrix representation, every B consecutive 1's in a column indicates an opportunity for merging information to reduce the communication cost, where B is a pre-specified positive integer called the *bandpass number*. Such a set of B consecutive 1's in a column of the matrix is said to form a *bandpass*. When counting the number of bandpasses in the present matrix, no two of them in the same column are allowed to share any common rows. The computational problem, the *Bandpass-B problem*, is to find an optimal permutation of rows of the input matrix $A_{m \times n}$ such that the total number of extracted bandpasses in the

---

J. Snoeyink, P. Lu, K. Su, and L. Wang (Eds.): FAW-AAIM 2012, LNCS 7285, pp. 351–358, 2012.
© Springer-Verlag Berlin Heidelberg 2012

resultant matrix is maximized [3,2,8]. Note that though multiple bandpass numbers can be used in practice, for the sake of complexities and costs, usually only one fixed bandpass number is considered [2].

The general Bandpass-B problem, for any fixed $B \geq 2$, has been proven to be NP-hard [8]. In fact, the NP-hardness of the Bandpass-2 problem can be proven by a reduction from the well-known *Hamiltonian path* problem [6, GT39], where in the constructed binary matrix $A_{m \times n}$, a row maps to a vertex, a column maps to an edge, and $a_{ij} = 1$ if and only if edge e_j is incident to vertex v_i. It follows that there is a row permutation achieving $m - 1$ bandpasses if and only if there is a Hamiltonian path in the graph.

On the approximability, the Bandpass-B problem has a close connection to the weighted B-set packing problem [6]. Given an instance I of a maximization problem Π, let $C^*(I)$ ($C(I)$, respectively) denote the value of the optimal solution (the value of the solution produced by an algorithm, respectively). The performance ratio of the algorithm on I is $\frac{C^*(I)}{C(I)}$. The algorithm is a ρ-approximation if $\sup_I \frac{C^*(I)}{C(I)} \leq \rho$. By taking advantages of the approximation algorithms designed for the weighted B-set packing problem [1,4], the Bandpass-B problem can be approximated within $O(B^2)$ [8]. Moreover, since the maximum weight matching problem is solvable in cubic time, the Bandpass-2 problem admits a simple maximum weight matching based 2-approximation algorithm [8].

In this paper, we present for the Bandpass-2 problem, or simply the Bandpass problem, the first improved approximation algorithm, with the worst case performance ratio proven to be at most $\frac{36}{19} \approx 1.895$. Our algorithm is still based on maximum weight matchings. While the algorithm is not too complex, our main contribution lies in the non-trivial performance analysis.

2   The Approximation Algorithm

A reduction from the Hamiltonian path problem has been used to prove the NP-hardness of the Bandpass problem. But the Bandpass problem does not readily reduce to the maximum traveling salesman problem (Max-TSP) [6] for approximation algorithm design. The main reason is that, an instance graph of Max-TSP is *static*, in that all (non-negative) edge weights are given at the front, while in the Bandpass problem the number of bandpasses extracted between two consecutive rows in a row permutation is permutation dependent. Nevertheless, as shown in the sequel, our design idea is based on maximum weight matchings, the same as in approximating Max-TSP [10,7,5,9]. Formally, in Max-TSP, a complete edge-weighted graph is given, where the edge weights are non-negative integers, and the goal is to compute a Hamiltonian cycle with the maximum weight. Note that there are several variants of Max-TSP been studied in the literature. In our case, the input graph is undirected (or symmetric) and the edge weights do not necessarily satisfy the triangle inequality. The following Lemma 1 states the currently best approximation result for Max-TSP.

Lemma 1. [9] *The Max-TSP admits a $\frac{9}{7}$-approximation algorithm.*

In our Bandpass problem, since we can always add a row of all 0's if needed, we assume without loss of generality that the number of rows, m, is even.

Given the input binary matrix $A_{m \times n}$, let r_i denote the i-th row. We first construct a graph G of which the vertex set is exactly the row set $\{r_1, r_2, \ldots, r_m\}$. Between rows r_i and r_j, the *static* edge weight is defined as the maximum number of bandpasses that can be formed between the two rows, and is denoted as $w(i, j)$. In the sequel we use row (of the matrix) and vertex (of the graph) interchangeably.

For a row permutation $\pi = (\pi_1, \pi_2, \ldots, \pi_m)$, its i-th row is the π_i-th row in the input matrix. We call a maximal segment of consecutive 1's in a column of π a *strip* of π. The length of a strip is defined to be the number of 1's therein. A length-ℓ strip contributes exactly $\lfloor \frac{\ell}{2} \rfloor$ bandpasses to the permutation π. We use $S_\ell(\pi)$ to denote the set of all length-ℓ strips of π, and $s_\ell(\pi) = |S_\ell(\pi)|$. Let $b(\pi)$ denote the number of bandpasses extracted from the permutation π. We have

$$b(\pi) = \sum_{\ell=2}^{m} s_\ell(\pi) \left\lfloor \frac{\ell}{2} \right\rfloor = s_2(\pi) + \sum_{\ell=3}^{m} s_\ell(\pi) \left\lfloor \frac{\ell}{2} \right\rfloor. \tag{1}$$

Let $p(\pi)$ denote the number of pairs of consecutive 1's in the permutation π. We have

$$p(\pi) = \sum_{\ell=2}^{m} s_\ell(\pi)(\ell - 1) = s_2(\pi) + \sum_{\ell=3}^{m} s_\ell(\pi)(\ell - 1). \tag{2}$$

2.1 Algorithm Description

In our algorithm denoted as APPROX, the first step is to compute a maximum weight matching M_1 in graph G. Recall that there are an even number of rows. Therefore, M_1 is a perfect matching (even though some edge weights could be 0). Let $w(M_1)$ denote the sum of its edge weights, indicating that exactly $w(M_1)$ bandpasses can be extracted from the row pairings suggested by M_1. These bandpasses are called the bandpasses of M_1.

Next, every 1 involved in a bandpass of M_1 is changed to 0. Let the resultant matrix be denoted as $A'_{m \times n}$, the resultant edge weight between rows r_i and r_j be $w'(i, j)$ — which is the maximum number of bandpasses can be formed between the two revised rows — and the corresponding resultant graph be denoted as G'. One can see that if an edge (r_i, r_j) belongs to M_1, then the new edge weight $w'(i, j) = 0$. In the second step of APPROX, we compute a maximum weight matching M_2 in graph G', and let $w'(M_2)$ denote its weight or its number of bandpasses. It is noted that no bandpass of M_1 shares a 1 with any bandpass of M_2.

If an edge (r_i, r_j) belongs to both M_1 and M_2, then it is removed from M_2. Such a removal does not decrease the weight of M_2 as $w'(i, j) = 0$. Consider the union of M_1 and M_2. Note that every cycle of this union, if any, must be an even cycle with alternating edges of M_1 and M_2. The third step of APPROX is to break cycles, by removing for each cycle the least weight edge of M_2. Let M denote the final set of edges of the union, which form into disjoint paths.

In the last step, we arbitrarily stack these paths to give a row permutation π. The number of bandpasses extracted from π, $b(\pi)$, is at least the weight of M, which is greater than or equal to $w(M_1) + \frac{1}{2}w'(M_2)$.

2.2   Performance Analysis

Let π^* denote the optimal row permutation such that its $b(\pi^*)$ is maximized over all row permutations. Correspondingly, $S_2(\pi^*)$ denotes the set of length-2 strips in π^*, which contributes exactly $s_2(\pi^*)$ bandpasses towards $b(\pi^*)$. The key part in the performance analysis for algorithm APPROX is to estimate $w'(M_2)$, as done in the following.

First, we partition the bandpasses of $S_2(\pi^*)$ into four groups: B_1, B_2, B_3, B_4. Note that bandpasses of $S_2(\pi^*)$ do not share any 1 each other. B_1 consists of the bandpasses of $S_2(\pi^*)$ that also belong to matching M_1 (such as the one between rows r_a and r_b in Figure 1); B_2 consists of the bandpasses of $S_2(\pi^*)$, each of which shares (exactly) a 1 with exactly one bandpass of M_1, and the other 1 of the involved bandpass of M_1 is shared by another bandpass in B_2; B_3 consists of the bandpasses of $S_2(\pi^*)$, each of which shares (exactly) a 1 with at least one bandpass of M_1, and if it shares a 1 with exactly one bandpass of M_1 then the other 1 of the involved bandpass of M_1 is not shared by any other bandpass of B_2; B_4 consists of the remaining bandpasses of $S_2(\pi^*)$. Figure 1 illustrates some examples of these bandpasses.

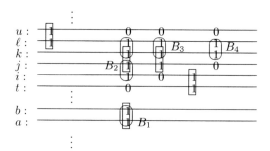

Fig. 1. An illustration of the bandpasses of $S_2(\pi^*)$ (in ovals) and the bandpasses of M_1 (in boxes) for grouping purpose. A horizontal line in the figure represents a row, led by its index. Rows that are adjacent in π^* and/or row pairs of M_1 are intentionally ordered adjacently. In this figure, rows r_a and r_b are adjacent in π^*, denoted as $(r_a, r_b) \in \pi^*$, and edge $(r_a, r_b) \in M_1$ as well; the bandpasses between these two rows in $S_2(\pi^*)$ thus belong to B_1. Edges $(r_t, r_i), (r_j, r_k), (r_\ell, r_u) \in M_1$, while $(r_i, r_j), (r_k, r_\ell) \in \pi^*$; the bandpasses between rows r_i and r_j and between rows r_k and r_ℓ in $S_2(\pi^*)$ shown in the figure have their group memberships indicated beside them respectively.

By the definition of partition, we have

$$s_2(\pi^*) = |B_1| + |B_2| + |B_3| + |B_4|. \tag{3}$$

From these "group" definitions, we know all bandpasses of B_1 are in M_1. Also, one pair of bandpasses of B_2 correspond to a distinct bandpass of M_1. Bandpasses of B_3 can be further partitioned into subgroups such that a subgroup of bandpasses together with a distinct maximal subset of bandpasses of M_1 form into an alternating cycle or path of length at least 2. Moreover, 1) when the path length is even, the number of bandpasses of this subgroup of B_3 is equal to the number of bandpasses of this subset of bandpasses of M_1; 2) when the path length is odd, 2a) either the number of bandpasses of this subgroup of B_3 is 1 greater than the number of bandpasses of this subset of bandpasses of M_1, 2b) or the path length has to be at least 5 and so the number of bandpasses of this subgroup of B_3 is at least $\frac{2}{3}$ of the number of bandpasses of this subset of bandpasses of M_1. It follows from 1), 2a) and 2b) that with respect to B_3, M_1 contains at least $\frac{2}{3}|B_3|$ corresponding bandpasses. That is,

$$w(M_1) \geq |B_1| + \frac{1}{2}|B_2| + \frac{2}{3}|B_3|. \tag{4}$$

Apparently, all bandpasses of B_4 are in graph G', while none of $B_1 \cup B_2 \cup B_3$ is in graph G'.

Note that the bandpasses of B_2 are paired up such that each pair of the two bandpasses share a 1 with a bandpass of M_1. Assume without loss of generality that these two bandpasses of B_2 are formed between rows r_i and r_j and between rows r_k and r_ℓ, respectively, and that the involved bandpass of M_1 is formed between rows r_j and r_k (see Figure 1). That is, in the optimal row permutation π^*, rows r_i and r_j are adjacent, and rows r_k and r_ℓ are adjacent; while edge $(r_j, r_k) \in M_1$. We remark that these four rows are distinct. We conclude that edge $(r_i, r_\ell) \notin M_1$. The proof is simple as otherwise in the particular column a bandpass would be formed between rows r_i and r_ℓ, making the two bandpasses of B_2 lose their group memberships (*i.e.*, they would belong to B_3).

Lemma 2. *Assume edge $(r_j, r_k) \in M_1$, and that one bandpass of (r_j, r_k) shares 1 with (two) bandpasses of B_2. Then in G edge (r_j, r_k) is adjacent to at most four edges in the optimal row permutation π^*, at most two of which are incident at row r_j and at most two of which are incident at row r_k.*

Proof. The lemma is straightforward from the above discussion, and the fact that edge (r_j, r_k) does not belong to π^*.

Continuing with the above discussion, assuming that edge $(r_j, r_k) \in M_1$, and that one bandpass of (r_j, r_k) shares 1 with two bandpasses of B_2, which are formed between rows r_i and r_j and between rows r_k and r_ℓ, respectively (see Figure 1). We know that in graph G', between rows r_i and r_ℓ, in the same column there is a bandpass (which contributes 1 towards the edge weight $w'(i, \ell)$). We call bandpasses constructed in this way the *induced* bandpasses. From Lemma 2, edge (r_j, r_k) is adjacent to at most two edges of π^* incident at row r_j. It follows that in graph G', row r_ℓ can form induced bandpasses with at most four other rows. In the other words, the subgraph of G' induced by the edges containing induced bandpasses, denoted as G'_s is a degree-4 graph.

Lemma 3. G'_s is a degree-4 graph, and its weight $w'(G'_s) \geq \frac{1}{2}|B_2|$.

Proof. The first half of the lemma is a result of the above discussion. Since every pair of bandpasses of B_2 leads to an induced bandpass, all the edge weights in G'_s sum up to at least $\frac{1}{2}|B_2|$, which is the number of bandpass pairs in B_2.

Lemma 4. *The weight of matching M_2 is* $w'(M_2) \geq \max\{\frac{1}{10}|B_2|, \frac{1}{2}|B_4|\} \geq x\frac{1}{10}|B_2| + (1-x)\frac{1}{2}|B_4|$, *for any* $x \in [0,1]$.

Proof. Vizing's Theorem [11] states that the edge coloring (chromatic) number of a graph is either the maximum degree Δ or $\Delta + 1$. Note that all edges of the same color form a matching in the graph. We conclude from Lemma 3 that, even in graph G'_s there is a matching of weight at least $\frac{1}{5}w'(G'_s) \geq \frac{1}{10}|B_2|$. As G'_s is a subgraph of G' and M_2 is the maximum weight matching of G', $w'(M_2) \geq \frac{1}{10}|B_2|$.

On the other hand, graph G' contains all bandpasses of B_4. Therefore, $w'(M_2) \geq \frac{1}{2}|B_4|$ as well. The last inequality in the lemma then follows trivially,

$$\max\left\{\frac{1}{10}|B_2|, \frac{1}{2}|B_4|\right\} \geq x\frac{1}{10}|B_2| + (1-x)\frac{1}{2}|B_4|,$$

for any $x \in [0,1]$.

Theorem 1. *Algorithm* APPROX *is a cubic time* $\frac{36}{19}$-*approximation for the Bandpass problem.*

Proof. The running time of algorithm APPROX is dominated by the computing for two maximum weight matchings, which can be done in cubic time. Since M_1 is the maximum weight matching in graph G, from Eq. (2) we have

$$w(M_1) \geq \frac{1}{2}p(\pi^*) \geq \frac{1}{2}\left(s_2(\pi^*) + \sum_{\ell=3}^{m} s_\ell(\pi^*)(\ell-1)\right). \tag{5}$$

Combining Eqs. (4) and (5), we have for any $y \in [0,1]$,

$$w(M_1) \geq y\frac{1}{2}\left(s_2(\pi^*) + \sum_{\ell=3}^{m} s_\ell(\pi^*)(\ell-1)\right)$$
$$+ (1-y)\left(|B_1| + \frac{1}{2}|B_2| + \frac{2}{3}|B_3|\right). \tag{6}$$

The permutation π produced by algorithm APPROX contains $b(\pi) \geq w(M_1) + \frac{1}{2}w'(M_2)$ bandpasses, as indicated at the end of Section 2.1. From Lemma 4, we have for any $x \in [0,1]$,

$$b(\pi) \geq w(M_1) + x\frac{1}{20}|B_2| + (1-x)\frac{1}{4}|B_4|. \tag{7}$$

Together with Eqs. (3) and (6), the above Eq. (7) becomes,

$$
b(\pi) \geq w(M_1) + x\frac{1}{20}|B_2| + (1-x)\frac{1}{4}|B_4|
$$

$$
\geq y\frac{1}{2}\left(s_2(\pi^*) + \sum_{\ell=3}^{m} s_\ell(\pi^*)(\ell-1)\right)
$$

$$
+(1-y)\left(|B_1| + \frac{1}{2}|B_2| + \frac{2}{3}|B_3|\right) + x\frac{1}{20}|B_2| + (1-x)\frac{1}{4}|B_4|
$$

$$
= \frac{y}{2}\left(s_2(\pi^*) + \sum_{\ell=3}^{m} s_\ell(\pi^*)(\ell-1)\right)
$$

$$
+(1-y)|B_1| + \left(\frac{1-y}{2} + \frac{x}{20}\right)|B_2| + \frac{2(1-y)}{3}|B_3| + \frac{1-x}{4}|B_4|
$$

$$
\geq \frac{5}{12}\left(s_2(\pi^*) + \sum_{\ell=3}^{m} s_\ell(\pi^*)(\ell-1)\right) + \frac{1}{18}|B_1| + \frac{1}{9}s_2(\pi^*), \tag{8}
$$

where the last inequality is achieved by setting $x = \frac{5}{9}$ and $y = \frac{5}{6}$. Note that for all $\ell \geq 3$, $(\ell-1) \geq \frac{3}{2}\lfloor\frac{\ell}{2}\rfloor$. It then follows from Eqs. (8) and (1) that

$$
b(\pi) \geq \frac{19}{36}\left(s_2(\pi^*) + \frac{15}{19} \times \frac{3}{2}\sum_{\ell=3}^{m} s_\ell(\pi^*)\left\lfloor\frac{\ell}{2}\right\rfloor\right) \geq \frac{19}{36}b(\pi^*). \tag{9}
$$

That is, the worst-case performance ratio of algorithm APPROX is at most $\frac{36}{19}$.

3   Conclusions and Future Work

In this paper, we presented a $\frac{36}{19}$-approximation algorithm for the Bandpass problem, which is the first improvement (≈ 1.895) over the maximum weight matching based 2-approximation algorithm. Our algorithm is still based on maximum weight matchings, similar to tackling the closely related Max-TSP. Though our algorithm description is not too complex, its performance analysis appears non-trivial.

It is noted that our algorithm applies to the Max-TSP as well, achieving a worst case performance ratio $\frac{8}{5}$. We are not sure whether a better analysis would narrow the gap between $\frac{8}{5}$ and $\frac{36}{19}$ on the Bandpass problem. For the Max-TSP, Serdyukov presented a $\frac{4}{3}$-approximation algorithm based on the maximum weight *assignment* (or called *cycle cover*) and the maximum weight matching [10], which is further improved to the currently best $\frac{9}{7}$-approximation algorithm in Lemma 1. But this assignment idea could not be easily adapted for the Bandpass problem. Hassin and Rubinstein gave the currently best randomized approximation algorithm for the Max-TSP with expected performance ratio $\frac{33}{25}$ [7] (which was subsequently de-randomized in [5]). It would be interesting to design a randomized approximation for the Bandpass problem too, with better than $\frac{36}{19}$ expected performance ratio.

Acknowledgement. This research was supported in part by NSERC.

References

1. Arkin, E.M., Hassin, R.: On local search for weighted packing problems. Mathematics of Operations Research 23, 640–648 (1998)
2. Babayev, D.A., Bell, G.I., Nuriyev, U.G.: The bandpass problem: combinatorial optimization and library of problems. Journal of Combinatorial Optimization 18, 151–172 (2009)
3. Bell, G.I., Babayev, D.A.: Bandpass problem. In: Annual INFORMS Meeting, Denver, CO, USA (October 2004)
4. Chandra, B., Halldórsson, M.M.: Greedy local improvement and weighted set packing approximation. In: ACM-SIAM Proceedings of the Tenth Annual Symposium on Discrete Algorithms (SODA 1999), pp. 169–176 (1999)
5. Chen, Z.-Z., Okamoto, Y., Wang, L.: Improved deterministic approximation algorithms for Max TSP. Information Processing Letters 95, 333–342 (2005)
6. Garey, M.R., Johnson, D.S.: Computers and Intractability: A Guide to the Theory of NP-completeness. W. H. Freeman and Company, San Francisco (1979)
7. Hassin, R., Rubinstein, S.: Better approximations for Max TSP. Information Processing Letters 75, 181–186 (2000)
8. Lin, G.: On the Bandpass problem. Journal of Combinatorial Optimization 22, 71–77 (2011)
9. Paluch, K., Mucha, M., Mądry, A.: A 7/9 - Approximation Algorithm for the Maximum Traveling Salesman Problem. In: Dinur, I., Jansen, K., Naor, J., Rolim, J. (eds.) APPROX 2009. LNCS, vol. 5687, pp. 298–311. Springer, Heidelberg (2009)
10. Serdyukov, A.I.: An algorithms for with an estimate for the traveling salesman problem of the maximum. Upravlyaemye Sistemy 25, 80–86 (1984)
11. Vizing, V.G.: On an estimate of the chromatic class of a p-graph. Diskretnogo Analiza 3, 25–30 (1964)

Partial Degree Bounded Edge Packing Problem

Peng Zhang

Shanghai Key Laboratory of Trustworthy Computing, East China Normal University
arena.zp@gmail.com

Abstract. In [1], whether a target binary string s can be represented from a boolean formula with operands chosen from a set of binary strings W was studied. In this paper, we first examine selecting a maximum subset X from W, so that for any string t in X, t is not representable by $X \setminus \{t\}$. We rephrase this problem as graph, and surprisingly find it give rise to a broad model of edge packing problem, which itself falls into the model of forbidden subgraph problem. Specifically, given a graph $G(V, E)$ and a constant c, the problem asks to choose as many as edges to form a subgraph G'. So that in G', for each edge, at least one of its endpoints has degree no more than c. We call such G' partial c degree bounded. This edge packing problem model also has a direct interpretation in resource allocation. There are n types of resources and m jobs. Each job needs two types of resources. A job can be accomplished if either one of its necessary resources is shared by no more than c other jobs. The problem then asks to finish as many jobs as possible. For edge packing problem, when $c = 1$, it turns out to be the complement of dominating set and able to be 2-approximated. When $c = 2$, it can be 32/11-approximated. We also prove it is NP-complete for any constant c on graphs and is $O(|V|^2)$ solvable on trees. We believe this partial bounded graph problem is intrinsic and merits more attention.

1 Introduction

An elementary problem of set operations is studied in [1]. Given two binary strings of the same length, namely s_1, s_2, let $s_1 \wedge s_2$ (resp. $s_1 \vee s_2$) be the binary string produced by bitwise AND $\wedge$ (resp. OR $\vee$) of s_1 and s_2. Given a set of m bits long binary strings, namely, $W = \{s_1, s_2, \cdots, s_n\}$, $s_i \in \{0, 1\}^m$, if there is a formula ϕ which calculates s, with operators in $\{\wedge, \vee\}$ and operands in some subset of W, then we say the target string s is representable by (or expressible from) W via formula ϕ, or simply s is representable.

A natural variant of this problem is finding a maximum subset, in which each string is not representable by the others. We call this variant *Maximum Expressive Independent Subset* (MEI) problem and examine the restricted case on strings with exactly two ones. Surprisingly, this is equivalent as maximum edge packing under partial degree bounded by 2.

This paper is structured as follows. We study the hardness of edge packing bounded by 1, by 2 and by a constant less than $\Delta(G)$ on graph in section 2. Then we study the general edge packing on trees in section 3. In section 4,

J. Snoeyink, P. Lu, K. Su, and L. Wang (Eds.): FAW-AAIM 2012, LNCS 7285, pp. 359–367, 2012.

approximation algorithms for bounded 1 and bounded 2 edge packing are presented. Some conclusions are given in section 5. Since the problem only concerns edges selecting, we assume the graph we deal with is free of isolated vertex.

1.1   Related Work

The decision problem of edge packing bounded by 1 turns out to be a parametric dual of the well known *Dominating Set* (DS) problem. The parametric dual means that for graph $G(V, E)$, a k sized dominating set implies a $|V| - k$ sized edge packing, and vice versa. The parametric dual of DS was studied in [2], in which the edges packed are called pendant edges. Further, the dual was well studied under the framework of parameterized complexity by Frank Dehne, etc in [3]. They coined the dual as NONBLOCKER problem and showed a linear kernel of $5/3 \cdot k_d + 3$, where k_d is the solution size.

2   Maximum Expressible Independent Subset

At first, we introduce some notations used in [1]. Let x denote any binary string, b_i^x denote the ith bit of x. So, $x = b_1^x b_2^x \cdots b_m^x$. Also, we define a function Zero : $\mathsf{Zero}(x) = \{i | b_i^x = 0\}$, from a binary string to a set of natural numbers which denotes the indices of bits with value 0 in the binary string. Similarly, $\mathsf{One}(x)$ denotes the indices of 1 valued bits of x. Also, $\mathbf{0}$ (resp. $\mathbf{1}$) denotes a binary string with no 1 (resp. 0) valued bits. Let N_i denote the set of strings whose i^{th} bit is 1, i.e, $N_i = \{y | b_i^y = 1, y \in W\}$. In addition, T_i denotes the set of binary strings in W whose i^{th} bit value is 0, i.e., $T_i = \{x \in W | b_i^x = 0\}$. Let $t_i = \bigvee_{x \in T_i} x$.

Definition 1 (Expressible Independent Set (EI)). *A set X of binary strings is expressible independent if and only if for each binary string $x \in X$, x is not expressible from $X \setminus \{x\}$.*

Then the Maximum Expressible Independent Subset (MEIS) problem is defined as follows. Given a set W of binary strings, MEIS asks to find a maximum expressible independent subset of W. The decision version with parameter k is denoted as MEIS(W, k).

2.1   MEIS on 2-Regular Set

We first pay attention to a restricted case of MEIS, when each binary string has the same number of bits valued 1. And we refer to the following theorem 1 from [1].

Theorem 1. *Given (W, s) where $s \neq \mathbf{1}$, then s is expressible from W if and only if $\forall i \in \mathsf{Zero}(s)$, $\mathsf{One}(s) \subseteq \mathsf{One}(t_i)$.*

Definition 2 (c-regular set). *A binary string is c-regular if and only if it contains exactly c one bits. A set of binary strings is c-regular if and only if each element is c-regular.*

Lemma 1. *Given a 2-regular set W and a 2-regular string $x \notin W$, $\mathsf{One}(x) = \{i, j\}$, then x is expressible from W if and only if $|N_i| \geq 2$ and $|N_j| \geq 2$.*

Proof. Sufficiency: We prove its contrapositive. By symmetry, suppose that $|N_i| \leq 1$. If $|N_i| = 0$, then $i \notin \mathsf{One}(t_l), l \in \mathsf{Zero}(x)$. If $N_i = \{y\}$, assume that $\mathsf{One}(y) = \{l, i\}$, then $i \notin \mathsf{One}(t_l)$. In both cases, $\mathsf{One}(x) \nsubseteq \mathsf{One}(t_l)$, thus x is not expressible from W according to Lemma 1.

Necessity: If $|N_i| \geq 2$ and $|N_j| \geq 2$, we assume $\{a, b\} \subseteq N_i$ and $\{c, d\} \subseteq N_j$. It is easy to check that $(a \wedge b) \vee (c \wedge d) = x$.

Definition 3 (Partial Degree Bounded Graph). *An undirected graph $G(V, E)$ is partial c bounded (PcB) if and only if $\forall_{e(u,v) \in E}(d_u \leq c \bigvee d_v \leq c)$. d_u is the degree of u.*

Given a graph G, the Maximum Partial c Degree Bounded Graph problem asks to find a PcB subgraph G' of G with maximum edges. The decision version with parameter k is denoted as $PcB(G, k)$. In the setting of resource allocation, each vertex stands for a resource, each edge stands for a job. And an optimum PcB subgraph maps to an optimum resource allocation.

Now, we will rephrase $\mathrm{MEIS}(W, k)$ on 2-regular set as $\mathrm{P2B}(G, k)$. Let $W \subseteq \{0, 1\}^m$, we construct the corresponding graph $G(V, E)$ as follows. Vertex $v_i \in V$ corresponds to the ith bit of string. Each edge $(v_i, v_j) \in E$ corresponds to a string $x \in W$ whose $\mathsf{One}(x) = \{i, j\}$. According to Lemma 1, $\mathrm{MEIS}(W, k)$ has a solution if and only if $P2B(G, k)$ has a solution. Just select the corresponding edges in G, and select the corresponding strings in W vice versa. The reduction can be done in the reverse way. So it is just a rephrasing. The following lemma was proved in [2], and we gave its proof here to make the paper more readable.

Lemma 2. *$P1B(G, k)$ is NP-complete.*

Proof. Given a graph $G(V, E)$, $P1B(G, k)$ is in NP trivially because we can check in $O(|E|)$ time that whether the given subgraph G' is partial 1 bounded. We prove its NP-completeness by showing that, there is a a k sized partial 1 bounded subgraph G' if and only if there is a $n - k$ sized dominating set D of G, $n = |V|$. Note that, any partial 1 bounded graph is a set of node-disjoint stars.

Necessity: If $D = \{v_1, \cdots, v_k\}$ is a dominating set, then we can construct a k node-disjoint stars as following, which is a partition of G. Let $P_i(V_i, E_i)$ denote the ith star being constructed. For each vertex $u \in V \setminus D$, if u is dominated by v_i, add u into V_i and (v_i, u) into E_i. To make the stars node-disjoint, when u is dominated by more than one vertices, break the ties arbitrarily. Note that, E_i may be empty, that is, the P_i only contains an isolated vertex. So $\sum_{i \leq k} |E_i| = \sum_{i \leq k} |V_i| - k = n - k$. Thus $\bigcup_{v_i \in D} P_i$ is a $n - k$ sized partial 1 bounded graph.

Sufficiency: If there is a G' with $|E_{G'}| = n - k$. Suppose that G' contains n_0 stars without leaf (i.e., isolated vertices) and n_1 stars with at least one leaf. It is easy to see, $n - k = (n - n_0) - n_1$. Thus $n_0 + n_1 = k$, so we just select the isolated vertices and the internal node of the n_1 stars. They make up a k sized dominating set.

Lemma 3. $P2B(G, k)$ is NP-complete, so is $MEIS(W, k)$ on 2-regular set.

Proof. This problem is trivial to be in NP. We show its NP-completeness via a reduction from $P1B(G, k)$. Given a $P1B(G, k)$ instance $G(V, E)$, $n = |V|$, we construct a $P2B(G', n+k)$ instance $G'(V', E')$ as follows. Adding a distinguished vertex u into V, i.e., $V' = V \cup \{u\}$ and $E' = E \cup \{(u, v)|v \in V\}$.

Necessity: If M is a k sized partial 1 bounded subgraph in G, then adding the n additional edges, i.e., $E' \setminus E$ into M will produce a $n + k$ sized partial 2 bounded subgraph M'.

Sufficiency: Let M' be a $n + k$ sized partial 2 bounded subgraph in G', and let $d_v^{M'}$ be the degree of v in M'. We prove it case by case. **Case 1:** When $(E' \setminus E) \subseteq E_{M'}$, then deleting all the n additional edges will make each node's degree in M' decrease 1, thus the remaining subgraph is a k sized partial 1 bounded subgraph. **Case 2:** When there exists an edge $(u, v_i) \notin E_{M'}$, if $d_{v_i}^{M'} < 2$, then we could just replace an arbitrarily edge $(v_j, v_k) \in E_{M'}$ by (u, v_i). If $d_{v_i}^{M'} \geq 2$, let $(v_i, v_j) \in E_{M'}$, then we could replace (v_i, v_j) by (u, v_i). Repeating this swap, we eventually arrive at a $n + k$ sized partial 2 bounded subgraph M' which contains all the n addition edges, that is Case 1.

Theorem 2. $PcB(G, k)$ is NP-complete.

Proof. We can easily generalize the proof technique of to any parameter c. Given a $P1B(G, k)$ instance $G(V, E)$, $n = |V|$, we add $c - 1$ additional vertices and $(c - 1)|V|$ edges connecting each of the vertex in V to each additional vertex, resulting a graph G'. Then $P1B(G, k)$ is a YES instance if and only if $PcB(G', k + (c-1)|V|)$ is a YES instance. Note that, it is trivial to select all the edges when $c \geq \Delta(G)$, where $\Delta(G)$ stands for the maximum degree of G.

It is clear that $PcB(G, k)$ is a case of forbidden subgraph of G, which asks to find a maximum subgraph G' of G so that G' does contain a subgraph which is isomorphic with the forbidden H. Here H is a tree with 2 internal nodes with degree $c + 1$ and $2c$ leaf nodes. Each internal node has incident edges to c leaf nodes and the other internal node.

3   Maximum Partial c Bounded Subgraph Problem on Tree

Due to the NP-hardness of P2B on the general graph, we would first consider it on some restricted structures, such as tree. In the scenario of Maximum Expressible Independent Subset, this corresponds to restricted instances where for any subset $A \subseteq W$, $|\bigcup_{x \in A} \mathsf{One}(x)| > |A|$.

Lemma 4. $PcB(G, k)$ is solvable in $O(n^2)$ for any parameter c via a dynamic programming, where $n = |V|$.

Proof. The sketch of the algorithm is bottom-up for the tree as a whole, and left to right knapsack like dynamic programming for selecting a vertex's children.

Let $T(V, E)$ denote the tree, and $v_1, \cdots, v_n$ be a breadth first ordering of vertices of V. Further, let d'_u be the number of u's children and let $T(u, i)$ be the subtree induced by u, u's first i children and all their descendants. Thus, $T(u, d'_u)$ is the subtree rooted at u and $T(u, 0)$ contains u alone. Let $f(u, i, q)$ denote the maximum PcB subtree in $T(u, i)$ under the condition that u has q neighbors in $T(u, i)$. Let $g_1(u)$ denote the maximum PcB subtree in $T(u, d'_u)$ under the condition that u has less than c neighbors, and $g_2(u)$ for exactly c neighbors and $g_3(u)$ for more than c neighbors respectively. So only $g_1(u)$ and $g_3(u)$ can be extended to have an edge connecting to u's parent when we are working upward. For simplicity, we abuse $f(u, i, q)$ and $g_i(u)$ to denote their edge cardinalities. Let $\mathsf{MAX}\{\cdots, a_i, \cdots\} = \arg max_i\{a_i\}$.

Algorithm 1. Solving Partial c Bounded Subgraph on Trees

1: **for all** u, u is a leaf **do**
2:     $f(u, 0, 0) = 0$
3: **for all** u, all subtrees rooted at u's children have been calculated **do**
4:     **for all** i from 1 to d'_u **do**
5:         Let v be the ith child of u counting from left to right
6:         **for all** q from 1 to i **do**
7:             $f(u, i, q) = f(u, i - 1, q) + \mathsf{MAX}\{g_1(v), g_2(v), g_3(v)\}$
8:             **if** $q \le c$ **then**
9:                 $f(u, i, q) = \mathsf{MAX}\{f(u, i, q), f(u, i - 1, q - 1) + 1 + \mathsf{MAX}\{g_1(v), g_3(v)\}\}$
10:            **else**
11:                $f(u, i, q) = \mathsf{MAX}\{f(u, i, q), f(u, i - 1, q - 1) + 1 + g_1(v)\}$
12:        update $g_1(u)$, $g_2(u)$ and $g_3(u)$

The algorithm above is correct because it *enumerates* every possible edge selection by a knapsack like way. Lines 1-2 take $O(|V|)$. It is important to note that lines 3-12 take only $\Sigma(d'_i)^2 \le (\Sigma d'_i)^2 \le (2|V|)^2$. So the running time of the algorithm is $O(|V|^2)$.

4   Approximation Algorithms for P1B and P2B

In this section, we are going to present two approximation algorithms. The first one for partial 1 bounded subgraph runs in $O(|E|)$ with approximation ratio 2, and the one for partial 2 bounded runs in $O(|V|)$ with ratio $\frac{32}{11}$ in expectation. In analyzing both algorithms, we only use upper bounds of the optimum solutions, without exploring deep relationships between the optimum and the solution our algorithm returned.

4.1 A 2-Approximation Algorithm for Partial 1 Bounded Subgraph

Given a graph $G(V, E)$, we first greedily calculate a dominating set with no more than $|V|/2$ vertices and then construct a partial 1 bounded subgraph M with no less than $|V|/2$ edges. Because the maximum P1B subgraph of G has less than $|V|$ edges, M is a 2 ratio approximation solution. The process is shown in algorithm 2.

Algorithm 2. Approximation Algorithm for maximum partial 1 bounded subgraph of $G(V, E)$

1: $D \subseteq V$, $A \subseteq V$, initiate $A = D = \emptyset$
2: **for all** $u \in V$ **do**
3:     **if** u is not dominated by any vertex $v \in D$ **then**
4:         $D = D \bigcup \{u\}$
5: **if** $|D| > |V|/2$ **then**
6:     $D = V \setminus D$
7: **for all** $u \in D$ **do**
8:     **for all** edge $(u, v) \in E$ **do**
9:         **if** $v \in D$ or $v \in A$ **then**
10:             delete (u, v)
11:         **if** $v \notin D$ and $v \notin A$ **then**
12:             $A = A \bigcup \{v\}$
13: G is a P1B graph

Lines 1-4 in algorithm 2 obtains a minimal dominating set (DS) D in $O(|E|)$. Then $V \setminus D$ is also a minimal DS and lines 5-6 obtains a minimal DS with no more than half vertices. Lines 7-12 obtains a P1B subgraph of G, this is proved in lemma 2.

4.2 A 32/11-Approximation Algorithm for Partial 2 Bounded Subgraph

We first present an upper bound of the optimum value and then give a randomized algorithm with expectation larger than $\frac{11}{32}$ times the upper bound. Eventually we show the process of derandomization. Let $N(u) = \{v | (u, v) \in E\}$ denote the neighbors of u in graph G. Also, if A is a set of vertices, then let $N(A) = \bigcup_{u \in A} N(u)$. In the sequel, $n = |V|$.

Lemma 5. If $G(V, E)$ is a maximum partial c bounded graph on n vertices, then $\forall_{(u,v) \in E}(d_u \geq c \bigwedge d_v \geq c)$.

Proof. If there is an edge (u, v) dissatisfies $(d_u \geq c \bigwedge d_v \geq c)$, i.e., $(d_u < c \bigvee d_v < c)$, we assume $d_u < c$. Let $X = V \setminus N(u, v)$, then $X \neq \emptyset$ because u and v can have at most $2c - 2$ neighbors. Otherwise, we can construct a graph with more edges. We do case by case proof as follows. **Case 1:** If $\exists_{x \in X} d_x > c$, then we apply an edge addition as $E = E \bigcup \{(u, x)\}$. This edge addition preserves G's property as a PcB, we call it *valid*. **Case 2:** If $\forall_{x \in X} d_x \leq c$, then $d_x \leq c$, we apply the same edge addition as in Case 1. Both edge additions contradict the fact that G is a maximum PcB. So the lemma holds.

According to lemma 5, $\forall_{u \in V}(d_u \geq c)$. By definition of P$c$B, $\forall_{(u,v) \in E}$ $(d_u \leq c \wedge d_v \leq c)$, then at least one endpoint has degree c. So the following corollary is correct.

Corollary 1. *If $G(V, E)$ is a maximum partial c bounded graph on n vertices, then*
$$\forall_{(u,v) \in E}(d_u = c \vee d_v = c).$$

Lemma 6. *If $G(V, E)$ is a maximum partial c bounded graph on n vertices, then* $\forall_{(u,v) \in E}(d_u = c \wedge d_v > c)$.

Proof. If there is an edge (u, v) dissatisfies $(d_u = c \wedge d_v > c)$, then according to lemma 5 and corrolary 1, the only possibility is $(d_u = c \wedge d_v = c)$. Let $X = V \setminus N(u, v)$, then $X \neq \emptyset$. We do case by case proof as follows. **Case 1:** If $\exists_{x \in X} d_x > c$, then we can apply an *valid edge augmentation* as $E = (E \setminus \{(u, v)\}) \bigcup \{(u, x), (v, x)\}$. **Case 2:** If $\forall_{x \in X} d_x = c$, then we can choose an x arbitrarily and apply the same *valid edge augmentation* as in Case 1. However, both edge augmentations contradict that G is a maximum PcB. So the lemma holds.

Theorem 3. *For any partial c bounded graph $G(V, E)$, $|E| \leq c \cdot (|V| - c)$.*

Proof. Let $G(V, E)$ be a maximum PcB graph on n vertices. With the help of lemma 6, we can calculate the number of vertices having degree c. Let y denote this number. Suppose $y > n - c$, then there are less than c vertices with degree more than c. Thus for any $d_u = c$, u can only have less than c neighbors, which contradicts $d_u = c$. So $y \leq n - c$, and $|E| \leq c \cdot y \leq c \cdot (|V| - c)$. When $y = n - c$, we can easily construct a PcB graph with $c \cdot (|V| - c)$ edges. So the theorem holds.

We justify an assumption that $\forall_{u \in G}(d_u^G > 2)$ as follows. Let $E' = \{(u, v)|d_u^G \leq 2\}$ and $M(V_M, E_M)$ be a partial 2 bounded subgraph. If $E' \subseteq E_M$, our assumption holds because we only need to consider the graph with minimum degree larger than 2. Otherwise, we repeatedly do the following swap in and out operations till $E' \subseteq E_M$. Let $(u, v_i) \in E' \setminus E_M$ and $(v_i, v_j) \in E_M \setminus E'$, then we could $E_M = (E_M \setminus \{(v_i, v_j)\}) \cup \{(u, v_i)\}$, i.e., swap (v_i, v_j) out of M and swap (u, v_i) in M. The replaced M is also a P2B subgraph.

It is clear that $B(V_B, E_B)$ is a partial 2 bounded subgraph of G. Now we will analyze the size of E_B. Let $f(u)$ be the degree of u, $u \in L$, so $|E_M| = \Sigma_{u \in L} f(u)$. And let d_u be the degree of u in G, the expectation of $f(u)$ is $\mathsf{E}[f(u)] = \frac{1}{2}\left(1 \cdot \frac{d_u}{2^{d_u}} + 2 \cdot \left(1 - \frac{1+d_u}{2^{d_u}}\right)\right) = 1 - \frac{2+d_u}{2^{d_u+1}} \geq \frac{11}{16}$. Using the linearity of expectation, $\mathsf{E}[E_M] = \Sigma_{u \in V}\mathsf{E}[f(u)] \geq \frac{11}{16}n > \frac{11}{32}OPT$, where OPT denotes the optimum value. According to theorem 3 conditioned on $c = 2$, $OPT < 2n$ and the last inequality holds.

Algorithm 3. Randomized Algorithm for maximum partial 2 bounded subgraph

1: $B(V_B, E_B)$ is a bipartite graph, $V_B = L \bigcup R$, initiate $L = R = \emptyset$
2: **for all** $u \in V$ **do**
3:     add u into L or R with equal probability $1/2$
4: **for all** $e(u, v) \in E$ with $u \in L$ **do**
5:     **if** u has no more than 2 edges in E_B **then**
6:         add e into E_B

Lines 1-3 take up $O(|V|)$ time and lines 4-6 take up $O(|E|)$ time, so algorithm 3 takes $O(|E|)$ time. Because it is not direct to show whether the variables $\{f(u)|u \in V\}$ are independent or with small dependency, so we are not sure whether $|E_M|$ is sharply concentrated around its expectation in $O(|E|)$. But we can de-randomize algorithm 3, using conditional expectation to decide whether the next vertex should be put in L or not. And the cost for deterministic algorithm is $O(|E|^2)$.

5   Conclusion

This paper presents a new model of edge packing problem with partial degree bounded constraint and several results on it. The author is still trying to study more deep results in the following respects.

PcB in a Parameterized View
When $c = 1$, PcB is fixed parameter tractable (FPT) with respect to its solution size. Does this hold for general c? When c is a constant, i.e., $c = o(|V|)$, it is easy to show PcB is in $W[1]$ defined in [4]. For example, when $c = 2$, for each forbidden subgraph of P2B, we create a antimonotone clause $(\overline{e_1} \bigvee \overline{e_2} \bigvee \cdots \overline{e_5})$ where each literal $\overline{e_1}$ corresponds to an edge in the subgraph. Thus there is a PcB subgraph with k edges if and only if the weighted 5-CNF satisfiability has a valid truth assignment with k variables being set true. Because weighted 5-CNF satisfiability is $W[1]$-complete, so PcB is in $W[1]$.

According to theorem 3, the solution may be close to c times n which renders the solution size not a good parameter. For example, when $c = 2$, let k be the parameter. Suppose $k < \Delta$, let M be a maximum matching of G. Thus $k > |M| > \frac{n}{\Delta} > \frac{n}{k}$. So $k > \sqrt{n}$ and $\sqrt{n}$ is certainly not a good parameter.

Section 3 shows that PcB is in P on trees, whether PcB could be efficiently (though not in P) solved on tree-like graph? Tree decomposition in [5] is a measure for this. Courcelle's theorem in [6] asserts that if a graph problem could be described in monadic second order (MSO) logic, then it could be solved in linear time with respect to its treewidth. Luckily, PcB is in MSO and thus establishes its FPT with respect to the treewidth. The author is trying to design a PcB specific algorithm with improved efficiency.

PcB in a Approximation View
In section 4, we only show algorithms which upper bounds optimum roughly. We might elaborate the analysis by correlate the optimum with the solution returned

by our algorithm. Also, both algorithms can not be extended when c increases. Constant ratio approximation algorithms for the general c or inapproximability results which exclude them would be really interesting.

Acknowledgements. Special thanks to Jukka Suomela and Chandra Chekuri for their valuable advice. Also, we would like to thank the anonymous referees for their suggestions in improving the readability of this paper.

References

1. Bu, T.M., Yuan, C., Zhang, P.: Computing on Binary Strings. In: arXiv:1112.0278v2 (2012)
2. Nieminen, J.: Two bounds for the domination number of a graph. Journal of the Institute of Mathematics and its Applications 14, 183–187 (1974)
3. Dehne, F., Fellows, M., Fernau, H., Prieto, E., Rosamond, F.: NONBLOCKER: Parameterized Algorithmics for MINIMUM DOMINATING SET. In: Wiedermann, J., Tel, G., Pokorný, J., Bieliková, M., Štuller, J. (eds.) SOFSEM 2006. LNCS, vol. 3831, pp. 237–245. Springer, Heidelberg (2006)
4. Downey, R.G., Fellows, M.R.: Parameterized complexity (1999)
5. Robertson, N., Seymour, P.: Graph minors. iii. planar tree-width. Journal of Combinatorial Theory, Series B 36(1), 49–64 (1984)
6. Courcelle, B.: The monadic second order theory of graphs i: Recognisable sets of finite graphs. Information and Computation 85, 12–75 (1990)

Erratum: The Approximability of the Exemplar Breakpoint Distance Problem

Zhixiang Chen[1], Bin Fu[1], and Binhai Zhu[2]

[1] Department of Computer Science, University of Texas-American, Edinburg,
TX 78739-2999, USA
{chen,binfu}@cs.panam.edu
[2] Department of Computer Science, Montana State University, Bozeman,
MT 59717-3880, USA
bhz@cs.montana.edu

Abstract. The paper "The Approximability of the Exemplar Breakpoint Distance Problem" [1], which appeared in AAIM 2006, contained several negative results and one positive result — a claimed $O(\log n)$-factor greedy approximation for the One-sided Exemplar Breakpoint Distance Problem. Here, we show that the analysis was incorrect and the approximation factor of the greedy algorithm could be $\Theta(n)$, where n is the size of the alphabet.

In Section 5 of [1], a greedy algorithm is presented for the One-sided Exemplar Breakpoint Distance Problem. The claimed approximation factor is $O(\log n)$. We show that the factor could be $\Theta(n)$ with an example. In our example, G is exemplar, so it satisfies the k-span condition. We start with a small $n = 9$.

$G = \langle 1, 2, 3, 4, 5, 6, 7, 8, 9 \rangle$, and
$H = \langle 9, 8, 7, 6, 1, 6, 2, 7, 3, 8, 4, 9, 5, 1, 2, 3, 4 \rangle$.

The optimal solution is to have $H^* = \langle 6, 7, 8, 9, 5, 1, 2, 3, 4 \rangle$. In other words, we will have two breakpoints between G and H^*.

The greedy algorithm would first select the NB-interval in H: $\langle 1, 2, 3, 4, 5 \rangle$. So the greedy algorithm would have a solution $H' = \langle 9, 8, 7, 6, 1, 2, 3, 4, 5 \rangle$. In other words, we will have four breakpoints between G and H'. (We thank Minghui Jiang for the idea regarding this example.)

By generalizing the alphabet to be $n = 2m + 1$, i.e., $|G| = 2m + 1$ and $|H| = 4m + 1$, the greedy algorithm would generate m breakpoints while the optimal solution only introduces two breakpoints. So the approximation factor of the greedy algorithm is $m/2 = \Theta(n)$.

It is an open question whether the One-Sided Exemplar Breakpoint Distance Problem admits a polynomial time $o(n)$-factor approximation. The only known negative result is the APX-hardness of the problem.

Reference

1. Chen, Z., Fu, B., Zhu, B.: The Approximability of the Exemplar Breakpoint Distance Problem. In: Cheng, S.-W., Poon, C.K. (eds.) AAIM 2006. LNCS, vol. 4041, pp. 291–302. Springer, Heidelberg (2006)

J. Snoeyink, P. Lu, K. Su, and L. Wang (Eds.): FAW-AAIM 2012, LNCS 7285, p. 368, 2012.
© Springer-Verlag Berlin Heidelberg 2012

Author Index

[1] Shandong University.
[2] East China Normal University.